REINFORCED CONCRETE SLABS

REINFORCED CONCRETE SLABS

SECOND EDITION

Robert Park
The University of Canterbury

William L. Gamble
University of Illinois at Urbana–Champaign

JOHN WILEY & SONS, INC.

New York · Chichester · Weinheim · Brisbane · Singapore · Toronto

Copyright © 2000 by John Wiley & Sons, Inc. All rights reserved.

Published simultaneously in Canada.

This publication is designed to provide accurate and authoritative information in regard to the subject matter covered. It is sold with the understanding that the publisher is not engaged in rendering professional services. If professional advice or other expert assistance is required, the services of a competent professional person should be sought.

Library of Congress Cataloging-in-Publication Data:

Park, R. (Robert), 1933–
 Reinforced concrete slabs / Robert Park, William L. Gamble. — 2nd ed.
  p.   cm.
  Includes bibliographical references.
  ISBN 0-471-34850-3 (cloth : alk. paper)
  1. Concrete slabs—Design and construction. 2. Concrete slabs—Testing. I. Gamble, W. L. (William Leo) II. TITLE.
 TA683.5.S6P36 2000
 624.1'8342—dc21   99-28984

10 9 8 7 6 5 4 3 2

CONTENTS

Preface   **xiii**

1. Introduction   **1**

  1.1 Scope and General Remarks, 1
  1.2 Types of Reinforced Concrete Slab Construction, 3
  1.3 Choice of Type of Slab Floor, 10
  1.4 Approaches to the Analysis and Design of Slab
      Systems, 11

      1.4.1 Complete Behavior of Slab Systems, 12
      1.4.2 Elastic Theory Analysis, 13
      1.4.3 Limit Analysis, 14
      1.4.4 ACI Building Code Method, 16
      1.4.5 Design Procedures, 18

  1.5 Notes on Units, 18
      References, 19

2. Basis of Elastic Theory Analysis   **21**

  2.1 Introduction, 21
  2.2 Classical Plate Theory, 21

      2.2.1 Lagrange's Equation, 21
      2.2.2 Equilibrium, 22
      2.2.3 Moment–Deformation Relationships, 24
      2.2.4 Shear–Deformation Relationships, 30
      2.2.5 Boundary Conditions, 31
      2.2.6 Reactions, 32
      2.2.7 Poisson's Ratio, 34
      2.2.8 Moments Acting at an Angle to the Coordinate
            Axes, 36
      2.2.9 Method of Solution, 37

  2.3 Elastic Models, 42
  2.4 Finite Difference Methods, 42
  2.5 Finite Element Methods, 46
  2.6 Approximate Methods, 49
  2.7 Examples of Elastic Theory Analysis, 51
      References, 57

3. **Results of Elastic Theory Analysis**   60

3.1 Introduction, 60
3.2 Moments in Interior Panels of Slabs, 68
    3.2.1 Effects of Relative Stiffness of Supporting
           Beams, 68
    3.2.2 Effects of Size of Supporting Column or
           Capital, 80
    3.2.3 Pattern Loadings, 81
3.3 Moments in Edge Panels of Slabs, 101
    3.3.1 Moments in the Span Parallel to the Edge of the
           Structure, 103
    3.3.2 Moments in the Span Perpendicular to the Edge
           of the Structure, 107
3.4 Moments in Corner Panels, 117
3.5 Special Cases of Loading and Geometry, 122
    3.5.1 Effects of Concentrated and Line Loads on
           Moments, 122
    3.5.2 Effects of Loads Varying Linearly Across One
           Span, 130
    3.5.3 Effects of Holes in Slabs, 133
    References, 139

4. **Background of 1971 and 1995 ACI Building Code
    Requirements for Reinforced Concrete Slab Design**   144

4.1 Introduction, 144
4.2 Static Moment and Structural Safety, 148
    4.2.1 Static Moment, 148
    4.2.2 Structural Safety and the Static Moment, 155
4.3 Equivalent Frame Method: Determination of Negative
     and Positive Slab Moments and Column Moments, 157
    4.3.1 General Comments and Manual Calculation
           Approach, 157
    4.3.2 Idealizations for Plane-Frame Computer
           Analysis, 174
4.4 Distribution of Moments Across Sections, 180
4.5 Direct Design Method, 195
    4.5.1 Negative–Positive Distribution of Moments, 195
    4.5.2 Requirements for Column Moments and
           Stiffnesses, 200
    References, 202

5. General Lower Bound Limit Analysis and Design   **205**

5.1 Introduction, 205
5.2 Governing Equations for General Lower Bound Limit
Analysis, 207

    5.2.1 Equilibrium Equation, 207
    5.2.2 Transformation of Moments to Different
Axes, 209
    5.2.3 Boundary Conditions, 210
    5.2.4 Yield Criterion, 211

5.3 Analysis of Slabs by General Lower Bound
Method, 217
5.4 Design of Reinforcement for Slabs in Accordance with a
Predetermined Field of Moments, 223

    5.4.1 General Approach, 223
    5.4.2 Reinforcement Arranged at Right Angles, 224

5.5 Comment on General Lower Bound Limit Design, 230
References, 230

6. Design by the Strip Method and Other Equilibrium Methods **232**

6.1 Introduction, 232
6.2 Simple Strip Method, 234

    6.2.1 Strip Action, 234
    6.2.2 Discontinuity Lines Originating from Slab
Corners, 238
    6.2.3 Discontinuity Lines Originating from Slab
Sides, 247
    6.2.4 Strong Bands, 254
    6.2.5 Skewed and Triangular Slabs, 255
    6.2.6 Comparison with the Yield Line Theory Ultimate
Load, 256
    6.2.7 Design Applications, 258

6.3 Advanced Strip Method, 277

    6.3.1 Types of Slab Element, 278
    6.3.2 Design Applications, 285

6.4 Segment Equilibrium Method, 287

    6.4.1 Flat Plates with Columns on a Rectangular
Grid, 287
    6.4.2 Flat Plates with Irregular Column Layouts, 293

6.5 Comparison with Test Results, 296
References, 300

7. Basis of Yield Line Theory   303

7.1   Introduction,   303
7.2   Slab Reinforcement, Section Behavior, and Conditions
       at Ultimate Load,   303

   7.2.1   Slab Reinforcement,   303
   7.2.2   Ductility of Slab Sections,   304
   7.2.3   Conditions at Ultimate Load,   305
   7.2.4   Yield Lines as Axes of Rotation,   306
   7.2.5   Ultimate Moments of Resistance at Yield
            Lines,   308
   7.2.6   Determination of the Ultimate Load,   310

7.3   Analysis by Principle of Virtual Work,   311

   7.3.1   Virtual Work Equation,   311
   7.3.2   Components of Internal Work Done,   314
   7.3.3   Minimum-Load Principle,   314

7.4   Analysis by Equations of Equilibrium,   321

   7.4.1   Equilibrium Equations,   321
   7.4.2   Statical Equivalents of Shear Forces Along a
            Yield Line,   322
   7.4.3   Magnitude of Nodal Forces,   323
   7.4.4   Method of Solution by the Equilibrium
            Equations,   332

7.5   Concentrated Loads,   339

   7.5.1   Types of Yield Line Patterns,   339
   7.5.2   Circular Fans,   341

7.6   Superposition of Moment Strengths for Combined
       Loading Cases,   346
7.7   Corner Effects,   349
7.8   Affinity Theorem,   354
7.9   General Cases for Uniformly Loaded Rectangular
       Slabs,   363

   7.9.1   Ultimate Moments of Resistance of the
            Slabs,   363
   7.9.2   Uniformly Loaded Orthotropic Rectangular
            Slabs with All Edges Supported,   363
   7.9.3   Uniformly Loaded Orthotropic Rectangular
            Slabs with Three Edges Supported and One
            Edge Free,   366
   7.9.4   Uniformly Loaded Orthotropic Rectangular
            Slabs with Two Adjacent Edges Supported and
            the Other Edges Free,   369

7.10 Composite Beam–Slab Collapse Mechanisms,   371

7.11   Beamless Floors,   377

    7.11.1   Folding Yield Line Patterns,   377
    7.11.2   Local Yield Line Patterns at Columns,   379
    7.11.3   Unbalanced Moment Transfer at Slab–Column
           Connections,   384
    7.11.4   Flat Slab Floors with Exterior Beams,   388
    7.11.5   Shear Strength of Slab–Column
           Connections,   390

7.12   Uniformly Loaded Rectangular Slabs with
      Openings,   390
7.13   Uniformly Loaded Circular and Ring Slabs,   399

    7.13.1   Circular Slab Supported on n Columns and
           Subjected to Uniform Loading,   399
    7.13.2   Ring Slabs,   401

7.14   Skew Slabs,   403
7.15   Approximate Yield Line Patterns for Uniformly Loaded
      Rectangular Slabs,   407

    7.15.1   Use of Approximate Yield Line Patterns,   407
    7.15.2   Uniformly Loaded Rectangular Slabs with All
           Edges Supported,   407
    7.15.3   Uniformly Loaded Rectangular Slabs with
           Three Edges Supported and One Edge
           Free,   409
    7.15.4   Uniformly Loaded Rectangular Slabs with Two
           Adjacent Edges Supported and the Remaining
           Edges Free,   413

7.16   Trial-and-Error Method for Approximate Yield Line
      Patterns,   415
7.17   Comparison with Test Results,   424

    7.17.1   Tests Conducted by the Deutscher Ausschuss
           für Eisenbeton,   424
    7.17.2   Tests Conducted by IRABA,   427
    7.17.3   Tests Conducted at the Technical University of
           Berlin,   428
    7.17.4   Tests Conducted at the TNO Institute for
           Building Materials and Structures,   428
    7.17.5   Tests Conducted at the University of
           Manchester,   430
    7.17.6   Tests Conducted at the University of
           Canterbury,   439
    7.17.7   Tests Conducted at the University of
           Illinois,   442

7.17.8   Tests Conducted at the Portland Cement
Association,   445
References,   445

8. Design by Yield Line Theory   **449**

8.1   Introduction,   449
8.2   Strength and Serviceability Provisions,   450
    8.2.1   Design Load and Moment of Resistance,   450
    8.2.2   Reinforcement Ratios,   451
    8.2.3   Reinforcement Arrangements,   452
    8.2.4   Serviceability Checks,   453
    8.2.5   Other Design Aspects,   454
8.3   Superposition of Loading,   455
8.4   Design of Uniformly Loaded Two-Way Slabs,   458
    8.4.1   Extent of Top Steel in Uniformly Loaded
Rectangular Slabs,   458
    8.4.2   Minimum-Weight Design,   468
    8.4.3   Design Examples,   474
8.5   Design of Beamless Slabs,   494
8.6   Design of Supporting Beams for Uniformly Loaded
Two-Way Slabs,   499
    8.6.1   Approach Based on Composite Beam–Slab
Collapse Mechanisms,   499
    8.6.2   Approach Based on Loading Transferred to the
Beams,   502
    8.6.3   Other Arrangements of Beams and
Columns,   509
    8.6.4   Summary of Design Method for Beams,   512
References,   513

9. Serviceability of Slabs   **515**

9.1   Introduction,   515
9.2   Deflections,   515
    9.2.1   General Comments on Deflections,   515
    9.2.2   Computation of Deflections,   521
    9.2.3   ACI Code Provisions for Deflection
Control,   526
9.3   Cracking,   530
    9.3.1   Need for Crack Control,   530
    9.3.2   Causes of Cracking,   532
    9.3.3   Computation of Width of Flexural Cracks in
One-Way Slabs,   533

9.3.4 Computation of Width of Flexural Cracks in
        Two-Way Slabs, 540
9.3.5 Code Provisions for Crack Control, 545
References, 547

10. Shear Strength of Slabs   **551**

10.1 Introduction, 551
10.2 Shear Strength of Slabs Transferring Uniform
       Shear, 554
       10.2.1 Mechanism of Shear Failure of Slabs Without
                Shear Reinforcement, 554
       10.2.2 ACI Code Approach to Shear Strength Without
                Shear Reinforcement, 560
       10.2.3 Truss Models for Shear Strength, 564
       10.2.4 ACI Code Approach to Shear Strength with
                Shear Reinforcement, 566
       10.2.5 Influence of Openings, Free Edges, and
                Service Ducts in Slabs, 572
       10.2.6 Special Problems with Pile Caps, 574
10.3 Shear Strength of Slab–Column Connections
       Transferring Shear and Unbalanced Moment, 579
       10.3.1 Behavior of Slab–Column Connections
                Transferring Shear and Unbalanced Bending
                Moment, 579
       10.3.2 Methods of Analysis and Design, 581
       10.3.3 ACI Code Approach, 584
       10.3.4 ASCE-ACI Committee 426 Suggested
                Approach, 594
       10.3.5 Alternative Beam Analogy Approach for
                Interior Connections, 601
       10.3.6 Truss Analogy Approach Results, 612
       10.3.7 Ductility of Slab–Column Connections, 613
10.4 Slab–Wall Connections, 615
10.5 Shear Capacity with High-Strength Concrete, 615
References, 617

11. Prestressed Concrete Slabs   **621**

11.1 Introduction, 621
11.2 Basis for Design, 622
       11.2.1 General Approach, 622
       11.2.2 Service Load Stresses, 622
       11.2.3 Flexural Strength, 628
       11.2.4 Shear Strength, 631

11.2.5   Concluding Comments,   633

11.3   Corrosion Concerns,   633
References,   634

12.   Membrane Action in Slabs   **636**

12.1   Introduction,   636

12.2   Uniformly Loaded Laterally Restrained Reinforced
Concrete Slabs,   636

12.2.1   General Behavior and Review of Past
Research,   636

12.2.2   Behavior in the Compressive Membrane
Range,   640

12.2.3   Behavior in the Tensile Membrane
Range,   679

12.3   Concentrated Loads on Laterally Restrained Reinforced
Concrete Slabs,   687

12.4   Slabs with Edges Free to Move Laterally,   687

12.5   Recent Computational Approaches,   689
References,   691

13.   Fire Resistance of Reinforced Concrete Slabs   **695**

13.1   Introduction,   695

13.2   Thermal Resistance,   697

13.3   Structural Fire Resistance,   698

13.3.1   Members Unrestrained Against Length
Change,   698

13.3.2   Members Restrained Against Length
Change,   707

13.4   Special Considerations for High-Strength
Concrete,   708
References,   709

Index   **711**

PREFACE

The book emphasizes the basic behavior of reinforced concrete slabs in both the elastic range and at the ultimate load. As such, it endeavors to give readers a thorough knowledge of the fundamentals of slab behavior. Such a background is essential for a complete and proper understanding of building code requirements and design procedures for slabs. The content and the treatment of the subject of reinforced concrete slabs in this book is intended to appeal to students, teachers, and practicing members of the structural engineering profession.

The book begins with a general discussion of slab analysis and design, and then treats at some depth the determination of the distribution of moments and shears using elastic theory. The *equivalent frame method* of the Building Code of the American Concrete Institute is explained, followed by limited coverage of the *direct design method* of that code. Next follows a detailed treatment of limit procedures for the ultimate load analysis and design of slabs using general lower bound theory, the strip method, and yield line theory. The behavior of slabs at the service load is then discussed, with emphasis on deflection and crack control. This is followed by an examination of the shear strength of slabs. Prestressed concrete slabs are discussed in an introductory manner near the end of the text, the earlier chapters having dealt specifically with reinforced concrete slabs. The effects of membrane action on the strength of slabs are reviewed. Finally, an introduction to the determination of the fire resistance of slabs is given.

The current building code of the American Concrete Institute (ACI 318-95) is one of the most widely accepted reinforced concrete codes. It has been adopted by some countries and has strongly influenced the codes of many others. For this reason, reference is made primarily to ACI provisions, although other building codes are also discussed. The book is not heavily code oriented, however. The emphasis is on why certain decisions should be made rather than on how to execute them. We believe that structural engineers should be capable of rationally assessing design procedures.

The book has grown from many years of experience in teaching slab theory and design, from significant involvement in slab research and design, and from association with design code committees. The chapters on yield line theory have been based on editions of seminar notes entitled *Ultimate Strength Design of Reinforced Concrete Structures,* Vol. 2, printed by the University of Canterbury. The book is also intended to complement a previously published text, *Reinforced Concrete Structures* by R. Park and T. Paulay (Wiley, 1975), which does not discuss slabs. The unusual combination of authors from New Zealand and the United States arises from complementary interests in aspects of slab behavior and the significant slab research that has been conducted at their two universities through the years.

An aspect of the book that distinguishes it from previous texts on slabs is that it attempts to give a full treatment of the background of most of the possible current approaches to reinforced concrete slab analysis and design. Previous texts have emphasized either elastic theory or the strip method or yield line theory but have not attempted to give a comprehensive treatment of all procedures, together with aspects of shear strength, serviceability, and membrane action. The authors have intentionally dealt almost entirely with reinforced concrete slabs and have given only an introduction to prestressed concrete slabs, since prestressed concrete is an extensive subject that deserves a book of its own. Also, the main consideration of the book is two-way floor systems. We hope that the book will serve as a useful text for teachers preparing advanced undergraduate or master's courses. It is also hoped that many practicing engineers, and research engineers, will find the book a useful reference.

The main changes in the design and analysis of slab structures, and indeed in all of structural engineering, which have occurred since the first edition was published may be summed up in one word: computers. The first edition was published slightly before the desktop computer era started. When the second edition was completed, many personal computers were more powerful than the mainframe machines existing at the time of the first edition. As computer power has grown, capabilities of the finite element method of analysis have greatly expanded, allowing solution of more and more complex slab problems. On a simpler level, the second author has made extensive use of spreadsheets for many of the routine tasks of slab design, such as selecting reinforcement for given moments, computation of stiffness and other coefficients needed for the analysis process, finding the critical arrangement of yield lines, and solution of frame analysis problems by the nearly archaic method of moment distribution. Various computer approaches are suggested in the text.

We would be grateful for any constructive comments or criticisms that readers may have and for notification of any errors that they will inevitably detect.

We have received a great deal of assistance, constructive comment, encouragement, and inspiration from numerous sources. Thanks are due our many colleagues at the University of Canterbury at Christchurch, at the University of Illinois at Urbana–Champaign, and in the profession in New Zealand and the United States. Particular thanks to colleagues in New Zealand are due to Professor H. J. Hopkins for his encouragement; to Professor T. Paulay, Dr. A. J. Carr, and Dr. P. J. Moss for comments; and to a number of dedicated graduate students and technicians who have been involved in test programs. Particular thanks to colleagues in the United States are due to the late Professor N. M. Newmark and Professor Emeritus C. P. Siess of the University of Illinois at Urbana–Champaign, Professor M. A. Sozen of Purdue University, and Professor B. Mohraz of Southern Methodist University for encouragement and much technical information, especially while the first edition was being written. Appreciation is also expressed to Professor Emeritus R. B. Peck of the University of Illinois for encouragement in deciding to undertake preparation of the manuscript of the first edition, and to various members of ACI-ASCE Committee 421, Reinforced Concrete Slabs, for encouragement to prepare the second edition.

Our thanks are also due to the following organizations for permission to reproduce copyrighted material: American Concrete Institute, American Society of Civil Engineers, Concrete Reinforcing Steel Institute, McGraw-Hill Book Co., and the University of Illinois at Urbana–Champaign, all in the United States; Cement and Concrete Association, Building Research Station, and Thames and Hudson of the United Kingdom; Institute of Engineers of Australia; Heron of the Netherlands; Springer-Verlag of Germany; and the Canadian Standards Association.

Finally, this undertaking could never have been completed without the patience and understanding of our wives, Kathie and Judy.

ROBERT PARK
WILLIAM L. GAMBLE

Christchurch, New Zealand
Urbana, Illinois
September 1999

1 Introduction

1.1 SCOPE AND GENERAL REMARKS

Reinforced concrete slabs are among the most common structural elements, but despite the large number of slabs designed and built, the details of the elastic and plastic behavior of slabs are not always appreciated or properly taken into account. This occurs at least partially because of the mathematical complexities of dealing with elastic plate equations, especially for support conditions which realistically approximate those in multipanel building floor slabs.

Because the theoretical analysis of slabs and plates is much less widely known and practiced than is the analysis of elements such as beams, the provisions in building codes generally provide both design criteria and methods of analysis for slabs, whereas only criteria are provided for most other elements. For example, Chapter 13 of the 1995 edition of the American Concrete Institute (ACI) *Building Code Requirements for Structural Concrete*,[1.1] one of the most widely used Codes for concrete design, is devoted largely to the determination of moments in slab structures. Once the moments, shears, and torques are found, sections are proportioned to resist them using the criteria specified in other sections of the same code.

Although the ACI Code approach to slab design is basically one of using elastic moment distributions, it is also possible to design slabs using plastic analyses (limit analyses) to provide the required moments. It is the intention of this book to provide some insight into both methods of design and analysis and the backgrounds to both.

To this end, the book may be viewed as being made of several separate, though interrelated parts. Basic information on the elastic analysis of slabs and plates and the moment distributions found using these analyses is provided in Chapters 2 and 3. The relationships between these data and the 1995 ACI Code provisions are described in Chapter 4.

Chapters 5 through 8 are concerned primarily with plastic design methods. Lower bound methods are discussed in Chapters 5 and 6. Chapter 5 also includes a scheme of reinforcing to resist a general field of moments. Use of the yield line method, an upper bound solution, for analysis and design purposes is described in Chapters 7 and 8.

Regardless of which design method is used, the resulting slab must be serviceable at the working load level, with deflections and cracking remaining within acceptable limits. These problems are discussed in Chapter 9. Slab design methods are concerned largely with flexure, but the shear forces may also be a limiting factor. The particular problems of shear in beamless slabs, especially when acting in combination with transfer of unbalanced moments from slab to columns, are discussed in Chapter 10.

Prestressed concrete slabs are important in some geographical areas and for some uses, and the field is so broad that the choice appears either to be to say very little or to write a separate book on them. The former path has been taken in Chapter 11, which provides introductory comments and literature sources for further study. Membrane action in slabs, which has been shown by some tests to result in a significant enhancement in the ultimate load of slabs, is discussed in Chapter 12.

Most buildings must be designed to some level of fire resistance in addition to the various structural requirements. The fire resistance of most reinforced concrete structures is inherently fairly high, but it must be checked for each specific case. There are many aspects to this process. In many cases, heat conduction through slabs sets a minimum slab thickness which may be greater than required for structural purposes, and information on this problem is presented in Chapter 13. In addition, the structural fire resistance must be adequate and Chapter 13 also gives basic information on this determination.

There are many other concerns in the field of fire safety which typically are not within the scope of the structural engineer. This include such items as detection and alarm systems, automatic extinguishing equipment, smoke control systems, exit travel distances, and protection openings by the use of doors with suitable fire resistance and closing and latching mechanisms. The minimum requirements are typically given in the general building code, with many details governed by codes published by the National Fire Protection Association.

The whole problem of slab analysis and design may be placed in better perspective through the following quotation, written by Hardy Cross[1.2] in 1929:

Perhaps it would be better not to mention slabs at all, for the principles here stated contribute only slightly to the study of continuous slabs. But these principles will, if judiciously applied, contribute something.

Tools available to the average engineer in thinking about slabs are the limitations imposed by statics upon the total moments, principles of symmetry and asymmetry, and mental pictures of the deflected slab as a means of judging of the variation of the moments along any given section. The last tool, though inexact, is very powerful. The writer finds the idea of distributing fixed-end moments useful in revising his mental pictures of the deflected slab when affected by continuity with other slabs or by discontinuities. It is consoling to realize that there is considerable evidence that these pictures need not be very exact and that if the total statical limitations are met, the assumed distribution of the moments need not conform precisely to results of mathematical theory.

The ideas expressed, the last sentence especially, should never be forgotten by designers and code writers.

1.2 TYPES OF REINFORCED CONCRETE SLAB CONSTRUCTION

Reinforced concrete slab floors have taken many forms since their introduction. Some of these were clearly direct imitations of earlier floors made entirely of wood or of wood supported on steel or iron beams. Others were just as clearly invented, with no recognizable ancestors, to suit the properties of the materials—steel bars and plastic concrete.

Economics and development of methods of construction, the suitability of particular slabs for particular sets of requirements, and advances in the methods of analysis of slabs have all joined to shape the current practice, and these factors will just as surely continue to change the types of slabs that are built.

Slabs may be divided into two general categories: beamless slabs and slabs supported on beams located on all sides of each panel. There are many hybrid variants, and many otherwise beamless slabs have beams at the edges of the structure and around large openings such as those made for elevators and stairways.

Beamless slabs are described by the generic terms *flat plates* and *flat slabs*. The *flat plate* is an extremely simple structure in concept and construction, consisting of a slab of uniform thickness supported directly on columns, as shown in Fig. 1.1. The flat plate is a direct development from the earlier *flat slab* structure, which was characterized by the presence of capitals at the tops of the columns and usually also by drop panels or thickened areas of the slab surrounding each column. The basic form of the flat slab is shown in Fig. 1.2. The most common subtypes are flat plates with drop panels and flat slabs

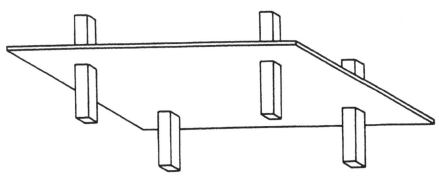

Figure 1.1 Flat plate.

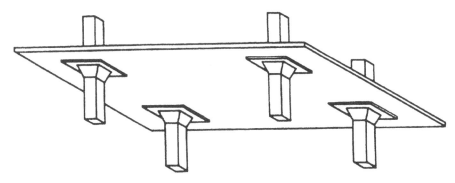

Figure 1.2   Flat slab.

without drop panels. Figure 1.3 is a photograph of a flat plate with drop panels.

The choice between the use of flat slabs and flat plates is largely a matter of the magnitude of the design loading and of the spans. The strength of the flat plate structure is often limited by the strength in punching shear at sections around the columns, and they are consequently used with light loads, such as are found in residential and some office construction, and relatively short spans. The column capital and drop panel provide the shear strength

Figure 1.3   Flat plate with drop panels. (Photo by W. L. Gamble.)

necessary for larger loads and spans, and the flat slab is often the choice for heavily loaded industrial structures and for cases where large spans are necessary.

Slabs supported on beams on all sides of each panel are generally termed *two-way slabs,* and a typical floor is shown in Fig. 1.4. This system is a development from beam-and-girder systems, as shown in Fig. 1.5. In a beam-and-girder system, it was quite easy to visualize the path from load point to column as being from slab to beam to girder to column, and from this visualization then to compute realistic moments and shears for the design of all members. This system is still used with heavy timber and steel frame construction, especially when the column spacing becomes large. Removal of the beams, except those on the column lines, results in the two-way slab structure. If the beam spacing is reduced to 2 to 3 ft (0.6 to 0.9 m), the *one-way* joist floor system is the result. One of the problems with earlier slab-and-girder construction was that the slabs were assumed to be one-way slabs, with no bending parallel to the beams. However, since the ends of the subpanels were supported, moments parallel with the beams developed near the ends of the subpanels and negative-moment cracking often developed where the slabs joined the girders. Since this had not been anticipated, no reinforcement was provided, and many buildings from 1930 or before will be found in which the locations of the girders can be traced by the large unrestrained cracks that opened on the tops of the slabs.

The *waffle slab,* a variant of the solid slab, may be visualized as a set of crossing joists, set at small spacings relative to the span, which support a thin top slab. The recesses in the slab, often cast using either removable or expendable forms, decrease the weight of the slab and allow the use of a large effective depth without the accompanying dead load. The large depth also leads to a stiff structure. Waffle slabs are generally used in situations demanding spans larger than perhaps about 30 ft (10 m).

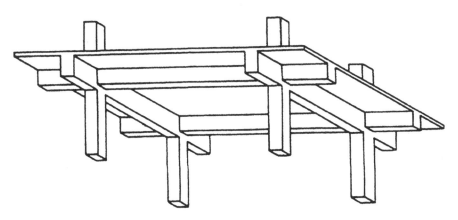

Figure 1.4 Two-way slab.

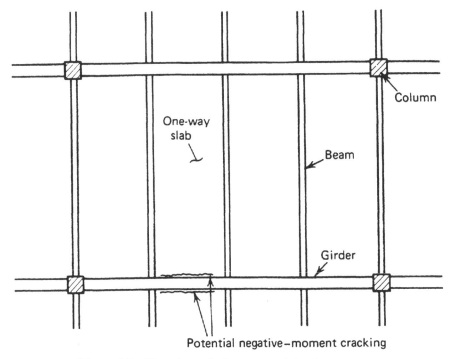

Potential negative–moment cracking

Figure 1.5   Plan view of a beam-and-girder floor system.

Waffle slabs may be designed as either flat slabs or two-way slabs, depending on just which recesses are omitted to give larger solid areas. Figure 1.6 shows two possible arrangements. The solid area near the column in the flat plate arrangement is comparable to a drop panel or column capital, as it provides a path for shear transfer and extra compression area in the highly stressed negative-moment regions surrounding the columns. The solid areas where the recesses have been omitted along the column lines in the two-way slab configuration are equivalent to beams since they are areas of concentrated flexural stiffness, even though they do not extend below the general lower surface of the slab.

The joists around each recess are designed as beams for the forces attributable to a width of slab equal to the joist spacing, and they may easily be reinforced for shear if necessary. The larger the number of joists per span, the better is the approximation of the behavior of a solid slab, and a minimum of six to eight joists would seem necessary if the slab is to be designed as a slab rather than as a series of crossing beams.

A further variation of a waffle slab is a hollow slab in which the holes are made by placing terra-cotta tiles, precast concrete boxes, or other fillers on the formwork, placing the reinforcing bars between the filler units, and cov-

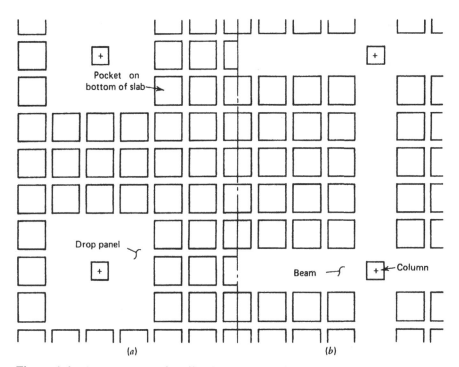

Figure 1.6 Arrangements of waffle slabs: (*a*) as a flat slab; (*b*) as a two-way slab.

ering the fillers with cast-in-place concrete. The lower surface of the slab may be either flat or have recesses, depending on the fillers used, and in some cases the fillers are the full depth of the slab. In a two-way configuration, the fillers along the column lines are omitted, and the beams may extend below the slab as well.

A comprehensive review of the history of the development of both slab types and of the building code regulations in the United States is given by Sozen and Siess,[1.3] and consequently, only abbreviated comments are given here. The first flat slab was built in 1906 in Minneapolis, Minnesota, by C. A. P. Turner. Since it was a completely new form of construction, and in addition no acceptable method of analysis was available, the building was built at Turner's risk and load-tested before the owner would accept it. The structure met its load test requirements, but acceptance tests were still required on a number of similar structures built in the next two years.

The flat slab was an instant commercial success, and more than 1000 structures were built in the United States within seven years of the first one. Unfortunately, there was no consistent way of analyzing flat slab structures because of the early stage of development of plate analysis methods, and no agreed-upon way of reinforcing the slabs once design moments had been determined. A 1910 study[1.4,1.5] compared the amounts of steel required for a given slab panel, having the same dead and live loads and spans, and the weight of steel required varied by a factor of 4 from lightest to heaviest.

The many schemes of arranging the reinforcement that were used in early flat slab construction, as described by Taylor et al.,[1.6] include two-way reinforcement in which all bars are placed parallel to the column lines approximately as in current practice; three-way steel; four-way schemes in which diagonal bands of bars are added to those parallel to the column lines; and the Smulski system, which used circular rings of steel, radial bars near the columns, and a few additional bars along both the column lines and panel diagonals. In light of current practice, the positive-moment reinforcement was terminated very far from the panel boundaries in all these systems.

The publication of a correct solution for the total moment for which a slab panel should be designed, by Nichols[1.7] in 1914, only added fuel to a raging controversy that persisted for many years. Nichols's analysis provided only a total moment and in current terminology would be recognized as a limit analysis approach. The distribution of this moment to positive- and negative-moment sections, and across these sections, was left undetermined until the publication of a classic study by Westergaard and Slater[1.8] in 1921.

Nichols's work was not widely accepted because it seemed to be so contrary to the findings of many load tests. However, these load tests were often wrongly interpreted to indicate that the strengths of the slabs were much higher than they really were because of two factors. The measured reinforcement strains were often very low, and since the influence of tensile stresses in the concrete in reducing the steel stresses was not generally recognized, it was assumed that the low strains implied the existence of low bending moments. However, the tensile stresses in the concrete are very important in reducing steel stresses at service load, especially when the reinforcement ratios are low, as is the case in most slab structures. The second factor was that most load tests had been conducted on single panels in buildings with many panels. In this situation, the adjacent panels can be of considerable assistance in resisting the load, and the failure load per unit area for a single panel is invariably much higher than for all panels loaded.[1.9]

The flat plate was developed as a direct simplification of the flat slab. Although Taylor et al.[1.6] mentioned flat slabs without column capitals in or before 1925, the structural type in the United States is primarily a post–World War II development accompanying the construction of high-rise apartment buildings.

The transition from beam-and-girder construction to the two-way slab is not well documented, but the single-panel test structure that was built and load tested in St. Louis in 1911[1.10] is certainly one of the earlier two-way slabs. Additional structures[1.11,1.12] plus laboratory specimens[1.13] were tested during the following decade. Moment distributions from plate theory solutions were reported by Westergaard and Slater[1.8] and a design method for two-way slabs was proposed by Westergaard[1.14] in 1926. The ACI Code requirements were developed from this proposal and another prepared by Di Stasio and van Buren[1.15] in 1936.

Unlike the flat slab, the two-way slab became a feasible structural system only after suitable analyses had been developed. However, the system remained a captive of the assumptions made in the analyses. It was assumed that the beams were nondeflecting, and the beams were then designed for the reactions the slab applied to the rigid supports. As a result of the magnitude and distribution of reactions computed, in combination with the additional restrictions imposed by the use of the working stress design method, the beams that were used in two-way slabs were either quite deep and stiff or had to be heavily reinforced in compression as well as tension. There was no intermediate ground that would allow the rational design of a slab supported on shallow beams until the 1971 ACI Code[1.16] was introduced.

Slabs supported on walls rather than beams are used in bearing-wall buildings. The walls are treated as flexurally rigid beams, and the walls may or may not resist moments transmitted from slab to wall, depending on both the wall construction and on the details of the connections. Many modern precast building systems are bearing-wall structures in which the slabs are supported by walls along most edges.

The prestressed concrete flat plate is a cast-in-place structure in which most of the reinforcement is in the form of post-tensioned tendons. The tendons are placed in a grid with steel parallel to the column lines in both directions. The chief advantage of this form of construction is in crack and deflection control. The prestressing forces cause compression stresses directly opposing the tensile stresses caused by dead and live loads, and as a result the slab will be nearly crack-free at service load levels. Since cracking is delayed, the full section remains effective, and as a consequence deflections at service load levels may be considerably less than in a reinforced concrete flat plate of the same thickness and span. The absence of cracking helps keep the slab watertight. In addition, the use of draped post-tensioned tendons, which cause an upward distributed loading within much of the spans, will reduce the deflections significantly. Proper selection of the steel profile and force may lead to a floor that undergoes only very small long-term deflection changes. The compression forces in the plane of the slab that result from prestressing also enhance the shear strength of the slab. Reinforcing bars are always used in addition to the post-tensioned tendons. These bars would be placed over the columns at the very least, and often in other locations as well. The tendons may be either bonded or unbonded, although they are nearly always unbonded in North American practice.

One-way floors, consisting of precast, prestressed concrete members placed side by side and spanning between beams or girders are also commonly used for floors. The precast members are typically of double- or single-tee section or a hollow-core plank, and a cast-in-place concrete topping slab is generally placed over the precast members to provide a smooth surface and additional stiffness and strength. Such one-way floors can be analyzed and designed using ordinary beam theory.

1.3   CHOICE OF TYPE OF SLAB FLOOR

The choice of type of slab for a particular floor depends on many factors. Economy of construction is obviously an important consideration, but this is a qualitative argument until specific cases are discussed, and is a geographical variable. The design loads, required spans, serviceability requirements, and strength requirements are all important.

For beamless slabs, the choice between a flat slab and a flat plate is usually a matter of loading and span. Flat plate strength is often governed by shear strength at the columns, and for service live loads greater than perhaps 100 lb/ft^2 (4.8 kN/m^2) and spans greater than about 20 to 24 ft (7 to 8 m) the flat slab is often the better choice. If architectural or other requirements rule out capitals or drop panels, the shear strength can be improved by using metal shear heads or some other form of shear reinforcement, but the costs may be high.

Serviceability requirements must be considered, and deflections are sometimes difficult to control in reinforced concrete beamless slabs. Large live loads and small limits on permissible deflections may force the use of large column capitals. Negative-moment cracking around columns is sometimes a problem with flat plates, and again a column capital may be useful in its control.

Deflections and shear stresses may also be controlled by adding beams instead of column capitals. If severe deflection limits are imposed, the two-way slab will be most suitable, as the introduction of even moderately stiff beams will reduce deflections more than the largest reasonable column capital is able to. Beams are also easily reinforced for shear forces.

The choice between two-way and beamless slabs for more normal situations is complex. In terms of economy of material, especially of steel, the two-way slab is often best because of the large effective depths of the beams. However, in terms of labor in building the floor, the flat plate is much cheaper because of the very simple formwork and less complex arrangement of steel. The flat slab is somewhat more expensive in labor than is the flat plate, but the forms for the column capitals are often available as prefabricated units, which can help limit costs. The real cost parameter is the ratio of costs of labor relative to material. Few two-way slabs are built in areas of high labor costs unless there are definite structural reasons, and many are built where steel is the most costly item. Hollow-tile slabs are still built in some places, but only where the cost of both steel and cement is very high relative to labor.

Local customs among builders, designers, and users should not be overlooked when selecting the slab type. There is a natural human tendency to want to repeat what one has previously done successfully, and resistance to change can affect costs. However, old habits should not be allowed to dominate sound engineering decisions.

If a flat plate or flat slab is otherwise suitable for a particular structure, it will be found that there is the additional benefit of minimizing the story height. In areas of absolute height restrictions, this may enable one to have an additional floor for approximately each 10 floors, as compared with a two-way slab with the same clear story heights. The savings in height lead to other economies for a given number of floors, since mechanical features such as elevator shafts and piping are shorter. There is less outside wall area, so wind loadings may be less severe and the building weighs less, which may bring cost reductions in foundations and other structural components. There are other cost savings when the ceiling finishes can be applied directly to the lower surfaces of the slabs.

Beamless slabs will be at a disadvantage if they are used in structures that must resist large horizontal loads by frame action rather than by shear walls or other lateral bracing. The transfer of moments between columns and a slab sets up high local moments, shears, and twisting moments that may be hard to reinforce for. In this situation, the two-way slab is the more capable structure because of the relative ease with which its beams may be reinforced for these forces. In addition, it will provide greater lateral stiffness because of both the presence of the beams and the greater efficiency of the beam–column connections.

The possible choice of a precast one-way floor system, consisting of prestressed concrete members placed side by side and spanning between the beams, girders, or walls and generally covered by a cast-in-place concrete topping slab, should not be overlooked.

1.4 APPROACHES TO THE ANALYSIS AND DESIGN OF SLAB SYSTEMS

There are a number of possible approaches to the analysis and design of reinforced concrete slab systems. The various approaches available are elastic theory, limit analysis theory, and modifications to elastic theory and limit analysis theory as in the ACI Code.[1.1] Such methods can be used to analyze a given slab system to determine either the stresses in the slabs and the supporting system or the load-carrying capacity. Alternatively, the methods can be used to determine the distribution of moments and shears to allow the reinforcing steel and concrete sections to be designed.

In the following sections the complete behavior of slab systems when loaded from zero load to ultimate load is considered briefly to illustrate the behavior at the various load stages, and then elastic theory, limit analysis theory, and the ACI Code method are discussed briefly to introduce the reader to the various possible approaches to analysis and design.

1.4.1   Complete Behavior of Slab Systems

The bending and torsional moments, shear forces, and deflections of slab systems, with given dimensions, steel content, and material properties, at any stage of loading from zero to ultimate load, can be determined analytically using the conditions of static equilibrium and geometrical compatibility if the moment–deformation relationships of the slab elements, and the yield criteria for bending and torsional moments and shear force when the strength of slab elements is reached, are known. In such an analysis of the complete behavior of slab systems, difficulties are caused by the nonlinearity of the moment–deformation relationships of the slab elements at high levels of stress, and a step-by-step procedure with load increased increment by increment is generally necessary. A possible procedure for such an analysis is described below.

At low levels of load the slab elements are uncracked and the actions and deformations can be computed from elastic theory using the uncracked flexural rigidity of the slab elements. The slab elements are searched at each load increment to ascertain whether cracking of the concrete has taken place. When it is found that the cracking moment has been reached, the flexural rigidity of the element is recomputed on the basis of the cracked section value and the actions and deformations of the slab are recomputed. This procedure is repeated at the same load level until all the flexural rigidity values are correct. At higher load increments, when the stresses in one or more elements begin to enter the inelastic range, the flexural rigidities of those elements are reduced to that corresponding to the particular point on the moment–deformation relationship of the element. This requires the computation to be repeated at the load level until the flexural rigidity of each element is correct. Eventually, with further load increments, the strength of one or more elements is reached and if the elements are sufficiently ductile, plasticity will spread through the slab system with further loading. The ultimate load is reached when deflections occur without further increase in load and hence when further load cannot be carried.

It is evident that the full analysis to determine the complete behavior of a slab system at all levels of loading is lengthy and can be undertaken successfully only with the aid of a computer having large storage. For such a general computer program the input information necessary includes the slab system geometry, the section dimensions, steel contents, material properties, and the type of loading. The output would be the distribution of bending and torsional moments and shear force, and the deflections, at any load level up to ultimate load. Such a computer program would be a powerful analytical tool for evaluating or checking structural performance of slab systems over the full range of loading, including the behavior at service loads and ultimate load.

Many finite element programs have been developed for the analysis of slabs, and these programs have become more capable as computers have become more powerful. The earlier efforts which included nonlinear effects generally were for simple cases with idealized boundary conditions and were essentially research tools. Examples of such analyses are those of Bell and Elms,[1.17] Jofriet and McNeice,[1.18] and Hand et al.[1.19]

Current commercial and semicommercial programs that can deal with slabs with fairly general support conditions include the following nonexhaustive list: FINITE (University of Illinois at Urbana–Champaign), SAP2000 and SAFE (Computers and Structures, Inc., Berkeley, California), RISA-3D (RISA Technologies, Lake Forest, California), ABAQUS (BKS, Providence, Rhode Island), and PCA-Mats (Portland Cement Association, Skokie, Illinois). PCA-Mats is a specialized program adapted to the analysis of mat foundations and slabs on grade. Some of these programs include nonlinear effects such as cracking and yielding, and some include selection of reinforcement.

Code design procedures are usually based on elastic theory moments modified in the light of some moment redistribution, which has been shown by tests to be possible without excessive cracking or deflections at service loads. Elastic theory moments without modification, and moments obtained from methods of limit analysis, form alternative design approaches that are recommended by some codes of practice.

1.4.2 Elastic Theory Analysis

Classical elastic theory analysis applies to isotropic slabs that are sufficiently thin for shear deformations to be insignificant and sufficiently thick for in-plane forces to be unimportant. Most floor slabs fall into the range in which classical elastic theory is applicable. The distribution of moments and shears found by elastic theory is such that:

1. The equilibrium conditions are satisfied at every point in the slab.
2. The boundary conditions are complied with.
3. Stress is proportional to strain; that is, bending moments are proportional to curvature.

The governing equation is a fourth-order partial differential equation in terms of the deflection of the slab at general point (x,y) on the slab, the loading on the slab, and the flexural rigidity of the slab section. This equation is difficult to solve in many realistic cases, particularly when the effects of deformations of the supporting system are to be taken into account. However, many analytical techniques have been developed to obtain solutions. In particular, the use of electronic computers and finite difference or finite element methods enables elastic theory solutions to be obtained for slab systems with any loading or boundary conditions. Standard computer programs and charts and tables are available to assist the designer. The solution gives the distributions of bending and torsional moments and shear forces throughout the slab. The actions in the supporting system can also be determined. Works by Timoshenko and Woinowsky-Krieger,[1.20] Westergaard,[1.8,1.21] and other authors give thorough treatments of elastic theory.

1.4.3   Limit Analysis

Limit analysis recognizes that because of plasticity, redistribution of moments and shears away from the elastic theory distribution can occur before the ultimate load is reached. Such redistribution of moments occurs because for typical reinforced concrete sections there is little change in moment with curvature once the tension steel has reached the yield strength. Thus, when the most highly stressed sections of a slab reach the yield moment they tend to maintain a moment capacity that is close to the flexural strength with further increase in curvature, while yielding of the slab reinforcement spreads to other sections of the slab with further increase in load. Limit analysis computes the ultimate load of the slab, and the distribution of moments and shears at the ultimate load, assuming that the slab sections are sufficiently ductile to enable the required redistribution of bending moments to occur. To determine the ultimate load, or the distribution of moments and shear at that load, of a given slab system, either a lower bound or an upper bound method may be used.

The *lower bound method* postulates a distribution of moments in the slab system at the ultimate load such that:

1. The equilibrium conditions are satisfied at all points in the slab system.
2. The yield criterion defining the strength of the slab sections is not exceeded anywhere in the slab system.
3. The boundary conditions are complied with.

The ultimate load is calculated from the equilibrium equation and the postulated distribution of moments. For a given slab system the lower bound method gives an ultimate load that is either correct or too low; that is, the ultimate load is never overestimated.

The *upper bound method* postulates a collapse mechanism for the slab system at the ultimate load such that:

1. The moments at the plastic hinges are not greater than the ultimate moments of resistance of the sections.
2. The collapse mechanism is compatible with the boundary conditions.

A collapse mechanism is composed of portions of the slab separated by lines of plastic hinging, and the ultimate load is calculated from the postulated collapse mechanism. However, the portions of the slab between the lines of plastic hinging are not examined to ensure that the moments there do not exceed the available ultimate moments of resistance of those sections. The ultimate moments of resistance between the plastic hinges are exceeded if an incorrect collapse mechanism is postulated. For a given slab system the upper bound method gives an ultimate load that is either correct or too high, but if all the possible collapse mechanisms for the slab system are examined, the mechanism giving the lowest ultimate load is the correct one. It is evident that care must be taken to ensure that the correct collapse mechanism is used; otherwise, the ultimate load will be overestimated.

The difference between the two methods can be illustrated for the simple case of the uniformly loaded one-way slab fixed at the supports shown in Fig. 1.7a. The slab has ultimate moments of resistance per unit width of m'_u for the negative moment and m_u for positive moment. For a lower bound solution, the limit moment diagram of Fig. 1.7b could be chosen. This distribution of moments is a possible solution because the ultimate moments of resistance are not exceeded anywhere in the slab. As is evident from statics, the ultimate load per unit area w_u calculated for the moment diagram is too low because the ultimate moments of resistance are not reached at all (three) critical sections. The moment diagram needs to be modified to allow the moments at all three critical sections to reach ultimate capacity before the correct ultimate load is calculated. Alternatively, for an upper bound solution the collapse mechanism shown in Fig. 1.7c could be chosen. It is evident that the ultimate load given by this possible mechanism is incorrect and predicts a load that is too high, as can be seen from the resulting limit moment diagram of Fig. 1.7d, in which the ultimate positive moment of resistance is exceeded. The correct collapse mechanism and ultimate load is given when plastic hinges form at the supports and at midspan. The foregoing example of a simple slab system may appear trivial because the correct solution is evident by inspection. However, the limit analysis solutions of slabs with more complex shapes and boundary conditions is more difficult, and it is not always evident whether the correct solution has been found.

The most commonly used lower bound approach is Hillerborg's *strip method*.[1.22,1.23] This method obtains the distribution of moments and shears by replacing the slab by systems of strips running in two directions, which share the load. Strip action is a valid lower bound procedure because if the loading is carried entirely by flexure, and therefore satisfies the requirements of statics, no assistance from torsion is necessary. Wood[1.24] has shown that alternative lower bound solutions, including torsion, are difficult to obtain in many cases. The upper bound method for slabs is the *yield line theory,* which was due primarily to Johansen.[1.25] Publications by Wood[1.24,1.27] and Jones[1.26,1.27] give more recent treatments of the theory. If yield line theory is used, the designer must examine all possible collapse mechanisms to ensure that the load-carrying capacity of the slab is not overestimated. The correct collapse mechanisms in nearly all cases are well known, however, and therefore the designer is not often faced with the uncertainty of whether further alternatives exist.

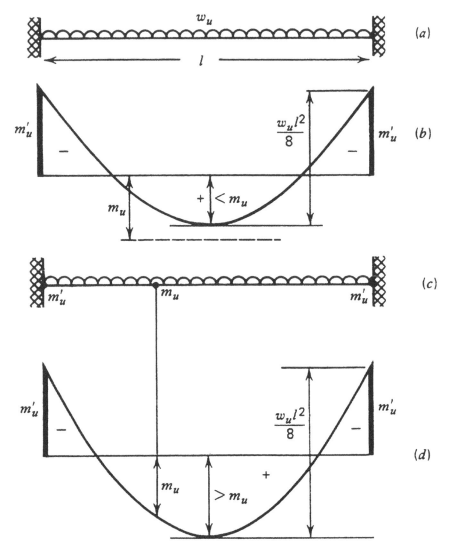

Figure 1.7   Lower and upper bound limit analysis solutions: (*a*) Uniformly loaded fixed-end one-way slab; (*b*) Possible limit moment diagram for lower bound solution; (*c*) Possible collapse mechanism; (*d*) limit moment diagram associated with possible collapse mechanism.

1.4.4   ACI Building Code Method

The ACI Code[1.1] contains procedures for the design of uniformly loaded reinforced concrete slab floors. The procedures are based on analytical studies of the distribution of moments using elastic theory and of strength using yield line theory, the results of tests on model and prototype structures, and experience of slabs built. Two detailed approaches are given in the code: the direct design method and the equivalent frame method. The methods apply both to floors with beams and floors without beams.

The direct design method may be used only for slab systems with three or more continuous spans in each direction, rectangular panels with aspect ratio not greater than 2, successive spans not differing by more than 33% of the longer span, columns not offset by more than 10% of the span length in the direction of the offset from centerlines of successive columns, loads due to gravity only and distributed uniformly over a whole panel, live loads not more than two times the dead load, and when beams are present, reasonable ratios between the beam flexural stiffnesses in each direction. The design static moment for a span is determined for a width of slab extending between the centerlines of the adjacent panels on each side of an interior column line. For exterior columns, only the slab width to one side is considered. The sum of the maximum negative and positive moments acting across the width of slab must for equilibrium be equal to the static moment $M_0 = wl_2l_n^2/8$, where w is the load per unit area, l_n the length of clear span between faces of supports in the direction the moments are being determined, and l_2 the width of slab considered. For interior spans the negative and positive design moments are taken as $0.65M_0$ and $0.35M_0$, respectively. For end spans the distributions of negative and positive design moments are tabulated for several support conditions. Alternatively, moments are computed which depend on the ratio of flexural stiffness of the equivalent column to the combined flexural stiffness of the slabs and beams at the end joint in the direction of the moments. The equivalent column includes the effect of the torsional stiffness of the edge beam of the floor. The determined negative and positive moments are shared between the column and middle strip regions of the floor, the distribution depending on the flexural stiffnesses of the beams (if any) and slabs, and the span lengths in the two directions. The moment in the column strip is shared between the beam (if any) and the adjacent regions of slab, the proportion of moment taken by each depending on the flexural stiffnesses of the beam (if any) and slab and the span lengths. The distribution of the static moment between the critical sections is based mainly on elastic theory, to ensure that the steel stresses are not excessively high at service load so that crack widths and deflections remain within acceptable limits.

The equivalent frame method allows consideration of a more general range of slab systems. The structure is considered to be made up of equivalent frames on column lines taken longitudinally and transversely through the building. Each equivalent beam is assumed to consist of the beam (if any) and the slab within the centerlines of the panels on each side of the column line. Each equivalent column is assumed to consist of the actual column plus attached torsional members at each floor level extending transversely to the panel centerlines each side of the column. The bending moments in the equivalent frame due to the design loading are found by linear elastic structural analysis and are distributed between the columns and middle strips, including the beams (if any), according to the rules used for the direct design method.

It is likely that these two explicit analysis methods will be removed from the Code early in the twenty-first century, leaving statements of general principles.

1.4.5 Design Procedures

The required reinforcement for slab systems can be designed from the distribution of moments at the factored (ultimate) loads using flexural strength theory for sections, or from the distribution of moments at the service loads using the working stress (elastic theory) method for sections.

The ACI Code[1.1] states that "A slab system may be designed by any procedure satisfying conditions of equilibrium and geometrical compatibility if shown that the design strength at every section is at least equal to the required strength . . . and that all serviceability conditions, including specified limits on deflections, are met." The direct design method and the equivalent frame method both comply with the code statement provided that deflections are checked. However, it may be necessary to seek alternative methods of design for cases of slab systems of irregular layout, or for unusual loading conditions, or for other reasons.

It is evident that the distribution of moments given by elastic theory for any design case can be used to satisfy the strength and serviceability requirements of the ACI Code. On the other hand, limit design, for example, by yield line theory or the strip method, will lead to the required strength but may be suspect from the point of view of serviceability. Thus, if limit design methods are used, checks of crack widths and deflections at service load may be necessary. However, limit design has been used widely in Europe and elsewhere. For example, it is specifically allowed by the building code of the United Kingdom[1.28] and New Zealand[1.29] and the recommendations of the CEB-FIP.[1.30] It is likely that with more experience of slab systems in service that have been designed by limit design methods, the approach will gain more acceptance in North America and elsewhere.

1.5 NOTES ON UNITS

Most of the examples where units are necessary are presented in English units, sometimes referred to as *U.S. Customary Units*. In many cases metric equivalents, in *SI units*, are given immediately following the English units. That pattern that has been used is that SI units which have been codified are enclosed in square brackets: for instance, $2 \sqrt{f_c'}$ lb/in^2 [$\sqrt{f_c'}/6$ N/mm^2]. On the other hand, metric equivalents that are simple conversions made by the authors are enclosed in parentheses: for instance, a bar spacing is stated as 6 in. (150 mm).

The SI units have one consistent departure from the strictly correct usage. Stresses and pressures are given in units that are clearly force per unit area. Thus, a steel yield stress is cited as 420 N/mm^2 rather than the equivalent 420 MPa. A distributed load on a floor is given as 5 kN/m^2 rather than the equivalent 5 kPa. These changes have been made as an aid to the student in understanding that stress times area equals force.

REFERENCES

1.1   *Building Code Requirements for Structural Concrete*, ACI 318-95, and *Commentary*, ACI 318R-95, American Concrete Institute, Farmington Hills, Mich., 1995, 371 pp.

1.2   H. Cross, "Continuity as a Factor in Reinforced Concrete Design," *Proc. ACI*, Vol. 25, 1929, pp. 669–708.

1.3   M. A. Sozen and C. P. Siess, "Investigation of Multi-panel Reinforced Concrete Floor Slabs: Design Methods—Their Evolution and Comparison," *Proc. ACI*, Vol. 60, August 1963, pp. 999–1028.

1.4   A. B. MacMillan, "A Comparison of Methods of Computing the Strength of Flat Reinforced Plates," *Eng. News*, Vol. 63, No. 13, March 13, 1910, pp. 364–367.

1.5   L. F. Brayton, "Methods for the Computation of Reinforced Concrete Flat Slabs," *Eng. News*, Vol. 64, No. 8, August 25, 1910, pp. 210–211.

1.6   F. W. Taylor, S. E. Thompson, and E. Smulski, *Concrete: Plain and Reinforced*, Vol. 1, *Theory and Design of Concrete and Reinforced Concrete Structures*, 4th ed., Wiley, New York, 1925.

1.7   J. R. Nichols, "Statical Limitations upon the Steel Requirement in Reinforced Concrete Flat Slab Floors," *Trans. ASCE*, Vol. 77, 1914, pp. 1670–1681; "Discussion," pp. 1682–1736.

1.8   H. M. Westergaard and W. A. Slater, "Moments and Stresses in Slabs," *Proc. ACI*, Vol. 17, 1921, pp. 415–538.

1.9   J. O. Jirsa, M. A. Sozen, and C. P. Siess, "Tests of a Flat Slab Reinforced with Welded Wire Fabric," *J. Struct. Div., ASCE*, Vol. 92, No. ST3, June 1966, pp. 199–224.

1.10 "Test of a Concrete Floor Reinforced in Two Directions," *Proc. ACI*, Vol. 8, 1912, pp. 132–157.

1.11 W. A. Slater, A. Hagner, and G. P. Anthes, "Tests of a Hollow Tile and Concrete Floor Slab Reinforced in Two Directions," *Technol. Pap. Bur. Stand.*, Vol. 16, 1923, pp. 727–796.

1.12 L. J. Larson and S. N. Petrenko, "Loading Tests of a Hollow Tile and Reinforced Concrete Floor of Arlington Building, Washington, D.C.," *Technol. Pap. Bur. Stand.*, Vol. 17, 1924, pp. 405–445.

1.13   C. Bach and O. Graf, "Versuche mit allseitig aufliegenden, quadratischen, und rechteckigen Eisenbetonplatten," *Dtsch. Ausschuss Eisenbeton (Berlin)*, Vol. 30, 1915.

1.14   H. M. Westergaard, "Formulas for the Design of Rectangular Floor Slabs and the Supporting Girders," *Proc. ACI,* Vol. 22, 1926, pp. 26–46.

1.15   J. Di Stasio and M. P. van Buren, "Slabs Supported on Four Sides," *Proc. ACI,* Vol. 32, January–February 1936, pp. 350–364.

1.16   ACI Committee 318, *Building Code Requirements for Reinforced Concrete*, ACI 318-71, American Concrete Institute, Detroit, 1971, 78 pp.

1.17   J. C. Bell and D. G. Elms, "A Finite Element Approach to Post-elastic Slab Behavior," in *Cracking, Deflection and Ultimate Load of Concrete Slab Systems*, ACI SP-30, American Concrete Institute, Detroit, 1971, pp. 325–344.

1.18   J. C. Jofriet and G. M. McNeice, "Finite Element Analysis of Reinforced Concrete Slabs," *J. Struct. Div., ASCE*, Vol. 97, No. ST3, March 1971, pp. 785–806.

1.19   F. R. Hand, D. A. Pecknold, and W. C. Schnobrich, "Nonlinear Layered Analysis of RC Plates and Shells," *J. Struct. Div., ASCE*, Vol. 99, No. ST7, July 1973, pp. 1491–1505.

1.20   S. Timoshenko and S. Woinowsky-Krieger, *Theory of Plates and Shells,* 2nd ed., McGraw-Hill, New York, 1959.

1.21   H. M. Westergaard, "Computation of Stresses in Bridge Slabs Due to Wheel Loads," *Public Roads,* Vol. 11, No. 1, March 1930, pp. 1–23.

1.22   A. Hillerborg, "Theory of Equilibrium for Reinforced Concrete Slabs," *Betong,* Vol. 41, No. 4, 1956, pp. 171–182. (Translated by Building Research Station, DSIR, Garston, England, 1962.)

1.23   A. Hillerborg, *Dimensionering av armerade betongplattor enligt strimlemetoden,* Almqvist & Wiksell, Stockholm, 1974, 327 pp. (*Strip Method of Design*, translated by Cement and Concrete Association, London, 1975, 256 pp.)

1.24   R. H. Wood, *Plastic and Elastic Design of Slabs and Plates,* Thames and Hudson, London, 1961, 344 pp.

1.25   K. W. Johansen, *Brudlinieteorier,* Jul. Ojellerups Forlag, Copenhagen, 1943. (*Yield Line Theory,* translated by Cement and Concrete Association, London, 1962, 181 pp.)

1.26   L. L. Jones, *Ultimate Load Analysis of Reinforced and Prestressed Concrete Structures,* Chatto & Windus, London, 1962, 248 pp.

1.27   R. H. Wood and L. L. Jones, *Yield-Line Analysis of Slabs,* Thames and Hudson, Chatto & Windus, London, 1967, 400 pp.

1.28   *Structural Use of Concrete*, BS 8110, Part 1:1997, *Code of Practice for Design and Construction*, British Standards Institution, London, 1997, 121 pp.

1.29   *The Design of Concrete Structures*, NZS 3101:1995, Standards New Zealand, Wellington, New Zealand, 1995.

1.30   *International System of Unified Standard Codes of Practice for Structures*, Vol. II, *CEB-FIP Model Code for Concrete Structures*, Comité Euro-International du Béton/Fédération Internationale du Précontrainte, Paris (English translation), April 1978, 348 pp.

2 Basis of Elastic Theory Analysis

2.1 INTRODUCTION

The purpose of this chapter is to give some background into the methods and problems of elastic theory analysis of plate structures. It is not intended to show the reader how to solve any but the simplest problems, but rather, to show how some problems have been solved and to direct the reader to more specific work on various topics and methods.

The derivation of the governing differential equations for a plate is given in Section 2.2, along with a few remarks on the classical methods of solution of these equations. Since the basic equation is a fourth-order partial differential equation that becomes difficult to solve for many realistic problems, many alternative methods of obtaining the moment and shear distributions required for design have been developed, and some are described in the following sections.

Section 2.3 includes some information on the use of models. The numerical methods of using finite difference equations and the finite element technique are described briefly in Sections 2.4 and 2.5, respectively. Approximate methods of solution have been important over a long period of time, and a few are outlined in Section 2.6.

2.2 CLASSICAL PLATE THEORY

2.2.1 Lagrange's Equation

Elastic deformations of isotropic plates loaded normal to their plane are governed by a fourth-order partial differential equation:

$$\frac{\partial^4 w}{\partial x^4} + 2\frac{\partial^4 w}{\partial x^2\,\partial y^2} + \frac{\partial^4 w}{\partial y^4} = \frac{q}{D} \tag{2.1}$$

where w is the deflection of plate in direction of loading at point (x,y), q the loading imposed on the plate per unit area, a function of x and y, and D the flexural rigidity of the plate $= Eh^3/12(1 - \mu^2)$, where E is Young's modulus of plate material, h the plate thickness, and μ is Poisson's ratio. This equation, derived from considerations of equilibrium and strain compatibility before 1820, applies specifically to medium-thick plates (i.e., plates which are thin enough that shear deformations are not important but thick enough that in-plane or membrane forces are not important). Most reinforced and prestressed concrete floor slabs fall within these limits at service load levels.

The derivation of Eq. 2.1 and some applications are covered in the following sections. The works by Timoshenko and Woinowsky-Krieger[2.1] and Westergaard and Slater[2.2] give a more complete treatment of elastic theory. Reference 2.1 is probably the most important current reference on the elastic analysis of slabs. Reference 2.2 includes a history of the solutions and attempts at solutions during the first century after Eq. 2.1 was derived.

2.2.2 Equilibrium

Equation 2.1 may be derived following a two-step process, the first considering equilibrium and the second strain compatibility. The forces acting on a small element of slab are shown in Fig. 2.1. All the slab actions are per unit width and have been multiplied by the element dimensions to obtain the forces acting on the element. For clarity the forces have been separated into two groups: the surface and shear forces (Fig. 2.1a) and the flexural and twisting moments (Fig. 2.1b). Note that the shear force V_x acts on the same face of the element as the moment m_x. This moment requires reinforcement extending in the x-direction. The moments are shown in vector form, using the right-hand screw rule. The vector represents the moment acting on a face, with the moment tending to produce rotation about the axis of the arrow. The length of the vector represents the magnitude of the moment, and vectors can be added graphically or otherwise. An alternative representation of the moments is shown in Fig. 5.2. Both the m_x and m_y moments shown in Fig. 2.1b are positive moments, producing compression at the top surface of the slab.

Equilibrium must be satisfied, and $\Sigma F_z = 0$ leads to

$$\frac{\partial V_x}{\partial x} + \frac{\partial V_y}{\partial y} + q = 0 \tag{2.2}$$

Summation of moment of all forces about the x-axis, and neglecting higher-order terms, leads to

$$\frac{\partial m_x}{\partial x} + \frac{\partial m_{xy}}{\partial y} = V_x \tag{2.3}$$

Similarly, with the assumption that $m_{xy} = m_{yx}$ since $\tau_{xy} = \tau_{yx}$ in every horizontal plane at any distance z from the neutral surface, one finds that

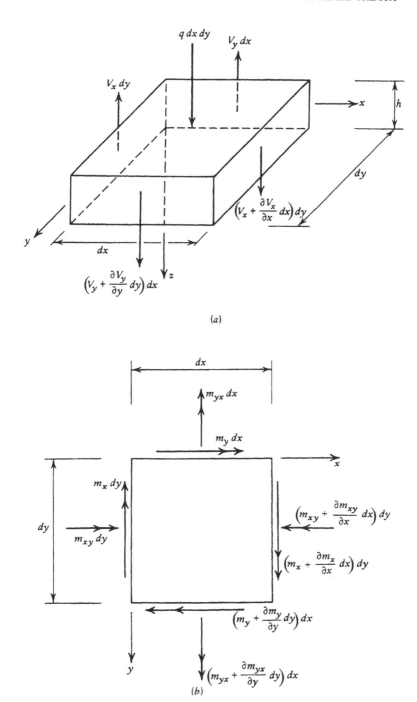

Figure 2.1   Forces acting on small element of slab: (*a*) Surface and shear forces; (*b*) flexural and twisting moment vectors (as viewed from above).

$$\frac{\partial m_y}{\partial y} + \frac{\partial m_{xy}}{\partial x} = V_y \qquad (2.4)$$

Substitution of Eqs. 2.3 and 2.4 into Eq. 2.2 leads to the plate equilibrium equation,

$$\frac{\partial^2 m_x}{\partial x^2} + 2\frac{\partial^2 m_{xy}}{\partial x\,\partial y} + \frac{\partial^2 m_y}{\partial y^2} = -q \qquad (2.5)$$

This equation is simply one of equilibrium, and is independent of the state of elasticity or plasticity, Poisson's ratio, or whether the plate is isotropic or orthotropic.

2.2.3   Moment–Deformation Relationships

The addition of the strain-compatibility part of the problem requires the following assumptions:

1. Material obeys Hooke's law and is isotropic.
2. The deflection is small relative to the thickness of the slab.
3. Any straight line perpendicular to the middle surface of the slab before bending remains straight and normal to the middle surface after bending.
4. Direct stresses normal to the middle surface are negligible.

The force–deformation relationships may be formulated as follows. Consider the small element of slab shown in Fig. 2.2, subjected to the moments m_x. The moments bend the element into a cylindrical segment with curvature $\phi_x = 1/R_x$, where R_x is the radius of curvature in the x-direction. The curvature may also be expressed as

$$\phi_x = \frac{\varepsilon_x}{z} \qquad \text{rad/unit length} \qquad (2.6)$$

where ε_x is the strain in x-direction in a fiber located at the distance z from the neutral surface.

The curvature is also defined as

$$\phi_x = \frac{\partial^2 w}{\partial x^2} \qquad (2.7)$$

This combined with Eq. 2.6 leads to the expression

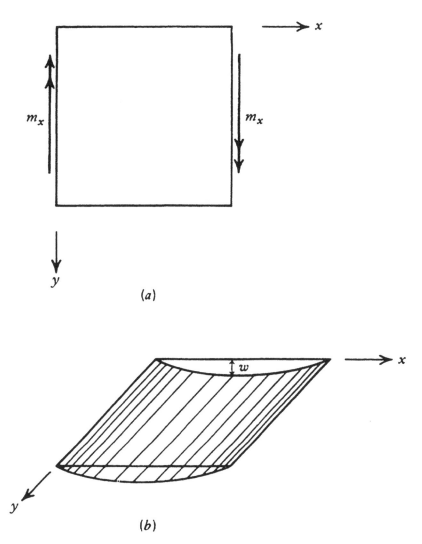

Figure 2.2 Applied moments and deformed shape for bending in x-direction: (a) applied moments per unit width; (b) deformed shape.

$$\varepsilon_x = -z\,\frac{\partial^2 w}{\partial x^2} \tag{2.8}$$

This expression also applies when moment m_y is acting on the element as well as m_x, provided that $-\partial^2 w/\partial x^2$ is the curvature in the x-direction due to both m_x and m_y. Similarly,

$$\varepsilon_y = -z \frac{\partial^2 w}{\partial y^2} \qquad (2.9)$$

which also applies to the general case where m_x and m_y are acting. The general two-dimensional stress–strain relationships for a linearly elastic material are given by any strength-of-materials text as

$$f_x = \frac{E}{1 - \mu^2} (\varepsilon_x + \mu\varepsilon_y) \qquad (2.10a)$$

$$f_y = \frac{E}{1 - \mu^2} (\varepsilon_y + \mu\varepsilon_x) \qquad (2.10b)$$

$$\tau_{xy} = \gamma_{xy} \frac{E}{2(1 + \mu)} \qquad (2.10c)$$

Thus, the stress f_x in a fiber at distance z from the neutral surface may be expressed in terms of the deformed shape of the plate as

$$f_x = -z \frac{E}{1 - \mu^2} \left(\frac{\partial^2 w}{\partial x^2} + \mu \frac{\partial^2 w}{\partial y^2} \right) \qquad (2.11)$$

The moment per unit width in the x-direction may be obtained by integrating the first moment of f_x about the neutral surface over the depth of the plate:

$$
\begin{aligned}
m_x &= \int_{-h/2}^{h/2} f_x z \, dz \\
&= \int_{-h/2}^{h/2} -\frac{E}{1 - \mu^2} \left(\frac{\partial^2 w}{\partial x^2} + \mu \frac{\partial^2 w}{\partial y^2} \right) z^2 \, dz \\
&= \frac{Eh^3}{12(1 - \mu^2)} \left(\frac{\partial^2 w}{\partial x^2} \mu \frac{\partial^2 w}{\partial y^2} \right)
\end{aligned}
$$

Introducing $D = Eh^3/12(1 - \mu^2)$ as the plate rigidity factor then gives

$$m_x = -D \left(\frac{\partial^2 w}{\partial x^2} + \mu \frac{\partial^2 w}{\partial y^2} \right) \qquad (2.12)$$

Similarly,

$$m_y = -D \left(\frac{\partial^2 w}{\partial y^2} + \mu \frac{\partial^2 w}{\partial x^2} \right) \qquad (2.13)$$

The application of twisting moments to a small element of slab, as shown in Fig. 2.3a, produces the anticlastic or saddle-shaped deformations shown in Fig. 2.3b. The twisting moments must always be equal (i.e., $m_{xy} = m_{yx}$) and have either the directions shown in Fig. 2.3a or all must be reversed; otherwise, equilibrium of shear stresses cannot be maintained. This means that the torques acting on any two adjacent faces must both tend to either lift or depress the common corner. It is not possible to have an arrangement of torques where one tries to lift a corner and the other tries to push it back down.

The twisting moment–distortion relationship can be derived in a number of different ways. A convenient way is to replace the system shown in Fig. 2.3a with the equivalent system shown in Fig. 2.3c. This replacement system (which is discussed later in this section) has the advantage of being subjected to only flexural moments, which are also the principal moments. The flexural moment–deformation relationships have already been derived and can be written as

$$m_{x'} = -D\left(\frac{\partial^2 w}{\partial x'^2} + \mu\,\frac{\partial^2 w}{\partial y'^2}\right) \tag{2.14}$$

$$m_{y'} = -D\left(\frac{\partial^2 w}{\partial y'^2} + \mu\,\frac{\partial^2 w}{\partial x'^2}\right) \tag{2.15}$$

These two equations can be solved to give the curvatures in the x'- and y'-directions, noting that $m'_x = m_{xy}$, $m_{y'} = -m_{xy}$, and $m'_x = m_y$:

$$\frac{\partial^2 w}{\partial x'^2} = \frac{-(m_{x'} - \mu m_{y'})}{D(1 - \mu^2)} = -\frac{m_{xy}}{D(1 - \mu)} \tag{2.16}$$

$$\frac{\partial^2 w}{\partial y'^2} = \frac{-(m_{y'} - \mu m_{x'})}{D(1 - \mu^2)} = -\frac{m_{xy}}{D(1 - \mu)} \tag{2.17}$$

The distortion relative to the x–y axes in terms of the curvatures relative to the x'–y' axes may be expressed (see below for proof) as

$$-\frac{\partial^2 w}{\partial x\,\partial y} = \frac{1}{2}\left(\frac{\partial^2 w}{\partial y'^2} + \frac{\partial^2 w}{\partial x'^2}\right) \tag{2.18}$$

Substituting Eqs. 2.16 and 2.17 into Eq. 2.18 and solving for m_{xy} gives the relationship

$$m_{xy} = -\frac{\partial^2 w}{\partial x\,\partial y}\,D(1 - \mu) \tag{2.19}$$

The term $\partial^2 w/\partial x\,\partial y$ represents the distortion or twist, that is, the rate of change of slope in the x-direction as one moves in the y-direction.

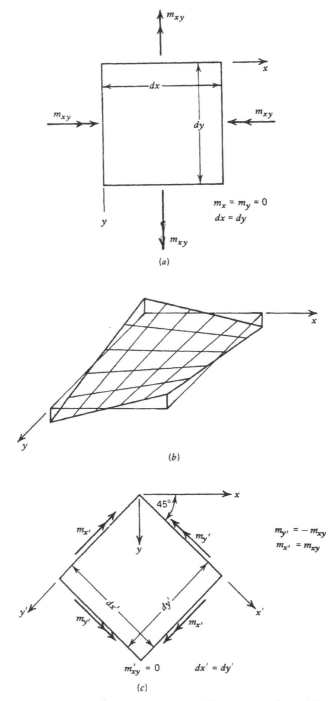

Figure 2.3 Applied twisting moments and resulting deformations: (*a*) applied twisting moments per unit width; (*b*) deformed shape; (*c*) equivalent system.

Substitution of Eqs. 2.12, 2.13, and 2.19 into the equilibrium equation, Eq. 2.5, leads to the Lagrange or biharmonic equation, Eq. 2.1.

The transformation equation given as Eq. 2.18 is a special case for a 45° shift of axes, simplified from a more general expression. The general expression, plus the transformations for flexural curvatures, may be derived as follows, considering the deformations of the small element of slab shown in Fig. 2.4. It is desired to find the curvatures and distortions relative to the $x'-y'$ axes, assuming that they are known relative to the $x-y$ axes.

First, the difference in deflection between points a and b is written as

$$dw = \frac{\partial w}{\partial x} dx + \frac{\partial w}{\partial y} dy \qquad (2.20)$$

The slope in the x'-direction is then

$$\frac{\partial w}{\partial x'} = \frac{\partial w}{\partial x} \frac{dx}{dx'} + \frac{\partial w}{\partial y} \frac{dy}{dx'} = \frac{\partial w}{\partial x} \cos \beta + \frac{\partial w}{\partial y} \sin \beta \qquad (2.21)$$

From this it is noted that

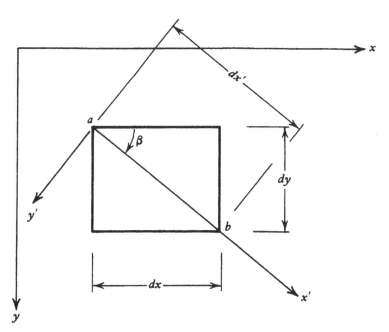

Figure 2.4 Slab element showing new orientation of coordinate axes.

$$\frac{\partial}{\partial x'} = \frac{\partial}{\partial x} \cos \beta + \frac{\partial}{\partial y} \sin \beta \qquad (2.22)$$

A second partial derivative with respect to x' gives

$$\frac{\partial^2 w}{\partial x'^2} = \left(\frac{\partial}{\partial x} \cos \beta + \frac{\partial}{\partial y} \sin \beta \right) \left(\frac{\partial w}{\partial x} \cos \beta + \frac{\partial w}{\partial y} \sin\beta \right)$$

$$= \frac{\partial^2 w}{\partial x^2} \cos^2\beta + \frac{\partial^2 w}{\partial y^2} \sin^2\beta + 2 \frac{\partial^2 w}{\partial x \, \partial y} \sin \beta \cos \beta \qquad (2.23)$$

The slope in the y'-direction may be found by substituting $\beta + 90°$ for β in Eq. 2.21, giving

$$\frac{\partial w}{\partial y'} = -\frac{\partial w}{\partial x} \sin \beta + \frac{\partial w}{\partial y} \cos \beta \qquad (2.24)$$

Taking the second partial derivative with respect to y' gives

$$\frac{\partial^2 w}{\partial y'^2} = \frac{\partial^2 w}{\partial x^2} \sin^2\beta + \frac{\partial^2 w}{\partial y^2} \cos^2\beta - 2 \frac{\partial^2 w}{\partial x \, \partial y} \sin \beta \cos \beta \qquad (2.25)$$

The distortion relative to the $x'-y'$ axes may be found by taking a partial derivative of Eq. 2.24 with respect to x':

$$\frac{\partial^2 w}{\partial x' \, \partial y'} = \left(\frac{\partial}{\partial x} \cos \beta + \frac{\partial}{\partial y} \sin \beta \right) \left(-\frac{\partial w}{\partial x} \sin \beta + \frac{\partial w}{\partial y} \cos \beta \right)$$

$$= \left(\frac{\partial^2 w}{\partial x^2} - \frac{\partial^2 w}{\partial y^2} \right) \sin \beta \cos \beta + \frac{\partial^2 w}{\partial x \, \partial y} (\cos^2\beta - \sin^2\beta)$$

$$= \frac{1}{2} \sin 2\beta \left(\frac{\partial^2 w}{\partial x^2} - \frac{\partial^2 w}{\partial y^2} \right) + \cos 2\beta \frac{\partial^2 w}{\partial x \, \partial y} \qquad (2.26)$$

For the case considered above, $2\beta = 90°$, which eliminates the second term for Eq. 2.26. Interestingly enough, $(\partial^2 w/\partial x'^2 + \partial^2 w/\partial y'^2)$ is independent of the angle β and the sum is sometimes referred to as the *average curvature*.

2.2.4   Shear–Deformation Relationships

Substitution of the bending and twisting moments from Eqs. 2.12, 2.13, and 2.19 into Eqs. 2.3 and 2.4 gives the following shear forces, per unit width:

$$V_x = -D\left(\frac{\partial^3 w}{\partial x^3} + \frac{\partial^3 w}{\partial x\,\partial y^2}\right) \tag{2.27}$$

$$V_y = -D\left(\frac{\partial^2 w}{\partial y^3} + \frac{\partial^3 w}{\partial x^2\,\partial y}\right) \tag{2.28}$$

2.2.5 Boundary Conditions

At a fixed edge there is neither deflection nor rotation. If the edge extends in the y-direction, the boundary conditions may be expressed mathematically as

$$w = 0$$
$$\frac{\partial w}{\partial x} = 0 \tag{2.29}$$

At a simply supported edge, there is no deflection, but the edge is free to rotate and therefore there is no flexural moment perpendicular to the edge. If the edge extends in the y-direction, the boundary conditions may be expressed as

$$m_x = 0 = -D\left(\frac{\partial^2 w}{\partial x^2} + \mu\,\frac{\partial^2 w}{\partial y^2}\right)$$

where $\partial^2 w/\partial y^3 = 0$ by inspection, leading to

$$w = 0$$
$$\frac{\partial^2 w}{\partial x^2} = 0 \tag{2.30}$$

If the edge is free (unsupported), there can be neither moment nor reaction, although deflection and rotation may occur. The boundary condition statement would appear to be

$$m_x = 0$$
$$m_{xy} = 0$$
$$V_x = 0$$

However, this has been demonstrated to be one too many conditions,[2.1] and it has also been shown that the second and third statements may be combined to give the requirement that the reaction is zero, the reaction being a function of the edge twisting moment and the edge shear force. The twisting moment produces a reaction component, V_x', which can be determined by replacing the twisting moment m_{xy} (which is a result of horizontal shearing stresses in the slab) acting over some length dy by an equivalent couple of vertical forces, as shown in Fig. 2.5. These give a net resultant force, per unit length, of

$$V_x' = \left[-\left(m_{xy} + \frac{\partial m_{xy}}{\partial y}\, dy \right) + m_{xy} \right] \frac{1}{dy} = -\frac{\partial m_{xy}}{\partial_y} \qquad (2.31)$$

The total reaction is then the shear less the component from the twisting moment:

$$R_x = V_x - V_x' = V_x + \frac{\partial m_{xy}}{\partial y} \qquad (2.32)$$

Then, equating $m_x = 0$ and $R_x = 0$ gives the boundary conditions:

$$\frac{\partial^2 w}{\partial x^2} + \mu\, \frac{\partial^2 w}{\partial y^2} = 0$$

$$\frac{\partial^3 w}{\partial x^3} + (2 - \mu)\, \frac{\partial^3 w}{\partial x\, \partial y^2}\, 0 \qquad (2.33)$$

At an elastically supported and restrained edge, Eqs. 2.33 become compatibility equations, and when multiplied by $-D$ are set equal to functions of the beam flexural and torsional rigidity and spans rather than to zero.

2.2.6   Reactions

On the basis of the discussion in Section 2.2.5, it is evident that the reaction at a supported edge is the shear plus (or minus) the contribution of the twisting moment, as follows:

$$R_x = V_x + \frac{\partial m_{xy}}{\partial y} = -D\left[\frac{\partial^3 w}{\partial x^3} + (2 - \mu)\, \frac{\partial^3 w}{\partial x\, \partial y^2} \right] \qquad (2.34)$$

$$R_y = V_y + \frac{\partial m_{xy}}{\partial x} = -D\left[\frac{\partial^3 w}{\partial y^3} + (2 - \mu)\, \frac{\partial^3 w}{\partial x^2\, \partial y} \right] \qquad (2.35)$$

At a fixed edge in the y-direction, there is no twisting moment since $\partial w / \partial x = 0$ along the border and consequently $\partial^2 w / \partial x\, \partial y = 0$. The twisting moment makes a definite contribution to the reaction at a simply supported edge, and may either increase or decrease the force per unit length. As a right-angle corner where two simply supported edges intersect is approached, R_x and R_y approach zero. However, the twisting moments reach a maximum

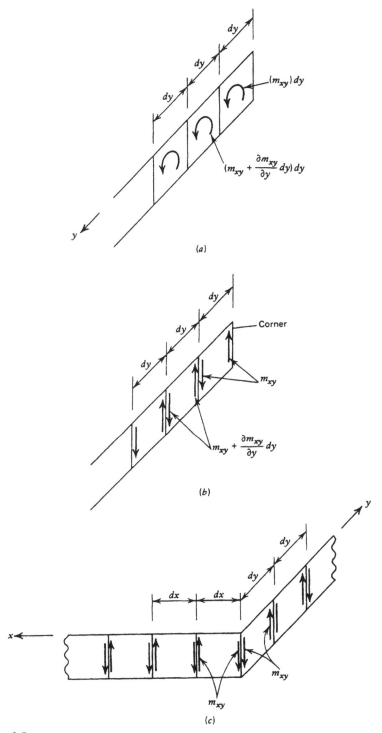

Figure 2.5 Replacement of twisting moments by equivalent couples: (*a*) twisting moments; (*b*) equivalent couples; (*c*) forces acting at corner of slab.

at the corner, and as a result there is an unequilibrated force m_{xy} acting vertically at the end of the side of the slab, as shown in Fig. 2.5b. There are effectively two such forces acting just at the corner of the slab since one is contributed by each of the intersecting sides, as shown in Fig. 2.5c, giving a corner force of

$$R_0 = 2m_{xy} = -2D(1 - \mu) \frac{\partial^2 w}{\partial x\, \partial y} \qquad (2.36)$$

In a uniformly loaded rectangular slab this corner force is upward when the loading acts downward, and if the corner is not positively held down by a force R_0, it will lift off the support. Note that if the corner lifts off, the distribution of moments and other forces throughout the panel changes from those calculated previously. Whether the corner is free to lift or not, the associated forces require special reinforcement details at the corner.

The term involving the first derivative of the twisting moment does not create additional shears at sections away from the edge of the plate. At any internal section, the sections on each side of any arbitrary cutting line are subjected to the twisting moments or equivalent couples shown in Fig. 2.5, but the torques on mating faces are equal and opposite and consequently cancel out.

2.2.7 Poisson's Ratio

Poisson's ratio appears as a significant variable in many plate equations, although it has often been taken as zero for work with reinforced concrete slabs. As is discussed below, this is reasonable for a cracked slab, and it should be borne in mind that all slabs become cracked as the collapse loads are approached. Poisson's ratio has little influence on the strength of the structure, although not necessarily on the stresses at lower load levels.

The influence of Poisson's ratio on the behavior of a cracked reinforced concrete slab could reasonably be expected to be less than in the case of a homogeneous plate, since the tensile stress element in the slabs consists of crossing, discrete, steel bars which are not connected together, nor necessarily even touching. Consequently, stress in the steel in the x-direction can affect the stress in the steel in the y-direction only by being transmitted through the concrete, and this would not be expected to be an efficient transfer because of the presence of cracking and the large differences in the values of Young's modulus of the materials. Poisson's ratio would have some influence on the concrete compressive stresses, but these are not controlling stresses in most slab systems. There is a minor influence of Poisson's ratio on slab stiffness, as reflected in the term D in Eq. 2.1, but this is much less than the expected variations in Young's modulus of concrete. Poisson's ratio could be important in a slab reinforced with a steel plate rather than bars.

The "true" value of μ for a slab undoubtedly varies over the area of the slab, depending on the severity of cracking, but making $\mu = 0.15$ as for concrete would be correct only for an uncracked slab, and then presumably only for the concrete stresses. It must approach zero as the slab approaches the fully cracked state. An experiment to determine the most appropriate value of μ would have to be very carefully planned and carried out, because small variations in more important properties such as E could easily mask the results.

However, an analytical study by Jofriet[2.3] suggests an experiment that might have the necessary sensitivity. He found that the moment m_x in the short-span direction in a typical interior rectangular panel of a beamless slab, in the location marked in Fig. 2.6, was very sensitive to μ. For $\mu = 0$, this is a negative moment, but for $\mu = 0.05$ it is zero, and for all larger values of μ the moment is positive. That this is possible can be seen with the aid of Eq. 2.12 and consideration of the fact that the curvature in the y-direction would be much larger than that in the x-direction, and of opposite sign. However, such an experiment might raise still other questions, since a normal slab design would average about four times as much steel per unit width in the y-direction as in the x-direction, and the slab could not strictly be considered as isotropic.

Disregarding the effect of μ on D, as it is small, the specific value of Poisson's ratio selected has an influence on the solution of Eq. 2.1 only if it enters the plate boundary conditions. It has no influence in the cases of simply supported and fixed slabs, but is of some consequence in slabs with free or elastically supported and restrained edges, as indicated by Eq. 2.33. In those cases where the boundary conditions do not include Poisson's ratio, moments for one value of μ may readily be determined from moments for another value, and it is convenient to solve for the case of $\mu = 0$ and assemble moments for other values of μ as required. If m_{x0}, m_{y0}, m_{xy0}, and w_0 denote moments and deflections when $\mu = 0$, the same quantities for finite values of μ are

$$m_{x\mu} = m_{x0} + \mu m_{y0}$$

$$m_{y\mu} = m_{y0} + \mu m_{x0}$$

$$m_{xy\mu} = (1 - \mu)m_{xy0} \qquad (2.37)$$

$$w_\mu = (1 - \mu^2)w_0$$

The shear forces are not affected by μ. Equation 2.37 can be derived starting with Eqs. 2.1, 2.12, 2.13, and 2.19, provided that it is remembered

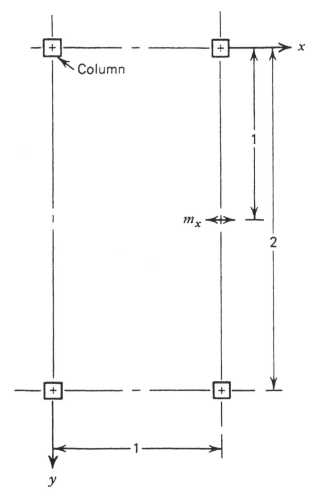

Figure 2.6 Layout of interior rectangular panel, showing location of m_x, where Poisson's ratio is critical.

that w and consequently the derivatives of w are smaller by $1 - \mu^2$ when μ is finite than when $\mu = 0$.

2.2.8 Moments Acting at an Angle to the Coordinate Axes

All of the previous discussion has been concerned with finding moments in the x- and y-directions, together with the corresponding twisting moments. In some instances it is necessary to find moments and twisting moments acting in other directions, and to determine principal moments, that is, the orthogonal moments that act on faces of a slab element which are free of twisting moments.

Consider the slab element shown in Fig. 2.7a, with the bending and twisting moments acting as shown by the vectors. The intensities of the moments acting on a section inclined at the angle α from the y-axis may be found by isolating the free body shown in Fig. 2.7b and summing components of m_x, m_y, and m_{xy} in the n and t directions to find the bending moment m_n and the twisting moment m_t acting at the inclined section. This calculation is exactly similar to the solution of a general two-dimensional stress problem, and results in the same equations, with normal and twisting moments substituted for direct and shear stresses. The moments per unit width are related by the following equations:

$$m_n = m_x \sin^2\alpha + m_y \cos^2\alpha + 2m_{xy} \sin \alpha \cos \alpha$$

$$= m_x \sin^2\alpha + m_y \cos^2\alpha + m_{xy} \sin 2\alpha \tag{2.38a}$$

$$m_t = (m_x - m_y) \sin \alpha \cos \alpha + m_{xy}(\sin^2\alpha - \cos^2\alpha)$$

$$= \frac{m_x - m_y}{2} \sin 2\alpha - m_{xy} \cos 2\alpha \tag{2.38b}$$

The angle α from the y-axis to the plane of the principal moment may be found from

$$\tan 2\alpha = \frac{2m_{xy}}{m_x - m_y} \tag{2.38c}$$

These equations are probably most conveniently solved through the use of a Mohr's circle construction, as shown in Fig. 2.7c. Figure 2.7d shows the specific Mohr's circle associated with the transformation used in Fig. 2.3. The transformations of curvatures and twisting distortions to different axes, as discussed in connection with the derivation of Eqs. 2.23, 2.25, and 2.26, can also be accomplished through the use of a Mohr's circle construction.

Shears relative to other coordinate axes rotated through the angle β from the x-axis may be found using the following equations, which were derived from consideration of the vertical equilibrium of the slab elements shown in Fig. 2.8:

$$\begin{aligned} V_{x'} &= V_x \cos \beta + V_y \sin \beta \\ V_{y'} &= V_x \sin \beta + V_y \cos \beta \end{aligned} \tag{2.39}$$

2.2.9 Method of Solution

The general procedure for solving an elastic slab problem is to determine the equation for the deflected shape of the slab and then combine the various derivatives of the deflection to obtain the internal forces. The deflected shape

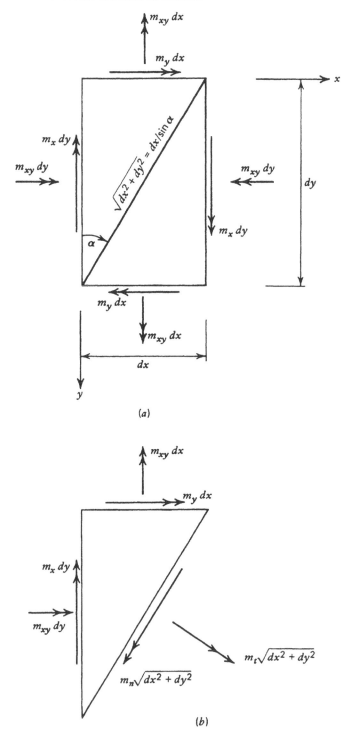

Figure 2.7 Computation of moments acting at angle α from y-axis: (a) slab element; (b) free body at angle a; (c) Mohr's circle for moments in slab; (d) Mohr's circle for slab element in Fig. 2.3.

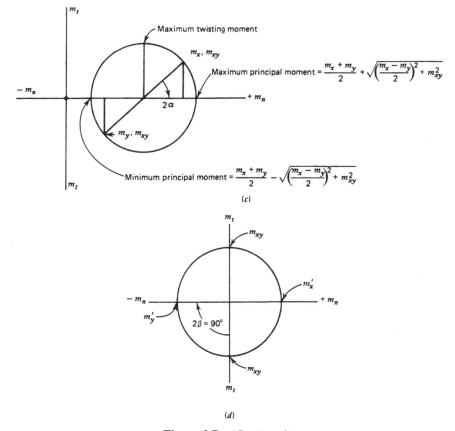

Figure 2.7 (*Continued*)

must satisfy both the Lagrange equation, Eq. 2.1, and the appropriate boundary condition equations. Although this procedure is exactly the same as that required for the solution of the general differential equation for a beam,

$$\frac{d^4 w}{dx^4} = \frac{q}{EI}$$

(2.40)

plus the associated boundary conditions, a practical slab problem can seldom be solved by the direct integration of Eq. 2.1.

Equation 2.40 is directly analogous to Eq. 2.1, and many analogies between the development of the slab and beam equations may be found. The student may find it helpful to look for these parallels to help develop a better physical understanding of various aspects of the slab problem.

Every civil engineer solves Eq. 2.40 routinely using various simple techniques, but the solution of the Lagrange equation is unfortunately not as

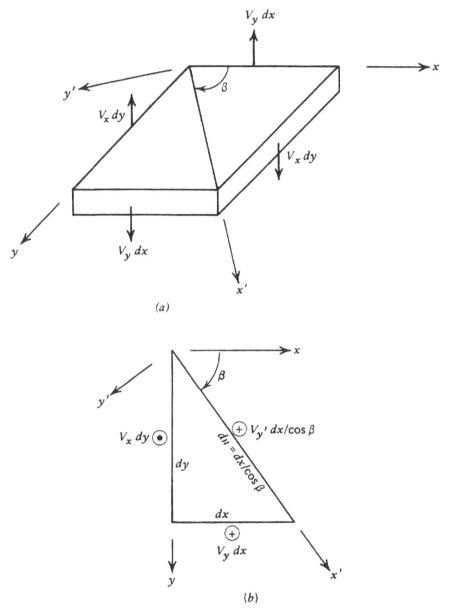

Figure 2.8 Shears acting on sections at the angle β from the coordinate axes: (*a*) slab element with shear forces; (*b*) free body of slab element, showing shear forces.

straightforward. Closed-form solutions of Eq. 2.1 have been found for a very limited number of slabs. Circular plates with axisymmetric loads, some elliptical plates, and some triangular plates have been solved, but the list is short and does not help structural engineers very much. For circular plates, the Lagrange equation can be formulated in polar coordinates[2.1] and directly integrated to find the deflected shape for axisymmetric loads and plates.

The first solutions of the Lagrange equation were obtained, by Navier in 1820, using doubly infinite Fourier series to describe the deflections and loading of simply supported rectangular plates under arbitrary loadings. The same general scheme was used by Lavoinne to solve a steam-boiler problem equivalent to an interior panel of a flat slab floor. The general form of the deflection function was

$$ w = qC \sum_{m=1}^{\infty} \sum_{n=1}^{\infty} A_{mn} \sin \frac{m\pi x}{a} \sin \frac{m\pi y}{b} \tag{2.41} $$

where C is a constant and A_{mn} is a variable depending on the integers m and n and on the ratio of the sides of the panel, a and b.

Such series solutions generally converge relatively quickly except in the case of concentrated loads, where they do not converge. For concentrated loads, various special techniques must be used, such as that developed by Westergaard.[2.4] A more general solution, using a singly infinite sine series with hyperbolic coefficients, was found by Levy for a rectangular slab with two parallel edges simply supported and with the other two having any arbitrary support conditions. Levy's method was used by Newmark[2.5] as the basis for a moment distribution scheme for slabs simply supported at two parallel edges and continuous across beams of any arbitrary stiffnesses in the other direction. This method has been used extensively in highway bridge analysis,[2.6] as well as for building slab design studies.

The concept of energy solutions for plates was developed by Ritz, based on the principle that the total energy of a deformed plate is a minimum when equilibrium exists. The solutions are usually in the form of a series solution, but more freedom in selection of the series to be used exists, as long as the functions satisfy the boundary conditions of the problem. Coefficients for the successive terms in the series are selected to minimize the total energy in the system. The Galerkin method of solution falls within this general class. The terms in the series may be polynomial or trigonometric terms, and series such as Bessel and Hankel functions[2.7] have been used to obtain solutions of particular problems. The discovery of suitable series that both satisfy the boundary conditions of the problem and approximate the deflected shape has been generally difficult, and consequently, other methods of solution have been developed.

From the point of view of the development of the 1995 ACI Code[2.8] provisions for the design of slab structures, the work by Sutherland et al.[2.9] was significant. They used the Ritz minimum energy approach, with Duncan S-functions[2.10] as the polynomial series. They studied the effects of the relative stiffness of supporting beams on the distribution of moments within interior slab panels, and the results obtained are related directly to values used in all editions of the ACI Code since 1971.

Unfortunately, as slabs become more complex and more representative of building slabs used in actual construction, it becomes more and more difficult to find suitable deflection functions that satisfy the boundary conditions of moment, slope, and reaction. No new Hardy Cross has yet arrived to put simple, straightforward plate analysis procedures within the easy grasp of every structural designer in the way that moment distribution transformed beam and frame analysis.

Because of the complexity of the "exact" methods of plate analysis referred to above (exact in the sense that any required provision can be attained if enough terms of a suitable series or polynomial are evaluated), much effort has been spent on the development of approximate methods of analysis.

2.3   ELASTIC MODELS

One traditional method of determining the moment distribution in a plate has been the use of elastic models, which may be made of a wide range of materials, including plaster, glass, plastic, metals, and hard rubber. These have been used for both research and design studies. Both strains and deformed shapes, and especially slopes, have been measured and related to the internal force distributions. Further information is contained in Refs. 2.11 to 2.15, for example.

However, many complex slabs that might have been designed starting with an elastic model only a few years ago can now be analyzed using numerical procedures and modern electronic computers. Since obtaining precise, quantitative information on the distribution of bending moments from models had to wait on the development of modern strain-measuring devices, elastic models made significant contributions to plate and slab research and design for only a short period before being largely supplanted by numerical techniques.

Reinforced concrete models have also been widely used, more in research than in design, and their role is discussed later. They cannot be viewed as being elastic, but they have been important in the development of design procedures.

2.4   FINITE DIFFERENCE METHODS

The method of finite differences, or difference equations, was introduced by Nielsen[2.16] in 1920 as an alternative method of solution of plates. His results were used by Westergaard and Slater,[2.2] with modifications to correct for circular column capitals, to derive the moment distributions used in ACI Codes for flat slab structures from that time on through to the 1963 version.

The finite difference method replaces Lagrange's fourth-order partial differential equation with a series of simultaneous linear algebraic equations for the deflections of a finite number of points on the slab surface. Once the deflections have been found, the moments and shears are found using the appropriate relationships between deflections of groups of points. The derivation of the finite difference operators and the determination of internal forces are covered by Timoshenko and Woinowsky-Krieger.[2.1] The method was used extensively to develop information on elastic moments in slabs throughout the first half-century after its introduction. Much basic information on the use of the method has been given by Jensen[2.17] in a report on an investigation of skew slab bridges with stiffened edges. Wood[2.18] also gives very helpful basic information on the method, including the difficult cases of free and elastically supported edges. In the use of the method, a slab or slab panel is divided into a suitable number of square subdivisions (other meshes are also possible, including rectangles, parallelograms, and spiderweb shapes), as shown in Fig. 2.9. The deflection of each interior node is to be determined, giving 40 points for the slab shown. If the load is uniformly distributed and the slab of uniform thickness, symmetry can be used to reduce the number of unknown deflections to 12 for this particular case.

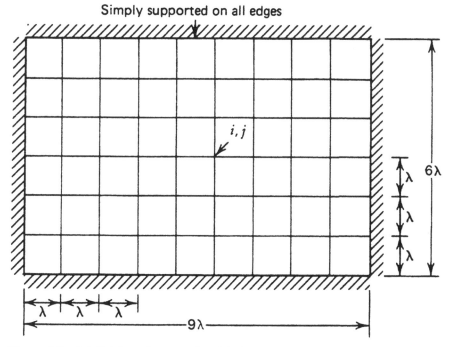

Figure 2.9 Subdivision of rectangular slab for finite difference or finite element solution.

Difference equations are then written for each point of intersection of the grid. This is readily done manually through the use of the difference equation operator, or molecule, shown in Fig. 2.10a at each point. The equation, in terms of deflections, w, for point i, j is shown in Fig. 2. 10b. One equation is written for each unknown deflection point, and the group of equations is then solved simultaneously for the unknown deflections.

The number of grid points to be considered is a matter of judgment, trial, and computational effort. Slabs with relatively small moment gradients, such as the simply supported slab shown in Fig. 2.9, require fewer points to obtain a given precision than slabs with large moment gradients, such as fixed slabs or flat slabs supported only on columns. Special techniques, such as the use of fine grids in areas of large shears and moment gradients and coarser grids in other areas of a plate, have been developed to cope with such problems.[2.19,2.20]

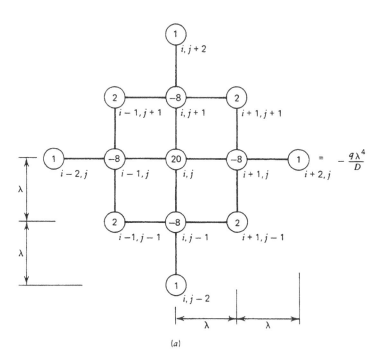

(a)

$$20w_{i,j} - 8(w_{i,j+1} + w_{i+1,j} + w_{i,j-1} + w_{i-1,j})$$
$$+ 2(w_{i+1,j+1} + w_{i+1,j-1} + w_{i-1,j-1} + w_{i-1,j+1})$$
$$+ w_{i,j+2} + w_{i+2,j} + w_{i,j-2} + w_{i-2,j} = -\frac{q\lambda^4}{D}$$

(b)

Figure 2.10   Finite difference operator for typical interior point: (a) finite difference operator for typical interior point; (b) finite difference equation for point i, j.

As points adjacent to the borders of a plate are reached, it will be found that some elements of the operators are out in space rather than on the slab. If the border is not able to deflect, imaginary values for the positions outside the boundary may usually be found from considerations of a symmetric or antisymmetric imaginary plate extending beyond the real boundary.

Unfortunately, if there is deflection of the plate boundary, it becomes more difficult to determine the form of the difference operators of the points at and near the boundary. The problem of an edge supported by a beam having finite flexural and torsional stiffnesses is an example of a practical problem for which the formulation of the difference equations becomes difficult. Other examples of practical problems that are difficult to handle are those of a multipaneled slab supported on columns with finite flexural stiffnesses and of a plate with either holes or reentrant corners.

To overcome some of these problems, a quite different approach to the formulation of the difference equations has been widely used. Rather than starting with the differential equation, a structure consisting of a series of orthogonal rigid bars connected by various flexural and torsional springs is imagined to replace the plate structure. This replacement structure is then analyzed. Newmark's plate analog[2.21] is probably the most widely used physical model, and by proper choice of the various spring constants the response to load of the actual slab is reproduced. Complete descriptions of the properties and use of the plate analog model are given by Ang and Newmark[2.22] in a description of a successful moment distribution procedure for slabs, and by Ang and Prescott.[2.23] The plate analog model is analyzed point by point, and a set of equations exactly like the difference equations are derived, and solved in the same way. The method has the capacity to take very complex boundary conditions into account, and was used by Simmonds[2.24] to solve for moments, shears, and deflection in a series of nine-panel slabs supported on flexible beams connected to flexible columns.

The finite difference equation methods are approximate procedures, but the quality of the approximation can be improved by considering a finer grid. A moment at a particular grid point represents an average moment acting over a width equal to the grid spacing and will be a close approximation of the elastic moment at that point if it is in a region of low moment gradient, and a less accurate representation if in a region of rapidly changing moments. Analyses by finite differences ordinarily satisfy the static requirements on moments quite precisely, and static checks are important parts of the verification of results.

One significant advantage of the physical analog methods of developing plate analysis problems is that the springs of the physical model can be made to obey elastoplastic stress–strain or moment–curvature relationships. Ang and Lopez[2.25] used this method to predict and follow the spread of yielding through various areas of slab panels. They developed a complete load–deflection curve for a simply supported square slab, for example, and obtained a plastic collapse load approximately the same as found by Jones and Wood[2.26] from a sophisticated upper bound procedure. Their solution effectively used two plate analog models, one offset from the other by half a grid spacing in both the x- and y-directions. This was required because of the necessity of defining the flexural and twisting moments at the same point in the plastic analysis procedure.

The finite difference method, either directly or with the aid of the plate analog, is a good method for finding numerical answers to otherwise complex plate problems, provided that the boundary conditions are simple. The number of points to be included in a difference equation grid does not necessarily have to be extremely large to give answers adequate for design purposes. For example, Timoshenko and Woinowsky-Krieger[2.1] showed that dividing a simply supported square plate by lines on each quarter and midspan line, giving nine unknown deflections but reduced to three by symmetry, resulted in a midspan deflection less than 1% smaller than the theoretical value, and the midspan moment was about 4.5% too small. As a consequence of this sort of precision, finite difference solutions of plate problems were widely used even in the days before the first generation of electronic computers.

In the early computer era of the late 1950s and early 1960s, finite difference- and plate analog–based methods were widely used in studies of such details as the distribution of stresses around columns in beamless slabs, effects of beam flexural and torsional stiffness, and effects of column stiffness, on slab moment distributions. Many of the detailed provisions of all ACI Codes since the 1971 version for the design of slab structures can be traced, directly or indirectly, back to such studies. The method has been larely supplanted by the finite element method.

2.5   FINITE ELEMENT METHODS

The finite element technique is another, quite different numerical approach to problems of plate analysis. In this method, a plate is divided into a number of rectangular, triangular, or quadrilateral areas, or elements. Figure 2.9 is again appropriate. The small areas partitioned off are of interest rather than the grid.

Each small element of plate has bending deformation properties that are either known or can be closely approximated. The general method of analysis is to concentrate the loads at the corners, or nodes, of the separated elements, and then restore continuity of slope and deflection at each node point, and sometimes at intermediate points as well, so as to satisfy equilibrium and boundary condition requirements.

The finite element method of analysis is a major contributor to the recent knowledge explosion, and several hundred references could probably be quoted. The earliest paper on a plate bending element was perhaps that by Melosh,[2.27] and the problem was stated in the current matrix terminology by Zienkiewicz and Chung.[2.28] Many later developments are also cited by Zienkiewicz.[2.29] The textbook by Cook et al.[2.30] contains a useful chapter on the application of the finite element method to slabs.

At the present state of development of the finite element technique, there are many elements of art mixed into the methods of science. The choice of the element shape, its particular deformation characteristics, the orientation of the elements, and the number of elements all influence the final results of a solution. As a consequence, every study of plates using this method includes (or should include) a certain amount of experimentation with these factors, and substantial agreement with known solutions reached before working on new ones. Comparisons with values required by static equilibrium are also important, as there is no guarantee that statics will be satisfied. Static equilibrium may be a reasonably serious problem in some instances. Once an element type and basic arrangement of elements has been selected, the number of elements still has to be decided. As was the case with finite difference solutions, in general the finer the grid, the better the answer but the greater the cost in terms of computation time. Element arrays with small elements in areas of high moment gradients and larger elements in other areas will improve the solution with reasonable computational effort. The optimum element size and combination of different size elements will be found by experimentation. A number of examples are given by Zienkiewicz,[2.29] illustrating the effects of these variables on the moments and deflections in square plates. The line dividing finite element and finite difference solutions sometimes gets very thin, as illustrated by Salonen.[2.31] The finite difference technique is used to find the required deformation properties of the various finite elements in some instances, so the two methods must be viewed as being complementary.

As was the case with the plate analog–based difference equation solutions, the finite element method can be taken into the inelastic range by various techniques. Jofriet,[2.3,2.32] for example, used the finite element technique to include the effects of flexural cracking and successfully predicted the short-term deflections of flat plate floor slabs at loads up to approximately the initiation of yielding of reinforcement. Scanlon[2.33] and Scanlon and Murray[2.34,2.35] have used finite element techniques to predict time-dependent deflections and stress changes in slabs. The finite element technique can accept almost any boundary condition, using relatively simple procedures, and so is a powerful analysis tool. Slabs supported on beams and columns having finite flexural stiffness, for example, have been studied by Jofriet and McNeice [2.32,2.36] and Pfaffinger.[2.37]

The development of finite element methods proceeded at the same time and pace as development of the large digital computers necessary for the use of the method. Several of the general structural analysis programs available include finite element analysis capability, and consequently, the method would be the logical choice for the analysis of a complex slab system. An example can be used to demonstrate the capability of the finite element method in solving for moments in a slab that would otherwise be very difficult to analyze. Figure 2.11*a* shows the plan view of a slab that is simply supported on two sides, fixed on two sides, and has free edges as may occur at the edge of a hole. The slab is subjected to a uniformly distributed load of 200 lb/ft^2

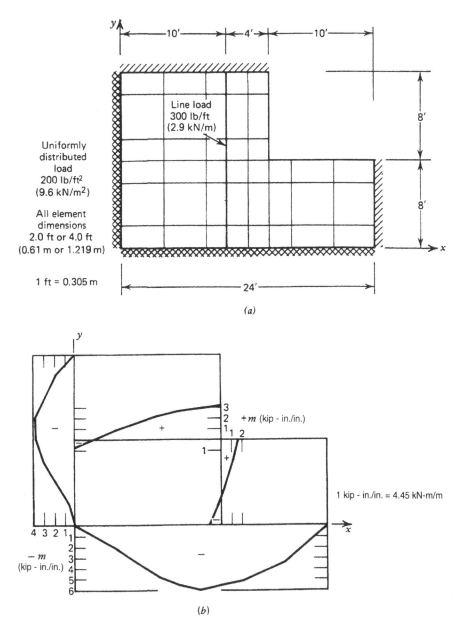

Figure 2.11   Example slab solved by finite elements: (*a*) slab plan, showing finite element array and loading; (*b*) moments acting across various sections of slab.

(9.6 kN/m²), including the weight of the slab, plus a line load of 300 lb/ft (2.9 kN/m) located as shown. The finite element mesh used for the analysis is also shown, resulting in 39 elements. The problem was solved using a computer program known as Finite,[2.38] developed at the University of Illinois at Urbana–Champaign. Plate elements with 16 degrees of freedom were used.

Some of the results of the analysis are shown in Fig. 2.11*b*, where moments acting across four different lines have been plotted. The results are quite reasonable, although the theoretical moment at the reentrant corner is infinite. The use of smaller elements would have led to higher computed moments in this area, but probably would have had little effect at other sections. The moments plotted are the average of the moments at the corners of the elements meeting at particular nodes. The output from the computer program includes the values of m_x, m_y, and m_{xy} at each corner of each element. The m_x and m_y moments computed for each of the corners at a particular node were generally not equal, but the differences between the several values were not large, except in the vicinity of the reentrant corner. This is a result of the method of computation of the moments from the deflected shape of the plate, and different analysis programs may treat this particular problem slightly differently.

2.6 APPROXIMATE METHODS

Numerous approximate methods of analyzing slabs have been proposed, developed, and used over the years. Many of these were developed before the computer era, although some are still being proposed and used, and many are useful for giving insights into the probable behavior of a slab system even if they are not capable of producing valid numerical information.

The replacement of a slab by a series of orthogonal crossing beams—perhaps only one beam in each direction—is probably the oldest such device. The flexural moments so calculated can differ considerably from the actual elastic theory distribution because of the omission of the twisting moments acting on each element of slab, which is comparable to the omission of the cross-differential term in the slab equilibrium statement, Eq. 2.5. This approach is, in fact, a lower bound limit design method and is discussed more fully in Chapter 6. A distribution of moments found this way would normally need moment redistribution due to inelastic behavior to achieve the ultimate load.

A moment-distribution procedure for multipanel slabs supported on nondeflecting beams was developed by Siess and Newmark.[2.39] The procedure is directly comparable to the Cross moment-distribution procedure for beams, except that the moment carryover factors are approximate, and when all the negative moments are balanced, another set of approximate factors is used to calculate the positive moments. Torsional stiffnesses of the supporting beams can also be taken into account. As long as the beams are flexurally rigid, the method can give reasonably accurate moment values—certainly good enough for design purposes. The main problem is that few beams are actually rigid, although bearing-wall buildings would satisfy this limitation, and the moments in edge and corner panels turn out to be particularly sensitive to small support deflections. Within the limits imposed by the nondeflecting supports, this was a useful tool for analysis and design.

Vanderbilt et al.[2.40] presented an approximate method of computing deflections of continuous slabs, with and without beams, which combines various aspects of frame analysis and known solutions of simply supported rectangular plates. The effects of cracking on stiffness were taken into account empirically. The procedure could probably also be used to calculate approximate moments, but this was apparently never fully explored, presumably because of developments in equivalent frame analysis methods that were occurring at the same time.

Frame analysis methods have been applied to slab design for many years. The 1963 and earlier ACI Codes described an *equivalent frame analysis* for flat plate and flat slab structures. In the 1971 and later ACI Codes[2.8,2.41] the method was revised considerably and extended to include two-way slabs with beams as well. These methods view a strip of a slab, including beams and columns, as being a two-dimensional frame. The frame might be either a single story, with the columns above and below considered fixed at the far ends, or the height of the structure, as if it had been sliced top to bottom along two successive centerlines between columns. The frame analysis allows a determination of the positive–negative moment distribution, taking into account the variations in lengths and loadings of various spans and support stiffnesses, but does not give any information on the lateral distributions of forces across the slab or to the beams.

The frame analysis method in the current ACI Code[2.8] is based on work by Corley[2.42] and Corley and Jirsa.[2.43] Their contributions include a rational evaluation of the effective stiffness of column–slab–beam combinations, which are so important in slab structures, especially at the edges of the structure. Once the positive–negative moment distributions are determined, these moments are distributed to the beams and the slab column and middle strips following additional guides and requirements. Frame analysis methods are discussed in Chapters 3 and 4.

Nichols[2.44] in 1914 used statics to establish the minimum possible value of the static moment, defined as the absolute sum of the positive plus average negative moments in a slab panel. This was the first rational treatment of what had been a highly empirical design procedure. His short 12-page paper attracted 55 pages of not entirely rational comments, including many from people with threatened vested economic interests. He proposed this equation:

$$M_0 = 0.125Wl\left(1 - \frac{2}{3}\frac{c}{l}\right)^2 \tag{2.42}$$

where W is the total uniformly distributed load on the panel, l the span of panel (assumed square by Nichols), and c the diameter of column capital. This formed the basis of flat plate and flat slab design for many editions of ACI Codes with the 0.125 changed to 0.09, ". . . which has puzzled engineers for over half a century . . . "[2.45] until the 1971 ACI Code[2.41] came into effect. Nichols's analysis did not give any information on the distribution of the bending moments, but rather, gave a total moment to be carried. In this sense, it was a limit analysis approach, although the term was not then in use.

The slab design provisions in the ACI Codes apply only to structures in which the supporting columns are at or very near the corners of a rectangular grid. However, in dwelling and some other uses, this is not always a convenient arrangement of supports from the point of view of the use of the structure, and it may be desirable to conceal columns in walls, the backs of closets, and so on. This is a problem primarily in the case of flat plate structures, since they are especially suitable for the light loads for which apartments are designed. Of the analyses discussed previously, only the finite difference and finite element techniques are capable of handling such a problem, and either requires the use of electronic computer. However, at least two approximate methods have been developed and proposed.

Van Buren[2.45] proposed a scheme for the analysis of slabs with similar or mirror-image parallelogram-shaped panels, termed as cases with staggered columns. The method is a reasonably sophisticated crossing-beam solution which gives quite satisfactory results. Wiesinger[2.46] presented a quite different scheme for the design of slabs with random column spacings. An equilibrium approach is used that has most of the elements of a limit analysis and could result in quite acceptable design moments for slabs when used with the stated rules for the positive–negative distributions of moments.

2.7 EXAMPLES OF ELASTIC THEORY ANALYSIS

Example 2.1. Compute the moments and deflections in a simply supported rectangular slab that is subjected to a sinusoidal loading pattern, as shown in Fig. 2.12a.

SOLUTION. The loading per unit area may be expressed as

$$q = q_0 \sin \frac{\pi x}{a} \cdot \sin \frac{\pi y}{b} \tag{2.43}$$

It is necessary to find a deflection function satisfying both the Lagrange equation, Eq. 2.1, and the boundary conditions, Eq. 2.30. The boundary conditions require that $w = 0$ for $x = 0$ and a and $y = 0$ and b. In addition, it is required that $\partial^2 w / \partial x^2 = 0$ for $x = 0$ and a and that $\partial^2 w / \partial y^2 = 0$ for $y = 0$ and b.

It is obvious that the deflection function

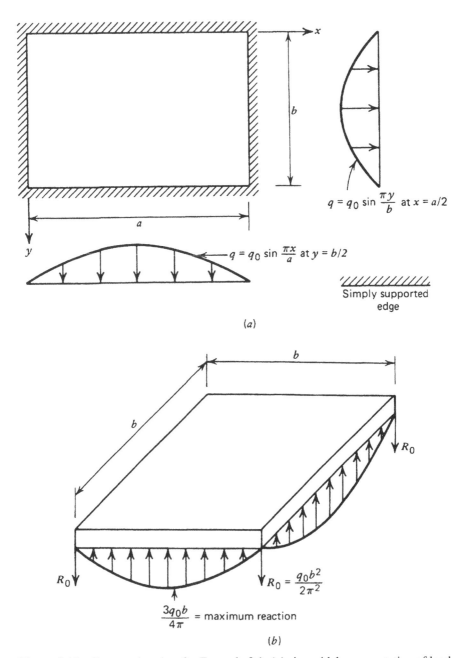

$q = q_0 \sin \dfrac{\pi y}{b}$ at $x = a/2$

$q = q_0 \sin \dfrac{\pi x}{a}$ at $y = b/2$

Simply supported
edge

(a)

R_0

$R_0 = \dfrac{q_0 b^2}{2\pi^2}$

$\dfrac{3 q_0 b}{4\pi}$ = maximum reaction

(b)

Figure 2.12   Rectangular plate for Example 2.1: (*a*) sinusoidal representation of loading; (*b*) reactions on simply supported square plate of Example 2.1, $\mu = 0$.

$$w = C \sin \frac{\pi x}{a} \sin \frac{\pi y}{b} \tag{2.44}$$

will satisfy both Eq. 2.1 and the boundary conditions since both the deflection function and all second and fourth partial derivatives with respect to x or y are equal to zero at the edges of the slab. The problem then is to find the constant C, and once this is done, to substitute w into the various expressions for the internal forces and reactions.

Equation 2.1 may be restated as

$$\frac{\partial^4 w}{\partial x^4} + 2 \frac{\partial^4 w}{\partial x^2 \partial y^2} + \frac{\partial^4 w}{\partial y^4} = \frac{q_0}{D} \sin \frac{\pi x}{a} \sin \frac{\pi y}{b} \tag{2.45}$$

Then noting the following derivatives of the deflection function:

$$\frac{\partial^4 w}{\partial x^4} = C \left(\frac{\pi}{a}\right)^4 \sin \frac{\pi x}{a} \sin \frac{\pi y}{b}$$

$$\frac{\partial^4 w}{\partial y^4} = C \left(\frac{\pi}{b}\right)^4 \sin \frac{\pi x}{a} \sin \frac{\pi y}{b}$$

$$\frac{\partial^4 w}{\partial x^2 \partial y^2} = C \frac{\pi^4}{a^2 b^2} \sin \frac{\pi x}{a} \sin \frac{\pi y}{b}$$

and substituting into Eq. 2.45 leads to

$$C \left(\frac{\pi^4}{a^4} + \frac{2\pi^4}{a^2 b^2} + \frac{\pi^4}{b^4}\right) \sin \frac{\pi x}{a} \sin \frac{\pi y}{b} = \frac{q_0}{D} \sin \frac{\pi x}{a} \sin \frac{\pi y}{b}$$

This is then solved to find the value of C:

$$C = \frac{q_0}{\pi^4 D} \frac{1}{1/a^4 + 2/(a^2 b^2) + 1/b^4} = \frac{q_0 b^4}{\pi^4 D} \frac{1}{[(b/a)^2 + 1]^2} \tag{2.46}$$

The deflection of the slab is, consequently,

$$w = \frac{q_0 b^4}{\pi^4 D} \frac{1}{[(b/a)^2 + 1]^2} \sin \frac{\pi x}{a} \sin \frac{\pi y}{b}$$

The deflection is thus a function of the load; the fourth power of the span; the flexural rigidity of the slab, as in the case of a beam; and the shape of the slab. The maximum deflection is at the center of the slab and the profile varies as a sine curve in both directions.

Internal forces can now be determined. For example, m_x is given by Eq. 2.12 and for this case becomes

$$
\begin{aligned}
m_x &= -D\left[-C\left(\frac{\pi}{a}\right)^2 \sin \frac{\pi x}{a} \sin \frac{\pi y}{b} - \mu C\left(\frac{\pi}{b}\right)^2 \sin \frac{\pi x}{a} \sin \frac{\pi y}{b} \right] \\
&= +q_0 \frac{b^4}{\pi^2[(b/a)^2 + 1]^2} \left(\frac{1}{a^2} + \mu \frac{1}{b^2}\right) \sin \frac{\pi x}{a} \sin \frac{\pi y}{b}
\end{aligned}
$$

Similarly, m_y is found by substituting into Eq. 2.13:

$$
m_y = +q_0 \frac{b^4}{\pi^2[(b/a)^2 + 1]^2} \left(\frac{1}{b^2} + \mu \frac{1}{a^2}\right) \sin \frac{\pi x}{a} \sin \frac{\pi y}{b}
$$

The twisting moment requires the cross partial differential:

$$
\frac{\partial^2 w}{\partial x\, \partial y} = C\frac{\pi^2}{ab} \cos \frac{\pi x}{a} \cos \frac{\pi y}{b}
$$

and substituting this into Eq. 2.19 gives the twisting moment:

$$
m_{xy} = -q_0 \frac{b^3(1 - \mu)}{\pi^2 a} \frac{1}{[(b/a)^2 + 1]^2} \cos \frac{\pi x}{a} \cos \frac{\pi y}{b}
$$

The bending moments are maximum at the center of the slab and go to zero as a corner is approached, while the twisting moments reach their maximum values in the corners. For a square slab under a sinusoidal loading, the maximum bending and twisting moments have the same numerical values. (If the load is uniformly distributed, the twisting moments are the larger.) As a corner is approached, the moment state reaches that shown by the Mohr's circle of Fig. 2.7d, and the twisting moments may be transformed into principal bending moments acting along and across the diagonal of the corner. These principal moments require reinforcement parallel to the diagonal in the top of the slab (i.e., for negative moments) and reinforcement across the diagonal in the bottom of the slab. The alternative is layers of steel in both the x- and y-directions both top and bottom.

The shear V_x is found by substituting into Eq. 2.27, giving

$$
V_x = q_0 \frac{b^4}{\pi} \frac{1/a^3 + 1/ab^2}{[(b/a)^2 + 1]^2} \cos \frac{\pi x}{a} \sin \frac{\pi y}{b}
$$

Similarly, the reaction is obtained using Eq. 2.34:

$$R_x = q_0 \frac{b^4}{\pi} \frac{1/a^3 + (2 - \mu)(1/ab^2)}{[(b/a)^2 + 1]^2} \cos \frac{\pi x}{a} \sin \frac{\pi y}{b}$$

where $x = 0$ or $x = a$ but not any intermediate values. Similar expressions are obtained for V_y and R_y.

It may be demonstrated that the total distributed reaction on the four sides of the slab exceeds the applied load. The total applied load is $4q_0(ab/\pi^2)$ for the sine loading considered. For a square slab with $\mu = 0$, it is found that $R_x = 3q_0(b/4\pi)$ at $y = b/2$, $x = 0$, and the total reaction from $y = 0$ to $y = b$ at $x = 0$ or a is $6q_0(b^2/4\pi^2)$. Consequently, the total reaction on the four sides of the slab is $6q_0(b^2/\pi^2)$. The discrepancy between the total load and total distributed reaction is due to the corner forces, R_0, and the solution of $R_0 = 2m_{xy}$ gives $R_0 = -q_0 (b^2/2\pi^2)$ at each corner, or a total of $-2q_0(b^2/\pi^2)$. This exactly balances the excess reaction. The reactions are shown in Fig. 2.12b.

All moments are proportional to the span squared, and shears and reactions are proportional to the span, as would be expected from experience with beams. In addition, all quantities vary as either a sine or cosine curve along any section parallel to the panel boundaries.

Example 2.2. Consider the generalization of the sinusoidal loading to any distribution of loading acting on the simply supported slab shown in Fig. 2.12a.

SOLUTION. From the case of the single sine-wave loading, it can be seen that a loading of q per unit area of the form

$$q = q_{mn} \sin \frac{m\pi x}{a} \sin \frac{n\pi y}{b} \tag{2.47}$$

plus a deflection function of the form

$$w = C_{mn} \sin \frac{m\pi x}{a} \sin \frac{n\pi y}{b} \tag{2.48}$$

will also satisfy both the Lagrange equation and the boundary conditions. This suggests the possibility of using a Fourier series approach to approximate a uniformly distributed loading, or any other loading, and this is, in fact, the approach used by Navier in the first solution of the plate problem. The Fourier series solution had conveniently been developed a decade earlier.

Following the same procedure as for the plate in Example 2.1, the multiple sine-wave loading gives

$$C = \frac{q_{mn}}{\pi^4 D} \frac{1}{[(m/a)^2 + (n/b)^2]^2} \qquad (2.49)$$

Therefore, the deflection due to a multiple sine-wave loading is

$$w = \frac{q_{mn}}{\pi^4 D} \frac{1}{[(m/a)^2 + (n/b)^2]^2} \sin\frac{m\pi x}{a} \sin\frac{n\pi y}{b} \qquad (2.50)$$

The double Fourier series describing a uniformly distributed load q per unit area is expressed as

$$q = \sum_{m=1}^{\infty} \sum_{n=1}^{\infty} q_{mn} \sin\frac{m\pi x}{a} \sin\frac{m\pi y}{b} \qquad (2.51)$$

where q_{mn} is to be determined, and has the function of controlling the relative values of successive terms of the series so that the summation approaches a constant value at all combinations of x and y.

The value of the coefficient q_{mn} can be determined following relatively well known procedures such as described by Timoshenko and Woinowsky Krieger[2.1] or by nearly any text on differential equations, such as Salvadori and Schwarz.[2.47]

It is found that $q_{mn} = 16q/\pi^2 mn$ for m and n odd, or equal to zero if m and n are even, and substitution in Eq. 2.50 and summarizing the terms leads to the expression

$$w = \frac{16q}{\pi^6 D} \sum_{m=1,3,5,\ldots}^{\infty} \sum_{n=1,3,5,\ldots}^{\infty} \frac{1}{mn[(m/a)^2 + (n/b)^2]^2} \sin\frac{m\pi x}{a} \sin\frac{n\pi y}{b} \qquad (2.52)$$

The terms m and n are odd-numbered integers only, as the even terms produce deflections that are not symmetrical about midspan. The equation is the explicit form of Eq. 2.41 for a simply supported rectangular slab subjected to a uniformly distributed load.

Once the deflected shape is known, internal forces are determined by combining the appropriate derivatives of Eq. 2.52, always retaining the double summation. Thus, the moment m_x may be expressed as

$$m_x = \frac{16q}{\pi^4} \sum_{m=1,3,5,\ldots}^{\infty} \sum_{n=1,3,5,\ldots}^{\infty} \frac{(m/a^2) + \mu(n/b)^2}{mn[(m/a)^2 + (n/b)^2]^2} \sin\frac{m\pi x}{a} \sin\frac{n\pi y}{b} \qquad (2.53)$$

It will be found that all the expressions converge relatively rapidly, and not very many terms in the series will have to be evaluated. It should also be noted, however, that the convergence rates of different expressions may be different and that they may be different at different locations on the slab, and consequently, a certain precision rather than a certain number of terms should be used to determine the end of a particular summation.

REFERENCES

2.1 S. Timoshenko and S. Woinowsky-Krieger, *Theory of Plates and Shells*, 2nd ed., McGraw-Hill, New York, 1959, 580 pp.

2.2 H. M. Westergaard and W. A. Slater, "Moments and Stresses in Slabs," *Proc. ACI*, Vol. 17, 1921, pp. 415–538.

2.3 J. C. Jofriet, "Analysis and Design of Concrete Flat Plates," Ph.D. thesis, University of Waterloo, 1971, 466 pp.

2.4 H. M. Westergaard, "Computation of Stresses in Bridge Slabs Due to Wheel Loads," *Public Roads*, Vol. 11, No. 1, March 1930, pp. 1–23.

2.5 N. M. Newmark, *A Distribution Procedure for the Analysis of Slabs Continuous over Flexible Beams*, Bulletin 304, University of Illinois Engineering Experiment Station, Urbana, Ill., 1938, 118 pp.

2.6 N. M. Newmark and C. P. Siess, *Moments in I-Beam Bridges*, Bulletin 336, University of Illinois Engineering Experiment Station, Urbana, Ill., 1942, 148 pp.

2.7 J. F. Brotchie, "General Elastic Analysis of Flat Slabs and Plates," *Proc. ACI*, Vol. 56, No. 8, August 1959, pp. 127–152.

2.8 *Building Code Requirements for Structural Concrete*, ACI 318–95 and Commentary, ACI 318R-95, American Concrete Institute, Farmington, Hills, Mich., 1995, 371 pp.

2.9 J. G. Sutherland, L. E. Goodman, and N. M. Newmark, "Analysis of Plates Continuous over Flexible Beams," *Civil Engineering Studies*, Structural Research Series 42, Department of Civil Engineering, University of Illinois, Urbana, Ill., 1953, 76 pp.

2.10 W. J. Duncan, *Normalized Orthogonal Deflexion Functions for Beams*, Reports and Memorandum 2281, (British) Aeronautical Research Council, 1951, 23 pp.

2.11 F. K. Ligtenberg, "The Moire Method—A New Experimental Method for the Determination of Moments in Small Slab Models," *Proc. Soc. Exp. Stress Anal.*, Vol. 12, No. 2, 1955, pp. 83–98.

2.12 M. W. Huggins and W. L. Lin, "Moments in Flat Slabs," *Trans. ASCE*, Vol. 123, 1958, pp. 824–841.

2.13 R. S. Fling, "Flat Plate Analysis of Oletangy River Dormitories," *Proc. ACI*, Vol. 64, September 1967, pp. 568–574.

2.14 R. C. Elstner, *Tests of Elastic Models of Flat Plate and Flat Slab Floor Systems*, ACI SP-24, Symposium on Models for Concrete Structures, American Concrete Institute, 1970, pp. 289–320.

2.15 A. J. Durelli, K. Chandrashekhara, and V. J. Parks, "In-Plane Moire Strain Analysis of Bent Plates," *Int. J. Solids Struct.*, Vol. 6, 1970, pp. 1277–1285.

2.16 N. J. Nielsen, *Bestemmelse af spaendinger i plader ver anvendelse at differensligninger*, J. Jorgensen & Co., Copenhagen, 1920.

2.17 V. P. Jensen, *Analysis of Skew Slabs,* Bulletin 332, University of Illinois Engineering Experiment Station, Urbana, Ill., 1941, 110 pp.

2.18 R. H. Wood, *Plastic and Elastic Design of Slabs and Plates,* Ronald Press, New York, 1961, 344 pp.

2.19 A. H.-S. Ang and C. P. Siess, "Bond in Flat Slabs," *Proc. ACI,* Vol. 57, May 1961, pp. 1512–1519.

2.20 D. N. de G. Allen, *Relaxation Methods in Engineering and Science,* McGraw-Hill, New York, 1954, pp. 69–71.

2.21 N. M. Newmark, "Numerical Methods of Analysis of Bars, Plates, and Elastic Bodies," in *Numerical Methods of Analysis in Engineering,* edited by L. E. Grinter, Macmillan, New York, 1949, pp. 138–168.

2.22 A. H.-S. Ang and N. M. Newmark, "A Numerical Procedure for the Analysis of Continuous Plates," *Proceedings of the 2nd Conference on Electronic Computation,* American Society of Civil Engineers, Pittsburgh, Pa., 1960, pp. 379–413.

2.23 A. H.-S. Ang and W. Prescott, "Equations for Plate-Beam Systems in Transverse Bending," *J. Eng. Mech. Div., ASCE,* Vol. 87, No. EM6, December 1961, pp. 1–15.

2.24 S. H. Simmonds, "The Effects of Column Stiffness on the Moments in Two-Way Slabs," Ph.D. thesis, University of Illinois at Urbana–Champaign, 1962, 216 pp. Also issued as *Civil Engineering Studies,* Structural Research Series 253, Department of Civil Engineering, University of Illinois, Urbana, Ill.

2.25 A. H.-S. Ang and L. A. Lopez, "Discrete Model Analysis of Elastic-Plastic Plates," *J. Eng. Mech. Div., ASCE,* Vol. 94, No. EM1, February 1968, pp. 271–293.

2.26 L. L. Jones and R. H. Wood, *Yield-Line Analysis of Slabs,* American Elsevier, New York, 1967, 401 pp.

2.27 R. J. Melosh, "A Stiffness Matrix for the Analysis of Thin Plates in Bending," *J. Aero. Sci.,* Vol. 28, 1961, pp. 34–42.

2.28 C. Zienkiewicz and Y. K. Chung, "The Finite Element Method for Analysis of Elastic Isotropic and Orthotropic Slabs," *Proc. Inst. Civ. Eng.,* Vol. 28, August 1964, pp. 471–488.

2.29 O. C. Zienkiewicz, *The Finite Element Method in Engineering Sciences,* 2nd ed., McGraw-Hill, London, 1971, 521 pp.

2.30 R. D. Cook, D. S. Malkus, and M. E. Plesha, *Concepts and Applications of Finite Element Analysis,* 3rd ed., Wiley, New York, 1989, 656 pp.

2.31 E.-M Salonen, "A Rectangular Plate Bending Element the Use of Which Is Equivalent to the Use of the Finite Difference Method," *Int. J. Numer. Methods Eng.,* Vol. 1, 1969, pp. 261–274.

2.32 J. C. Jofriet, "Short Term Deflections of Concrete Flat Plates," *J. Struct. Div., ASCE,* Vol. 99, No. ST1, January 1973, pp. 167–182.

2.33 A. Scanlon, *Time Dependent Deflections of Reinforced Concrete Slabs,* Structural Engineering Report 38, Department of Civil Engineering, University of Alberta, Edmonton, Alberta, Canada, December 1971.

2.34 A. Scanlon and D. W. Murray, "An Analysis to Determine the Effects of Cracking in Reinforced Concrete Slabs," *Proceedings of the Specialty Conference on*

Finite Elements in Civil Engineering, McGill University, Montreal, Quebec, Canada, June 1972.

2.35 A. Scanlon and D. W. Murray, "Time-Dependent Reinforced Concrete Slab Deflections," *J. Struct. Div., ASCE,* Vol. 100, No. ST9, September 1974, pp. 1911–1924.

2.36 J. C. Jofriet and G. M. McNeice, "Pattern Loading on Reinforced Concrete Flat Plates," *Proc. ACI,* Vol. 68, December 1971, pp. 968–972.

2.37 D. D. Pfaffinger, "Column–Plate Interaction in Flat Slab Structures," *J. Struct. Div., ASCE,* Vol. 98, No. ST1, January 1972, pp. 307–326.

2.38 L. A. Lopez, "FINITE: An Approach to Structural Mechanics Systems," *Int. J. Numer. Methods Eng.,* Vol. 11, No. 5, 1977, pp. 851–866.

2.39 C. P. Siess and N. M. Newmark, *Moments in Two-Way Concrete Floor Slabs,* Bulletin 385, University of Illinois Engineering Experiment Station, Urbana, Ill., 1950, 122 pp.

2.40 M. D. Vanderbilt, M. A. Sozen, and C. P. Siess, "Deflections of Multiple-Panel Reinforced Concrete Floor Slabs," *J. Struct. Div., ASCE,* Vol. 91, No. ST4, August 1965, pp. 77–101.

2.41 *Building Code Requirements for Reinforced Concrete,* ACI 318–71, American Concrete Institute, Detroit, 1971, 78 pp.

2.42 W. G. Corley, "The Equivalent Frame Analysis for Reinforced Concrete Slabs," Ph.D. thesis, University of Illinois at Urbana–Champaign, 1961, 166 pp. Also issued as *Civil Engineering Studies,* Structural Research Series 218, Department of Civil Engineering, University of Illinois, Urbana, Ill.

2.43 W. G. Corley and J. O. Jirsa, "Equivalent Frame Analysis for Slab Design," *Proc. ACI,* Vol. 67, November 1970, pp. 875–884.

2.44 J. R. Nichols, "Statical Limitations upon the Steel Requirement in Reinforced Concrete Flat Slab Floors," *Trans. ASCE,* Vol. 77, 1914, pp. 1670–1681; "Discussion," pp. 1682–1736.

2.45 M. P. Van Buren, "Staggered Columns in Flat Plates," *J. Struct. Div., ASCE,* Vol. 97, No. ST6, June 1971, pp. 1791–1797.

2.46 F. P. Wiesinger, "Design of Flat Plates with Irregular Column Layout," *Proc. ACI,* Vol. 70, February 1973, pp. 117–123; "Discussion," August 1973, pp. 597–598.

2.47 M. G. Salvadori and R. J. Schwarz, *Differential Equations in Engineering Problems,* Prentice-Hall, Englewood Cliffs, N.J., 1954.

3 Results of Elastic Theory Analysis

3.1 INTRODUCTION

The distributions of elastic theory bending moments within various kinds of panels of slab floors are presented and discussed in this chapter, with information on interior panels, edge panels, and corner panels presented. Before making these presentations, however, the various parameters affecting the moment distributions are presented and discussed. The notation and terminology will follow that of the ACI Building Code[3.1] wherever possible.

Each slab panel is assumed to be rectangular, with sides of l_1 and l_2, measured center to center of supporting columns, as shown in Fig. 3.1. The span l_1 is always the span being considered, while l_2 is the transverse span. The clear span face to face of supporting columns, l_n is very important in the development of the code provisions discussed in Chapter 4 but is not used extensively in this chapter, although the effects of the size of the support, c_1 and c_2, are discussed here. Panels ranging from square to those with a long side twice the short side are primarily discussed. Some comments are also made on slabs with greater aspect ratios.

Several stiffness parameters are required to define the distribution of moments within panels and the structure as a whole. For an interior panel surrounded by similar panels supporting the same distributed loads, the stiffness of the supporting beams, relative to the slab stiffness, is the controlling factor. The beam relative stiffness in direction 1 is defined as

$$\alpha_1 = \frac{E_{cb}I_{b1}}{l_2 D} = \frac{E_{cb}I_{b1}}{E_{cs}I_s} \tag{3.1}$$

where $E_{cb}I_{b1}$ is the flexural rigidity of beam in direction 1, D the slab flexural rigidity (see Eq. 2.1), and $E_{cs}I_s$ the flexural rigidity of slab of width l_2. Similarly,

$$\alpha_2 = \frac{E_{cb}I_{b2}}{l_1 D} \tag{3.2}$$

Other definitions of relative stiffness, such as $\alpha_1 = E_{cb}I_{b2}/l_1 D$, could also be used and are equally valid, but the definition adopted has the conceptual

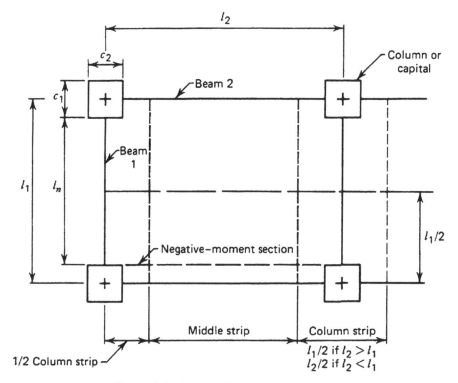

Figure 3.1   Layout of typical interior panel.

advantage of comparing the rigidities of the beam and the width of slab directly supported by the beam.

The range of values for beam stiffness α_1 and α_2 is obviously from zero to infinity, ranging from the case of no beams to the case of the slab supported on walls. In more practical terms, for square panels and concrete beams, it is unlikely that α_1 and α_2 will be much less than 1.0. For beams supporting the slab and without other major loads, it is unlikely that α_1 and α_2 will exceed 4 or 5, which is about the normal beam stiffness used in two-way slabs with beams designed using the 1963 and earlier ACI Codes. The use of bare steel beams, although not recognized by the ACI Code, is possible and may result in beam stiffness values considerably lower than 1.0, especially for noncomposite designs.

Values of beam stiffnesses α_1 and α_2 are ordinarily calculated using uncracked, gross section moments of inertia of both slab and beam. The use of transformed cracked or uncracked sections would generally result in somewhat higher values of α_1 and α_2, since the reinforcement ratios of the beams are always higher than those of the slabs. The beam cross sections to be considered in calculating the I_{b1} and I_{b2} values are shown in Fig. 3.2.

At the edge of the structure, the value of beam stiffness is computed as follows:

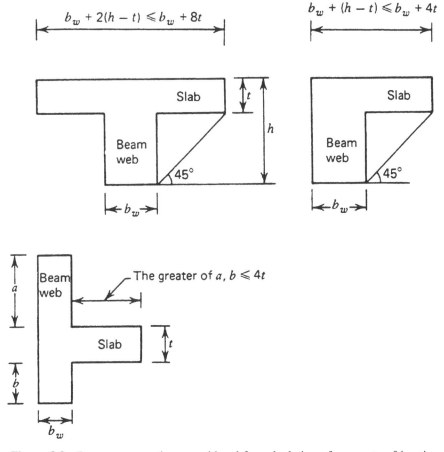

Figure 3.2   Beam cross sections considered for calculation of moments of inertia.

$$\alpha_1^e = \frac{E_{cb}I_{b1}}{(l_2/2)D} \tag{3.3}$$

Thus, the edge beam relative stiffness is computed considering half of the width of the slab that exists on one side of the beam.

In edge and corner panels, and in interior panels with unequal spans or loadings, the flexural stiffnesses of the supporting columns or walls and the torsional stiffness of the beams become very important in determining the distribution of moments in the panel.

The relative torsional stiffness of a beam spanning in direction 1 may be defined as

$$\beta_t = \frac{G_{cb}C}{l_1 D} = \frac{E_{cb}C}{2l_1 D} \tag{3.4}$$

where G_{cb} is the shear modulus of beam concrete $= E_{cb}/2$, C the torsional constant of cross section, and E_{cb} the Young's modulus of beam concrete. The value of C for a section made of rectangular components may be calculated by subdividing the section into rectangles and applying the approximate expression from the 1995 ACI Code:

$$C = \sum \left(1 - 0.63 \frac{x}{y}\right) \frac{x^3 y}{3} = \sum x^4 \left(\frac{y}{3x} - 0.21\right) \tag{3.5}$$

where x is the shorter side of a rectangular area and, y is the longer side of the same area (see Fig. 3.3). The subdivision should be done so as to minimize the length of common borders between segments.[3.2] A more sophisticated analysis for C may be found in most advanced strength-of-materials texts, such as that by Seely and Smith.[3.3]

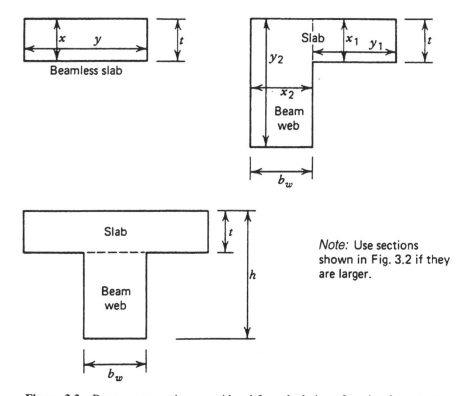

Figure 3.3 Beam cross sections considered for calculation of torsional constants.

The definition of the flexural stiffness of the columns, especially of exterior columns, is somewhat more difficult. The basic problem is that just making the column rigid does not mean that the slab then has a fixed edge, since the rotation of the edge of the slab panel has been prevented only at the location of the column and not across the entire panel width. As a consequence, moment can "leak" around the column and affect the conditions in the next span even though the column is rigid, and a discontinuous edge can be fixed only if both the column is flexurally rigid and the edge beam is torsionally rigid.

Two definitions of column stiffness were used in the 1977 ACI Code.[3.4] Although neither appears in the 1995 Code, both may be useful for certain purposes. The first, and simplest, simply views the slab–beam–column system as a frame and defines the relative column stiffness as

$$\alpha_c = \frac{\Sigma\, K_c}{\Sigma(K_s + K_b)} \tag{3.6}$$

where K_c, K_s, and K_b are the flexural stiffness, in terms of moment per unit rotation, of each column, slab, or beam framing into the joint, respectively. In this definition, the slab is merely viewed as a wide beam. It is appropriate to observe that the expression $\alpha_c/(1 + \alpha_c)$ is equivalent to the combined moment distribution factor to the columns at the joint.

The term α_c is a perfectly valid variable for use in examining the effects of changes of stiffness of various elements on the moment distribution, but it has some difficulties, especially when applied to a design rather than to an analysis situation. This may be illustrated by imagining two structures that are identical except for the torsional stiffnesses of the spandrel beams. The end span moments (corner panel or span perpendicular to the edge in an edge panel) in a structure with torsionally very stiff edge beams would be quite different from those in a similar structure having torsionally flexible edge members, even though the α_c values would be the same. This torsional stiffness problem exists at all joints between columns and slab, even though it is most important at the edges of the structure.

The concept of reducing the stiffness of the column to reflect the influence of the torsional rotation of the beams, or a portion of the slab if there are no beams, in the span perpendicular to the one considered was developed by Corley[3.5] and Corley and Jirsa[3.2] and has been incorporated in the Equivalent Frame Method of the 1995 ACI Building Code.

Rather than using the stiffness of the columns, the stiffness of the assemblage of columns and transverse beams shown in Fig. 3.4 is used. The triangular distribution of moment (for a case with no beam in the direction of the span being considered) is assumed as being reasonable. The actual analysis of this assemblage is more conveniently done in terms of flexibilities or rotations than in terms of stiffnesses, as follows.

Here $I_{\text{slab+beam}}$ is the moment of inertia of slab of width l_2, including the effects of beam as a stiffening rib (see Fig. 3.5b), and I_{slab} is the moment of inertia of slab of width l_2 (see Fig. 3.5a).

Adding the rotations of the column plus torsional member then gives

$$\theta_{ec} = \frac{1}{K_{ec}} = \frac{1}{\Sigma K_c} + \frac{1}{K_t} \tag{3.13}$$

where K_{ec} is the flexural stiffness of the equivalent column and K_t is replaced by K_t' if a beam exists in direction l_1.

A variety of assumptions have obviously been made in the development of this equivalent column concept, but the results of analyses of a number of structures were checked against experimentally obtained values, with satisfactory agreement being obtained.[3.2]

The use of the equivalent column stiffness, K_{ec}, in the computation of the relative column stiffness, α_{ec}, does not have the mathematical neatness of the computation of relative column stiffness, α_c, using K_c because of the approximations made in the derivation. However, it has some conceptual advantages due to the combining of two strongly interdependent variables and consequently is a useful parameter. The equivalent column relative stiffness is defined, similar to Eq. 3.6, as

$$\alpha_{ec} = \frac{K_{ec}}{\Sigma(K_s + K_b)} \tag{3.14}$$

The range of possible and practical values of K_c, K_{ec}, α_c, and α_{ec} is obviously quite large. Columns approaching zero relative flexural stiffness are possible, as in the case of a flat plate supported on relatively light steel pipe columns.[3.7] The K_c values for the lower floor columns in large buildings may

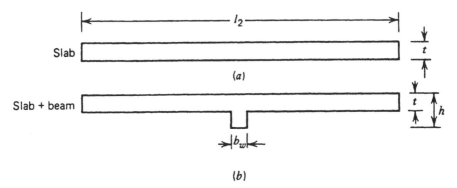

Figure 3.5 Cross sections considered for computation of modified torsional member stiffness.

be extremely high, and in such a case the degree of restraint is primarily dependent on the torsional stiffness of the transverse members.

The moments in the following sections are presented in terms of moments in beams, column strips, and middle strips in most cases. The strip widths are defined as in Fig. 3.1, where the column strip is shown as half the panel width but not more than half the shorter span of the panel. For panels with $l_2 > l_1$, this results in a column strip width less than $l_2/2$, but this definition has resulted in somewhat more consistent relationships between column and middle strip moments than if a column strip width of $l_2/2$ is always used. The middle strip occupies the space between column strips.

In most cases the moments in the following sections are presented as percentages of the total static moment for a panel, where the static moment is defined as in the ACI Code:

$$M_0 = \frac{wl_2l_n^2}{8} = \frac{Wl_1}{8}\left(1 - \frac{c_1}{l_1}\right)^2 \tag{3.15}$$

where w is the uniformly distributed load on panel per unit area; l_2 the transverse span, center to center of supports; l_n the clear span in direction considered $= l_1 - c_1$; W the total load on panel $= wl_1l_2$; l_1 the span in direction considered, center to center of supports; and c_1 the column or capital width in direction of span l_1. Stating the moments in terms of the static moment for each span emphasizes the static aspects of the problem, and that all the static moment must be accounted for in determining the moment distributions within a panel.

In a few cases in the following sections, moments are presented in terms of coefficients of moments per unit width, in the form m/wl_1^2. This form is used when examining the distribution of moments across or along sections. The elastic moments are for the case of Poisson's ratio $\mu = 0$ in nearly all cases. The accuracy of this assumption for reinforced concrete was discussed in Section 2.2.7.

3.2   MOMENTS IN INTERIOR PANELS OF SLABS

3.2.1   Effects of Relative Stiffness of Supporting Beams

The variations in moments caused by changes in beam relative stiffness are plotted in Figs. 3.6 to 3.10 for rectangular panels of five different ratios of side lengths. Total negative moments at supports and positive moments at midspan are shown. The basic data were developed by Sutherland et al.[3.8] The data are also tabulated and discussed by Appleton.[3.10] Beam, column strip, and middle strip moments are plotted versus the beam relative stiffness factor, $\alpha_1 l_2/l_1$. A linear scale is used for the beam stiffness factor, with a discontinuity indicated to show the moments for the rigid beam case, so that the

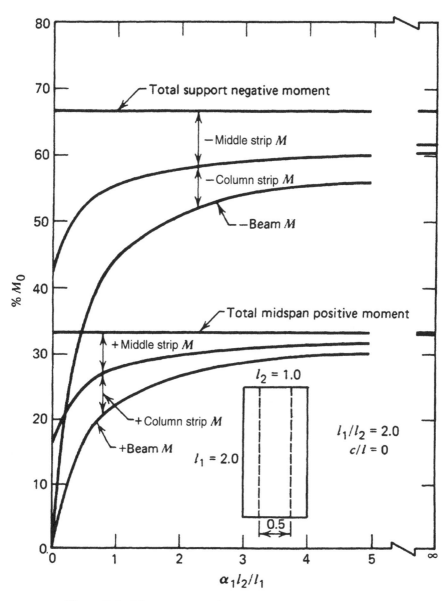

Figure 3.6 Moments versus beam stiffness for $l_1/l_2 = 2.0$.[3.9]

relative sensitivity of the moments to the beam stiffness in the low stiffness range, and the relative insensitivity for $\alpha_1 l_2/l_1$ values greater than 2 to 3, can easily be seen. The moments are for the case of a typical interior panel, surrounded by similar panels, with all panels supporting the same uniformly distributed load. The panel boundary conditions could also be stated as al-

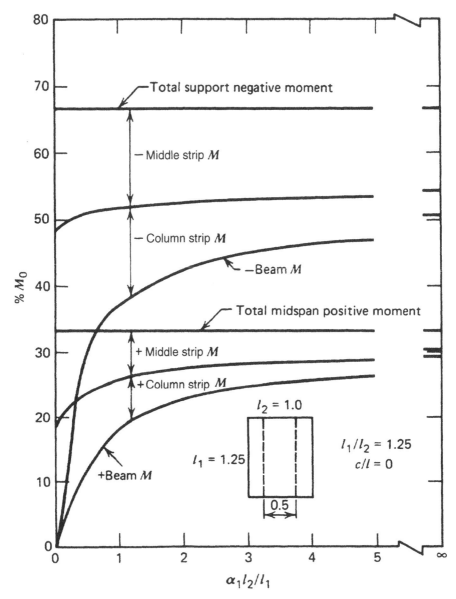

Figure 3.7 Moments versus beam stiffness for $l_1/l_2 = 1.25$.[3.9]

lowing deflection of the panel borders but maintaining the slope across the borders at zero. The supporting columns have no width; that is, c_1 and c_2 tend to zero.

The data of Figs. 3.6 to 3.10 are for the specific combination of beam flexural stiffness given by the expression

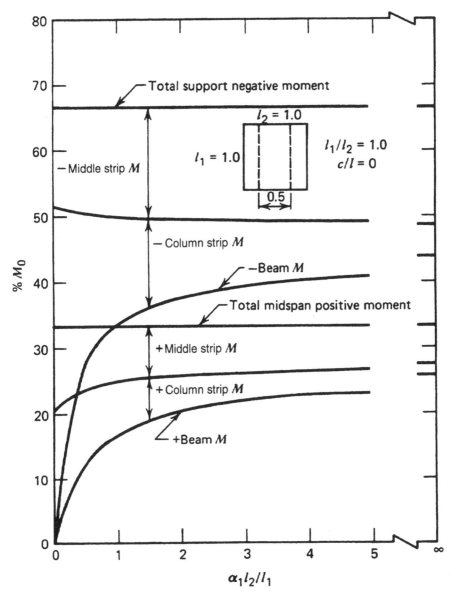

Figure 3.8 Moments versus beam stiffness for $l_1/l_2 = 1.0$.[3.9]

$$\frac{E_{cb}I_{b1}}{E_{cb}I_{b2}} = \frac{l_1}{l_2} \tag{3.16}$$

In other words, the moments of inertia of the beams are proportional to their spans. Another case is discussed later.

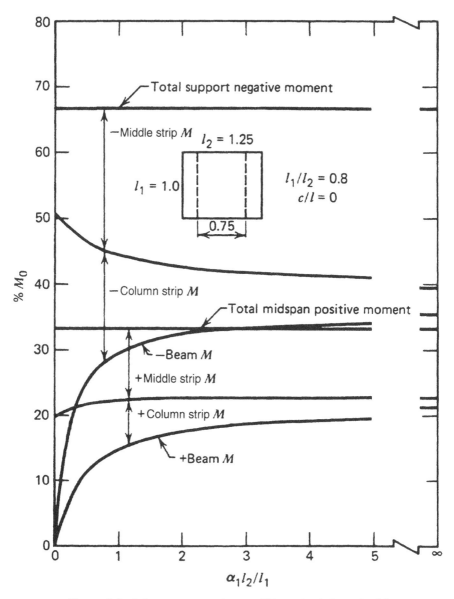

Figure 3.9 Moments versus beam stiffness for $l_1/l_2 = 0.8$.[3.9]

The graphs are best discussed with reference to Fig. 3.6, for $l_1/l_2 = 2$, where the separation of the various curves is greatest. As long as the supporting columns have no width, the support negative moments total $\frac{2}{3}M_0$ and the midspan positive moments $\frac{1}{3}M_0$ regardless of panel shape or beam stiffness, where M_0 is the static moment. The total negative, or positive, moment is shared between beam and slab, with the beam moment shown by the band

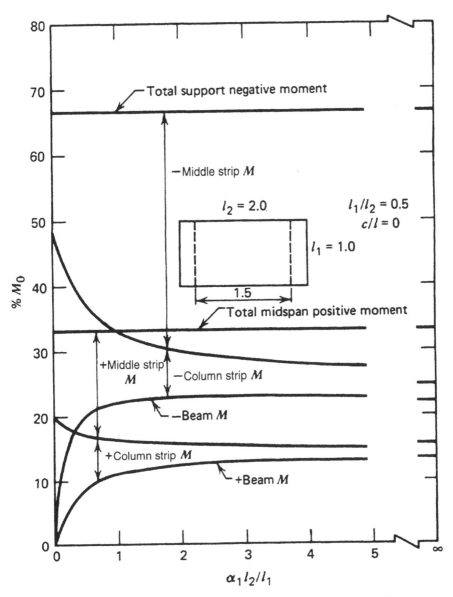

Figure 3.10 Moments versus beam stiffness for $l_1/l_2 = 0.5$.[3.9]

below the lower curve, the slab column strip moment by the intermediate band, and the slab middle strip moment by the upper band located below the straight, horizontal total section moment line.

The beam moment is most sensitive to beam stiffness, as would be expected. An increase in beam stiffness always causes a decrease in slab column strip moment, and there may be some merit in considering the slab column strip and beam as a unit. Increasing beam stiffnesses causes reductions in middle strip moments in the long span of rectangular panels and increases in the middle strip moments in the short spans, with the latter trend most marked for the negative moments. The increased moments in the short-span middle strips with increasing beam stiffnesses are a result of the greater support of the slab by the beams.

In most cases, the moment distributions for the case of $\alpha_1 l_2 / l_1 = 5$ are about the same as for rigid beams. The important exception is the case of the long span of a panel with $l_1 / l_2 = 2.0$ (Fig. 3.6). Here, with rigid beams the slab midspan positive moment is less than 0.5% of the static moment. If the beam stiffness is $\alpha_1 l_2 / l_1 = 5$, the slab midspan positive moment is more than 3% of M_0. In this case the difference between beam moments for the two cases are not important, but the differences in slab moments are quite large, even though in a practical case the minimum reinforcement requirements would probably eliminate any real differences unless the design loads were extremely heavy. If the beams are rigid, the slab midspan positive moment is not the maximum; the maximum is located about 0.2 of the long span from the end, and if the aspect ratio is 2.0 has a value of about 2.3 times the midspan moment. At an aspect ratio of 2, the slab approaches a one-way slab condition, but only if the supporting beams are rigid, and only for the positive moments, as the negative slab moments in the long-span direction remain appreciable.

The moments in panels with aspect ratios of 2 or more which are supported on nondeflecting supports can best be examined by visualization of the deflected shape of the panel. The panel tends to deflect to a trough shape, and if it is long enough relative to width, the deflection will be constant for some distance along the length of the slab. Constant deflection implies zero curvature and hence zero moment in the direction of the long span. If the aspect ratio is 2.0, the deflection curve does not quite include a portion with constant deflection, but the midspan curvature parallel to the long edge of the panel is considerably smaller than the curvature at points nearer the ends of the panel. The negative moments in the direction of the long span remain large for all aspect ratios, and in absolute terms remain about the same as the negative moments in a square panel having the same span as the short span of the rectangular panel.

To illustrate these points, moments acting in the direction of the long span in a uniformly loaded clamped slab having an aspect ratio of 2.0 are plotted in Fig. 3.11. The moments are from Ref. 3.11 and are specifically for the case of Poisson's ratio = 0. The moment diagram would be somewhat different in shape if Poisson's ratio = 0.3 or if the long-span supports are able to deflect even a very slight amount.

The case of beam stiffnesses $\alpha_1 l_2 / l_1 = 1.0$ is an interesting one, at least from a theoretical viewpoint, as at this particular beam stiffness the slab moments are uniformly distributed across the full width of slabs. Additional implications of this are that there are no twisting moments in the slabs, and

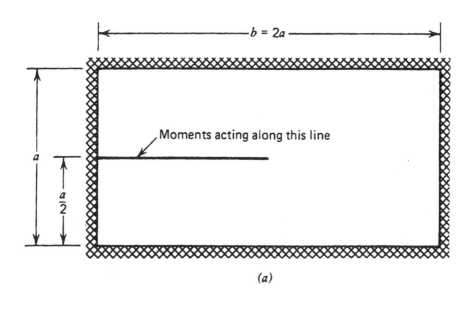

(a)

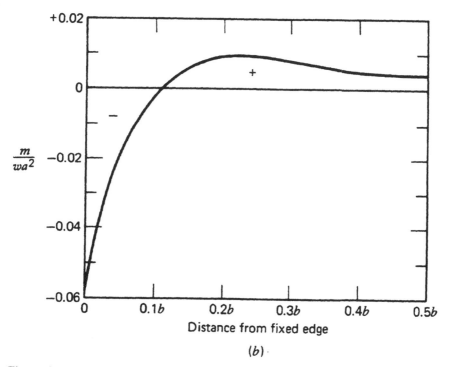

(b)

Figure 3.11   Moments in uniformly loaded clamped slab having aspect ratio = 2.0, $\mu = 0$: (a) uniformly loaded slab, fixed edges; (b) distribution of m along longitudinal centerline.

that the reactions of the slabs on the beams are uniformly distributed. In terms of the Lagrange equation (Eq. 2.1), the term $\partial^4 w / \partial x^2 \, \partial y^2$ vanishes, which leads to a simple expression for the deflected shape and a closed-form solution for the plate and beam moments. Wood[3.12] investigated a single panel simply supported on flexible beams and found a similar critical beam stiffness at which the twisting moments vanished. In addition, the nature of the reaction applied to the beams by the slab changes at this particular beam stiffness, as shown in Fig. 3.12, for an interior beam supporting a uniformly loaded continuous plate of square panels.

The foregoing information is for cases in which the beam relative stiffnesses in the two directions were related in a specific way, with EI values being proportional to span. This is probably a reasonable enough assumption if the two spans are greatly different, but if the aspect ratio is only slightly different from 1.0, it is probably more reasonable to make the beams in the two directions the same. Unfortunately, not very many such cases have been reported. Sutherland et al.[3.8] computed moments for square panels with many different combinations of beam stiffness in the two directions, but for rectangular panels solved only the cases with EI proportional to span. Vanderbilt et al.[3.13] developed some solutions using difference equations, for the cases of $l_1 / l_2 = 0.8$ and 1.25 and with the same beam rigidity in the two directions, and some of their results are plotted in Fig. 3.13 for comparison. Of the various beam and slab sections, the middle strip support negative moment

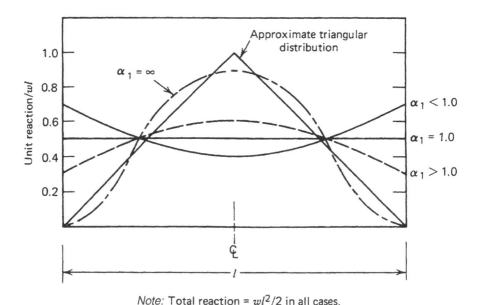

Note: Total reaction = $wl^2/2$ in all cases.

Figure 3.12 Approximate distributions of reaction on beams supporting uniformly loaded square interior panels.

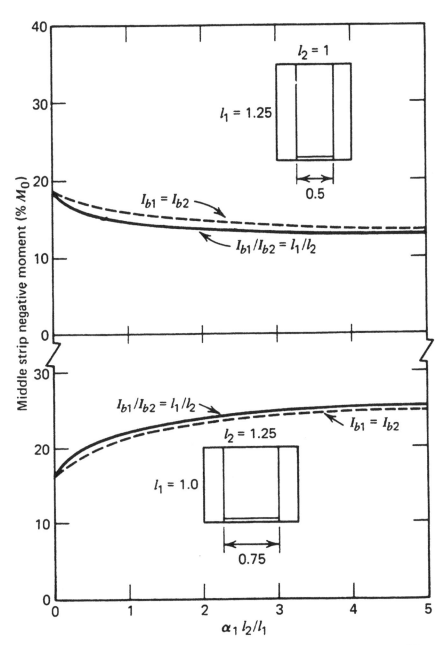

Figure 3.13 Effects of beam stiffness on middle strip negative moment.[3.9]

was the most sensitive to the stiffness of the transverse beam, but Fig. 3.13 makes it clear that this is not an important problem. Evidently, the beam stiffness in the direction considered is the most important; the transverse beam stiffness can be considerably different from being in direct proportion to the span without altering the moment distribution significantly.

A related problem is the case of a slab with beams in one direction only. The question then becomes: When does this structure become a one-way slab? At one limit, it is a beamless slab (a flat plate) spanning in two directions. At the other limit, the beams carry all the moment in one direction. The slab spanning from beam to beam may be divided into conventional middle and column strips. It should be obvious that at low beam stiffnesses the column strip moments are considerably larger than the middle strip moments, and Fig. 3.14 has been prepared to illustrate this, using Sutherland's data.[3.8] With these data, one is in a position to make a judgment about what constitutes the minimum beam stiffness for the use of uniformly distributed reinforcement in the slab. If one is willing to accept a support negative column strip average steel stress 20% greater than the panel average, and a support negative middle strip average steel stress 20% less than the panel average, a beam with $\alpha_2 = 1.1$ is sufficient. For the positive moment, no beam stiffness is required. If the overstress that can be allowed is 10%, a reasonable value, a beam stiffness of $\alpha_2 = 3$ is required for support negative moments, and the midspan positive moment column strip overstress will be very low. Thus, from the point of view of the slab in the span between the parallel beams, a beam stiffness value of $\alpha_2 = 3$ is adequate for nearly uniform distribution of the positive and negative slab reinforcement.

The span parallel to the beams is quite different. The moments in the beam, slab column strip, and slab middle strip are plotted versus beam stiffness in Fig. 3.15. The slab sections resist an appreciable portion of the total moment for all finite beam stiffness values. For example, if $\alpha_1 = 2$, the beam resists about 79% of M_0, as the sum of midspan positive and support negative moments, and the slab the remaining 21%. A conventional structure designed by working stress design methods would ordinarily have relatively stiff beams, but a structure designed using strength design concepts and the minimum possible beam dimensions could have beams small enough to cause large stresses in the slab spans parallel to the beams. The same is true if the beams are wide and shallow, as in a structural form sometimes labeled the *slab-band building*. The ACI and other codes require some slab steel to be placed parallel to the beams in one-way slabs, and this is ordinarily placed in the bottom of the slab, where it is available, and probably adequate, to resist the slab midspan positive moments. However, there ordinarily would not be top steel parallel to the beams, and hence in such a case there would be no control of the cracking that may occur due to negative moments.

This is not a structural safety problem, as the beams have adequate strength capacity to resist the entire static moment, since they are designed assuming they resist that moment. It is rather a serviceability problem in that potential

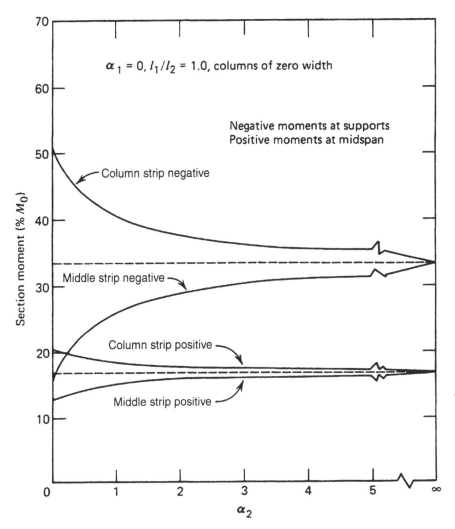

Figure 3.14 Slab moments versus beam stiffness in one-way slab, span perpendicular to beams.

slab negative moment cracks are not controlled. This can be reduced to the question of how much moment an unreinforced slab section should be expected to resist. At least the minimum steel should be provided as negative-moment reinforcement, parallel to the beams and across the lines connecting column centers, unless the beam relative stiffnesses are extremely high.

These data are for the case of slab continuous over parallel beams. It is not unreasonable to expect the same trends in cases where the slabs are simply supported on parallel beams and that about the same numerical values of beam stiffness would be required for a given uniformity of positive-moment

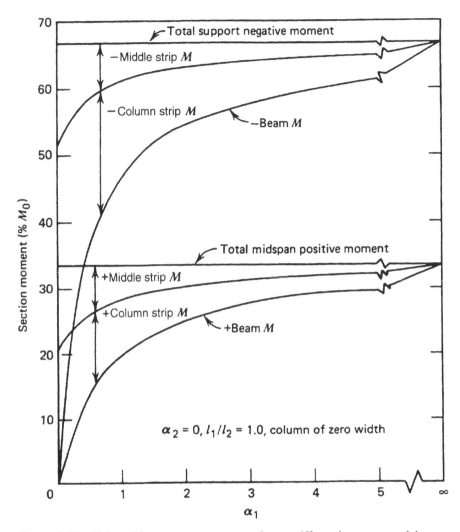

Figure 3.15 Slab and beam moments versus beam stiffness in one-way slab, span parallel to beams.

distribution, and reduction of slab moments parallel to the beams to reasonably small values.

3.2.2 Effects of Size of Supporting Column or Capital

The influence of the size of the supporting columns on the static moment has long been recognized. A larger column obviously results in a smaller clear or effective span. However, the question of what influence the support size has on the moment distribution within a panel has not received much attention.

Westergaard and Slater, in their classic 1921 paper,[3.14] studied the distribution of moments in flat slab panels supported on circular column capitals, with c/l values ranging up to 0.3, where c is the column capital diameter and l the slab span. They expressed their results in terms of percentages of the static moment, M_0, using the value of static moment proposed by Nichols (Eq. 2.42). Within the range of $c/l = 0.15$ to 0.3, the midspan positive moment was essentially invariant at 35% of Nichols's static moment. This information then formed the basis for the design of flat slabs in the ACI Codes until the 1971 edition was adopted.

The 1971 ACI Code introduced a new static moment definition, based on the clear span, face to face of the supports. If Westergaard's moments are compared to the M_0 value from the 1971 Code (including the replacement of the round columns with square ones of the same area for the purpose of calculation of the static moment), it is found that there is a significant change in the ratio of negative to positive moment as the c/l ratio is increased. As c/l increases from 0 to 0.3, the total midspan positive moment in the slab increases from $0.33M_0$ to $0.42M_0$, as is plotted in Fig. 3.16. This is a significant redistribution which should be taken into account in the design of slabs but which is currently ignored.

Additional information on this redistribution with increasing column size is given in Fig. 3.17, based on data developed by Vanderbilt et al.[3.13] in connection with their study of deflections. This figure indicates that regardless of beam relative stiffness, the total midspan positive moment increases from $M_0/3$ to about $0.4M_0$ as c/l increases from 0 to 0.3. These figures are for square panels supported on square columns, and the total values for the beamless case compare quite favorably with Westergaard's values for slabs on circular capitals. There are not sufficient data to subdivide Westergaard's support face negative moments between column and middle strips. The relative values of the midspan positive column and middle strip moments do not change with changes in c/l.

There are some rather strange changes in the beam moments with increasing c/l, especially in the negative-moment regions. These are due to transfer of reactions and moments directly to the capital rather than through the beam as the capitals become large. However, their practical significance is very small, since slabs supported on beams will seldom have c/l values greater than about 0.1.

3.2.3 Pattern Loadings

The basic proportioning of most floor slabs is carried out assuming that all panels are uniformly loaded by live load, but it is also necessary to check to see whether some serious serviceability problems may exist if only part of the panels carry live load. The strength of the slab is ordinarily not a problem under partial loadings, but excessive cracking and/or deflection may be. Two different types of partial loadings have been considered, *checkerboard loading* and *strip loading*. Each has significance in different types of structures, de-

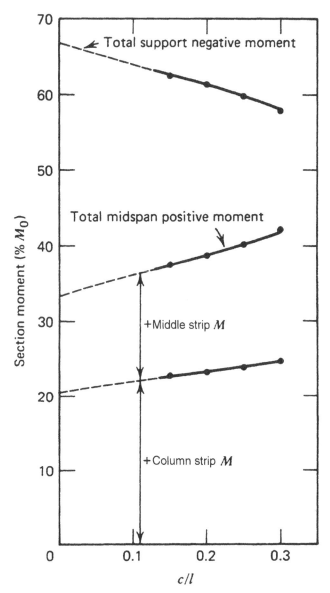

Figure 3.16   Effects of circular capital size on slab moments in square panels.

pending on the panel aspect ratio, beam flexural stiffness, beam torsional stiffness, and column flexural stiffness. The following discussion is concerned entirely with service loads rather than factored design loads.

Arrays of loaded and unloaded panels, for cases producing maximum slab moments, are shown in Fig. 3.18. The two loadings for maximum slab positive

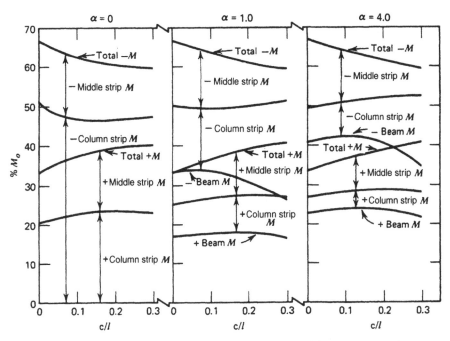

Figure 3.17 Effects of column or capital size on slab and beam moments in square panels, square capitals.[3.9]

moment (Fig. 3.18a and b) are the ones that have been considered in the literature, and solutions are readily attainable, at least for the case of zero flexural stiffness of columns. The loadings producing the theoretical maximum slab negative moment (Fig. 3.18c and e) have in general not been analyzed because of mathematical problems, and alternative patterns producing slightly lower maximum moments (Fig. 3.18d and f) have been used instead. However, this is not a serious limitation, since the increases in slab positive moments are much more important than any potential increases in slab negative moments. The loadings shown (Fig. 3.18b, e, and f) also produce the maximum positive moments in the beams, and the strip loads (Fig. 3.18c and d) produce the maximum negative beam moments. The checkerboard loading for negative beam moment is not shown but would start with a cluster of four loaded panels surrounded by a checkerboard pattern.

In addition to the general structural characteristics, the ratio of live load to dead load, or movable load to total load, is very important in determining the significance of pattern loads. Pattern loads are obviously of much greater potential importance in a structure in which the live load is several times the dead load than in a structure in which the live load is only part of the dead load. The ratio of live load to dead load may be taken into account by the following expression for the effective moment ratio for pattern loads:

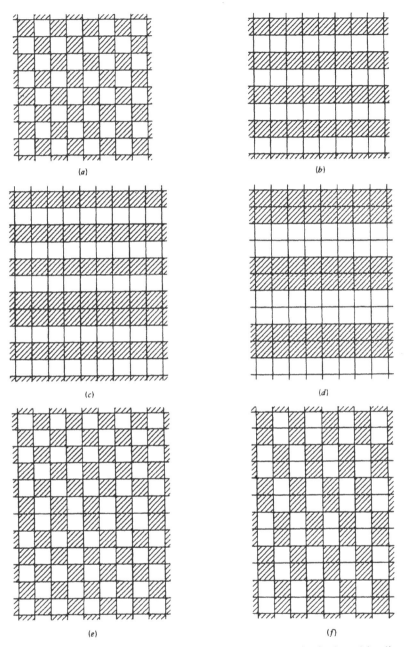

Figure 3.18 Loading patterns for maximum moments: (*a*) checkerboard loading for maximum slab positive moment; (*b*) strip loading for maximum slab and beam positive moments; (*c*) strip loading for theoretical maximum slab and beam negative moments; (*d*) strip loading used for maximum slab and beam negative moments; (*e*) checkerboard loading for theoretical maximum slab negative and beam positive moments; (*f*) checkerboard loading used for maximum slab negative and beam positive moments.

$$\gamma = \frac{\text{MAXM}}{\text{DLM} + \text{LLM}} = \frac{\gamma_1 + \beta_a}{1 + \beta_a} \tag{3.17}$$

where MAXM is the dead load moment plus maximum pattern live load moment, DLM the dead load moment, LLM the live load moment with all panels loaded, γ_1 the maximum pattern live load moment/LLM, and β_a the DLM/LLM = dead load/live load.

Although continuous beams and frames have ordinarily been designed for the maximum moments taking into account pattern loading, or moments closely related to those maximum moments recognizing some potential inelastic redistribution, slab structures and especially beamless slabs have generally been designed considering all panels to be loaded. Consequently, a pattern loading may be expected to cause some overstress, and this has been accepted as long as it is not too large. The 1977 ACI Code[3.4] contained provisions based on the implicit assumption that an overstress of 33%, or $\gamma = 1.33$, may be permitted in the slab sections of the structures. Those provisions led to combinations of minimum beam and column stiffness, as required to limit the overstress to about 33%. Those requirements were dropped from the code when other changes made them unnecessary. The increases in beam moments were not considered separately in that code.

The loading patterns specified in the ACI Code, Sec. 13.7.6.3, using 75% of the live load in the pattern loadings, lead to variable values of γ, with the maximum values ranging from about 1.21 for $\beta_a = \frac{4}{3}$ to 1.38 for $\beta_a = \frac{1}{3}$, both for cases with very low column stiffness. Smaller values result as the column stiffness increases. This conclusion is based on a value of $\gamma_1 = 2.0$, as is true for a typical interior panel under the strip loading of Fig. 3.18b. Using 100% of the live load leads to $\gamma = 1.43$ and 1.75, respectively, for the same two cases.

The most extensive study of the effects of partial loadings on moments in slab structures is that reported by Jirsa et al.[3.15,3.16] This work was based on solutions developed by Jirsa[3.16] and Gamble,[3.17] which in turn were based on work reported in a number of the other references cited in Chapters 2 and 3. Important work has also been reported by Jofriet and McNeice.[3.18] Much of Jirsa's work was concentrated on the case of typical interior panels and includes the effects of the full range of beam stiffnesses. Jofriet and McNeice's was concerned with beamless structures with five spans in one direction and a large number in the other. Pioneering work was also reported by Westergaard and Slater.[3.14]

One basic computational method, which works relatively easily as long as the columns have no flexural stiffness and the beams have no torsional stiffness, is the superposition of symmetric and antisymmetric loadings, as shown in Fig. 3.19. The computational advantage of the negative-moment loadings using alternate pairs of spans strip loaded rather than the arrangement giving the theoretical maximum is obvious. The error in the slab maximum negative

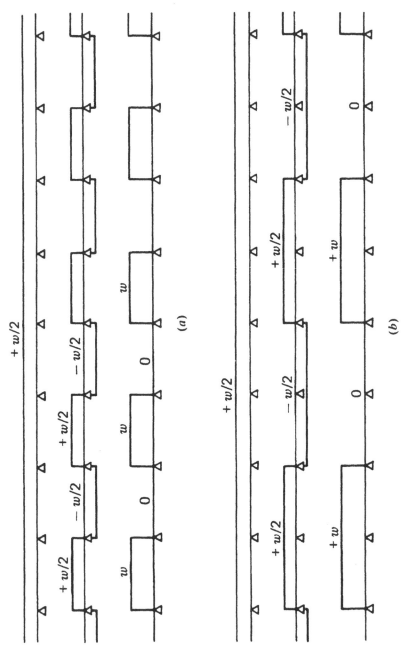

Figure 3.19 Superposition of symmetric and antisymmetric loadings to produce maximum moments: (*a*) loadings for maximum positive moments; (*b*) loadings for maximum negative moments.

moment from the checkerboard loading cannot readily be assessed but would not be very significant. For the strip loading case, with all live load movable and zero column flexural stiffness, the increase in the total negative moment in slab plus beam is 25% (γ_1 = 1.25) above that produced by live load on all panels. The theoretically correct loading for maximum negative moment causes an increase of about 37% (γ_1 = 1.37), but no rigorous solution has been made and it can only be assumed that the distribution of moment between beam and slab and across the slab sections would be the same for the two cases. In any event; this is not an important problem, since the values of γ will be acceptable for any realistic combination of dead load, live load, and finite column stiffness. Because of the low values of γ and γ_1 at the negative-moment sections, the remainder of the discussion of the effects of partial loadings will be concentrated on the positive moments.

The strip loading patterns are more important for positive moment than the checkerboard patterns for two reasons. First, they produce larger increases in positive moment unless the beam stiffnesses are large; second, the probability of getting a distribution of loading in a building that approximates the strip loading must be much greater than the probability of getting a checkerboard loading, especially in the case of structures with large live loads such as warehouses. In addition, the strip loading cases produce the maximum beam moments.

The slab positive-moment ratios are plotted versus beam stiffness in Figs. 3.20a to 3.22a for three different aspect ratios of panels. For the present, considering only the case of zero column flexural stiffness and zero beam torsional stiffness, the strip loading produces the maximum moments in a square panel (Fig. 3.21a) when $\alpha_1 < 3$ [i.e., $\alpha_1/(1 + \alpha_1) < 0.75$], and the checkerboard loading causes the larger moments for greater stiffnesses. Other aspects of these figures are discussed later.

For zero column stiffness, moment analysis shows that the strip loading doubles the total of the slab plus beam positive moment. Since Figs. 3.20a to 3.22a show that the slab moment ratio always reduces as the beam stiffness increases, the moment ratios for the beams must be larger than 2.0. Beam moment ratios for several aspect ratios are plotted against beam stiffness in Fig. 3.23 and show values of γ_1 ranging from 2.0 to 2.9. (The curves become undefined as α_1 approaches zero, since the moments then vanish.) These increases are so large in some cases that it seems necessary to take them into account separately in the assessment of the effects of partial loadings.

It does not appear that much work has been done that would give information on the effects of column stiffness on controlling the increases in beam moments. Morrison[3.19] considered a three-span structure with rigid columns, square panels and $c/l = 0.1$, and various beam stiffnesses. For interior span beams he found γ_1 values of 1.07 and 1.09 for cases of $\alpha_1 = 0.5$ and 2.0, respectively, with $\beta_t = 0.5$ in both cases. No similar case with flexible columns was analyzed, but considering the basic value of $\gamma_1 = 3$ for the interior span of a three-span continuous beam structure and the values given in Fig.

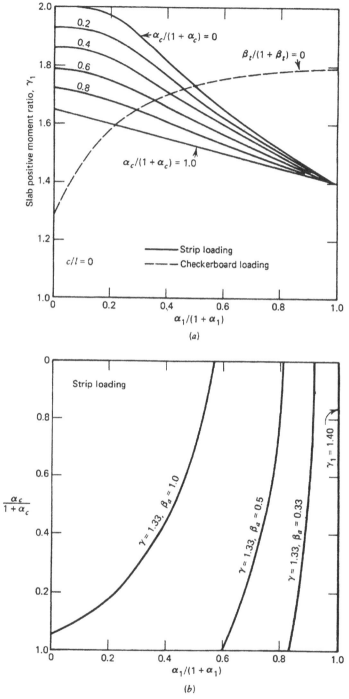

Figure 3.20 Effects of pattern loadings on positive moments in slab, $l_1/l_2 = 0.5$: (a) γ_1 versus support stiffness; (b) permissible β_a versus support stiffness.

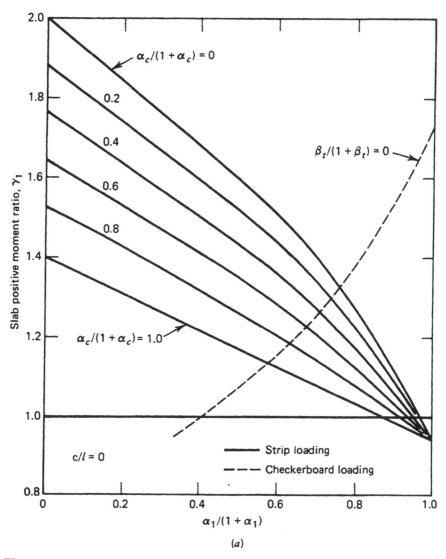

Figure 3.21 Effects of pattern loadings on positive moments in slab, $l_1/l_2 = 1.0$: (*a*) γ_1 versus support stiffness; (*b*) permissible β_a versus support stiffness.

3.23 for typical interior spans, it seems likely that $\gamma_1 > 3.5$ for the beam when $\alpha_c = 0$. The only conclusion possible is that columns appear to be considerably more effective in controlling beam moments than in controlling slab moments. The beam negative moment ratios are not a problem. The loading shown in Fig. 3.18*d* causes no values of γ_1 greater than 1.4 for any finite beam stiffness. The maximum value is 1.45 for rigid beams, $l_1/l_2 = 0.5$, and $\alpha_c = 0$.

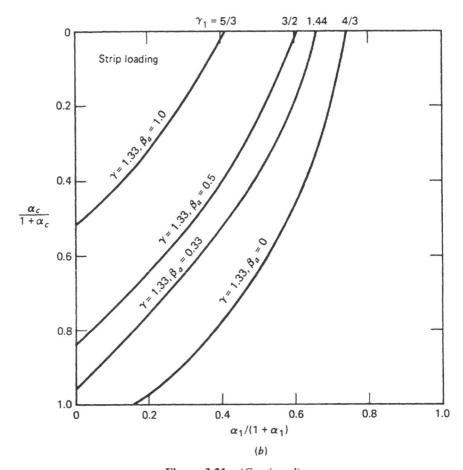

Figure 3.21   (*Continued*).

In addition to the beam flexural stiffness, the beam torsional stiffness and the column flexural stiffness influence the maximum moments produced by partial loadings. For the checkerboard loadings, however, only the beam torsional stiffness is important, as no moments are induced in the columns by the loading shown in Fig. 3.18a. For the strip loading, only the column flexural stiffness is important, since, to a reasonable first approximation, the strip loading of Fig. 3.18b produces uniform rotation along each column line at each edge of a loaded area, and consequently, produces no significant torsional rotations in the beams, especially at low column stiffness values.

The moment increases caused by checkerboard loadings will generally be held within acceptable limits by the torsional stiffness of the beams. If $\beta_t = 1.0$, the increases in moment will be half or less of those occurring when $\beta_t = 0$. β_t will be this large in many cases where slabs are supported on reinforced concrete beams with flexural stiffnesses large enough to make the checkerboard loadings important in the first place.

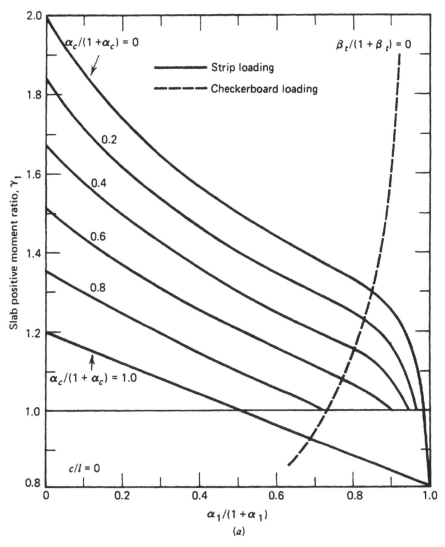

Figure 3.22 Effects of pattern loadings on positive moments in slab, $l_1/l_2 = 2.0$: (a) γ_1 versus support stiffness; (b) permissible β_a versus support stiffness.

To provide some additional information for slabs on beams with high flexural stiffness but low torsional stiffnesses, the following values are given in Table 3.1. For values of β_t greater than zero, a linear reduction in γ_1 from the value given to 1.0 as the term $\beta_t/(1 + \beta_t)$ increases from 0 to 1.0 may be assumed. The very high values of γ_1 given for positive moments in long spans of rectangular panels are applied to very small moments, and the increased moments are still so small that minimum reinforcement will often govern. These moment ratios are different in some cases from those given by

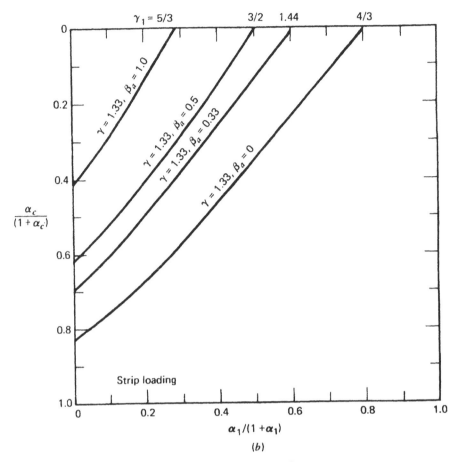

Figure 3.22   (*Continued*).

Jirsa et al.[3.15] in their study of the effects of torsional stiffness of the beams because of the sources of the basic moment data. The values of γ_1 reported here were obtained by superposing various relatively precise plate theory solutions for the loading patterns shown in Fig. 3.18a and f. Those reported by Jirsa were obtained from the analysis of 25 panel slabs using the moment distribution scheme described in Ref. 3.11, as was necessary to take the torsional stiffness into account. However, that analysis procedure is an approximate method that results in more accurate negative moments than positive moments. The procedure gives positive moments, for the case of all panels loaded, which are higher than the more exact theoretical moments, especially in the long spans of rectangular panels, and for these same cases it also underestimates the checkerboard loading moments.

The problem of the influence of column flexural stiffness on the effects of strip loadings has vexed designers and research workers for three-quarters of

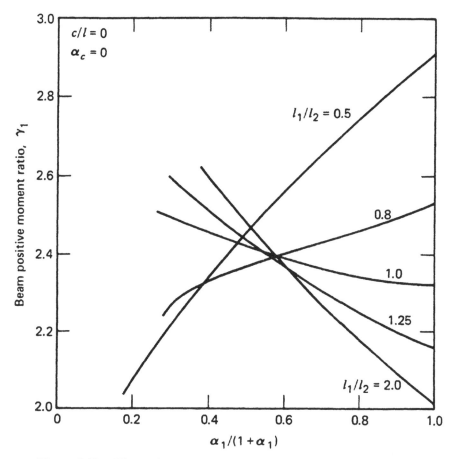

Figure 3.23   Effects of strip loadings on beam positive moments, $\alpha_c = 0$.

a century, and no completely satisfactory answer has yet been produced. Fortunately, there are some reasonably valid analogies between continuous slabs and continuous beams that help set some limits on the effects. In a typical interior panel of a slab, the total of the slab plus beam positive moment is doubled by the strip loading if the columns have no flexural stiffness and

TABLE 3.1   Values of γ_1 for Span in Direction l_1 for Slabs Under Checkerboard Loadings, $\alpha_1 = \infty$, $\beta_t = 0$

l_1/l_2	0.5	0.8	1.0	1.25	2.0
Positive moment	1.79	1.65	1.73	2.61	11.4
Negative moment	1.31	1.36	1.47	1.58	1.68

Source: After Ref. 3.16.

$c/l = 0$; the positive moment in an interior span of a continuous beam is also doubled. However, as the stiffness of the supporting columns is increased, the beam approaches the case of an isolated, fixed span; the slab panel is not completely fixed even if the columns are rigid. The degree of fixity for a given column stiffness depends on the panel aspect ratio, the beam stiffnesses, and the size of the support or capital.

Westergaard and Slater[3.14] studied the effects of the capital size in slabs with square, beamless panels. Data from that study are plotted, along with the point at $\gamma_1 = 2$ for the case of $c/l = 0$ and $\alpha_c = 0$, in Fig. 3.24. The case of $c/l = 0$ and $\alpha_c = \infty$ was not considered, but it appears that a value of $\gamma_1 = 1.35$ to 1.40 is reasonable for rigid columns, with an approximately linear decrease to $\gamma_1 = 1.0$ for $c/l = 0.3$. An interesting aspect of the data is the sharp increase in γ_1 as c/l increases for the case of zero column stiffness. This increase can be explained easily by use of a continuous beam analogy. If the end sections of each span of a continuous beam are made stiffer than

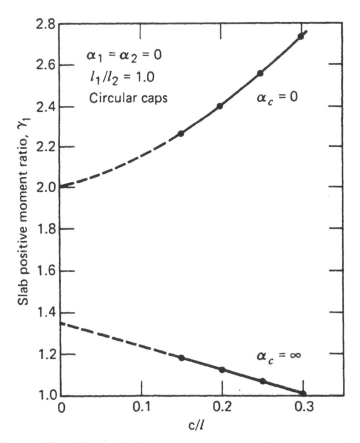

Figure 3.24   Effects of column capital size on positive-moment ratios.

the remainder of the span, the negative moments increase, with a consequent decrease in positive moments, for the case of all spans loaded. On the other hand, the positive moments produced by the antisymmetric loading that is superposed with the uniform loading are not altered by the increased end stiffnesses, which leads to larger values of γ_1. (This is not in conflict with the increase in positive moment with increasing capital size shown in Fig. 3.16, since those moments are expressed in terms of the ACI Code static moment.)

Westergaard provided a method for computing the moments when the columns had intermediate stiffness values, using a frame analysis approach combined with a modified column stiffness reflecting the "leakage" of moment around the columns, the capital diameter, and the capital height plus slab thickness relative to the story height. He thus anticipated the equivalent column of the 1971 and later ACI Codes by more than 50 years.

Jofriet and McNeice[3.18] studied the effects of strip loadings in five-span structures having no beams but having columns of finite flexural stiffness. Although the center span of a five-span structure is not the same as a typical interior span, their study of the effects of panel shape, with l_1/l_2 values of 0.5, 1.0, and 2.0, is quite valuable. They also considered the influence of the c/l ratio, although the results are not given in their very short paper. The positive-moment ratio for a typical interior span is 2.0, while it is 1.86 for the central span of a five-span continuous beam and about 1.85 for their slabs for the case of zero column stiffness.

Values of the positive-moment ratio for the three panel shapes[3.18] and a beam are given in Table 3.2 for the end, first interior, and central spans, of five-span structures, for the cases of completely flexible and rigid columns.

TABLE 3.2 Values of Positive Moment γ_1 for Five-Span Slabs[3.18] and Beams, $\alpha_1 = 0$, $c/l = 0$

l_1/l_2	$\alpha_c = 0$	$\alpha_c = \infty$
Interior span		
0.5	1.90	1.62
1.0	1.85	1.40
2.0	1.83	1.20
Beam	1.86	1.00
First interior span		
0.5	2.60	1.76
1.0	2.37	1.45
2.0	2.33	1.20
Beam	2.40	1.00
End span		
0.5	1.26	1.19
1.0	1.33	1.17
2.0	1.35	1.08
Beam	1.36	1.00

The slab and beam values are similar in most cases where the column stiffness is zero, although one must wonder why the value for the first interior span when $l_1/l_2 = 0.5$ is so high, and so low for the end span. The values for the rigid column cases illustrate very well the leakage of moment around the columns. From the various values it appears reasonable to suggest that the following γ_1 values are appropriate for the case of rigid columns, $c/l = 0$ and no beams, for any interior panel:

l_1/l_2	0	0.5	1.0	2.0	∞
γ_1	2.0	1.65	1.4	1.2	1.0

The values of γ_1 for the first interior and center spans are so similar when the columns are rigid that it seems reasonable to assume that the values found for the center span are representative of those for a typical interior span. The variations in γ_1 with the column stiffness, $\alpha_c/(1 + \alpha_c)$, are plotted in Fig. 3.25 for the center span. In all cases the variations have the same basic shapes. The relatively high value of $\gamma_1 = 2.4$, shown for the first interior of span of the beam in Table 3.2, is a consequence of having a small moment when all spans are loaded rather than of having an unusually large moment under the strip loading. The strip loading produces a slightly smaller maximum moment in the first interior span than in the central span or in a typical interior span. Similarly, the low value of $\gamma_1 = 1.36$ for the end-span moment is a consequence of both a high moment when all panels are loaded and of changing the conditions of restraint only at one end of the span by the application of the strip loading.

There apparently have been no rigorous analytical studies of the influence of the flexural stiffness of the columns on the moment ratios caused by strip loadings in structures with beams. To overcome this problem, Jirsa et al.[3.15,3.16] developed a method of making a reasonable estimate of the effects. This can best be explained with reference to Fig. 3.21a, considering the square panel case. The top curve for the strip loading case, for zero column stiffness, is derived from precise analytical solutions and shows that the moment ratio decreases from 2.0 to 0.94 as the beam stiffness increases from zero to infinity. The only points available when the columns are rigid are those at $\gamma_1 = 1.4$ for zero beam stiffness, as suggested above, and 0.94 for rigid beams, since the column stiffness is not relevant if the beams are rigid. It is not unreasonable then to connect these two points with a straight line, giving the lower curve.

The next step is to provide some transition from completely flexible to rigid columns. In this figure it has been assumed that the transition is a linear function of the value of the column stiffness parameter, $\alpha_c/(1 + \alpha_c)$. Figure 3.25 indicates that the linear transition is not quite valid for zero beam stiff-ness but that it is always conservative in that it predicts a slightly larger value

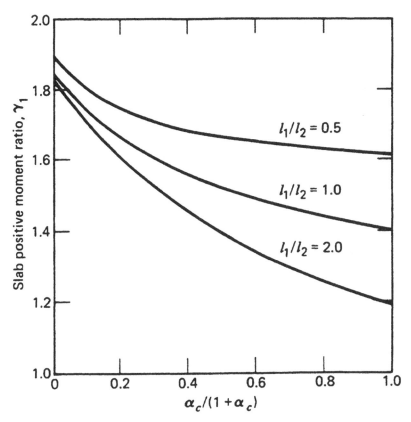

Figure 3.25 Positive moment ratios for center span of five-span slab versus column stiffness, $c/l = 0.$[3.18]

of γ_1 than actually exists. There is no evidence of any kind about the nature of the transition for larger beam stiffnesses, but it is believed that a linear transition is adequate and safe.

The three graphs shown in Figs. 3.20a to 3.22a are similar to, but not the same as, those prepared by Jirsa et al.[3.15] In each case the upper curves, for zero column stiffnesses, are the same, but the lower straight lines for the rigid column cases are somewhat different. Jirsa assumed, for lack of any better data, that the columns in beamless slabs were ineffective in isolating a panel if $l_1/l_2 = 0.5$, that rigid columns were completely effective if $l_1/l_2 = 2.0$, and that $\gamma_1 = 1.2$ for a square panel with rigid columns. These figures have been prepared using values of γ_1 of 1.65, 1.4, and 1.2 for l_1/l_2 ratios of 2.0, 1.0, and 0.5, respectively, as was suggested earlier, and are specifically for the case of $c/l = 0$.

The next step, still following Jirsa's lead, was the preparation of the three figures comprising Figs. 3.20b to 3.22b. These are contour graphs for equal values of γ_1 plotted against both beam relative stiffness and column relative stiffness. In addition, each contour is labeled with the permissible value of β_a for the case of $\gamma = 1.33$, which was assumed in development of the ACI Code. For any given ratio of dead load to live load and any allowable over-stress, the required combination of beam and column stiffness may readily be identified. These graphs differ from Jirsa's[3.15] only in detail. The square and long-span panels would require slightly greater column stiffnesses than in Jirsa's version, while the short span of the rectangular panel, $l_1/l_2 = 0.5$, would require less stiff columns for a given degree of control of the effects of partial loadings.

No complete picture of the effects of the capital size, or the c/l ratio, can yet be assembled. Such studies have been only on beamless slabs, which is fairly reasonable since only beamless slabs are likely to have values of c/l much larger than 0.1. However, enough data can be assembled to at least allow some speculation about the effects in the case of a square panel. The information available has been assembled in Fig. 3.26a for the case of $c/l = 0.1$. The dashed lines give the limiting information for $c/l = 0$, from Fig. 3.21. For rigid columns, both Westergaard's data (Fig. 3.24) and Morrison's as reported by Jirsa et al.[3.15] indicate $\gamma_1 = 1.25$, approximately, and Morrison's work provides two more points for rigid columns and low beam torsional stiffnesses, enabling construction of the lowest curve. Westergaard's data for zero column stiffness indicate that $\gamma_1 = 2.15$, and starting from that point and following the shape of the curve for $c/l = 0$, the upper curve was drawn as a reasonable approximation. Then the intermediate curves could be approximated, and using Fig. 3.26a, the contour plot of Fig. 3.26b was prepared.

Relative to the data for $c/l = 0$, the use of columns with $c/l = 0.1$ tends to require slightly stiffer beams at very low column stiffnesses but somewhat less stiff columns at very low beam stiffnesses, for a given degree of control of the effects of partial loadings. This figure is very similar to the one presented by Jirsa et al.[3.15] This information must be regarded as a fairly crude approximation, but the trends should be reasonably well forecast, and the same trends would be expected for the case of rectangular panels.

Jofriet and McNeice[3.18] studied the effect on the slab positive moment ratio of the capital size in beamless slabs, and presented the following equation to approximate the combined influence of the c/l ratio and the column stiffness ratio. The equation is rewritten to bring the notation into conformance with the remainder of this section.

$$\gamma_1 = \frac{1.85}{1 - (c_1/l_1)} - \frac{l_1/l_2}{1 + (l_1/l_2)}\left[1.75\,\frac{\alpha_c}{1 + \alpha_c} - 0.8\left(\frac{\alpha_c}{1 + \alpha_c}\right)^2\right]$$

$$- 2.8\,\frac{\alpha_c}{1 + \alpha_c}\frac{c/l_1}{1 - (c/l_1)} \qquad (3.18)$$

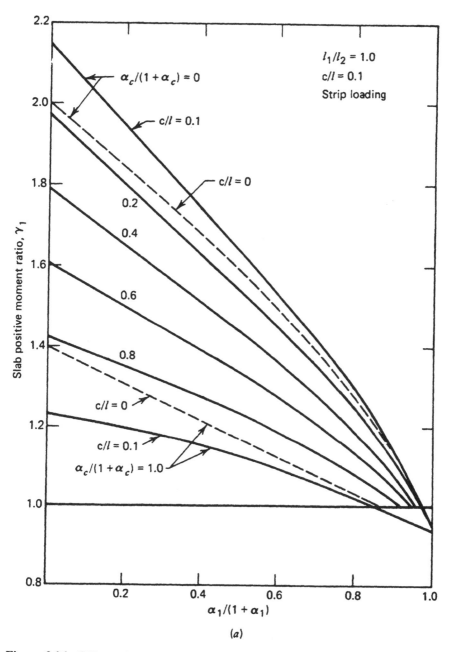

Figure 3.26 Effects of pattern loadings on positive-moment ratios when $c/l = 0.1$, $l_1/l_2 = 1.0$: (a) γ_1 versus support stiffness; (b) permissible β_a versus support stiffness.

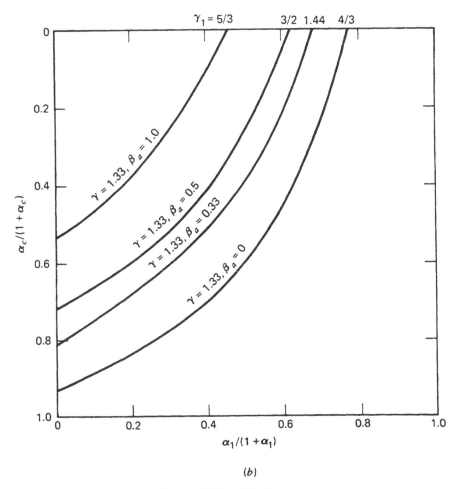

(b)

Figure 3.26b   (*Continued*).

This equation is in reasonable agreement with the values shown in Figs. 3.20 to 3.26, although the various coefficients need minor changes so that the equation converges to $\gamma_1 = 2.0$ when $\alpha_c = 0$ and $c/l = 0$. Otherwise, it gives a good picture of the increase in live load moments due to partial loading on beamless slabs.

In closing it is worthwhile repeating the opening philosophy of this section. Partial loading on slabs may cause serviceability problems, but they should not cause safety or loss-of-strength problems, as slabs will in general be able to redistribute moments caused by high overloads. The serviceability problems would principally arise from greater cracking because of higher steel stresses than under uniformly distributed loads, and larger deflections because of reduced restraints at the ends of various spans.

3.3 MOMENTS IN EDGE PANELS OF SLABS

All of the variables influencing the distribution of moments in interior panels also affect the moments in edge panels. In addition, the flexural stiffness of the columns and the torsional stiffness of the beams are always important variables instead of having influence primarily under pattern loadings.

The edge panel is probably best thought of as being two separate systems: one spanning parallel to the direction of the edge of the structure, and the other spanning perpendicular to the edge. The system spanning parallel to the edge of the structure is very similar to a typical interior panel, and the moments are relatively insensitive to the stiffness of the edge columns and the torsional stiffness of the spandrel beams. The moments in the span perpendicular to the edge of the structure are sensitive to both the flexural stiffness of the edge columns and the torsional stiffness of the edge beams; these moments are very similar to those in corner panels, as discussed later. Of the moments in the span perpendicular to the edge of the structure, the negative moments at the exterior end of the span are extremely sensitive to the edge column stiffness and the midspan positive moments are relatively sensitive; the interior negative moments are somewhat less sensitive than a frame analogy would indicate, as is reasonable since a rigid column cannot produce a fixed end. For a square panel, the moments in the span perpendicular to the edge of the structure are generally larger than those in the span parallel to the edge, just as end span moments are larger than interior span moments in beams.

The last statement is contrary to some past design practice, as Methods 1 and 3 of the 1963 ACI Code[3.20] for the design of slabs supported on beams indicated that slab moments in the span parallel to the edge of the structure were considerably higher than those in the perpendicular span. Those design moments were based on results of analyses of panels having three edges fixed against deflection and rotation and the fourth side simply supported, and in this case the parallel span moments are indeed significantly higher than the perpendicular span moments, as shown in Fig. 3.27. However, this model is a fairly unrealistic representation of an edge panel, since real beams are seldom rigid, and once some degree of flexibility of beams and columns is introduced, the perpendicular span slab moments become the larger ones.

The elastic moment distributions in edge panels have received much less attention than in interior panels, at least partially because edge panels are more complex and the means of investigating some of the variables simply did not exist in the precomputer era. A number of different kinds of structures have been analyzed to give information on edge panels. Appleton[3.10] used finite differences to analyze a structure having two spans in one direction and an infinite number in the other direction. He considered several combinations of beam flexural stiffness, but did not include edge beam torsional stiffness and considered the columns to be rigid. Jofriet[3.21,3.22] and Jofriet and McNeice[3.18] considered semi-infinite structures with one, three, or five spans in

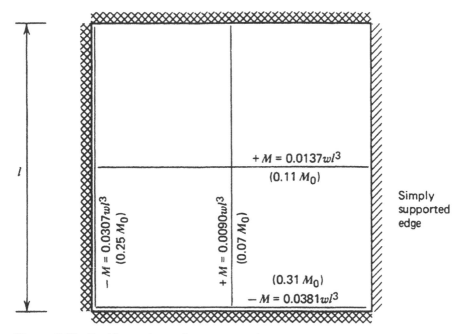

l

$+ M = 0.0137wl^3$

$(0.11 M_0)$

Simply
supported
edge

$- M = 0.0307wl^3$
$(0.25 M_0)$

$+ M = 0.0090wl^3$
$(0.07 M_0)$

$(0.31 M_0)$

$- M = 0.0381wl^3$

Figure 3.27   Positive and negative moments in a square plate fixed on three sides and simply supported on the fourth, $\mu = 0.$[3.11]

one direction using finite elements. They varied the column stiffnesses and considered slabs with various edge beams but no interior beams. In the study of the influence of the torsional stiffness of flexurally rigid beams, Jirsa et al.[3.15] considered the panel at the center of one edge of a 25-panel structure, arranged five by five. The moment-distribution scheme developed by Siess and Newark[3.11] was used.

Most of the other available information comes from analyses of nine-panel slabs with square panels. Ang and Newmark[3.23] and Morrison[3.19] considered cases with different beam stiffness for rigid columns having $c/l = 0.1$. Simmonds[3.24] considered approximately the same range of beam stiffness variables, except the case of zero beam stiffness, and included the effects of column flexural stiffness, but with $c/l = 0$. There are some obvious problems with the edge panels of such slabs, because of the influence of the adjacent end spans on the moments in the span parallel to the edge of the structure causing high negative moments, but the interior panel has the same difficulty and the moments in the two cases can be compared directly.

In the direction perpendicular to the edge, the number of spans is of more importance, as it has a modest potential influence on the interior negative moment. This moment is highest for a two-span slab, and for the most severe case of zero column stiffness, any slab with more than two equal spans would have an interior negative moment 80 to 85% of that in the two-span case.

The important relative stiffness parameters were discussed in Section 3.1. However, the edge beam relative stiffness requires further discussion. The ACI Code in effect defines the beam relative stiffness factor, α_1, as the flexural rigidity of the beam cross section divided by the flexural rigidity of the width of slab supported by the beam. Consequently, Eq. 3.1 for an interior beam should be restated for the specific case of the edge beam:

$$\alpha_1^e = \frac{E_{cb}I_{b1}}{E_{cs}I_s} = \frac{E_{cb}I_{b1}}{(l_2/2)D} \tag{3.19}$$

as in Eq. 3.3. Thus, if the EI of the edge beam is half that of the adjacent interior beam, the α_1 values will be the same for the edge beam and the adjacent interior beam.

3.3.1 Moments in the Span Parallel to the Edge of the Structure

The similarities of moments in the span parallel to the edge and interior span moments can be shown in several ways. Figure 3.28 illustrates the relationships between slab moments and beam relative stiffnesses for interior and edge panels of slabs with nine square panels. Figure 3.28a is for the case of rigid columns, and Fig. 3.28b is for the case of columns whose EI/h values

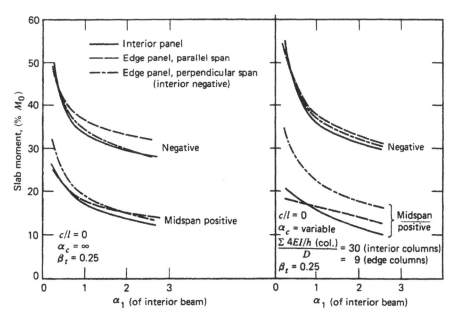

Figure 3.28 Edge panel slab moments versus α_1, square panel from nine-panel slab.[3.24]

remain constant as α_1 varies, resulting in a variable α_c. The columns are of moderate stiffness.

The values shown in Fig. 3.28 show how unrealistic the moment distributions for panels on rigid beams (Fig. 3.27) are. If the columns are rigid but the beams have flexibility, the midspan positive moments become essentially equal in the two spans of an edge panel. The addition of flexibility to the columns ensures that the midspan positive moments in the perpendicular span of the edge panel are considerably greater than in the parallel span.

The values used in Fig. 3.28 are from the analyses by Simmonds.[3.24] In these structures, the *EI* of the edge beams was 0.625 of that of the interior beams, as was the torsional stiffness. The *EI* of the edge columns was 0.3 of that of the interior columns, considering bending perpendicular to the edge of the structure. The α_1 and β_t values used here are for the interior beams. The α_c, α_{ec}, and *EI/h* values for columns specified are for the edge columns, with bending perpendicular to the edge of the structure unless noted otherwise.

For the range of column and beam stiffnesses illustrated, the moments in the interior panel and the parallel span of the edge panel never differ by more than about 20% of the smaller moment. Other cases investigated indicated that higher values of β_t lead to smaller differences. The work reported by Morrison,[3.19] also on slabs with nine square panels, but with rigid columns having $c/l = 0.1$, indicated the same range of maximum differences, including cases of slabs with no edge beams.

Comparisons can also be made between moments in typical interior panels, as reported earlier in this chapter, and the edge panel moments found by Appleton[3.10] for structures two panels wide by many panels in length. Figure 3.29 shows average column strip, middle strip, and wall strip moments for square panels and rectangular panels with aspect ratios of 2.0. Both midspan positive and support negative moments are shown. The wall strip is the column strip adjacent to and parallel to the edge of the structure. The moments are shown in terms of coefficients of m/wl_1^2, where m is the moment per unit width of slab, and are for the three cases of no interior beams, rigid beams, and the intermediate flexural stiffness value of $\alpha_1 = 1.0$ for the square panels and $\alpha_1 l_2/l_1 = 0.5$ for the rectangular panels. The edge panels designated as having zero beam stiffness actually have edge beams with the same flexural stiffnesses as in the intermediate cases, but in all cases the edge beam torsional stiffness is zero. Moments in both the parallel and perpendicular spans of the edge panels are presented. The latter case is considered in the next section.

Comparisons can be made directly for the cases with beams on all sides of the panels. For slabs with no interior beams, the edge beam alters the moments primarily in the wall strip, and the wall strip moment should be compared with the interior panel column strip moment for the beam stiffness corresponding to that of the edge beam rather than for a $\alpha_1 = 0$.

In most cases, the moments in interior and edge panels are reasonably similar. In the case of the rectangular panel, $l_1/l_2 = 2.0$, which has a spandrel

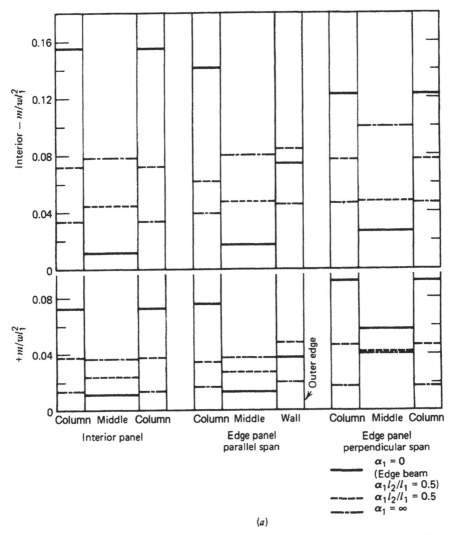

Figure 3.29 Comparisons of distributions of moments across critical sections of interior and edge panels: (a) $l_1/l_2 = 0.5$, $\beta_t = 0$, $K_c = \infty$; (b) $l_1/l_2 = 1.0$, $\beta_t = 0$, $K_c = \infty$; (c) $l_1/l_2 = 2.0$, $\beta_t = 0$, $K_c = \infty$.

beam along one long side of the panel, the isolated edge beam appears to reduce the moment across the entire width of the panel, which is not unreasonable considering the shape of the panel. This is consistent with the results of Jofriet's analyses,[3.21] where an edge beam was found to be generally more heavily loaded than an interior beam with the same α_1.

It seems reasonable to conclude that for purposes of design, both the magnitude and distribution of moments in edge panels in the span parallel to the edge of the structure may safely be taken as the same as for interior panels

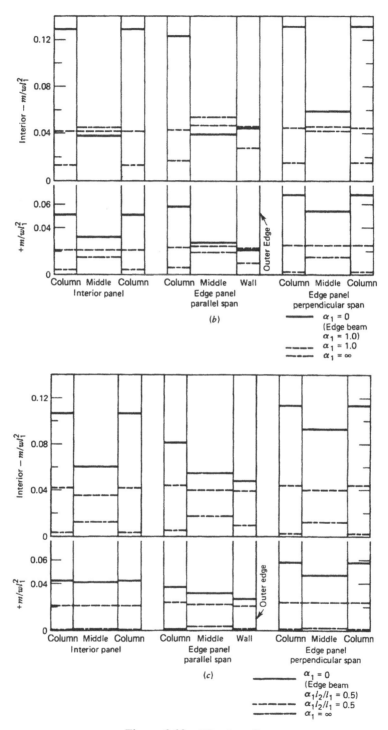

Figure 3.29 (*Continued*).

of the same shape and having the same relative stiffnesses of beams. This approximation becomes even better when the edge beams have reasonable torsional stiffnesses.

3.3.2 Moments in the Span Perpendicular to the Edge of the Structure

In the span perpendicular to the edge of the structure, both the total moments and their distributions across the sections must be determined. The magnitudes of the total moments, for a given span and loading, depend strongly on the relative stiffness of the edge columns of the structure, and this relative stiffness must include the effects of both the column flexibility and the torsional rotations of the edge beams of the structure. The equivalent column concept, described earlier, provides a convenient tool for this purpose.

Once the total section moments are determined, they must be divided between slab middle and column strips and the beams. At the interior negative- and positive-moment sections, the beam relative stiffness α_1 and the l_1/l_2 ratio appear to be the most significant factors, as is the case in interior panels. At the exterior negative-moment section, the flexural and torsional stiffnesses of the edge beam, the flexural stiffness of the beams perpendicular to the edge of the structure, and the l_1/l_2 ratio are all factors that must be taken into account. The magnitudes of the section moments as influenced by the column and edge beam stiffnesses have been the subject of few rigorous analyses, and theoretical information is quite limited even though the designer faces the problem every day.

The work by Simmonds[3.24] shows quite clearly that for square panels, neither the column flexural stiffness nor the edge beam torsional stiffness has a significant influence on the shape of the moment distribution curves at sections away from the edge of the structure. Figure 3.30 illustrates the effects of different column stiffnesses on the distribution of midspan positive mo-

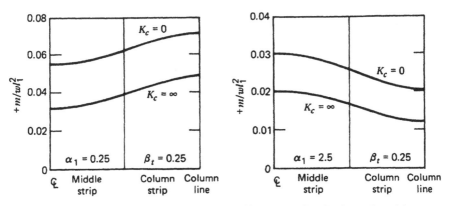

Figure 3.30 Influence of exterior column stiffness on distribution of positive moments in perpendicular span of edge panel.[3.24]

ments across the slab sections for two values of beam flexural stiffness α_1. It was also found that the edge beam torsional stiffness, β_t, influences the magnitudes of the total section moments appreciably only when the flexural stiffness of the beam perpendicular to the edge of the structure is very low and the column stiffnesses are high, as is consistent with the results of the use of the equivalent column concept. Simmonds apparently provides the only information available on slabs with interior beams. Jofriet[3.21,3.22] and Jofriet and McNeice[3.18] considered flat plate structures with various column and edge beam stiffnesses and panel aspect ratios, and this work is discussed later.

Midspan and support slab and beam moments are plotted versus edge column stiffness in Fig. 3.31 for two of Simmonds's slabs. The exterior negative moments are controlled almost completely by the edge column stiffness, and the slab and beam moments plot as linear functions of $\alpha_{ec}/(1 + \alpha_{ec})$. This finding supports the general validity of the equivalent column concept. These moments are not quite linear functions of $\alpha_c/(1 + \alpha_c)$, which is the stiffness parameter accounting for column stiffness alone, although the deviations from straight lines are small.

The other moments do not, in general, follow such straight lines when plotted against either column stiffness parameter. This is reasonable since the stiffnesses of the interior columns also changed. The EI values of the edge columns are 0.3 of those of the interior columns. However, the equivalent column stiffnesses are related by a variable ratio, depending on the exact combination of column and beam flexural stiffness and beam torsional stiffness considered.

The variable α_c is a mathematically explicit definition of column stiffness. The definition of α_{ec} includes the approximations discussed earlier in this chapter and has some additional problems in this particular case as it is being applied to an idealized structure consisting of slabs with no thickness and beams and columns without depth, width, or thickness. Consequently, the overhanging portions of slabs shown in Figs. 3.2 and 3.3 are not included in the determination of the cross-sectional properties.

With either definition, $\alpha/(1 + \alpha)$ is equivalent to the combined moment-distribution factor for the columns framing into the joint considered. While the range of $\alpha_c/(1 + \alpha_c)$ is from 0 to 1.0, the range of $\alpha_{ec}/(1 + \alpha_{ec})$ is limited to some maximum values less than 1.0, depending on the edge beam torsional stiffnesses. The low maximum stiffness of the edge equivalent columns for the case of $\beta_t = 0.25$ is very obvious in Fig. 3.31.

It should be noted that the absolute sum of the average support negative plus midspan positive moments shown in Fig. 3.31 is slightly more than the static moment for one panel. This is the consequence of the panel being at the edge of a structure with three panels each direction; the beams are subjected to slightly more than the tributary loading expected from similar edge panels. The total static moment for the three end-span panels, one edge panel plus two corner panels, is satisfied by the solutions.

The straight-line variation of the exterior negative moments with $\alpha_{ec}/(1 + \alpha_{ec})$ strongly supports the validity of the equivalent column concept. However,

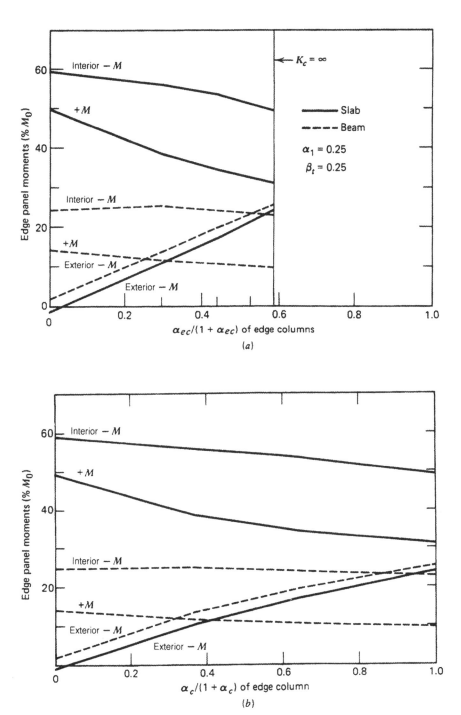

Figure 3.31   Edge panel slab and beam moments versus relative stiffness of edge columns.

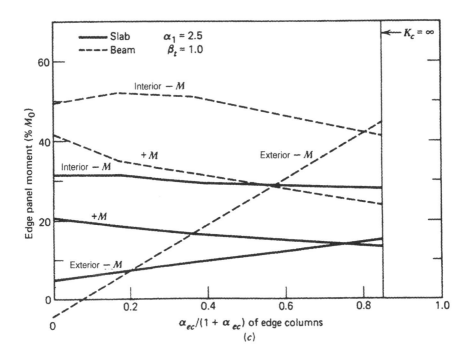

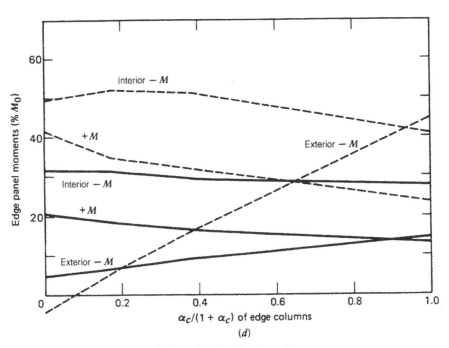

Figure 3.31   *(Continued)*.

a more crucial test is whether it can be used to predict exterior negative moments correctly when used in conjunction with a frame analysis. The slabs investigated by Simmonds[3.24] were analyzed taking into account the equivalent column stiffnesses. For the slabs supported on rigid columns ($K_c = \infty$ but $K_{ec} < \infty$), the results were as shown in Table 3.3. Both the first approximations to the moment, $[\alpha_{ec}/(1 + \alpha_{ec})]$FEM, where FEM is the fixed-end moment, and the moment from analyses of three span frames are given as percentages of the fixed-end moments. The fixed-end moments were taken as $wl_1^2 l_2/12$, which is the exact value of the fixed-end moment when $c/l = 0$ and both ends of the span considered have the same support conditions. In this case, however, α_2 for the edge beam is 1.25 times the interior beam value, which would lead to slightly different fixed-end moments at the two ends of the span. The error introduced is not known but should not be significant. For the case of rigid columns, moments were evaluated for the full three-panel width of the slab, since the α_{ec} values of the edge and corner columns turned out to be the same for each value of β_t. If the comparisons of Table 3.3 had been made using the sum of the moment in the edge panel plus the moment in one beam, the plate analysis moments would have been 2 to 3% higher than those shown.

The first approximations to the exterior negative moments shown in Table 3.3 are very close to the moments resulting from the frame analyses. The approximation would be less precise at lower column stiffnesses but should always be a satisfactory design approximation. Although the frame analysis gives for a constant β_t, the same moments for all values of α_1, while Simmonds's analyses gave slightly different moments, the frame analysis gives a very good approximations for these moments at the larger values of β_t. It appears that the equivalent column stiffness was reduced too much when the correction for the torsional stiffness of the edge beam was made for $\beta_t = 0.25$.

However, this may be largely a problem of interpretation in the application of the equivalent column to an extremely idealized structure. An arbitrary addition to the torsional member of the small portion of slab necessary to raise the first approximation of the moment to 70% of the fixed-end moment

TABLE 3.3 Exterior Negative Moments Across Width of Structure, Percent of Fixed-End Moment, Rigid Columns

β_t	$\dfrac{\alpha_{ec}}{1 + \alpha_{ec}}$ FEM	Frame Analysis with Equivalent Columns	Plate Analysis[3.24] α_1		
			0.25	1.0	2.5
0.25	58.4	60.0	71.3	76.1	79.2
1.0	84.9	85.2	82.3	87.5	88.1
2.5	93.4	93.5	89.5	91.3	94.7

when $\beta_t = 0.25$ has only a very small influence for the larger values of β_t. The addition of this portion of slab results in an increase in moment of about 2% of the fixed-end moment when $\beta_t = 1.0$, and the increase is less than 0.5% when $\beta_t = 2.5$.

The first approximations to the sum of the slab plus interior beam exterior negative moments for all column stiffnesses are plotted against the theoretical moments from Simmonds' analyses in Fig. 3.32. Only the values for the slabs with low torsional stiffnesses of the beams, $\beta_t = 0.25$, vary much from the line of perfect agreement, and these points define a distinct line of first approximation computed moments 25% lower than the theoretical values found by Simmonds. The same adjustments to the effective torsional stiffness sug-

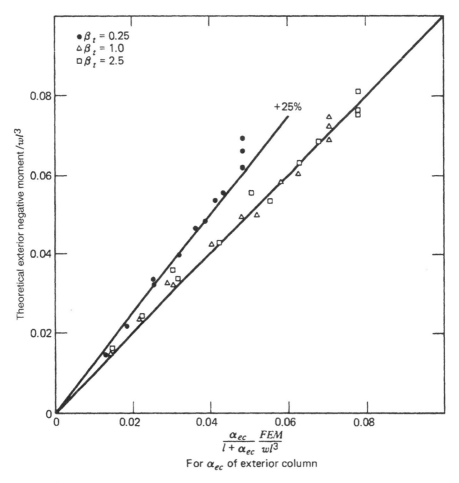

Figure 3.32 Comparison of theoretical exterior negative moments with moments computed using equivalent column concept.

gested previously would greatly improve the agreement. The general good agreement shown is a very strong confirmation of the validity and usefulness of the equivalent column as a design and analysis tool.

Jofriet[3.21] found good agreement between his analytical results and those from equivalent frame analysis, considering both one- and three-span structures with and without edge beams. There were no interior beams. The results of his work are a little harder to evaluate, however, since he used fixed-end moments calculated assuming that K_c and $K_t = \infty$ using the finite element method, and all moments were presented in terms of percentages of M_0 rather than as numerical values. It cannot readily be determined how the original fixed-end moments compare with those computed following the ACI 318 method for taking c_1/l_1 and c_2/l_2 into account, but it does not appear that the values are greatly different.

The distributions of interior negative and positive moments across the widths of the cross sections are shown for several cases in Fig. 3.29. The simplest generalization is that the moment distributions at interior sections in edge panels in the span perpendicular to the edge of the structure have about the same relative proportions as in interior panels of the same shape and having beams with the same relative stiffnesses.

Although this is a fair generalization, the moments shown in Fig. 3.29 indicate a definite tendency for the middle strip moments to be somewhat larger, relative to the column strip moments, in beamless slabs. This is especially true in the positive moments for $l_1/l_2 = 0.5$ with one long edge of the panel supported by a rather flexible beam having zero torsional stiffness and for the interior negative moments for $l_1/l_2 = 2.0$.

Simmonds' moment distributions across sections in square interior panels and the interior sections of edge panels appear to be identical. His solutions were all for slabs having beams on all column lines, and all beams had at least some torsional stiffness. The torsional stiffness of the edge beams may be of special importance in controlling the moment distributions in edge panels.

The distribution of the exterior negative moments between the slab middle and column strips and the beam is controlled by a fairly complex interaction of the effects of the flexural stiffnesses of the column and the interior and edge beams and torsional stiffness of the edge beam. It is difficult to be very general since, for example, the effects of increasing the interior beam stiffness are very much different for cases with relatively flexible columns and quite stiff columns, as shown in Fig. 3.33.

The results of Simmonds's[3.24] analyses do indicate that increasing the torsional stiffness of the edge beams always has the same effect on the exterior negative moments. It causes a significant increase in the middle strip moments, reductions in the beam moments, and relatively small increases or no change in the column strip moment when the moments are expressed as percentages of the total exterior negative moment. This is shown graphically for three typical cases in Fig. 3.34.

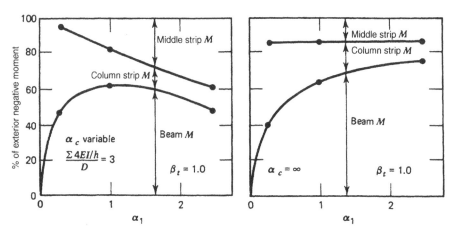

Figure 3.33   Distribution of exterior negative moments versus beam relative stiffness.[3.24]

The column strip moment depends primarily on the flexural stiffness of the beam perpendicular to the edge of the structure, as shown in Fig. 3.35, where the limiting moments are plotted versus α_1 for all values of β_t and K_c studied. The potential influence of β_t and K_c on the column strip moment is obviously quite limited.

The column strip moment does not tend to zero when $\beta_t = 0$. It might be argued that it should approach zero if the column had no width and there was no beam perpendicular to the edge of the structure. However, real columns always have finite widths. In a numerical solution of plate theory problems, a width of plate equal to the grid spacing in a finite difference solution, or to approximately an element width in a finite element solution, is always available to resist the moment applied by even a zero-width column, and so there is always a finite width of plate available to resist the edge moments.

The middle strip moments, which must be taken as zero when $\beta_t = 0$, are of course strongly dependent on the edge beam torsional stiffness, but the interior beam and column flexural stiffnesses are also important. There is one common point. When $K_c = \infty$, only the torsional stiffness matters and the middle strip moments are about 6, 14, and 20% of the exterior negative moments for $\beta_t = 0.25$, 1.0, and 2.5, respectively. The same percentages apply reasonably well regardless of the values of K_c when $\alpha_1 = 1.0$. However, at lower values of α_1, reducing the column stiffness reduces the share of the moment in the middle strips, while the same change increases it for higher values of α_1. This is illustrated in Fig. 3.36, where the middle strip moments are plotted versus $\alpha_{ec}/(1 + \alpha_{ec})$ for two values of β_t and three of α_1. The beam moments show trends approximately opposite those shown by the middle strip moments, as can be seen in Fig. 3.37.

No theoretical information is available on structures with rectangular panels and interior beams, but work by Jofriet[3.21] includes structures with edge

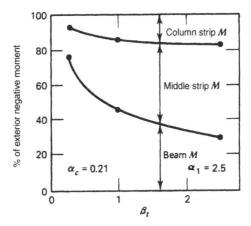

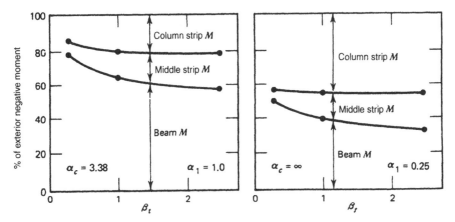

Figure 3.34 Distribution of exterior negative moment versus beam torsional stiffness.[3.24]

beams only. His extensive study leads to the conclusion that the column strip moments, expressed as a percentage of the total exterior negative moment, decrease linearly from 100% to 50% as $\beta_t/(1 + \beta_t)$ increases from 0 to 1.0. This applies for all panel shapes up to aspect ratios of 2.0 and all combinations of column and edge beam flexural stiffness.

Salvadori[3.25] has made an interesting contribution to the analysis of the effect of the torsional stiffness of the spandrel beams. If a spandrel beam is eccentric (i.e., centered either above or below the midplane of the slab), torsional rotations are necessarily about an axis at the level of the slab rather than about the midheight of the beam. He showed that by so restraining the free torsional rotation of the beam, the torques developed in it may be substantially reduced even though the reduced rotation leads to slightly larger negative slab moments acting perpendicular to the beam. A step-by-step pro-

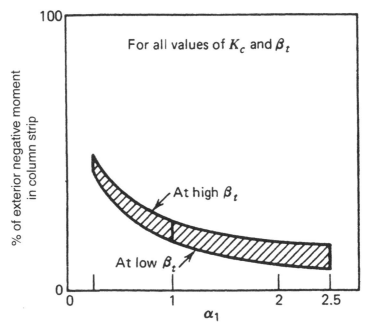

Figure 3.35 Exterior negative column strip moments versus beam relative stiffness.[3.24]

cedure to take into account the torsional stiffness of the edge beam relative to the slab stiffness, the eccentricity of the beam axis from the slab midplane, the panel shape, the beam flexural stiffness, and the column flexural stiffness

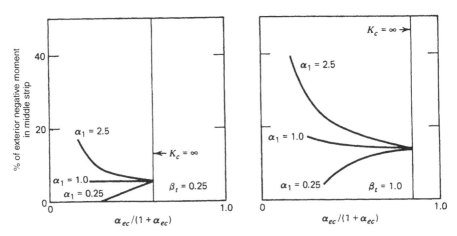

Figure 3.36 Exterior negative middle strip moments versus equivalent column relative stiffness.[3.24]

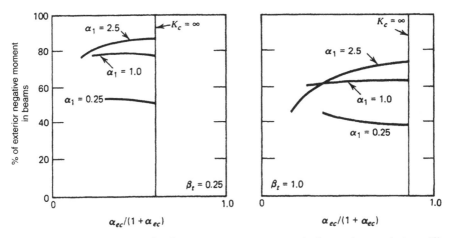

Figure 3.37   Exterior negative beam moments versus equivalent column relative stiffness.[3.24]

and width was presented for the calculation of both beam torque and slab moments. The analysis applies strictly to a semi-infinite slab which is either fixed or simply supported along an edge one span from the spandrel beam and which does not have interior beams, but the conclusions should be reasonably valid for any slab without interior beams. Slabs without spandrel beams are also considered.

There is a minor difficulty with Salvadori's analysis which can be shown by referring to a slab with rigid columns, $K_c = \infty$. As the first step in the calculation, K_t is assumed to be infinite, and a uniformly distributed fixed-end slab moment, m_0 per unit width, is found. The next step is to relax the torsional stiffness of the beam, and a slab moment that varies from a maximum at the columns to a minimum midway between the columns may be computed. For the case of rigid columns, the maximum slab moment remains equal to the fixed-end moment, m_0. This cannot be correct, however, as relaxing the beam torsional stiffness should produce some increase in slab moment at the columns, which become points of moment concentration. Simmonds's analyses,[3.24] for instance, always indicate this local increase in exterior negative slab moment at the columns as β_t is reduced, with the magnitude of the change depending on both the change in β_t and on α_1 perpendicular to the edge. Despite this problem, Salvadori's analysis method provides useful insights into the behavior of the edge of a slab structure.

3.4   MOMENTS IN CORNER PANELS

Corner panels present all the analysis complexities found in edge panels, and more, since moments and distributions depend on the column stiffnesses in two directions and on the torsional stiffnesses of two edge beams. Fortunately, however, it has been found that the moments in each span are influenced primarily by the relative stiffnesses in that span direction and are not greatly influenced by stiffnesses in the orthogonal direction. Consequently, the design problem can be approached one span at a time.

Because of the mathematical difficulties involved, nearly all the available information is from numerical analysis of slabs with square panels. Appleton[3.10] provides the only information on rectangular panels with flexible beams. Siess and Newmark[3.11] used classical techniques to determine moments in rectangular slabs that were clamped along two adjacent edges and simply supported along the other two.

There is no agreement on what kind of a panel should be analyzed. Appleton[3.10] considered slabs with four identical panels. Nine-panel slabs have been considered by several investigators[3.17,3.21,3.23,3.24] and 25-panel slabs by two others.[3.16,3.17] As a reasonably accurate generalization, it can be stated that the moments in corner panels are about the same as in the perpendicular span of an edge panel having the same aspect ratio and relative stiffness characteristics. In the slabs that Simmonds analyzed,[3.24] either the edge or corner panel slab moments might be the larger, depending on the values of α_1, β_t, and K_c, but the differences never exceed about 20% of the smaller moment and are less than 10% in most instances. In addition to having about the same total magnitude, the distributions across the widths of the panels have the same general shapes.

Moments in rectangular edge and corner panels with $l_1/l_2 = 2.0$ are shown in Figs. 3.38 and 3.39. All moments are taken from Appleton's analyses.[3.10] Figure 3.38 is for the case of panels with beams on all four sides. The relative stiffnesses are $\alpha_1(l_2/l_1) = \alpha_2(l_1/l_2) = 0.5$ for all beams. Figure 3.39 is for panels with only edge beams of the same relative stiffnesses. The corner panels are from four-panel structures and the edge panels from semi-infinite structures two spans wide, so the interior negative-moment restraint conditions are comparable. The columns are rigid, $c/l = 0$, and $\beta_t = 0$.

The beam and slab moments in Fig. 3.38 show that the slab moments are comparable in the corresponding sections of edge and corner panels. The interior beam of the corner panel supports slightly more load than the beam under the typical edge panel, and the edge beam moments in the corner panel are slightly less than half of those in the interior beams. The relative beam moment values are quite reasonable, especially for this case of zero torsional stiffness.

Slab moment intensities in the corner panel with only spandrel beams (see Fig. 3.39) are generally smaller than in the comparable edge panel. The corner panel spandrel beam is obviously effective in reducing moments across the entire panel width. The middle strip moments in the corner panel are approximately equal to the average of the middle strip moments in edge panels with and without interior beams, which is an important conclusion from the designer's point of view. The corner panel spandrel beam moments are consid-

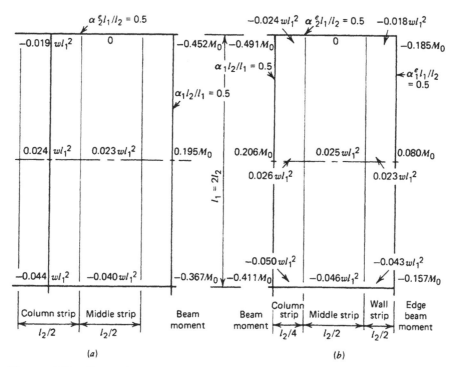

Figure 3.38 Comparison of moment distributions in rectangular edge and corner panels, with beams on all sides. $l_1/l_2 = 2.0$, $\alpha_1(l_2/l_1) = \alpha_2(l_1/l_2) = 0.5$: (a) edge panel; (b) corner panel.[3.10]

erably higher than those shown in Fig. 3.38, where there is a parallel interior beam.

The exterior negative moments are shown as average column strip moments distributed over a width of $l_2/2$. Since $\beta_t = 0$, however, the moment is actually concentrated in a very small width of slab at the column, with a theoretical intensity approaching infinity as c/l_2 approaches zero for this case of rigid columns.

Comparable data for the rather troublesome case of $l_1/l_2 = 0.5$ are unfortunately not available, but information on square edge and corner panels is given in Figs. 3.40 and 3.41. Where there are beams on all sides of the panels, the moments in the two cases are quite similar, with only the interior negative middle strip moments differing significantly. The moments are nearly uniformly distributed across the widths of the panels, as would be expected since $\alpha_1 = \alpha_2 = 1.0$. The beams resist about half the total moment.

When there are only spandrel beams (Fig. 3.41), the column strip moments are comparable, but the middle and wall strip moments in the corner panel are smaller than in the edge panel. Again, the spandrel beam moments in the l_1 direction are appreciably larger than when there is a parallel interior beam,

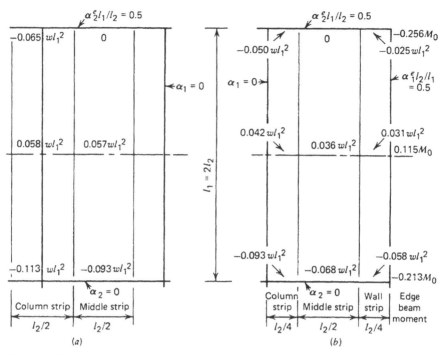

Figure 3.39 Comparison of moment distributions in rectangular edge and corner panels with spandrel beams, and without interior beams. $l_1/l_2 = 2.0$: (a) edge panel; (b) corner panel.[3.10]

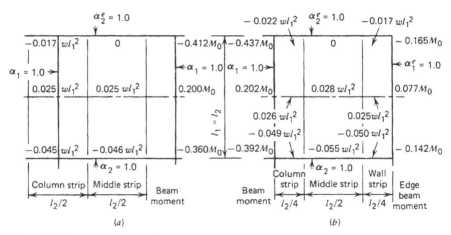

Figure 3.40 Comparison of moment distributions in square edge and corner panels, with beams on all sides, $\alpha_1 = 1.0$: (a) edge panel; (b) corner panel.

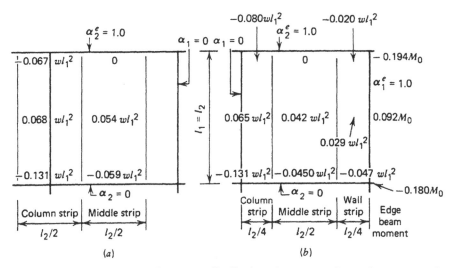

Figure 3.41 Comparison of moment distributions in square edge and corner panels with spandrel beams without interior beams: (a) edge panel; (b) corner panel.

although the difference is not as great as in the case of $l_1/l_2 = 2.0$. Jofriet's work[3.21] also shows spandrel beam moments to be larger when there is no parallel beam, although most of the increase appears to be in the negative-moment regions, while in Appleton's results both the positive and negative moments are higher.

As a note of caution, the moments discussed above were derived from finite difference solutions using relatively coarse mesh sizes. A reanalysis using a finer mesh or a fine grid finite element solution would be expected to change some of the numerical values, but it should not change the relative values or trends appreciably.

The addition of beam torsional stiffness would be expected to reduce the difference between edge and corner panel moments in cases where there are beams on all sides of the panels, and analyses by both Morrison[3.19] and Simmonds[3.24] provide evidence of this. It might also be speculated that in a structure with edge beams only, increasing β_t would tend to increase the differences between corner and edge panel moments by increasing the flexural moments in the edge beam in the l_1 direction, thereby further reducing the corner panel slab moments.

Results of the various analyses by Simmonds[3.24] and Jofriet[3.21] indicate that the equivalent column concept is as valid for predicting moments in corner panels as in edge panels. The distribution of the exterior negative moments is also controlled by the same interaction of beam flexural and torsional stiffness and column stiffness effects, and further discussion would largely repeat the findings of Section 3.3.

3.5 SPECIAL CASES OF LOADING AND GEOMETRY

There are a number of common design problems that have not received very much attention in the technical literature. Short discussions of some of these topics are contained in the following sections.

One of the occasionally serious problems is the treatment of heavy concentrated loads on building slabs. There is a reasonably large volume of data on the effects of concentrated loads on bridges slabs, but this is not always directly applicable to building slabs because of differences in support conditions and span/thickness ratios. This is particularly true of the beamless slabs—the flat plate and flat slab. Some aspects of this problem are discussed in Section 3.5.1.

Some slabs, especially those used as walls of various kinds in hydraulic structures, are subjected to loads that vary linearly along one of the spans of the slab. This is a large subject in its own right, and some information and sources of additional data are cited in Section 3.5.2. Many slabs have holes in them, and rule-of-thumb procedures for reinforcement details have been developed by various design agencies. However, the actual states of stress near openings and the effects of openings on the force distribution throughout the remainder of the structure are often not appreciated. This important topic is discussed briefly in Section 3.5.3.

3.5.1 Effects of Concentrated and Line Loads on Moments

Westergaard's 1930 paper[3.26] still appears to be one of the most important analytical investigations of the effects of wheel loads on bridge slabs, and some of his findings are worth reviewing as an introduction. A 1943 paper[3.27] has additional discussion and information.

One of the problems in the analysis of plates is that the ordinary theory of flexure of plates indicates that moments under a load approach infinity as the area over which the load acts approaches zero. In overcoming this difficulty, Westergaard used a "special theory" to compute the moments in the vicinity of the load and concluded that either the special theory or the use of a fictitious loaded area was necessary whenever the load was distributed over a circular area whose diameter, a, was less than $3.45t$, where t is the slab thickness. He also demonstrated that the span/depth ratio l_1/t was of some consequence for values of a/l_1 less than 0.1. These conclusions were for a slab simply supported over a span of l_1 and extending both ways to infinity in the l_2 direction, but the general conclusions should be valid for any slab structure. The layout of this structure is shown in Fig. 3.42.

Using Westergaard's approximation to a more complex expression, and for Poisson's ratio of $\mu = 0.15$, the moments per unit width under a midspan load can be calculated using the following expressions:

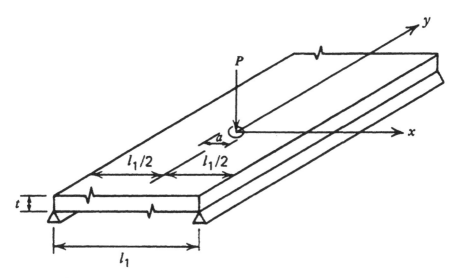

Figure 3.42 Layout of simply supported one-way bridge slab.[3.26]

$$m_x = \frac{P}{2.32 + 8(a/l_1)} \tag{3.20}$$

$$m_y = m_x - 0.0676P \tag{3.21}$$

where m_x and m_y are moments per unit width acting in the directions of the x- and y-axes, respectively. These expressions give moments very close to the exact moments for $l_1/t = 20$.

If the edges shown as simply supported are clamped instead, the moments per unit width under a midspan load become

$$m_{xf} = m_x - 0.0699P \tag{3.22}$$

$$m_{yf} = m_x - 0.1063P \tag{3.23}$$

The additional case of a square slab simply supported on all sides and subjected to a midspan load gave the following moments per unit width under the load:

$$m_{xsq} = m_{ysq} = m_x - 0.0490P \tag{3.24}$$

Starting with Westergaard's basic equations but considering that $\mu = 0$, it can be shown that the moments per unit width under the load in a simply supported one-way slab subjected to a midspan concentrated load can be approximated as follows:

$$m_x = \frac{P}{2.60 + 9.6(a/l_1)}$$
(3.25)

$$m_y = m_x - 0.0706P$$
(3.26)

In a simply supported one-way slab, the m_x moments are positive throughout the area of the slab. The m_y moments are positive throughout an area extending about $0.4l_1$ each way in the y-direction (see Fig. 3.42) from the load point. Beyond that distance very small negative m_y moments are produced, with a maximum intensity of about 4% of the positive m_y moment directly under a very small loaded area.

This moment distribution is quite different from that implied by the yield line analysis method discussed in Chapter 7, which indicates that one of the yield mechanisms for a point load includes a circular or elliptical negative moment yield line surrounding the load at some indeterminate distance from the load plus radial positive-moment yield lines originating from the load point, forming a collapse mechanism with a conical or nearly conical deflected shape.

For small values of a/l_1, the changes in support conditions do not produce large changes in the moments. For normal building slab span ranges, a load distributed over an area equivalent to that loaded by an automobile or truck tire would produce a maximum moment per unit width of $0.30P$ to $0.38P$, regardless of the support conditions.

Westergaard's paper also includes consideration of the effects of multiple loads and the derivation of an equivalent width of slab that may be treated as a beam when designing a bridge slab, a concept that is widely used by bridge designers. It should be noted that the maximum moments found exist only at the load location, and the unit moment intensities reduce very quickly as one moves away from the load point.

Although the bridge designer has commonly worked with equivalent widths of slab, the building designer faced with concentrated loading has often worked with equivalent loads (i.e., uniformly distributed loads producing about the same moment and shears as the concentrated loads). An example of this concept is the design load for an open parking structure for automobiles specified by the *BOCA National Building Code*[3.28] and ASCE 7-95.[3.29] The slab is designed for either a uniformly distributed load of 50 lb/ft² (2.4 kN/m²) or for a concentrated load of 2000 lb [8.9 kN] distributed over any arbitrarily located 20-in² [12900-mm²] area, whichever produces the greater moment and shear. Earlier editions of the BOCA Code had distributed the same concentrated force over an area which was 2.5 ft square (0.76 m square), which greatly lessened its effects.

Woodring[3.30] and Woodring and Siess[3.31] used a combination of fine mesh finite difference solutions plus the moment distribution scheme described in Ref. 3.23 to develop influence surfaces for moments in the interior panels of several nine-panel slabs. The structures had square panels, square rigid columns with $c/l = 0.1$, $\mu = 0$, and several different beam flexural and torsional stiffness combinations, including the important beamless slab case. The influence surfaces were then used in a study of the effects of concentrated and line loads on slab moments.

One of the conclusions from Woodring's analysis[3.30] was that the concentrated load required by the older BOCA and other building codes never governed when $l_1 > 20$ ft (6.1 m), regardless of where it is placed and regardless of the beam stiffness. A loaded area that is more realistically representative of an automobile tire print, as in the current loading requirements, would produce the governing midspan positive moments in many slab structures.

Woodring also found that the positive moment under a midspan concentrated load was not very sensitive to the support conditions at the edge of the panel. When $\alpha_1 = \beta_t = 0$, the moment per unit width in both directions was

$$m = 0.30P \tag{3.27}$$

and when $\alpha_1 = \beta_t = \infty$, it was reduced to

$$m = 0.28P \tag{3.28}$$

The averaging influence of the use of a finite difference grid of 20 spaces per span was such that the concentrated load was effectively spread over a circle of diameter $a = 0.015l_1$.

The influence of the size of the loaded area is illustrated in Fig. 3.43, where the positive moment per unit width under a midspan load distributed over a small square area is plotted versus the ratio of the side of the loaded area to the span. The curved lines giving the relationships for a beamless slab and a slab equivalent to a clamped panel are parallel.

If the concentrated load is located at the midspan of an interior column line in a beamless slab, positive slab moments per unit width under the load of

$$m_1 = 0.31P \tag{3.29}$$

$$m_2 = 0.25P \tag{3.30}$$

are produced, where m_1 acts in the direction of the span between columns and m_2 in the transverse direction. The m_2 moment may be particularly troublesome in short-span structures with low dead loads, since it is acting in opposition to negative moment from dead and distributed live loads and is in an area that is normally not adequately reinforced for positive moments.

The maximum negative moment per unit width at the column face in a beamless slab is produced by a load on the column line located at $0.25l$ from the center of the column (see Fig. 3.44). The maximum negative moment per unit width is

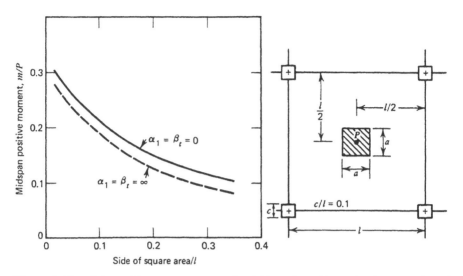

Figure 3.43 Slab positive moment per unit width under midspan load versus size of loaded area [3.30]

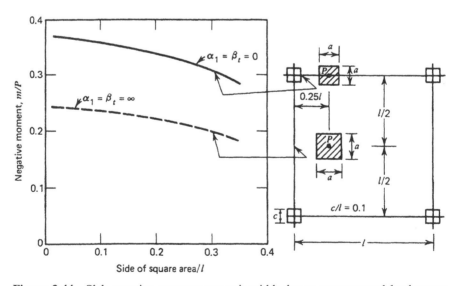

Figure 3.44 Slab negative moment per unit width due to concentrated load versus size of loaded area.[3.30]

$$m_c = 0.37P \tag{3.31}$$

This moment is at the center of one side of the column, and it would be sensitive to the details of the finite difference mesh near column. The moment at the corners of the columns would presumably be higher, but they are seldom taken into account in design. Introduction of beams rapidly reduces this moment and moves the most critical load point slightly toward midspan as α_1 of the beam increases.

As is shown in Fig. 3.44, enlarging the loaded area reduces the maximum negative moment only slowly from the maximum of $0.37P$. It is also obvious from the nature of the influence surface for this point[3.30,3.31] that the load can be moved at least $0.1l$ in any direction without reducing the moment at the face of the support significantly.

Figure 3.44 also shows the variation in negative moment at the middle of one edge of a clamped plate, $\alpha_1 = \beta_t = \infty$, with the size of the loaded area for a load centered $0.25l$ from midspan of the edge. Again, the moment is not very sensitive to the area actually loaded or to the exact position of the load.

Although the negative-moment magnitudes caused by concentrated loads are comparable to the positive moments, they are of less structural significance relative to the moments due to distributed loads since the negative moments caused by distributed loads are about twice the positive moments. As an illustration of this, Woodring[3.30] solved for the distributed loads equivalent to a 2000-lb (8.9-kN) load acting in a 2.5-ft (760-mm) square area on a beamless slab. When $l_1 = l_2 = 30$ ft (9.15 m), this load produces the same midspan positive moment as a load of 20 lb/ft² (0.96 kN/m²), but it causes negative moments equal to those produced by a distributed load of 10 lb/ft² (0.48 kN/m²).

When the load is located over the midspan point of a beam, the beam supports a reasonably large share of the load. When $\alpha_1 = 1.0$, the beam carries 40% of the total moment, with the positive and negative section moments about equal in this case of rigid columns. It seems reasonably safe to speculate that a concentrated load acting on a small area ($a < l_1/50$) anywhere within the area common to the two middle strips will produce positive moments per unit width of $0.25P$ to $0.30P$, regardless of the support conditions at the edges of the panel. It also appears that similar moments can be caused anywhere within a middle strip in a beamless slab. The influence of the size of the loaded area can be estimated using Fig. 3.43 or Eq. 3.25.

Multiple concentrated loads are a common design problem, and the effects can be studied through the use of Westergaard's[3.26] and Woodring's[3.30,3.31] influence surfaces. Westergaard found that to determine the moment at one load point due to other loads, it was sufficiently accurate to take the other loads as point loads rather than as loads distributed over small finite areas. This simplifies the use of the influence surfaces.

It is found in all cases that the positive moments caused at one location by a load at some other point decreases very rapidly as the distance between the points increases. For example, Westergaard's study showed that a load displaced by $0.2l_1$ toward the support of a simply supported one-way slab reduces the midspan moment in direction l_1 to one-third to one-fourth of the moment that the load would have caused had it been located at midspan. Moving the load along the y-axis (Fig. 3.42) requires a displacement of about $0.4l_1$ to bring the same reduction. Woodring's influence surfaces suggest similar or greater reductions in moment as the distance between the loads increases, with a reduction to two-thirds or less of the moment under the load for a load spacing of $0.2l_1$ in any direction.

A perhaps more serious multiple loading situation exists in the negative moments at columns in beamless slabs. For this case, placement of a load anywhere within a reasonably large area centered about the $\frac{1}{4}$-span point on the column line produces significant moments at the column, and hence the effects of multiple loads could be quite important.

Line loads, as may be applied by walls, are a special case of loads acting on small areas and were also studied in Refs. 3.30 and 3.31. It has apparently been customary to take the weight of interior walls into account by adding an additional uniformly distributed load of 30 lb/ft^2 (1.44 kN/m^2) to the dead load of the slab. Woodring[3.30] studied the effects of an arbitrarily placed concrete block wall weighing 304 lb/ft (4.43 kN/m) and extending an arbitrary length. The wall was considered to have zero stiffness and hence to apply a uniform line load to the slab regardless of the slab deflection.

Considering a single wall, the 30-lb/ft^2 allowance was more than adequate for any length and placement of the wall as far as the negative moments in beamless slabs were concerned. However, if the wall were placed at midspan and especially if it extends across several panels, the 30-lb/ft^2 allowance was inadequate for positive moments. When the wall extended across the full width of one panel, the required allowance to compensate properly for the effects on midspan moments acting in the direction perpendicular to the wall varied from 67 lb/ft^2 (3.2 kN/m^2) when l_1 = 20 ft (6.1 m) to 45 lb/ft^2 (2.1 kN/m^2) when l_1 = 30 ft (9.15 m). If the wall extended across more than one panel, the required equivalent loads would be slightly larger. Thus, it may be desirable, if inconvenient, to design the positive-moment sections with one equivalent load and the negative-moment sections for another lower equivalent load.

The equivalent uniformly distributed load is computed using the following equation, taken from Ref. 3.30:

$$w^* = \frac{\overline{w}C}{l_1} \tag{3.32}$$

where w^* is the equivalent uniformly distributed load per unit width; $\overline{w}$ the weight of wall per unit length; l_1 the slab of square panel; and C the concentration coefficient, where a positive number indicates moments with the same sign as are caused by distributed loads.

For a wall placed at midspan on a beamless slab, $C = 4.4$ for positive moments acting perpendicular to the wall and 1.7 for positive moments parallel to the wall. For a wall placed on the column line, the factor for middle strip negative moments perpendicular to the wall is $C = -2.2$, which indicates that positive moments are caused by the wall in a region where dead loads cause negative moments. This positive moment should seldom govern in structures of normal proportions, assuming that the wall is treated as dead load, subject to the same load factors as the slab itself. The concentration factors for column strip moments acting parallel to a wall on the column line are $C = 1.1$ and $C = 1.5$ for the positive- and negative-moment sections, respectively.

It may be argued that the wall can be partially self-supporting, especially when the slab deflects away from a very stiff wall. This may be true and if so could be expected to reduce the C values in some cases. However, it does not appear that this effect could significantly reduce the value of $C = 4.4$ for positive moments acting perpendicular to a midspan wall, since there is no possibility of redistributing the load to any other structural element; at the most, the distribution of the load applied to the slab at the base of the wall may change somewhat, but the total force cannot. On the other hand, a column line wall that is tightly fitted between columns could be expected to develop some arching action between the slab areas adjacent to the columns and relieve the midspan region of the slab of some of the load.

In slabs with beams, the midspan wall still results in approximately $C = 4.4$, requiring the same equivalent distributed loads, for all beam stiffnesses. In this case the beams effectively isolate a panel, and the extension of the wall across adjacent panels does not increase the slab moments, although two panels would have to be loaded to produce the maximum beam moments. It appears that all other values of C are less than 2.0 as long as the beams are not rigid.

Comparable information is not available for multiple-panel slabs supported on nondeflecting beams. However, a square clamped plate represents a limiting case for which information is available. The middle strip concentration factors for moments perpendicular to a midspan wall are $C = 4.4$ and $C = 2.1$ for the positive- and negative-moments sections, respectively. For moments parallel to the same midspan wall, the factors are $C = 1.9$ and $C = 4.2$ for positive and negative sections, respectively.

The foregoing discussion of the effects of both point and line loads is incomplete in that most of the information is for slabs with square panels. However, the good agreement generally found between the information on square slab panels and semi-infinite bridge slabs should give confidence that the maximum moments due to concentrated loads in rectangular panels will not be too different from those found in square panels. Reference 3.11 contains a brief discussion of the effects of concentrated loads on panels with rigid beams, $\alpha_1 = \infty$, with most of the emphasis on the negative moments that could be produced. The moment distribution procedure proposed in the reference was found to be reasonably accurate for the cases considered.

3.5.2   Effects of Loads Varying Linearly Across One Span

Walls of hydraulic structures such as rectangular tanks and counterfort retaining walls, which are subjected to loads varying linearly from zero or some minimum at the top to a maximum at the bottom, are common slab problems. Both the magnitude and distribution of bending and shear forces must be determined if safe, serviceable designs are to be produced.

Information on moment and shear distributions is available from a number of sources for plates with various suitable boundary conditions. Plates fixed at two parallel vertical edges, free at the top edge, and either simply supported or fixed at the lower edge, are the most common, although the free edge has also been replaced by a simply supported edge. Analyses with the vertical edges simply supported have been reported and are especially suited to precast structures. The few comments in this section are restricted to cases with the upper, horizontal edge of the plate free (unsupported). The layout of a typical wall panel is shown in Fig. 3.45.

Apparently, all the available analyses are for cases with rigid, nondeflecting supports. This is a reasonable assumption for tank walls that are supported by orthogonal tank walls and for walls that are counterforted with bracing of the usual counterfort proportions. The moments derived for these support conditions are generally not suitable for cases of walls with pilasters of usual

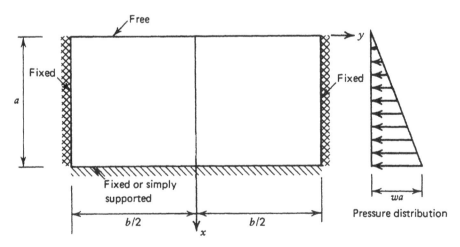

Figure 3.45   Layout and loading of typical tank wall panel.

dimensions, and in cases of doubt the value of the relative stiffness of the pilaster should be computed. As a rough rule of thumb, if $\alpha_1(l_2/l_1) > 5$, the pilaster can be assumed to be rigid as far as the moment distribution calculations are concerned.

Moment distributions for various cases are given by Timoshenko and Woinowsky-Krieger,[3.32] the Portland Cement Association,[3.33,3.34] Bares,[3.35] Moody,[3.36] and Jofriet.[3.37] A few comments about these sources are appropriate. A quick comparison of moments from any two sources will often show that there are disagreements about the magnitudes of moments in similar cases. In some instances the differences may be as large as 30% of the smaller moment, and these must be resolved before a design is prepared. The basic problem appears to be one of the precision in the original calculations of the moments, although one must also be certain that cases with the same assumed values of Poisson's ratio are being considered. The moments given in Table 45 of Ref. 3.32 were derived from finite difference solutions first reported in 1936. It must be assumed that the finite difference grid used was quite coarse, which always tends to mask peak values on distribution curves. This type of error is consistent with most of the differences between the references. The earlier Portland Cement Association bulletin on rectangular tanks[3.33] gave few details of the background of the calculations, but the moments were computed using an infinite series approach. The assumed value of μ was 0.2. The later, much enlarged edition[3.34] gives tabulated values derived from the finite element method. There appear to be anomalies in some of the solutions for horizontal negative moments ($-m_y$ in Fig. 3.45) at the sections near the top of the end supports. The discrepancies are not large, but the data do not lead to the smooth curves describing the moment distribution which are to be expected given the physical nature of the structure.

The book by Bares[3.35] reports moments for cases with several values of μ and also for triangular loadings extending only partway up the wall. The moments were apparently derived from finite difference solutions considering an 8 by 8 mesh within each panel, which is fine enough for most purposes. Moody's values[3.36] are also based on finite difference solutions considering a reasonably fine mesh. Considering these four references, the publications by the PCA and Moody give considerably more information on the distribution of moments and shears along and across sections than the other two, while the book by Bares has some information on a large number of combinations of support conditions and loadings. Jofriet's[3.37] solutions include values for larger ratios of length to height of wall, and also covers cases of tapered walls, where the base thickness is 1.5 times the thickness at the top edge.

It is not intended to present tabulations of moment values, but some of the information from the earlier PCA publication[3.33] is presented in Figs. 3.46 and 3.47 to illustrate the very great changes in both the magnitude and the distribution of moments as the aspect ratio of the wall changes. Note the different scales of moment values used in the figures. The moments are pre-

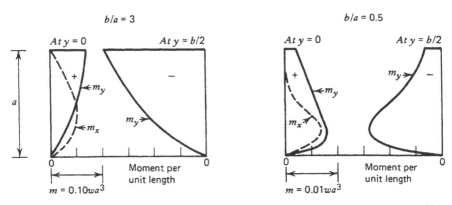

Figure 3.46 Distributions of moments in slabs with simply supported bases.[3.33]

sented in terms of wa^3, where w is the weight per unit volume of the liquid or equivalent liquid, and a positive moment produces tension on the face of the plate away from the load.

From the distributions of moments shown in Figs. 3.46 and 3.47, it is obvious that the m_y moments acting in the horizontal direction are more important in carrying the load when the base of the wall is simply supported than when the base is fixed. The change in the distribution of moments as the aspect ratio b/a is changed from 3.0 to 0.5, for either support condition, could be interpreted as a change from major two-way bending, where much of the load is transferred directly to the supports at the base, to a situation where the wall is predominantly acting as a one-way slab spanning between, and transferring the load to, the vertical supports.

This can be illustrated through the use of a static moment concept. A total load of $W = wa^2b/2$ must be resisted. Therefore, in the y-direction, a total

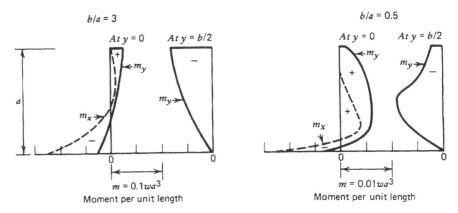

Figure 3.47 Distributions of moments in slabs with fixed bases.[3.33]

static moment of $M_0 = wa^2b^2/16$ must be accounted for. For the wall with the simply supported base, the sum of the positive and negative moments in the y-direction in the wall account for about 30% of M_0 when $b/a = 3$, but for about 88% when $b/a = 0.5$. The remaining moment must be effectively resisted by the base slab.

In the fixed-base wall, the span carries about 10% of M_0 in the y-direction when $b/a = 3.0$, indicating that the fixed base transfers even more of the load directly into the base slab. When $b/a = 0.5$, the slab carries about 65% of M_0 in the y-direction. In the x (vertical)-direction, a long cantilever wall without cross-walls or counterforts would be subjected to a base moment of $-wa^3/6$ per unit length. For the boundary conditions considered in Fig. 3.45, when $b/a = 3.0$ the m_x moment at the center of the fixed base is 75% of this limiting value, indicating the great importance of vertical wall bending for the case. When $b/a = 0.5$, the same m_x moment is reduced to about 7.5% of the limiting value.

The very large changes in the nature of the distribution of moments with changes in aspect ratio effectively warn the designer that it is not satisfactory to prepare a set of standard details of steel arrangement for all walls. The details of steel spacing and cutoff points must be determined for each aspect ratio.

As an additional note on the detailing of reinforcement, the corners of tanks subjected to internal liquid pressures deserve special attention. The moments tend to open the corner, and this presents special problems of anchorage of bars and provision for tension forces tending to tear the inside bars from the concrete mass. This already serious problem would be aggravated by the presence of the in-plane tensions in the walls, which must be transmitted through the joint. References 3.38 to 3.43 describe some of the recent work on this problem, although they are not all concerned specifically with tanks.

Some analytical and experimental work examining the moments in square tanks with rigid connections between the bases of walls and the floor slabs has been reported by Davies.[3.44] This paper points out some of the difficulties in determining the effective flexural stiffness of the base slab. These difficulties are related to the possible lifting of the slab from the subbase as the edges rotate because of the wall moments.

3.5.3   Effects of Holes in Slabs

There are many aspects to the problems of holes in slabs, and this discussion is necessarily limited to only a few problems. This section is concerned with the effects of holes on elastic moments in slab panels. It is specifically not concerned with the reduction in shear strength that may occur if the holes are too near columns in flat plate or flat slab construction; that problem is discussed in Chapter 10. It is also not concerned with the effects of holes on the strength or collapse loads, in a flexural mode; such problems are discussed in Chapters 6, 7, and 8, which deal with strength analysis and design procedures. Elastic moments have a strong influence on the initiation and extent of cracking, and consequently this section is concerned primarily with serviceability problems.

Holes, even large holes, do not necessarily cause major changes in the magnitudes or distributions of bending moments in a slab panel. There is always a large but local moment at a sharp corner of a hole, but the effects may not extend far from the corner. In other regions of the slab, the effects of removing the load from the area of the hole often outweigh the effects of removing the same area of the slab itself and may lead to either no change or to net reductions in bending moments.

Relatively few analyses of slabs with rectangular holes have been published, primarily because of mathematical difficulties with the analysis. All of the analyses known to the authors were done using numerical methods, generally finite difference equations, and no solutions using classical plate analysis techniques are known. Because of the general complexity of the problem, it appears that only single panels have been analyzed, and no solutions of multiple-panel slabs containing openings have been published.

Apparently, the work by Weidemann[3.45] is the earliest analysis of the effects of square holes on moments in square slabs. He used a relatively fine finite difference grid to analyze fixed and simply supported slabs having central square holes. The hole dimension was one-third of the span of the plate. There are some anomalies in the analysis, but the negative moment at the center of an edge of a fixed plate is slightly less than in a solid plate, although the average moment across the diagonal of a simply supported plate is slightly larger than in a solid plate. The moments near the corners of the holes are much larger than in a solid plate, but the region of high moments extends no more than about 0.1 of the span from the corner. The paper unfortunately did not include analyses of solid plates with the same boundary conditions. Such solutions would have been very helpful in providing comparisons and in evaluating the analysis procedure itself.

The only other major study was done at the University of Illinois,[3.46–3.48] where square clamped plates with various holes were analyzed using the Newmark plate analog to assemble the finite difference equations. Square and rectangular holes, and holes with stiffened edges, were studied. In addition to cases with uniformly distributed loads, line loads distributed around the edges of the openings were also examined.

Fluhr[3.46] separated the effects of the change in the total load on the panel from the effects of removing part of the slab by analyzing solid slabs subjected to uniformly distributed loads except in the area of a future hole, and then analyzing slabs with the holes cut out. It was clear that for the case of central square holes, most of the effects were due to unloading of the slab, and relatively few due to removal of the same area of the slab, especially when sections 0.2 of the span away from the hole were considered.

This is illustrated in Fig. 3.48a, which compares moments in the solid slab with and without the central area loaded, and with a hole of 0.4 of the span.

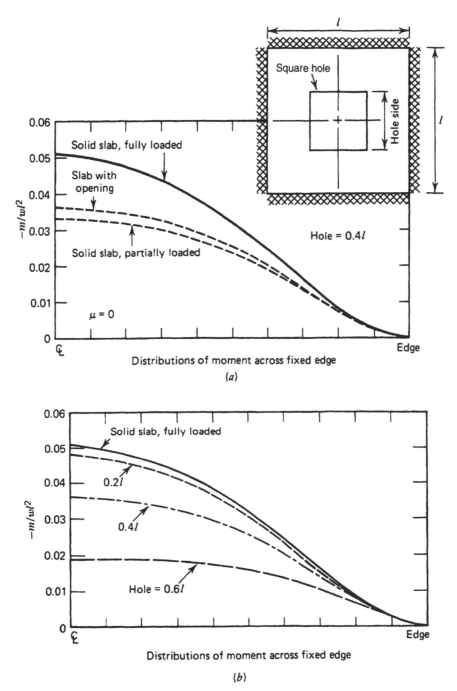

Figure 3.48   Effects of square central holes on negative moments:[3.46] (*a*) effects of removing load and removing slab; (*b*) effect of size of opening.

The negative moments at the edges of the slab are plotted. Effects of the size of the hole on the negative moment are shown in Fig. 3.48b. In the limiting case of a large hole, the negative moments at the center of the edge approach the moment to be expected in a cantilever slab, which would be $-0.02wl^2$ for the hole equal to $0.6l$.

At a sharp corner of a hole, the local moment theoretically approaches infinity in a very small area. In the real reinforced concrete slab, cracking and possibly some very localized yielding result, with little influence on the rest of the slab. A numerical analysis such as the finite difference method cannot handle the infinite moment either, and the moment computed at the corner of a hole corresponds to that for a corner having a fillet of some finite radius. The coarser the grid, the larger the equivalent fillet.

Most of Fluhr's analyses were with a grid of span/20. Although this is a fine grid in most respects, it does not give very large moments at corners of holes. To investigate this aspect of the problem, two cases were analyzed with a special fine grid of span/80 in the vicinity of the hole (and the normal grid in the rest of the panel). These analyses indicated sharply peaked moments at the corners of the holes; there were no significant changes at sections located at span/10 or more from the edge of the hole.

Moments near a central square hole of width $0.2l$ are shown in Fig. 3.49, where moments acting across and along a line through the edge of the hole are plotted. The influence of the grid size on the computed moment is clearly visible, and it is apparent that the hole is not a major problem. The corner moment is real enough, but it extends over such a small area that it should not be too difficult to provide reinforcement to control it.

Fluhr analyzed a large number of cases, including several with multiple holes, rectangular holes, and square holes located at the corner of the panel. As long as there were only uniformly distributed loads, the moments were usually no greater than, and often smaller than, those in solid slabs, except for the local effects at the corners of holes.

He also considered some loading cases in which the total load lost from the area of the hole was applied as a uniformly distributed line load about the perimeter of the hole, in addition to the uniformly distributed load acting on the remaining area of the slab. The moments resulting from this loading were generally slightly greater than in the solid slab. The concentrations of moments at the corners of openings were increased somewhat by this loading, and the moments in regions between multiple openings were often appreciably greater (by up to about 50%) than in either the solid slab or the similar slab with distributed loads only.

In a few instances the concentrated moments at the corners of holes were opposite in sign from the moments at the same location in a solid plate. In these cases, both moments were small and apparently well within the capacity of nominal reinforcement. In general, the computed elastic deflections were no greater for the slabs with holes than in fully loaded solid slabs, as long as the only loading was the uniformly distributed load on the remaining area.

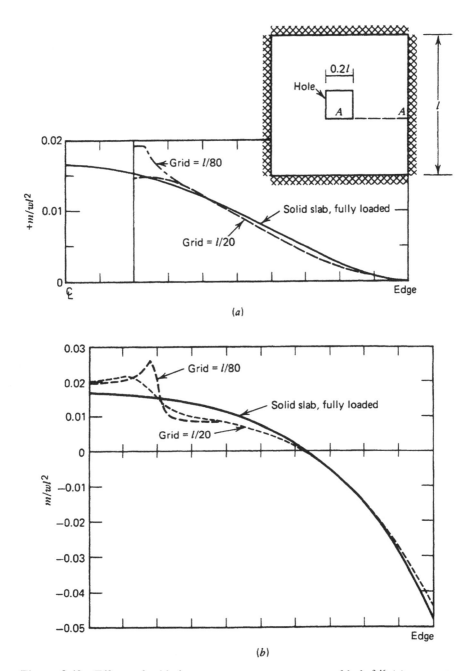

Figure 3.49   Effects of grid size on moments near corners of hole:[3.46] (a) moments acting across section A–A; (b) moments acting along section A–A.

The addition of the line load at the edge of the hole usually led to deflections larger than in the solid slab.

Although a reasonably broad range of cases of different size, shape, number, and location of holes was analyzed, all cases were for clamped square panels. There is no reason to expect greatly different results for simply supported square or for clamped or simply supported rectangular panels. Unfortunately, however, no continuous slabs of any kind have been analyzed to determine the effects of holes, so there must remain numerous questions to be considered, especially for beamless slabs.

Bhatti, et al.[3.49] analyzed a series of square slabs with central holes. Some slabs were simply supported and others were clamped. They used a finite element analysis, taking into account cracking of the concrete and eventual yielding of the reinforcement. When there were uniformly distributed loads, the hole made little difference to either the deflections or the expected collapse loads. The also considered a loading consisting of four concentrated loads, each located one-fourth of the span from the adjacent edges. For this loading, deflections increased and load capacity decreased, rapidly, as the size of the central hole was increased. This would be expected, since the loading must cause a nearly constant moment field in the central part of the plate. Removing material by making the hole larger reduces the resistance, but does not reduce the applied moment field.

A beamless slab appears to have fewer alternative load paths than either a multipanel slab supported on stiff beams or an isolated panel on nondeflecting supports. Consequently, it is reasonable to expect that moderate-sized midspan holes in a flat slab or flat plate, especially if the holes are in the column strips, will be more serious than in the cases analyzed so far.

Despite the unresolved questions, current design and construction practice is probably generally adequate and often quite conservative in the amount of reinforcement provided. Present practice is probably fairly represented by the following recommendation:

Holes in floors are frequent because of stairs, elevators, ducts, pipes, and the like. Reentrant corners are points of weakness and require supplementary reinforcement in the way of corner bars or overlapped side and end bars. One rough rule is to make the tensile resistance of the added bars at least equal to that of the concrete removed by the opening. Obviously, the stress bars must not be cut off, or the capacity reduced; rather the full amount of required steel must be grouped, some at either side of the opening. In addition, bars at right angles are needed to divert the stresses around the opening.[3.50]

The 1995 ACI Code,[3.1] Section 13.4, has a number of specific requirements on the treatment of holes, with limits on size depending on the location of the hole. The area common to two middle strips is allowed a large hole without a special analysis to justify the reinforcement arrangement; a much smaller hole is allowed in the region common to two column strips. The provisions for the quantity and arrangement of reinforcement are quite similar in character to the suggestions quoted above.

Prescott[3.47] and Ang and Prescott[3.48] studied slabs with central square holes having stiffened edges. Square clamped slabs were considered, with the size of the hole and the torsional and flexural stiffnesses of the edge beams being varied in the elastic analysis. Various combinations of flexural and torsional stiffness of the edge beams, relative to the slab stiffness, were found to be effective in reducing moments in the slab and increasing the overall stiffness of the panel.

Because of the basic deflected shape of the slab, the edge stiffeners were subjected to positive moments for their entire lengths, for cases of holes of up to 0.4 of the span. The positive moment at the end of a stiffener was largely resisted by torque induced in the stiffener on the adjacent edge of the hole. The flexural and torsional moments found in the stiffeners were relatively sensitive to the flexural stiffness of the beams at the edges of the holes; they were relatively insensitive to the torsional stiffness of these beams.

The magnitudes of the bending moments computed for a given assumed edge beam size appear reasonable and within the capability of a reinforced concrete beam. However, the torsional moments computed, even in cases of quite low torsional stiffness, appear to be generally beyond the torsional capacity of a concrete beam, no matter how carefully the beam may be reinforced. Increasing the torsional capacity by enlarging the cross section would not be effective, as this would also increase the flexural stiffness and hence also the applied torque.

Because of this torque problem, and also because of the difficulties in making simple-to-construct connections between beams at the corners of holes, the stiffening of edges of holes cannot now be recommended. If the magnitudes of loads applied at the edges of a hole are so great that stiffening of the opening appears necessary, complete framing with beams back to the supporting columns or walls would appear preferable.

Furr[3.51,3.52] presented an approximate analysis method for slabs and applied it to slabs with holes. He found reasonable agreement between his and Wiedemann's results.[3.45] A number of authors have looked at the effects of circular and elliptical holes on slab moments.[3.32,3.52–3.54] Circular slabs with central circular holes are relatively simple to analyze elastically[3.32] as long as the loads are axisymmetric. The other cases, triangular and rectangular slabs with circular or elliptical holes, were also solved by classical elastic procedures.[3.55] In most cases the holes produce only a local concentration of moments of no great consequence. The moment at the edge of a hole may be twice that existing without the hole,[3.56] but the concentration extends only a very short distance from the hole.

REFERENCES

3.1   *Building Code Requirements for Structural Concrete,* (ACI 318-95) and *Commentary,* (ACI 318R-95), American Concrete Institute, Detroit, 1995, 371 pp.

3.2   W. G. Corley and J. O. Jirsa, "Equivalent Frame Analysis for Slab Design," *Proc. ACI,* Vol. 67, November 1970, pp. 875–884.

3.3   F. B. Seely and J. O. Smith, *Advanced Strength of Materials,* 2nd ed., Wiley, New York, 1952, 680 pp.

3.4   *Building Code Requirements for Reinforced Concrete,* (ACI 318-77) American Concrete Institute, Detroit, 1977, 102 pp.

3.5   W. G. Corley, "The Equivalent Frame Analysis for Reinforced Concrete Slabs," Ph.D. thesis, University of Illinois at Urbana–Champaign, 1961, 166 pp. Also issued as *Civil Engineering Studies,* Structural Research Series 218, Department of Civil Engineering, University of Illinois, Urbana, Ill.

3.6   Discussion of Ref. 3.2, *Proc. ACI,* Vol. 68, No. 5, May 1971, pp. 397–401.

3.7   A. E. Cardenas and P. H. Kaar, "Field Test of a Flat Plate Structure," *Proc. ACI,* Vol. 68, No. 1, January 1971, pp. 50–59.

3.8   J. G. Sutherland, L. E. Goodman, and N. M. Newmark, "Analysis of Plates Continuous over Flexible Beams," *Civil Engineering Studies,* Structural Research Series 42, Department of Civil Engineering, University of Illinois, Urbana, Ill., 1953, 76 pp.

3.9   W. L. Gamble, "Moments in Beam Supported Slabs," Proc. *ACI,* Vol. 69, March 1972, pp. 149–157.

3.10   J. H. Appleton, "Reinforced Concrete Floor Slabs on Flexible Beams," Ph.D. thesis, University of Illinois at Urbana–Champaign, 1961, 221 pp. Also issued as *Civil Engineering Studies,* Structural Research Series 223, Department of Civil Engineering, University of Illinois, Urbana, Ill.

3.11   C. P. Siess and N. M. Newmark, *Moments in Two-Way Concrete Floor Slabs,* Bulletin 385, University of Illinois Engineering Experiment Station, Urbana, Ill., 1950, 122 pp.

3.12   R. H. Wood, "Studies in Composite Construction, Part II. The Interaction of Floors and Beams in Multistory Buildings," *National Building Studies,* Research Paper 22, Department of Scientific and Industrial Research, Building Research Station, Watford, Hertfordshire, England, 1955.

3.13   M. D. Vanderbilt, M. A. Sozen, and C. P. Siess, "Deflections of Multiple-Panel Reinforced Concrete Floor Slabs," *J. Struct. Div., ASCE,* Vol. 91, No. ST4, August 1965, pp. 77–101.

3.14   H. M. Westergaard and W. A. Slater, "Moments and Stresses in Slabs," *Proc. ACI,* Vol. 17, 1921, pp. 415–538.

3.15   J. O. Jirsa, M. A. Sozen, and C. P. Siess, "Pattern Loadings on Reinforced Concrete Floor Slabs," *J. Struct. Div., ASCE,* Vol. 96, No. ST6, June 1969, pp. 1117–1137.

3.16   J. O. Jirsa, "The Effects of Pattern Loadings on Reinforced Concrete Floor Slabs," Ph.D. thesis, University of Illinois at Urbana–Champaign, 1963, 145 pp. Also issued as *Civil Engineering Studies,* Structural Research Series 269, Department of Civil Engineering, University of Illinois, Urbana, Ill.

3.17   W. L. Gamble, "Measured and Theoretical Bending Moments in Reinforced Concrete Floor Slabs," Ph.D. thesis, University of Illinois at Urbana–Champaign, 1962, 322 pp. Also issued as *Civil Engineering Studies,* Structural Research Series 246, Department of Civil Engineering, University of Illinois, Urbana, Ill.

3.18 J. C. Jofriet and G. M. McNeice, "Pattern Loading on Reinforced Concrete Flat Plates," *Proc. ACI,* Vol. 68, December 1971, pp. 968–972.

3.19 G. C. Morrison, "Solutions for Nine-Panel Continuous Plates with Stiffening Beams," M.S. thesis, University of Illinois at Urbana–Champaign, 1961, 46 pp.

3.20 *Building Code Requirements for Reinforced Concrete,* (ACI 318-63), American Concrete Institute, Detroit, 1963, 144 pp.

3.21 J. C. Jofriet, "Analysis and Design of Concrete Flat Plates," Ph.D. thesis, University of Waterloo, 1971, 466 pp.

3.22 J. C. Jofriet, "Short Term Deflections of Concrete Flat Plates," *J. Struct. Div., ASCE,* Vol. 99, No. ST1, January 1973, pp. 167–182.

3.23 A. H.-S. Ang and N. M. Newmark, "A Numerical Procedure for the Analysis of Continuous Plates," *Proceedings of the 2nd Conference on Electronic Computation,* American Society of Civil Engineers, Pittsburgh, Pa., 1960, pp. 379–413.

3.24 S. H. Simmonds, "The Effects of Column Stiffness on the Moments in Two-Way Slabs," Ph.D. thesis, University of Illinois at Urbana–Champaign, 1962, 216 pp. Also issued as *Civil Engineering Studies,* Structural Research Series 253, Department of Civil Engineering, University of Illinois, Urbana, Ill.

3.25 M. G. Salvadori, "Spandrel–Slab Interaction," *J. Struct. Div., ASCE,* Vol. 96, No. ST1, January 1970, pp. 89–106.

3.26 H. M. Westergaard, "Computation of Stresses in Bridge Slabs Due to Wheel Loads," *Public Roads,* Vol. 11, No. 1, March 1930, pp. 1–23.

3.27 H. M. Westergaard, "Stress Concentrations in Plates Loaded over Small Areas," *Trans. ASCE,* Vol. 108, 1943, pp. 831–856; "Discussion," pp. 857–886.

3.28 *BOCA National Building Code,* 13th ed., Building Officials and Code Administrators International, Country Club Hills, Ill., 1996.

3.29 *Minimum Design Loads for Building and Other Structures,* ASCE 7-95, American Society of Civil Engineers, New York, 1995.

3.30 R. E. Woodring, "An Analytical Study of the Moments in Continuous Slabs Subjected to Concentrated Loads," Ph.D. thesis, University of Illinois at Urbana–Champaign, 1963, 151 pp. Also issued as *Civil Engineering Studies,* Structural Research Series 264, Department of Civil Engineering, University of Illinois, Urbana, Ill.

3.31 R. E. Woodring and C. P. Siess, "Influence Surfaces for Continuous Plates," *J. Struct. Div., ASCE,* Vol. 94, No. ST1, January 1968, pp. 211–226.

3.32 S. Timoshenko and S. Woinowsky-Krieger, *Theory of Plates and Shells,* 2nd ed., McGraw-Hill, New York, 1959, 580 pp.

3.33 *Rectangular Concrete Tanks,* Report IS003D, Portland Cement Association, Skokie, Ill., 1981, 15 pp.

3.34 J. A. Munshi, *Rectangular Concrete Tanks,* 5th ed., Portland Cement Association, Skokie, Ill., 1998, 182 pp.

3.35 R. Bares, *Tables for the Analysis of Plates, Slabs, and Diaphragms Based on the Elastic Theory,* Bauverlag GmbH, Wiesbaden, Germany, 1969, 579 pp.

3.36 W. T. Moody, *Moments and Reactions for Rectangular Plates,* Engineering Monographs 27, U.S. Bureau of Reclamation, Denver, Colo., 1963, 85 pp.

3.37 J. C. Jofriet, "Design of Rectangular Concrete Tank Walls," *Proc. ACI,* Vol. 72, July 1975, pp. 329–332.

3.38 G. Somerville and H. P. J. Taylor, "The Influence of Reinforcement Detailing on the Strength of Concrete Structures," *Struct. Eng.,* Vol. 50, No. 1, January 1972, pp. 7–19.

3.39 P. S. Balint and H. P. J. Taylor, *Reinforcement Detailing of Frame Corner Joints with Particular Reference to Opening Corners,* Technical Report 42.462, Cement and Concrete Association, London, February 1972, 16 pp.

3.40 Discussion of Ref. 3.38, *Struct. Eng.,* Vol. 50, No. 8, August 1972, pp. 309–321.

3.41 P. W. Birkeland et al.; N. W. Hanson; and R. A. Shawnn, Discussions of the paper "Experimental Study of Reinforced Concrete Frames Subjected to Alternating Sway Forces," *Proc. ACI,* Vol. 66, May 1969, pp. 441–444.

3.42 R. Park and T. Paulay, *Reinforced Concrete Structures,* Wiley, New York, 1975, 769 pp.

3.43 I. H. E. Nilsson and A. Losberg, "'Reinforced Concrete Corners and Joints Subjected to Bending Moments," *J. Struct. Div., ASCE,* Vol. 102, No. ST6, June 1976, pp. 1229–1254.

3.44 J. D. Davies, "Bending Moments in Square Concrete Tanks Resting on Rigid Supports," *Struct. Eng.,* Vol. 41, No. 12, December 1963, pp. 407–410.

3.45 E. Wiedemann, "Der Formanderungszustand emer quadratischer Platte mit quadratischer Offnung," *Ing. Arch.,* No. 7, 1936, pp. 56–70, 196–202.

3.46 W. E. Fluhr, "Theoretical Analysis of the Effects of Openings on the Bending Moments in Square Plates with Fixed Edges," Ph.D. thesis, University of Illinois at Urbana–Champaign, 1960, 148 pp. Also issued as *Civil Engineering Studies,* Structural Research Series 203, Department of Civil Engineering, University of Illinois, Urbana, Ill.

3.47 W. S. Prescott, "Analysis of Square Clamped Plates Containing Openings with Stiffened Edges," Ph.D. thesis, University of Illinois at Urbana–Champaign, 1961, 202 pp. Also issued as *Civil Engineering Studies,* Structural Research Series 229, Department of Civil Engineering, University of Illinois, Urbana, Ill.

3.48 A. H.-S. Ang and W. S. Prescott, "Equations for Plate–Beam Systems in Transverse Bending," *J. Eng. Mech. Div., ASCE,* Vol. 87, No. EM6, December 1961, pp. 1–15.

3.49 M. A. Bhatti, B. Lin, and J. P. Idelin Molinas Vega, "Effect of Openings on Deflections and Strength of Reinforced Concrete Floor Slabs," in *Recent Developments in Deflection Evaluation of Concrete,* ACI SP-161, edited by E. G. Nawy, American Concrete Institute, Detroit, 1996, pp. 149–164.

3.50 W. L. Gamble, "Reinforced-Concrete Design," Sec. 12 in *Structural Engineering Handbook,* 4th ed., edited by E. H. Gaylord, Jr., C. N. Gaylord, and J. E. Stallmeyer, McGraw-Hill, New York, 1996, pp. 12–54.

3.51 H. L. Furr, "Numerical Method for Approximate Analysis of Building Slabs," *Proc. ACI,* Vol. 56, December 1959, pp. 511–541.

3.52 H. L. Furr, "Approximate Elastic Analysis of Slabs with Openings," Ph.D. thesis, University of Texas at Austin, 1958, 143 pp.

3.53 W. G. Bickley, "The Effects of a Hole in a Bent Plate," *Philos. Mag.,* Ser. 6, Vol. 48, 1924, pp. 1014–1024.

3.54 J. N. Goodier, "The Influence of Circular and Elliptical Holes on the Transverse Flexure of Elastic Plates," *Philos. Mag.,* Ser. 7, Vol. 22, 1936, pp. 69–80.

3.55 H. R. Hicks, "Laterally Loaded Plates, Moment Distribution Around a Small Circular Hole," *Eng. (London),* Vol. 184, No. 4770, August 9, 1957, p. 175.

3.56 E. Reissner, "The Effects of Transverse Shear Deformation on the Bending of Elastic Plates," *J. Appl. Mech.,* Vol. 67, 1945, pp. A-68 to A-77.

4 Background of 1971 and 1995 ACI Building Code Requirements for Reinforced Concrete Slab Design

4.1 INTRODUCTION

The purpose of this chapter is to explain the development of the 1971 ACI Building Code[4.1] requirements for reinforced concrete slab floors and similar requirements in the 1995 ACI Code.[4.2] These requirements were based on many factors, including the results of analytical studies of elastic theory moments such as reported in Chapter 3, the results of tests of various structures at both model and prototype scales, the analysis of the strengths of various types of slab structures using the yield line methods described in Chapter 7, and experience of successes and failures with slab structures which has accumulated during close to a century since the first beamless slabs were built.

The 1971 ACI Code contained many changes from earlier editions, some major and some minor. The most significant change was the unification of the design methods for all slabs, with and without beams. As one result of this, all types of slabs will have about the same factors of safety, which is an important improvement from earlier codes under which a slab supported on beams was significantly stronger than a beamless slab even when both had the same dead and live loads and the same qualities of materials.

A short history of the development of the code provisions may give some perspective to the complexity of the task of completely revising just one section of the ACI Code. In 1956 a large research project was started at the University of Illinois at Urbana–Champaign, Department of Civil Engineering, under the sponsorship of the Reinforced Concrete Research Council and with funding from several U.S. government agencies and the construction materials industry. This led to the testing of five $\frac{1}{4}$ scale nine-panel slabs of various designs, plus several smaller slabs and many auxiliary specimens. The results of these tests have been described by Sozen and Siess,[4.3] plus additional papers that are referenced later. A $\frac{3}{4}$-scale nine-panel flat plate that was tested by the Portland Cement Association was an important complement to the test series.[4.4] An extensive analytical investigation was launched at the same time, and literally hundreds of slab structures were analyzed using numerical techniques that were just then becoming feasible because of the development of electronic computers. Many of the resulting papers and reports were referenced in Chapter 3. The bulk of the analytical and experimental work was completed by 1963, and much effort was expended in reconciling the differences and establishing the areas of agreement. The formulation of proposed building code requirements was going at the same time, using primarily the results of the analytical studies but with the very important correlations with experimental results always having a bearing on decisions about various factors. Joint Committee 421 of ACI and ASCE, on Reinforced Concrete Floor Slabs, had the general responsibility for development of Chapter 13 of the 1971 ACI Code, and the committee instigated and had helped plan many aspects of the investigation. Even so, it was 1964, eight years after the inception of the program, before the first recognizable ancestor of the provisions in the 1971 Code was presented to this group. Seven more years of development, of discussions and haggling between theoretical and pragmatic viewpoints, and between the need to be general and the need to be simple, ensued before publication of Chapter 13 as part of the 1971 Code. Chapter 13 of the ACI Code gives two design procedures for slab systems: the direct design method and the equivalent frame method.

The direct design method is for slab systems, with or without beams, loaded only by gravity loads and having a fairly regular layout meeting the following conditions:

1. There must be three or more spans in each direction.
2. Adjacent span lengths may differ by no more than one-third of the longer span.
3. Panels should be rectangular, and the long span may be no more than twice the short span.
4. Columns must be near the corners of each panel, with an offset from the general column line of no more than 10% of the span in each direction.
5. The live load should not exceed twice the dead load. (This limit need not apply in cases where the same live load must always be present in all panels at the same time.)
6. If there are beams, there must be beams in both directions, and the relative stiffnesses of the beams in the two directions must be related as follows

$$0.2 \le \frac{\alpha_1 l_2^2}{\alpha_2 l_1^2} \le 5$$

(For notation, see Chapter 3.) This is not a very restrictive condition. Although the code does not say so, this requirement should not be construed to mean that beamless slabs cannot have edge beams and still be designed using these provisions.

Slab systems meeting the foregoing restrictions may be designed following the direct design method of the 1995 ACI Code (Sec. 13.6). For slab systems loaded by both horizontal loads and uniformly distributed gravity loads, or not meeting the foregoing restrictions, the equivalent frame method of Sec. 13.7 of the 1995 ACI Code, or a more comprehensive analysis, must be used. Although Sec. 13.7 of the 1995 Code implies that the equivalent frame method may be satisfactory in cases with lateral as well as horizontal loads, the commentary cautions that additional factors may need to be considered. The method is probably adequate when lateral loads are small, but serious questions may be raised when major lateral loads must be considered in addition to the vertical loads. Additional information on this topic is included later in this chapter.

The direct design method gives rules for the determination of the total static design moment and its distribution between negative- and positive-moment sections. The equivalent frame method defines an equivalent frame for use in structural analysis to determine the negative and positive moments acting on the slab system. Both methods use the same procedure to divide the moments so found between the middle and column strips of the slab and the beams (if any).

In addition to the many detailed provisions for the two methods for the design of slab systems described above, the 1995 ACI Code makes provisions for cases that obviously lie outside the limitations of the code and gives a certain degree of license to the analytically minded designer. Section 13.5.1 of the code could be viewed as an escape clause from the specific requirements of the code. It states: "A slab system may be designed by any procedure satisfying conditions of equilibrium and geometrical compatibility, if shown that the design strength at every section is at least equal to the required strength considering Secs. 9.2 and 9.3 of the ACI Code, and that all serviceability conditions, including specified limits on deflections, are met." The methods of elastic theory moment analysis described in Chapter 2 satisfy the requirements of this clause. The limit design methods, for example the strip method and yield line theory (see Chapters 6, 7, and 8), alone do not satisfy these requirements, since although the strength provisions are satisfied, the serviceability conditions may not be satisfied without separate checks of the crack widths and deflections at service load levels.

The thickness of a floor slab must be determined early in the design process since the weight of the slabs is an important part of the dead load of the structure and hence influences nearly every other aspect of the design. The minimum thickness may be determined by any of several factors. The shear strength of a beamless slab is often a controlling factor, and the slab must be thick enough to provide adequate shear strength. Less often, the thickness is determined by flexural moment requirements, and this led to the equations for minimum thickness of beamless slabs that were in the 1963 ACI Code.[4.5] Fire resistance requirements may also place minimum requirements on both total slab thickness and on the concrete cover over the reinforcement. The most common thickness limits arise from the need to control deflections, although one must never be blind to the fact that strength requirements may indicate thicker slabs. Section 9.5.3 of the 1995 ACI Code gives a set of equations and other guides to slab thickness and indicates that slabs which are equal to or thicker than the computed limits should have deflections within acceptable ranges at service load levels. These thickness limitations are discussed in Section 9.2 of this book in connection with the discussion of deflections.

The code direct design and equivalent frame methods can be discussed conveniently in terms of a number of steps used in design. The determination of the total design moment is concerned with the safety (strength) of the structure. The remaining steps are intended to distribute the total design moment so as to lead to a serviceable structure in which no crack widths are excessive, no reinforcement yields until a reasonable overload is reached, and in which deflections remain within acceptable limits. These steps are discussed in the following sections.

In Section 4.2 we discuss mainly the determination of the total moments used in the direct design method. This first involves determination of the total static bending moment, M_0, which is the absolute sum of the positive plus average negative moments for which a panel must be designed in the usual case in which the effects of partial loadings are not too important. The relationship between M_0 and the failure load is also discussed, using data from five test structures.

In Section 4.3 we describe the equivalent frame method, which may be used in those cases where the slab layout is irregular and does not comply with the restrictions stated previously, where horizontal loading is applied to the structure, or where partial loading patterns are significant because of the nature of the loading and high live load/dead load ratios. Steps for the determination of the distribution of total moments throughout the frame are discussed in this section. The equivalent frame method includes some fairly complex assumptions about the stiffness of frame members. Comments about how to adapt the equivalent frame method for use in a ordinary plane-frame computer program are included in this section.

Once the total bending moment acting on a section across the full width of a panel is determined by either the direct design or equivalent frame methods, the distribution of the moment across the width of the section, including the portion of the moment assigned to the beam, if any, must be determined. The step from the information given in Chapter 3 to the code provisions is described in Section 4.4.

In Section 4.5 we present details of the direct design method, including the distribution of M_0 to the various critical sections and the determination of design moments for columns. There seems little reason to retain the direct design method in the ACI Code given its limitations and given the general use of computer programs, which make the equivalent frame method much simpler and more straightforward than in the past.

4.2   STATIC MOMENT AND STRUCTURAL SAFETY

4.2.1   Static Moment

The static moment for a slab panel, which is the absolute sum of the midspan positive plus average support negative moments, is defined in the ACI Code direct design method as

$$M_0 = \frac{wl_2 l_n^2}{8} \tag{4.1}$$

where w is the uniformly distributed load per unit area; l_2 the span, center to center of columns, or width of panel considered, in transverse direction; and l_n the clear span, face to face of supports, in direction being considered. This equation may be rewritten as

$$M_0 = \frac{Wl_1}{8} \left(1 - \frac{c_1}{l_1} \right)^2 \tag{4.1a}$$

where W is the total uniformly distributed load acting on panel $= wl_1 l_2$; l_1 the span, center to center of columns, in direction being considered; and c_1 the width of supporting rectangular column, capital or bracket, measured in direction of span being considered. Although this total moment is explicit in the direct design method, it is also implicit in the equivalent frame method for the basic loading case of all panels loaded. The average of the negative moments, corrected to the face of the supports, plus the midspan moment, as determined by the equivalent frame method should be equal to M_0.

Equation 4.1a is in a convenient form for comparisons with previously specified values of M_0. Equation 4.1 may be derived by considering the equilibrium of the slab free body of a typical interior panel shown in Fig. 4.1a and b, assuming that the three edges which are on the panel centerlines are free of shear forces and twisting moments since they are lines of symmetry. All shear forces act on the line A–A through the face of the column, and moments are summed about that line.

Example 4.1. As an example of the application of Eq. 4.1 for the direct design method, Fig. 4.2 shows the dimensions l_1, l_2, and l_n for a slab with nine square panels. For this slab there are four varieties of panels, designated I, II, III, and IV, for consideration in the design. Each of these panel cases contains two of the shaded areas shown in Fig. 4.1b used in the derivation of Eq. 4.1.

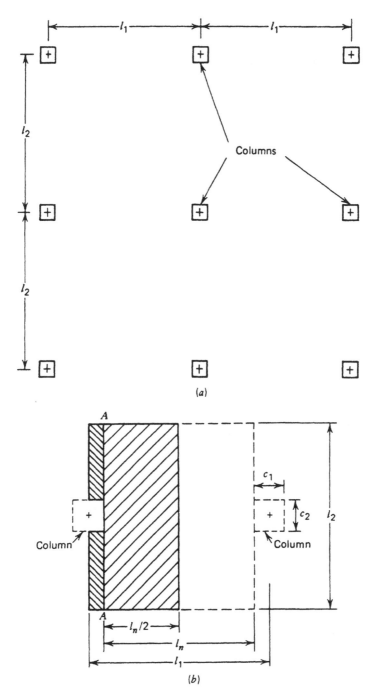

Figure 4.1 Slab sections considered for static moment computations. (*a*) plan of part of floor slab system; (*b*) slab sections considered in 1971 and 1977 ACI code static moment calculations; (*c*) slab section considered by Nichols.

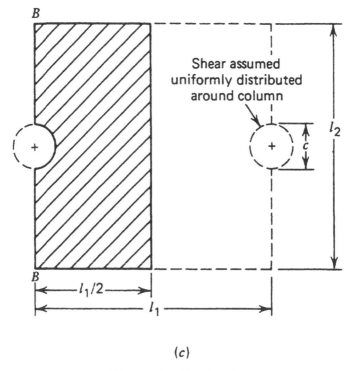

(c)

Figure 4.1 *(Continued).*

The static moment is calculated for each panel for the example slab, as follows. According to the 1995 ACI Code, the design load w_u = 1.7L + 1.4D = 1.7(70) + 1.4(94) = 251 lb/ft^2 = 0.251 kips/ft^2 (12.0 kN/m^2), where D and L are the service dead and live loads. Substitution into Eq. 4.1 gives:

For panel I: $M_0 = \dfrac{0.251 \times 20 \times 18^2}{8} = 203$ kip-ft (275 kN-m)

For panel IV: $M_0 = \dfrac{0.251 \times 10.67 \times 18.33^2}{8} = 112.4$ kip-ft (152.5 kN-m)

Similarly, for panels II and III, M_0 = 108.5 kip-ft (147 kN-m) and 211 kip-ft (286 kN-m), respectively. The moments due to the weight of the beam stem extending below the slab, and to any other loads applied directly to the beam, also have to be considered before the beam sections can be finally designed. The width of panels II and IV could have been taken as 10 ft 0 in. (3.05 m) width. The only difference in the end would be the additional dead load of the beam stem.

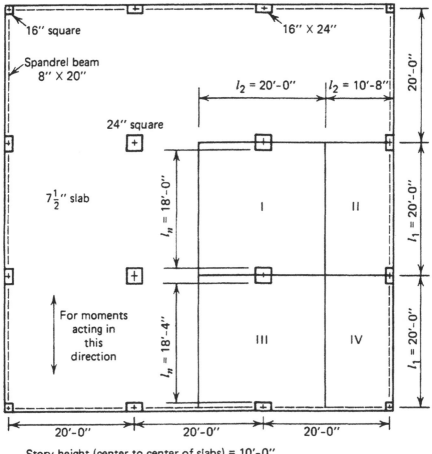

Story height (center to center of slabs) = 10'-0"

D = 94 lb/ft^2 (4.50 kN/m^2)   1 in. = 25.4 mm
L = 70 lb/ft^2 (3.35 kN/m^2)   1 ft = 0.305 m

Figure 4.2   Example slab floor, showing division into panels for static moment computations.

The 1963 and 1956 Codes[4.5,4.6] had the following value of M_0 for beamless slabs:

$$M_0 = 0.09Wl_1F \left(1 - \frac{2}{3}\frac{c_1}{l_1} \right)^2 \qquad (4.2)$$

where $F = 1.15 - (c_1/l_1) \geq 1.0$ and c_1 is the width of supporting column or capital measured in the direction of the span being supported. When strength design procedures were used with the 1963 Code, the 0.09 was replaced by 0.10.

The static moment expressions from the 1995 and 1963 Codes are plotted versus c_1/l in Fig. 4.3. Compared with the 1963 ultimate strength requirements, the 1995 Code (Eq. 4.1) requires a greater static design moment throughout the range of $c_1/l < 0.23$. Thus the 1995 ACI Code requires a moment that is about 7% greater than the 1963 ACI Code strength design method, for $c_1/l = 0.2$, and about 18% larger than the 1963 working stress design method, for the same load, w.

Equation 4.2 was based on the study by Nichols,[4.7] who considered the slab free body shown in Fig. 4.1c. However, Nichols' original study correctly led to the equation

$$M_0 = \frac{W l_1}{8} \left(1 - \frac{2}{3}\frac{c_1}{l_1}\right)^2 \tag{4.3}$$

where c is the diameter of supporting column or capital. Equation 4.3 is a

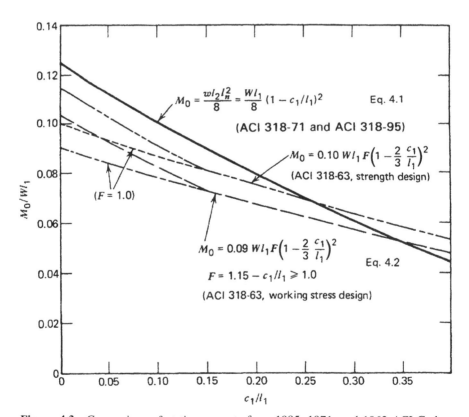

Figure 4.3 Comparison of static moments from 1995, 1971, and 1963 ACI Codes.

simplified form of a more complex expression. In deriving Eq. 4.3, Nichols summed moments about line B–B of Fig. 4.1c. It is thus obvious that the 1963 and 1956 Codes (Eq. 4.2) were specifying a static moment of only 72 or 80% (with $F = 1.0$) of Nichols's original moment (Eq. 4.3), if $c = c_1$. M_0 from Eq. 4.3 is also significantly larger than M_0 from Eq. 4.1, especially when $c/l > 0.2$ (i.e., for flat slabs with large column capitals).

The decision to use Eq. 4.2 rather than Eq. 4.3 in the earlier codes can perhaps be better understood if one reads the often bitter discussion of Nichols's paper.[4.7] Mistrust of Nichols's work, together with the then poorly understood concepts of plate theory and the results of a number of improperly interpreted tests of various early flat slab structures, contributed to this decision. Additional discussion, presented in a paper by Sozen and Siess,[4.3] adds more historical perspective to this vexing question.

Equation 4.3 was strictly correct only for circular columns but was commonly applied to slabs with rectangular columns as well. However, replacement of the $\frac{2}{3}$ factor by $\frac{3}{4}$ makes the expression suitable for use with square columns.[4.3] Equation 4.1 applies strictly to slabs supported on rectangular columns, with the additional provision that if other shapes are used, such as circular or polygonal, replacement square columns having the same cross-sectional area are to be considered when computing the clear span.

The static moment concept of Eq. 4.1 should not be blindly applied to slabs with columns or capitals with very large (or small) values of the ratio c_1/c_2. Simmonds[4.8] considered the case of a slab supported on very elongated columns with $c_1/c_2 = 8$ (or $\frac{1}{8}$). In such a case, the slab tends to behave much as a one-way slab in the direction of the small column dimension. However, in the direction of the long dimension of the columns, there are very high negative moments near the corners of the columns which may not be adequately accounted for unless a more complete analysis is used.

The decision to change from Eq. 4.2 and the various two-way slab provisions in the 1963 Code to Eq. 4.1 for all slabs in the 1971 Code is of considerable significance. The immediate result is to raise the design strength of any beamless slab somewhat and lower the design strength of slabs on beams. However, in practice the minimum reinforcement requirements often govern in lightly loaded slabs regardless of which expressions are used.

The choice of the clear span concept in the 1971 Code as opposed to the Nichols method of static analysis was based on perhaps four fairly persuasive arguments.

1. It is simple in concept and use.
2. It agrees well with tests to failure of beamless slabs in that the negative-moment yield lines tend to be straight lines intersecting the faces of the supporting columns. Also, theoretically the negative-moment yield lines must be straight, unless fan mechanisms develop adjacent to the columns. In a test the negative-moment yield lines may, in fact, bend back

toward the lines connecting the column centers somewhat, but they generally will not be offset the full distance.[4.9–4.11]

3. In beamless slabs, the elastic negative moments acting across the line connecting the column faces are of the same order or larger (much larger near the column) than the moments acting across the line connecting column centers, and consequently are more critical for service load conditions. This is illustrated in Fig. 4.4, which shows the distributions of negative moments across the two sections for a square interior panel with $c/l = 0.1$, where c is the column dimension and l is the span.

4. The increase in static moment using Eq. 4.3 over that obtained using Eq. 4.2 was believed to be greater than necessary to obtain the desired safety.

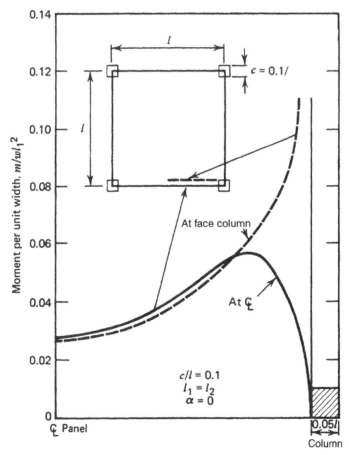

Figure 4.4   Distributions of negative moments across panel borders. $c/l = 0.1$[4.12]

The second and third factors are less persuasive for slabs supported on beams since the negative-moment yield lines in such slabs definitely tend to follow the face of the transverse beam and then be offset to the face of the column.[4.13,4.14] In addition, there is a definite moment gradient as the beam is approached, and the moments in the floor at the beam face are always greater than at slab sections slightly away from the beam. However, in practical terms these are usually not very important, for two reasons. First, beams, especially very narrow beams, are seldom used in combination with large values of c/l. Second, beams are usually assigned a relatively large portion of the total moment capacity, and beams would be designed for the clear span face to face of columns in any circumstances.

Van Buren[4.15] has considered the clear span concept for beamless slabs. He examined the conditions of equilibrium of both the major slab element shown lightly shaded in Fig. 4.1b and also the small slab element located between the lines connecting the column faces and the column centers, shown heavily shaded in the figure. He found that a reasonable set of assumptions could be made concerning the distribution of forces around the column that led to an alternative derivation of Eq. 4.1.

4.2.2 Structural Safety and the Static Moment

The decision to increase the static moment for beamless slabs and reduce it for slabs with beams was supported strongly by the results of the tests of five slabs at the University of Illinois[4.9–4.11,4.13,4.14] in addition to the results of analytical studies. The five slabs each had nine panels which were 5 ft (1.52 m) square. The beamless slabs, F1, F2, and F3, were 1.75 in. (44.5 mm) thick and those with beams, T1 and T2, were 1.5 in. (38 mm) thick. All slabs were supported on short columns extending below the slab. The nominal concrete strength was 3000 lb/in² (20.7 N/mm²) in compression and the design steel yield stress was 40,000 lb/in² (276 N/mm²). Four slabs were designed according to the 1956 ACI Code, using the working stress design method with allowable stresses of 1350 lb/in² (9.3 N/mm²) compression in the concrete and 20,000 lb/in² (138 N/mm²) tension in the steel; the fifth, T2, was designed using limit analysis concepts. All were designed at prototype scale with 20-ft (6.1-m) square panels, 7 in. (178 mm) thick for the F slabs and 6 in. (152 mm) thick for the T slabs, and then scaled down by reducing all dimensions to one-fourth of the prototype size. The $\frac{3}{4}$- scale PCA test slab[4.4] was based on the same prototype as slab F1.

The results of the tests to failure of these slabs are summarized in Table 4.1, where the different ratios of measured ultimate load to total service load are very evident. The most significant comparisons are in the bottom line of the table, which gives the ratio of measured ultimate load, normalized to a 40,000-lb/in² (276-N/mm²) yield strength for the steel to the total service load.

TABLE 4.1   Comparison of Ultimate and Design Loads[a]

Item	Fl Flat Plate[4.9]	F2 Flat Slab[4.10]	F3 Flat Slab, Structural Fabric[4.11]	T1 Two-Way Slab[4.13]	T2 Two-Way Slab, Shallow Beams[4.14]
Measured ultimate load,[b] w_u (lb/ft^2)	360	551	955	537	46
Total service load,[c] w (lb/ft^2)	155	285	285	145	145
Actual yield stress, f_y (kips/in^2)	36.7	42	70	42.0	47.6
$\dfrac{w_u}{w}$	2.3	1.9	3.4	3.7	3.2
$\dfrac{w_u}{w}\dfrac{40 \text{ kips/in}^2}{f_y}$	2.5	1.8	1.9	3.5	2.7

Source: After Ref. 4.3

[a] 1 lb/ft^2 = 0.0479 kN/m^2; 1 kip/in^2 = 6.89 N/mm^2.

[b] Maximum load carried by slab in tests.

[c] Sum of service dead and live loads.

Both flat slabs had normalized w_u/w ratios less than 2.0, which would have been the minimum value expected for other structures designed under the 1956 Code. These values were low enough to be regarded as serious problems. The ratio for the flat plate structure, Fl, was greater than 2.0 largely because the minimum reinforcement requirements governed in most middle strip sections, which increased the collapse load.

The ratios for the two-way slabs are obviously very high, and even if the value for T1 were normalized using the yield strength of the beam reinforcement, 50,000 lb/in^2 (344 N/mm^2), the ratio would still be 3.0. The load capacity of slab T1 was slightly enhanced by minimum reinforcement requirements, but this cannot have been a major factor since only a few sections were so governed, and in addition about two-thirds the total moment capacity was carried by the beams. The design of slab T2 was such that the minimum reinforcement requirements never controlled the amount of steel, and its rather high load capacity must have been due largely to strain hardening of the steel. The observed deformations were consistent with this.

The results of these tests, plus supporting studies which indicated that the specimens were in no way unusual slab structures, were used to justify the change to the use of Eq. 4.1 for determining the static design moment for all slab systems, with and without beams. A normalized value of $w_u/w = 1.8$ was low for the 1956 Code under which the flat slabs were designed, since the basic assumption of that code was that structures were to be capable of resisting at least 2.0 times dead plus live service loads before major yielding or collapse was reached. This is evident from the use of an allowable stress of 20,000 lb/in² (138 N/mm²) for steel having a minimum yield stress of 40,000 lb/in² (276 N/mm²).

Under the 1995 Code, the minimum ratio of w_u/w to be expected varies with the ratio of service live to dead loads, and can be expressed as

$$\frac{w_u}{w} \geq \frac{1.4D + 1.7L}{\phi(D + L)} \tag{4.4}$$

where D is the service dead load, L the service live load, and ϕ the strength reduction factor. For a flexural failure with $\phi = 0.90$, the minimum required value of w_u/w varies between 1.72 and 1.79 for the slabs considered above. Had slab F2 been designed using the 1995 Code load factors and Eq. 4.2 from the 1956 (and 1963) Code for determining the static moment, the value of w_u/w found from the test would have been about 1.6, which is obviously too low by current standards, just as 1.8 was too low by the 1956 standards.

It should be noted that the static moment requirement must be met in both span directions of each panel. This is an occasional source of confusion for the student. That the full static moment must be resisted in both directions can be seen by considering a simple example. A structure consisting of two parallel beams of span l_1 separated by the distance l_2 is shown in Fig. 4.5. A floor made of individual slab units separated by small gaps is supported by the beams and is subjected to a uniformly distributed load per unit area, w, from dead and live loads. Ignoring the weight of the beams, the total midspan moment to be shared by the two beams must be $wl_2l_1^2/8$. The sum of the midspan moments in the individual slab units must be $wl_1l_2^2/8$. If either moment capacity is less than indicated, the structure will collapse at a lower load than required. It is evident that the full static moment must be resisted in each direction.

4.3   EQUIVALENT FRAME METHOD: DETERMINATION OF NEGATIVE AND POSITIVE SLAB MOMENTS AND COLUMN MOMENTS

4.3.1   General Comments and Manual Calculation Approach

Slabs that do not meet the geometric or loading limitations for the direct design method listed in Section 4.1 may be analyzed using the equivalent frame method rather than by alternative, more complex rational analysis methods. Two common building arrangements that could be analyzed using the equivalent frame method are those having only two spans in one or both directions, and those having the "institutional" floor plan, with long end spans on each side of a short interior panel whose span length corresponds to the width of a corridor.

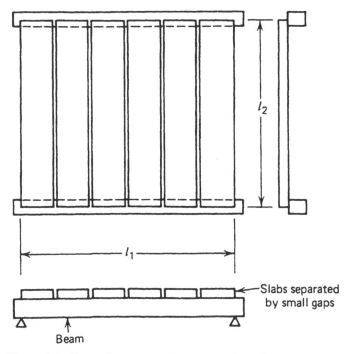

Figure 4.5 Example structure for computation of static moments.

In addition, slabs meeting all the limitations of the direct design method may also be designed using the equivalent frame method. Indeed, that would appear to be the logical method of analysis if a design firm has obtained or developed a fairly general computer program capable of handling the equivalent frame method. A separate alternative computer program for handling the direct design method is, of course, possible, but does not appear logical.

As the name implies, the equivalent frame method is a frame analysis scheme in which beam and column joint forces are found, and from these the moments and shears at the faces of the supports and the maximum positive moments may be determined. Any of the conventional frame analysis techniques may be used. However, since the structure being analyzed is not really a frame but a slab system, a number of special approximations must be made so that the frame analysis results will have some valid relationship to the slab. These approximations are concerned with the effective stiffnesses of the various members and the fixed-end moments to be used in the analysis. It will be seen that nearly all of the material in Sec. 13.7 of the 1995 ACI Code is concerned with these two questions. These sometimes complex definitions of stiffness have been criticized as making the slab provisions too complicated. However, these definitions are necessary if slabs are to be represented satisfactorily.

The application of the equivalent frame method is best explained through the use of an example. A slab with nine square panels, as shown in Fig. 4.6, will be considered. This is the same slab as was considered in connection with the direct design method example. The first step is to make two-dimensional frames from the three-dimensional structure. Each equivalent frame consists of either a row of interior columns plus the slab extending to the panel centerlines on each side of the column line, or a row of columns

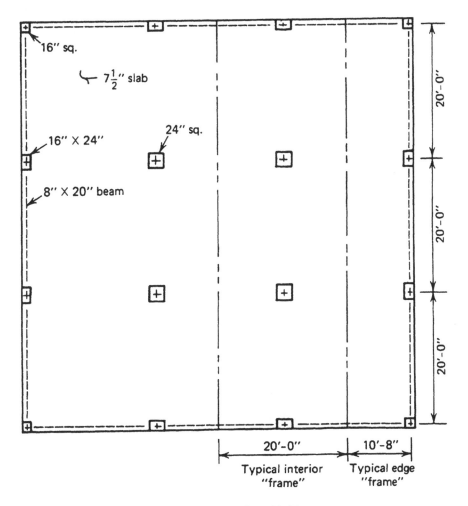

16" sq.

⌐ $7\frac{1}{2}''$ slab

24" sq.

16" X 24"

8" X 20" beam

20'-0"

20'-0"

20'-0"

20'-0"   10'-8"

Typical interior   Typical edge
"frame"   "frame"

Story height (center to center of slabs) = 10'-0"

D = 94 lb/ft^2 (4.50 kN/m^2)   1 in. = 25.4 mm
L = 70 lb/ft^2 (3.35 kN/m^2)   1 ft = 0.305 m

Figure 4.6   Example slab structure, showing division of slab into equivalent frames.

at the edge of the structure plus the slab extending from the edge to the panel centerline. As shown in Fig. 4.6, the division lines extend through the full height of the structure. The equivalent frames therefore consist of the columns and very wide beams.

One story of the resulting frame, with columns above and below the slab, is shown in Fig. 4.7. The code assumes that the stress analysis may be carried out by a hand-calculation method such as the moment distribution technique, and allows some simplifications to reduce the amount of labor. One of the permissive aspects is that either the entire frame or a single floor may be considered when only gravity loads act. The entire frame clearly should be considered if horizontal loads act and must be resisted by bending of the columns rather than by shear walls or other bracing. If only one intermediate

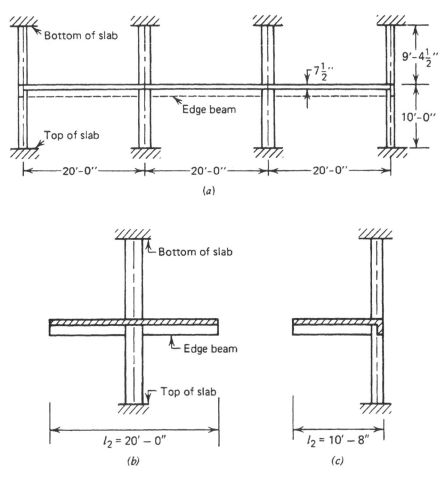

Figure 4.7 Elevation and sections of equivalent frames: (*a*) elevation of frame cut from slab structure; (*b*) cross section of interior frame; (*c*) cross section of edge frame.

story is considered at a time, the replacement structure is as shown in elevation in Fig. 4.7a, with the far ends of the columns both above and below fixed against rotation. Sections through the interior and edge frames are also shown in Fig. 4.7b and c. The top and bottom stories may have other support conditions.

As an additional permissible simplification, in a structure of several spans, a joint two or more spans away from a joint being considered may be taken as fixed if the structure extends beyond that joint. Both simplifications are holdovers from earlier codes and are common engineering practice for hand-calculation methods.

The assumptions to be made about the variation of EI on along the lengths of the members are carefully spelled out in Sec. 13.7 of the 1995 Code. (These assumptions vary in small details from those proposed by Corley and Jirsa.[4.16]) Determination of the relative stiffnesses of the sections may be carried out considering the gross, uncracked concrete sections. All variations in concrete cross section along a member that occur outside the joint regions should be taken into account. Therefore, a slab-beam (a slab that forms the beam of the equivalent frame) which has drop panels has a larger I value near the ends than in the central part of the span.

In addition, the parts of the slab—beam lying between the column center-line and column face; that is, the end segments, of length $c_1/2$, are effectively stiffer than the clear span portion of the member. The code approximation is that the I of this end section should be obtained by dividing the I of the slab–beam just outside the column by the quantity $[1 - (c_2/l_2)]^2$. The variation in moment of inertia that results is shown in Fig. 4.8 by the plot for $1/I$ for beam–slabs without and with drop panels. For these members the flexural stiffnesses are more than $4EI/l_1$ (considering I of the midspan section), the carryover factors are more than 0.5, and the fixed-end moments are more than $W(l_1/12)$. These necessary quantities may be computed from first principles, but they are tabulated for many combinations of c_1/l_1, including different values at each end of the span, in the commentary to the 1977 Code[4.17] and in Simmonds and Misic.[4.18]

For the columns, the moment of inertia is assumed to have an infinite value throughout the joint region, from the top of the slab to the bottom of the slab, drop panel, column capital, or beam. The variations of $1/I$ along the height of an interior and edge column from the example slab are shown in Fig. 4.9. The numerical dimensions shown in Figs. 4.8 and 4.9 are specifically for the example slab shown in Fig. 4.6.

The stiffness and carryover factors must account for the rigid portions of the members. These may conveniently be calculated using the column analogy method[4.19] or any of a number of other methods. In addition, useful tables have been published by various agencies and authors: for example, the Portland Cement Association[4.20] and Cross and Morgan.[4.21]

Various design aids have been produced to ease the way through the necessary calculations. Hoffman et al.[4.22] and Rice[4.23] give several graphs plus

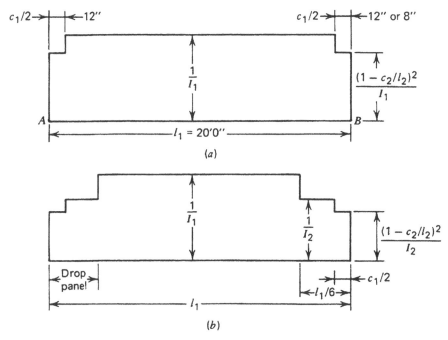

Figure 4.8 Typical $1/I$ diagrams for slab–beams. (a) $1/I$ diagram for slab–beam without drop panels; (b) $1/I$ diagram for slab–beam with drop panels.

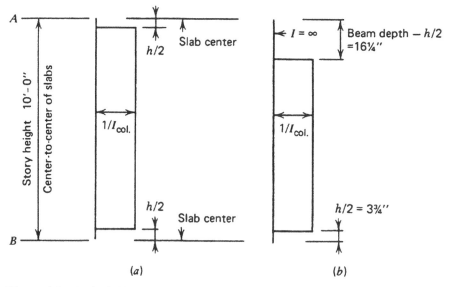

Figure 4.9 Typical $1/I$ diagrams for columns: (a) interior column in beamless slab; (b) column supporting slab on beams.

approximations for stiffness factors. The 1996 *CRSI Handbook*[4.24] also has some useful information on stiffness factors, in addition to extensive material on the selection of reinforcement once the forces and member sizes are known. Zweig[4.25] published an article giving design aids primarily for the direct design method, but some of the information is also applicable to the equivalent frame method.

Once the stiffnesses of the columns have been determined, the stiffnesses of the equivalent columns can be found, as explained in Section 3.1. The equivalent stiffness K_{ec} is found as

$$\frac{1}{K_{ec}} = \frac{1}{\Sigma K_c} + \frac{1}{K_t} \tag{4.5}$$

where ΣK_c is the sum of the flexural stiffnesses of the columns above and below the slab which frame into the joint being considered, and K_t is the torsional stiffness, computed as indicated in Section 3.1. The torsional constant, C, may be computed using Eq. 3.5, which is shown plotted in Fig. 4.10. Equation 3.5 is repeated here:

$$C = \Sigma \left(1 - 0.63 \frac{x}{y}\right) \frac{x^3 y}{3} = \Sigma x^4 \left(\frac{y}{3x} - 0.21\right) \tag{4.6}$$

The K_{ec} value found from Eq. 4.5 is then used in the frame analysis. K_{ec} is for a combination of two columns (one above and one below), and if it is

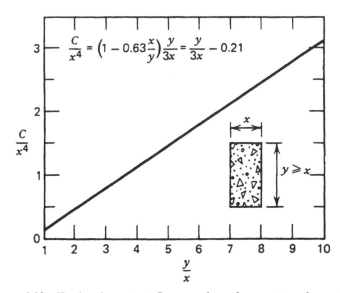

Figure 4.10 Torsional constant C versus shape for a rectangular section.

used to find a moment distribution factor, the factor applies to the total moment in the two columns. This is usually not satisfactory, as the individual column moments must be found for design purposes. The distribution factors to the two columns may be found by subdividing the total equivalent column factor K_{ec} in direct proportion to the individual K_c values.

Example 4.2. The design example shown in Fig. 4.6 will be continued to find the design moments using the equivalent frame method. The equivalent frames are shown in Fig. 4.7.

INTERIOR EQUIVALENT FRAME. The moments of inertia of the interior slab and column sections are:

Slab: $\qquad\qquad I_s = \dfrac{1}{12} \times 7.5^3 \times 240 = 8438$ in.^{3} $(3.512 \times 10^9$ mm$^4)$

Interior column: $\quad I_c = \dfrac{1}{12} \times 24^3 \times 24 = 27{,}648$ in.^{3} $(11.51 \times 10^9$ mm$^4)$

Edge column: $\qquad I_c = \dfrac{1}{12} \times 16^3 \times 24 = 8192$ in.^{3} $(3.410 \times 10^9$ mm$^4)$

(weak axis bending)

The fixed-end moment FEM, flexural stiffnesses K, and carryover factors COF for the interior frame beam and column members shown in Figs. 4.8a and 4.9 are as follows. For the interior span beam, from Table 13.1 of the 1977 Code commentary:

1. Interior span $[c_1/2 = 12$ in. (305 mm) at each end, $l_1 = 20$ ft (6.10 m)]:

$$\text{FEM} = 0.085Wl_1$$

$$K = 4.18E\,\frac{I_s}{l_1} = 4.18E \times \frac{8438}{240} = 147E \text{ kip-in./rad } (E \text{ in kips/in}^2)$$

$$\text{COF} = 0.513$$

2. End span $[c_1/2 = 12$ in. (305 mm) at end A, 8 in. (203 mm) at end B, $c_2 = 24$ in. at both ends, and $l_1 = 20$ ft (6.10 m)], values computed from a column analogy solution:

$$\text{FEM}_A = 0.085Wl \qquad\qquad \text{FEM}_B = 0.084Wl_1$$

$$K_{AB} = 4.17(I_s/l_1) \qquad\qquad K_{BA} = 4.13E(I_s/l_1)$$

$$= 147E \text{ kip-in./rad} \qquad\qquad = 145E \text{ kip-in./rad}$$

$$\text{COF}_{AB} = 0.508 \qquad\qquad \text{COF}_{BA} = 0.513$$

Tabulated values in various design aids are based on the assumption that the columns at both ends of the span are square, with c_1 assumed in both directions. Interpolated values from the table noted above are thus slightly different from those given here.

For the columns, the following values were obtained from column analogy solutions:

Interior column: $\qquad\qquad K_c = 4.71E(I_c/l_c)$

$$\text{COF} = 0.547$$

Edge column: $\quad K_{AB} = 7.52E(I_c/l_c) \qquad K_{BA} = 5.36E(I_c/l_c)$

$$\text{COF}_{AB} = 0.522 \qquad\qquad \text{COF}_{BA} = 0.733$$

where l_c is the length of column measured from center to center of slabs = 10 ft (3.05 m). The K_{ec} value for the edge column is computed as follows. The total stiffness of the columns above and below the slab is

$$\Sigma K_c = (7.52 + 5.36)\frac{E(8192)}{120} = 879E \text{ kip-in./rad}$$

where E has kips/in^2 units (1 kips/in^2 = 6.89 N/mm^2).

The computation of K_t requires a value of C for the edge beam. The beam cross section is shown in Fig. 4.11a, and $C = 3647$ in^4 (1.518 $\times$ 10^9 mm^4).

$$K_t = \Sigma \frac{9E_cC}{l_2(1 - c_2/l_2)^3} = 2 \times \frac{9E(3647)}{240[1 - (24/240)]^3} = 375E$$

K_{ec} for the edge column is computed using the values above.

$$\frac{1}{K_{ec}} = \frac{1}{\Sigma K_c} + \frac{1}{K_t} = \frac{1}{879E} + \frac{1}{375E}$$

$$K_{ec} = 263E \text{ kip-in./rad}$$

This is the total equivalent stiffness of the edge columns above and below

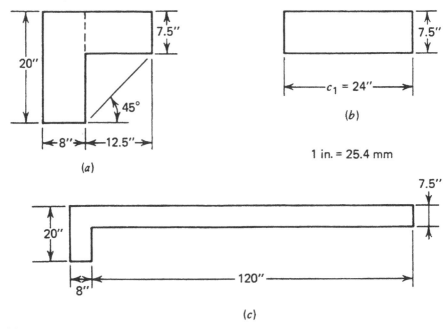

1 in. = 25.4 mm

Figure 4.11 Slab and beam cross sections considered when computing stiffnesses for the equivalent frame method: (*a*) edge beam cross section; (*b*) interior torsional member cross section; (*c*) slab-plus-beam cross section considered when modifying K_t.

the slab. The individual edge column equivalent stiffnesses may be found from the column stiffness:

Column above slab: $K_{ec}^* = 263E \dfrac{5.36}{5.36 + 7.53} = 109E$

Column below slab: $K_{ec}^* = 263E \dfrac{7.52}{5.36 + 7.52} = 154E$

where K_{ec}^* denotes the equivalent stiffness of an individual column where two columns frame into a joint.

The calculation of all the stiffness and other factors, starting from the basic dimensions of the members, is fairly easily accomplished using a simple computer program. It is readily adapted to a spreadsheet, for instance, or to a math–equation solver program.

Similarly, $\Sigma K_c = 2169E$ for the interior columns, and $C = 2711$ in^4 (1.128×10^9 mm^4), leading to $K_t = 279E$ for the torsional member consisting of a slab strip of width c_1 as shown in Fig. 4.11*b*. This gives a combined $K_{ec} = 247E$ for the interior columns, divided equally between the columns above and below the slab. The stiffnesses are summarized in Fig. 4.12, which

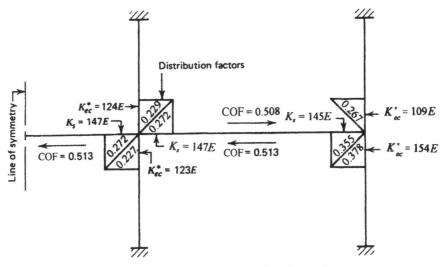

Figure 4.12 Summary of stiffnesses of interior equivalent frame.

also shows the moment distribution and carryover factors for interior equivalent frames.

The next step is the computation of the fixed-end moments. The total design load for each panel is $W = wl_1l_2$, where the design load per unit area is

$$w = 1.7L + 1.4D = 1.7(70) + 1.4(94) = 251 \text{ lb/ft}^2$$

$$= 0.251 \text{ kip/ft}^2 \ (12.0 \text{ kN/m}^2)$$

Therefore, the total design panel load

$$W = 0.251 \text{ kip/ft}^2 \times (20 \text{ ft})^2 = 100.4 \text{ kips } (447 \text{ kN})$$

The interior span fixed-end moment is

$$-M = 0.085Wl_1 = 0.085 \times 100.4 \times 20 = 171 \text{ kip-ft } (232 \text{ kN-m})$$

The static moment for each span is

$$M_{\text{static}} = \frac{Wl_1}{8} = 100.4 \times \frac{20}{8} = 251 \text{ kip-ft } (340 \text{ kN-m})$$

The fixed-end moments are shown in Fig. 4.13. The first step of the moment distribution process for the case of all panels loaded is shown, but the final moments given here are from a frame analysis program. Moments from

Figure 4.13 Fixed-end, balanced, and design moments for interior frame.

Line of Symmetry

(– – – – – – Signs – – – – – – +)

	Beam	Column Below	Column Above	Beam	Maximum + Moment	Beam	Column Below	Column Above	Notes
	0.272	0.228	0.228	0.272		0.355	0.378	0.267	Distribution factors
	+171	0	0	-171		+169	0	0	FEM, (kip-ft)
	--	--	--	-30.5		-59.5	-65.5	-46.0	Balance
	⋯	⋯	⋯	⋯		⋯	⋯	--	Carryover, etc.
	+176.0	+8.5	+8.5	-192.0		+112.2	-65.5	-46.7	Final Moments — kip-ft
	+237.3	+11.5	+11.5	-260.3		+152.1	-88.8	-63.3	kN-m
Midspan	+76				+101				(Maximum + Moment)
	+102				+137				
	-127	+8	+8	-141	+101	-83	-52	-44	Design Moments — kip-ft
	-174	+11	+11	-191	+137	-112	-71	-60	kN-m

a moment-distribution solution differed by no more than 1.0 kip-ft, with the differences consistent with those caused by differential shortening of the columns.

The balanced moments obtained from the distribution process are those at the centers of the supports. They must be reduced to the moments at the faces of the supports and used to find the maximum positive moments before reinforcement can be selected. This process is illustrated graphically in Fig. 4.14, where the static moment diagrams are plotted from the baseline, and the straight lines connecting the negative-moment values at each end of the spans are drawn over the parabolas. A simple spreadsheet program will facilitate the construction of such a graph. The design moments so found are also shown in Fig. 4.13. The slab–beam moments are shown with designer's signs, while the column moments are given with rotational signs. The column moments have been corrected from those at the center of the slab to values at the top of the slab and bottom of the beam or slab.

EDGE EQUIVALENT FRAME. The analysis of the edge frame is similar in most respects. The stiffness of the beam–slab is taken as the sum of the individual flexural stiffnesses of the beam section, as shown in Fig. 4.11a, plus the stiffness of the remaining width of slab to the panel centerline. Neither the 1995 Code nor the commentary is specific about the cross section to be considered, but the code background paper by Corley and Jirsa[4.16] states plainly that the stiffness should be that of the beam plus the remaining part of the slab. (The same reasoning applies to cases of interior frames where beams are present.)

For the edge equivalent frame of the slab shown in Fig. 4.6, $I_{beam} = 8082$ in^4 (3.364×10^9 mm^4) for the beam cross section shown in Fig. 4.11a and $I_{slab} = 3779$ in^4 (1.573×10^9 mm^4) for the 107.5 in. (2730 mm) width of slab remaining after the beam width is removed from l_2. Consequently, I for the slab–beam is $8082 + 3779 = 11,861$ in^4 (4.937×10^9 mm^4).

For the calculation of K for the slab-beam, the variation of I near the ends of the member should be taken as in Fig. 4.8a, following Sec. 13.7.3.3 of the ACI Code, taking I over the end $c_1/2$ as $I_1/[1 - (c_2/l_2)]^2$. It may appear more logical to compute K for the beam assuming that $I_{beam} = \infty$ for the length $c_1/2$ at each end, and K for the slab with $I_{slab} = I_1/[1 - (c_2/l_2)]^2$ for the length $c_1/2$ at each end, and to add the two K values to get the stiffness of the edge beam, but this is not in accordance with the code or the background paper. The second interpretation would present great difficulty when it came to computing the fixed-end moments.

Since there is a beam extending in the direction of the span being considered, the torsional stiffness K_t must be modified before computing K_{ec} as noted in Section 3.1. K_t is multiplied by the ratio of the EI of slab plus beam divided by EI of the slab alone, to give K_{ta} (see Eq. 3.12). If the materials are the same, only the I values need be considered. (The code actually specifies using I values, but the analysis is general enough to apply to such com-

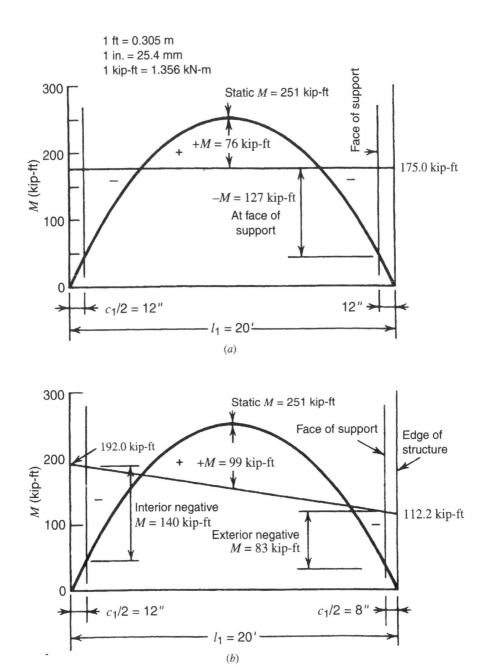

Figure 4.14   Graphical determination of design moments for interior equivalent frame from end moments: (*a*) interior span; (*b*) end span.

binations as concrete slabs composite with steel beams, and for these, EI values would be necessary.) The moment of inertia for the slab and beam combination is not that calculated previously, but is for the entire cross section taken as a slab with the beam as a stiffening rib, as shown in Fig. 4.11c. This moment of inertia is somewhat larger than the sum of the moments of inertia of the individual sections.

For the slab, K_t at the edge column of the edge frame would be, considering the slab section shown in Fig. 4.11b as the torsional member,

$$K_t = \frac{9E(2711)}{240[1 - (16/240)]^3} = 125.0E$$

In this particular case l_2 has to refer to the full distance from the center of the column to the center of the adjacent row of columns. This is not perfectly clear in the code, but it is necessary in keeping with the derivation of the equation for K_t. There is no summation since there is a torsional member on only one side of the column.

The I of the 10 ft 8 in. (3.25 m) width of slab is 4500 in⁴ (1.873×10^9 mm⁴) and of the slab plus beam section is 14,859 in⁴ (6.185×10^9 mm⁴). Therefore, K_{ta} becomes

$$K_{ta} = 125.0E \frac{14,859}{4500} = 413E$$

I for strong-axis bending of the edge column is 18,432 in⁴ (7.672×10^9 mm⁴) and

$$\Sigma K_c = \frac{(7.52. + 5.36)E(18,432)}{120} = 1978E$$

From this, the equivalent column stiffness is

$$\frac{1}{K_{ec}} = \frac{1}{\Sigma K_c} + \frac{1}{K_{ta}} = \frac{1}{1978E} + \frac{1}{413E}$$

$$K_{ec} = 342E$$

From the flexural stiffnesses of the columns, $K_{ec}^* = 200\,E$ for the edge column below and $K_{ec}^* = 142E$ for the edge column above.

The equivalent stiffness of the corner column is computed in the same manner. I for the 16-in. (406-mm) square column is 5461 in⁴ (2.273×10^9 mm⁴), which leads to $\Sigma\,K_c = 586E$. One spandrel beam forms the edge torsional member, with $C = 3647$ in⁴ (1.518×10^9 mm⁴), as before. From this,

$$K_t = \frac{9E(3647)}{240[1 - (16/240)]^3} = 168E$$

The multiplier to find K_{ta} is the same as for the edge column since the same beam in the l_1 direction frames into the corner column, giving $K_{ta} = 555E$. From these stiffnesses, the equivalent corner column stiffness is found to be $K_{ec} = 285E$, and the K^*_{ec} values for the corner columns below and above are $166E$ and $119E$, respectively. The stiffnesses and resulting moment distribution factors are summarized in Fig. 4.15.

The loaded area is 20 ft (6.1 m) by 10.67 ft (3.25 m), for a total panel load of 53.5 kips (238 kN) for each span of the edge frame. The fixed-end moments are calculated in the same manner as for the interior equivalent frame, with the same constants, and are shown in Fig. 4.16. The balanced end moments and the design moments at the faces of the columns and at midspan are also shown.

The loadings to be considered in connection with the equivalent frame method are specified in the 1995 ACI Code. If the service live load is less than 0.75 of the service dead load, only the case of all panels loaded needs to be considered. For larger service live loads, maximum moments due to pattern loads must also be considered. For maximum positive moments, the span being considered plus alternate spans are subjected to live loads. For maximum negative moments, only the two spans adjacent to the section being considered are loaded. For pattern loading cases, the slab is subjected to the design dead load (including load factor) plus 0.75 of the design live load (also including load factor). The use of the reduced design live load is a recognition

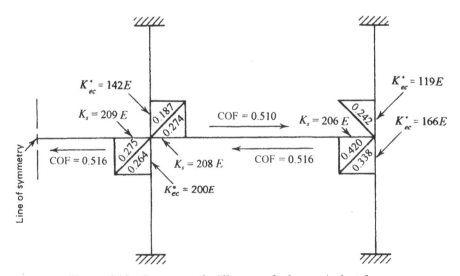

Figure 4.15 Summary of stiffnesses of edge equivalent frame.

— Line of Symmetry

------- Signs ------- +

Beam	Column Below	Column Above	Beam		Beam	Column Below	Column Above	Notes
0.275	0.264	0.187	0.274		0.420	0.338	0.242	Distribution factors
+91.1	0	0	-91.6		+90.1	0	0	FEM, (kip-ft)
+0.2	+0.1	+0.1	+0.2		-37.8	-30.5	-21.8	Balance
--			-19.5		-+0.1	--	--	Carryover, etc.
• • •	• • •	• • •	• • •		• • •	• • •	• • •	Final Moments
+93.2	+6.0	+4.3	-103.5		+54.6	-31.8	-22.8	kip-ft
+126.4	+8.1	+5.8	-140.3		+74.0	-43.1	-30.9	kN-m
				Maximum+Moment				
-68	+5	+4	-76	+56	-39	-25	-22	Design Moments kip-ft
-92	+7	+5	-103	+76	-112	-34	-30	kN-m

Midspan: +41 (kip-ft), +50 (kN-m)

Figure 4.16 Fixed-end, balanced, and design moments for edge frame.

173

of both the normally low probability of getting a particular loading pattern with full life load and of the fact that some overstress may be safely allowed at a few sections at one time. The final design moments must not be less than those caused by full design loading on all panels. If the loading planned for the structure includes a high probability of known patterns of loading, such as in some warehousing operations, they should of course be considered.

When the slab analyzed meets the requirements of the direct design method, the positive and average negative moments found by the equivalent frame method may be reduced in proportion until the sum equals M_0. The loadings specified in the code should be adequate for the determination of the slab and beam moments. However, when the design live load is small enough to allow the slab and beam moments to be found from only the case of all panels loaded, the column moments may be underestimated. Pattern loadings should still be considered to determine the controlling column moments. As an example of the potential problem, the interior frame of the example slab was analyzed with the design dead load in all panels, 0.75 of the design live load on each end span, and zero live load on the interior span. The interior column moments for the columns above and below the slab were each 23.9 kip-ft (32.4 kN-m), as compared with 8.25 kip-ft (11.2 kN-m) when all panels were subjected to the full design load. Each column moment is associated with a different axial load, and it cannot always be readily determined which combination is most critical before the column has been designed.

The 1995 Code does not address the question of the loadings for column moments using the equivalent frame method. The code specifies only that the columns shall be designed for the moments determined with the equivalent frame analysis. It would appear reasonable that the column moments and axial loads caused by pattern loadings involving design dead load plus 0.75 of the design live load be considered even for cases where the pattern loadings are not required for the slab-beam moments.

There is at present a conflict between the column moment provisions of the equivalent frame method and the direct design method. The pattern loadings, when required, for the equivalent frame analysis involve 0.75 of the design live load. An expression for the column moments for slabs designed by the direct design method is given in Sec. 13.6.9 of the ACI Code, and the moment is based on having 0.50 of the design live load on some panels and none on the remainder. This difference should be resolved. The distribution of the total section bending moments found at the negative- and positive-moment regions of the slab system to the column and middle strips of the panels, and to the beams (if any), is described in Section 4.4.

4.3.2 Idealizations for Plane-Frame Computer Analysis

The application of a plane-frame analysis to a slab structure spanning in two directions sounds like a contradiction in terms. That is, of course, exactly what the equivalent frame analysis of the ACI Codes has been doing for several decades, and the following are comments on how this can be done using whatever plane-frame program is available, rather than using a specialized slab program such as ADOSS (distributed by the Portland Cement Association), moment-distribution methods as described in the preceding section, or a finite element method. The material in this section is largely based on Ref. 4.26. Cano and Klingner[4.27] considered the same general problem and developed a different solution, using a three-dimensional frame program rather than a plane-frame approach.

Different assumptions are needed when considering vertical and lateral loading cases, and they must be treated differently. For vertical loads, the primary interest is in the *relative* stiffnesses of the members. For lateral loads, one must have information on both the *relative* and *absolute* stiffnesses if one is to assess both the force distributions and displacements. In addition, the *relative* stiffnesses apparently must be quite different in the two cases. The original calibrations between analyses and experimental values of moment distributions were based on uncracked, gross section, EI values, as is still commonly done for the vertical loading case. The lateral loading case requires more complex evaluations of EI if the results are to be reasonable.

Vertical Loading Cases

Slab–Beams. The ACI equivalent frame method assigns a variable moment of inertia to the slab–beam components. In a frame analysis, the slab–beam for a span is simply broken into several members as in Fig. 4.17. The moments of inertia are either I_{slab} or $I_{slab}/(1 - c_2/l_2)^2$, as is appropriate to the segment. No more searching for the stiffness, fixed-end moment, and carry-over factors for a nonprismatic beam with stiff end sections. And the moment at the node at the face of the column is the one that one wants for the slab

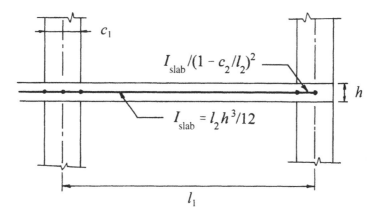

Figure 4.17 Stiffnesses of slab–beam elements.[4.26]

reinforcement design, after division between the column and middle strips (and beam).

A potential problem in this analysis is that the stiffness of the short end members is very much greater than of the central member. In some cases, this may introduce numerical instabilities into the results, and this must be checked for. In one case, a program worked properly when the dimensions and section properties were given in meters, but the same program failed to give correct answers when the dimensions were given in millimeter units. For that program, one solution to the problem was the "condensation" of the stiffnesses of the three members making up one beam span into a single equivalent member. In this particular case, the error was that the member end moments, at the centers of the supports, were all approximately equal to the fixed-end moments, even at the exterior columns where the moments should have been much smaller.

Equivalent Columns. The column stiffness is complex, on two bases. First, the end sections are assumed to be rigid for some distance. Second, a torsional member is interposed between the column and slab (for purposes of the analysis), and this member is thought of as being part of the column. The stiffness of this assembly is then taken as K_{ec}, as has already been discussed. The stiffness, K_{ec}, is of no help if you wish to use a plane-frame program, as opposed to a special-purpose slab analysis program or moment distribution. K_{ec} also has the problem of lumping together the combined stiffnesses of the columns above and below the slab, and these columns are of course often dissimilar. The subdivision of K_{ec} into K_{ec}^* was noted in the preceding section. However, instead of , the moment of inertia of an equivalent column, I_{ec}, is needed. This can be developed as follows.

The K_{ec}^* value for an individual column can be computed as

$$K_{ec}^* = K_{ec} \frac{K_c \text{ of column concerned}}{\Sigma K_c} \tag{4.7}$$

In a parallel calculation, the moment of inertia of an individual equivalent column, I_{ec}, can be computed as

$$I_{ec} = I_c \frac{K_{ec}^* \text{ of column concerned}}{\Sigma K_c} \tag{4.8}$$

This gives phoney values of I to be used in the elastic plane-frame program, in connection with also telling the program that the appropriate end lengths of the columns are rigid. These rigid lengths are also essential if the process is to give the correct answers. The I_{ec} calculation is easily accomplished with a computer program that has this calculation as its sole purpose. However, there is a snag in this process. If we consider the edge column shown in Fig.

4.18, we can compute I_{ec} values for each column at each joint. If the columns have the same cross section, the values I_{ec2} and I_{ec3} will be the same, even if the two stories have different heights and different rigid lengths. This is a reasonable conclusion. The snag comes at the top column, where you will find that $I_{ec3} \neq I_{ec4}$. Since the column is of constant section between the rigid end sections, this is not a reasonable answer. Nor is the solution to replace the column with a nonprismatic member with some transition in I_{ec} from one end to the other. The problem seems to be that the torsional member has a different effect on K_{ec} when attached to one column than when it is connected to two columns.

Although it may not be a scientifically pleasing solution, the practical engineering solution is to average the values of I_{ec3} and I_{ec4} for the upper column. The EI distribution suggested in Fig. 4.18 is based on this averaging. The answer, in a limited study, is that this will give about the same upper

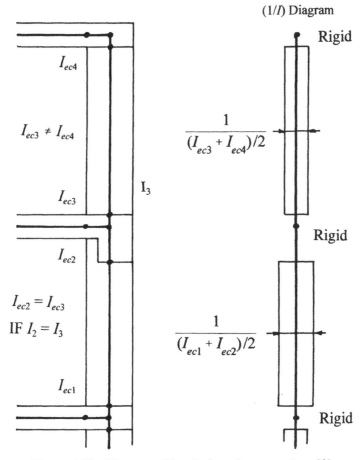

Figure 4.18 Moments of inertia for column members.[4.26]

joint moments as one gets from a moment distribution solution, and all other moments are less sensitive to the stiffness of this problem column. This is a reasonable comparison to make, considering the origins of the equivalent frame method. In simpler structures, such as the single-story frame shown in Fig. 4.7a, the use of this phoney I_{ec} in a plane-frame program gives numbers nearly identical with those from the moment distribution procedure, with the very small differences consistent with the expected consequences of differential shortening of columns.

Lateral Loading Cases. It is useful to start with a quote from Vanderbilt,[4.28] who was writing about an unbraced flat plate structure: "Its analysis presents a number of interesting problems, most of which center on the proper way to consider the behavior of the planar slabs." It also useful to emphasize that the problems in analysis being considered are those in beamless slabs. The two-way slab with substantial beams on all column lines will respond as a beam-and-column frame, with little participation from the slabs since most of the stiffness will be concentrated in the beams. Questions about states of cracking will remain important if drift is to be predicted properly, but the problems are less crucial than those in beamless slabs.

This introduction-conclusion has not changed in the last two decades. Vanderbilt went on to compare the results of analyses that were extensions of the equivalent frame method to the lateral load case (with the torsional members present), with analyses that used some reduced width of the slab as the beam member in a plane-frame program and with the few tests results that were available. His equivalent frame method (a program called EFRAME) had cases in which the torsional members acted with the columns (*equivalent columns*) and cases in which they acted with the beams (*equivalent beams*). These two approaches produced nearly the same reasonable results. Comparisons with the deflected shapes of a nearly uncracked eight-story model slab structure under various lateral loadings were quite favorable.

The second approach that Vanderbilt considered was the use of a reduced *EI* of the slab elements, based on an effective-width concept. In this frame, the columns have the *EI* of the columns, with no reduction to account for the torsional members. Effective widths had been determined earlier by several investigators, including at least Khan and Sbarounis,[4.29] Pecknold,[4.30] and Allen and Darvall,[4.31] using both analytical and experimental methods. Each study considered uncracked, elastic slabs. The effective width problem is, of course, the same one that led to the equivalent column with its attached torsional members—merely making a column rigid will not produce a fixed-end condition. The rigid column clamps a piece of slab equal to the width of the column, c_2. The rest of the width of panel, $(l_2 - c_2)$, rotates and deflects as it wishes. Vanderbilt used effective width $= \alpha l_2$ and cited values of α ranging from 0.25 to 0.67 from a literature survey. His comparisons based on matching deflected shapes of the same eight-story structure led him to use $\alpha = 0.5$ in one direction of the structure and 0.22 in the other. The difference between the α values is related to the panel shape, in which the long span/ short span ratio = 1.57 for the model considered. The larger α was in the longer span direction, where a greater portion of the narrower slab width was effective. The trends with panel shape are consistent with those found by Pecknold and Allen and Darvall. These papers contain information on a wide range of panel shapes and column sizes relative to panel size. Figure 4.19 shows the relationships between effective width, αl_2, and l_2 for a panel with a span of l_1 = 6.0 m and square columns, plotted using data from Allen and Darval. The important factor to note is that the effective width does not increase much as the transverse span increases, although the value of α changes greatly. For the smallest values of l_2, the effective width occupies most of the physical width, and the effective width increases somewhat for the square panel. However, there is little additional increase as the transverse span l_2 becomes larger than l_1.

An even more important factor is the state of cracking in the members of the structure. The slabs would usually be cracked more extensively than the columns, and lightly reinforced slabs would be expected to lose a significant amount of stiffness upon cracking. The reductions in stiffness that have to be accounted for will be found to be much more severe in seismic (reversed loading) cases than in wind or other more or less static lateral load cases. This question has been studied by Morrison[4.32] and is outside the scope of the present discussion.

This differential state of cracking is undoubtedly the source of ACI 318-95, Sec. 10.11.1(b), which lists the following reduced moments of inertia that may be used when investigating column length effects:

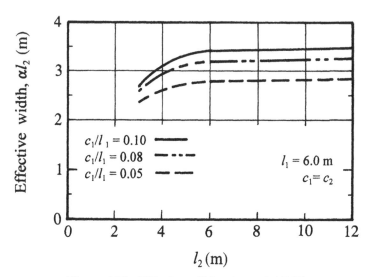

Figure 4.19 Effective width, αl_2 vs. l_2.[4.26,4.31]

Beams: $0.35I_g$
Columns: $0.70I_g$
Slabs: $0.25I_g$

Although these relative stiffness values were not necessarily intended for use with the frame analysis of an unbraced slab and column structure, this has apparently been done, perhaps widely. The $0.25I_g$ value may be thought of as a 50% reduction in I, which reflects the effective width concept, and another 50% reduction, which reflects the expected state of cracking. This probably is not far off for square panels but needs a modification to account for the differing equivalent widths of different panel shapes and column sizes. Rather than taking the blanket $0.25I_g$ value, this should be replaced by something involving the α noted earlier, so that the I to be used in the frame analysis, to one significant figure on the cracking reduction, becomes

$$I = 0.5\alpha I_{\text{slab}}$$

where I_{slab} is the gross moment of inertia of the slab–beam member. The 0.5 reduction may be very optimistic about the loss in stiffness at cracking, except that the greatest concentration of cracking in the lateral loading case will be near the columns, where the reinforcement ratios are also highest. Thus, the frame to be considered for lateral load may have the member stiffnesses shown in Fig. 4.20. The columns should be assumed to be rigid through the thicknesses of the floor slabs.

A remaining issue to be considered is the problem of distribution of the lateral load between the various parallel frames that may be sliced from a building. The interior and exterior frames are likely to have different stiffness characteristics. Figure 4.21 shows a scheme, suggested by Vanderbilt and Corley,[4.33] which imposes equal lateral deflections on the various frames. This is equivalent to an assumption that the floor system forms a rigid diaphragm. The lateral loads are then shared in proportion to lateral load stiffness, in a reasonable manner that is not too demanding of computational effort.

4.4   DISTRIBUTION OF MOMENTS ACROSS SECTIONS

Once the section moments have been determined, using either the direct design or equivalent frame methods, the distribution of these moments across the sections has to be found before the design, including the selection of reinforcement, can be completed. The distributions may be made following Sec. 13.6.4 of the 1995 Code in the case of slabs meeting the restrictions of the direct design method, and also for slabs analyzed using the equivalent frame method if the slab panels have aspect ratios of 2.0 or less and meet

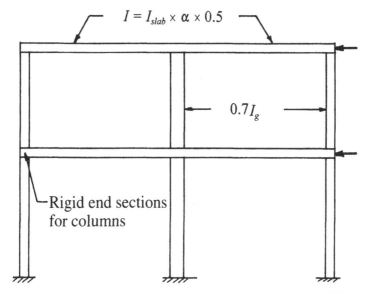

Figure 4.20 Member stiffnesses for lateral loading cases.[4.26]

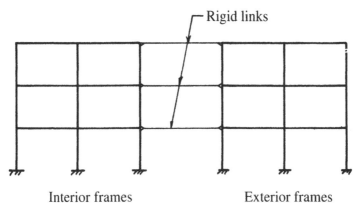

Interior frames   Exterior frames

Figure 4.21 Enforcing same lateral deflection, interior and exterior frames.[4.26]

the restrictions on relative beam stiffnesses in the two directions given for the direct design method.

The distributions at positive- and interior negative-moment sections are based on the theoretical elastic moments which are plotted versus the beam stiffness parameter $\alpha_1(l_2/l_1)$ in Figs. 3.6 to 3.10. In the ACI Code formulation, the curves were replaced by bilinear functions which were fairly good representations of the curves and were easy to describe in mathematical or tabular forms. The curves themselves were not used, since curves are subject to potential reading errors and hence are not suitable for use in a legal document such as the code.

The code moment distributions for positive and interior negative moments are plotted in Figs. 4.22 to 4.24, along with the theoretical moments, for a square interior panel and for both spans of a rectangular interior panel with an aspect ratio of 2.0. It can be seen that the bilinear relationships are fairly good representations. The largest significant differences occur when $\alpha_1(l_2/l_1)$ is about 1.0, where the code values for beam moments are generally too high.

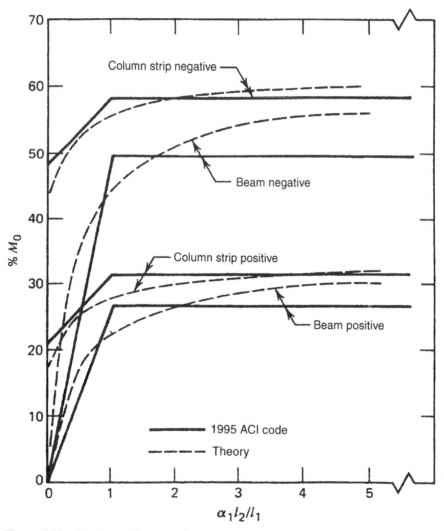

Figure 4.22   Design and theoretical moments versus beam stiffness for interior panels with $l_1/l_2 = 2.0$.

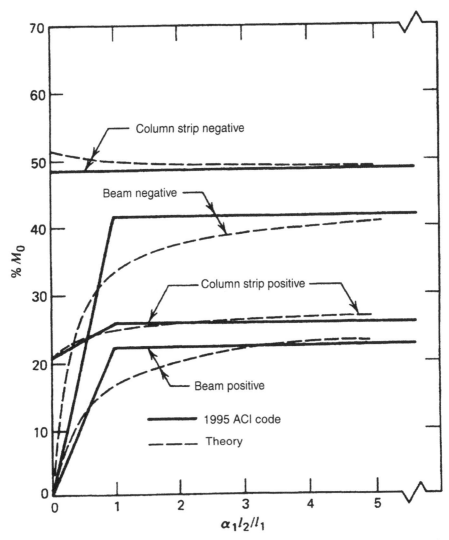

Figure 4.23 Design and theoretical moments versus beam stiffness for interior panels with $l_1/l_2 = 1.0$.

When $\alpha_1(l_2/l_1) \approx \frac{1}{2}$, the code beam moments are generally low, but such beams are generally too small to be practical, since they introduce most of the costs of beam formwork without any significant benefits of beams. Within the context of the total design moment for the panel, these differences are not large.

The code assigns the percentages of the total section design moments to the column strips shown in Tables 4.2, 4.3, and 4.4. The moments not assigned to the column strips are automatically assigned to the middle strips. The

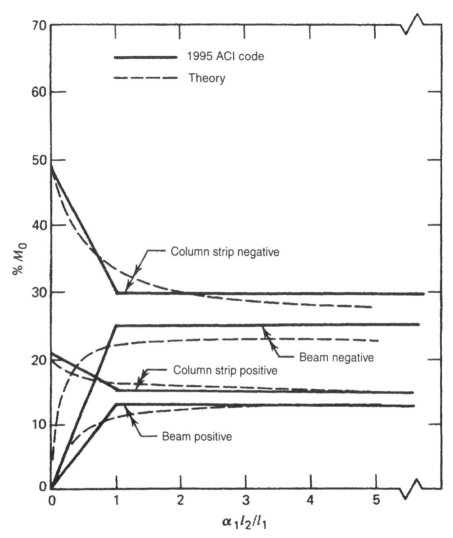

Figure 4.24   Design and theoretical moments versus beam stiffness for interior panels with $l_2/l_1 = 0.5$.

column strip moment includes the beam moment, with the division between beam and slab as determined later. The positive-moment and interior negative-moment distribution is a function only of the flexural stiffness of the beam α_1 and the aspect ratio l_2/l_1. The exterior negative-moment distribution is a function of the torsional stiffness of the edge beam, β_t; the flexural stiffness of any interior beam, α_1, framing into the edge column; and the aspect ratio l_2/l_1. Linear interpolations may be made between the values in each of Tables 4.2 to 4.4. These tables are summarized in Figs. 4.25 and 4.26, in which the

TABLE 4.2 Percentage of Positive Moment Assigned to Column Strip

l_2/l_1	0.5	1.0	2.0
$\alpha_1 \dfrac{l_2}{l_1} = 0$	60	60	60
$\alpha_1 \dfrac{l_2}{l_1} \geq 1$	90	75	45

Source: After Ref. 4.2.

percentage of moment assigned to the column strip is plotted against l_2/l_1 for both positive and negative sections, and for cases with and without beams.

The distributions of exterior negative moment (Table 4.4) can be related to the studies of edge and corner panel moments reported in Chapter 3, but they cannot be related to particular figures as was done above for interior positive and negative moments. The aim of Table 4.4 is to keep as much of the exterior negative moment as possible in the column strips unless the beam torsional stiffness is extremely high. If there is no edge beam, it becomes almost mandatory to place sufficient reinforcement within a bandwidth of $c_2 + 3h$ (i.e., the width of the column plus the 1.5 times the thickness of the slab on each side of the column) to transfer the design moment from slab to column. See Section 10.3 for more details on this question.

It will be found that in most cases minimum reinforcement requirements will govern in the middle strips of exterior negative-moment sections even in heavily loaded slabs. One important effect of this is to minimize the torsional moments that can be developed in the spandrel beams. Torsional distress had occurred at high overload levels in several of the slabs tested in the University of Illinois program. In addition, the decrease in torsional stiffness accompanying torsional cracking of reinforced concrete beams is so great that it virtually reduces the edge to a simply supported condition. Creep appears to reduce torsional stiffness considerably more than it reduces flexural stiffness,[4.34] which would lead to further reduction in torque and middle strip moments.

TABLE 4.3 Percentage of Interior Negative Design Moment Assigned to Column Strip

l_2/l_1	0.5	1.0	2.0
$\alpha_1 \dfrac{l_2}{l_1} = 0$	75	75	75
$\alpha_1 \dfrac{l_2}{l_1} \geq 1$	90	75	45

Source: After Ref. 4.2.

TABLE 4.4 Percentage of Exterior Negative Design Moment Assigned to Column Strip

l_2/l_1	0.5	1.0	2.0
$\alpha_1 \dfrac{l_2}{l_1} = 0$			
$\quad \beta_t = 0$	100	100	100
$\quad \beta_t \geq 2.5$	75	75	75
$\alpha_1 \dfrac{l_2}{l_1} \geq 1$			
$\quad \beta_t = 0$	100	100	100
$\quad \beta_t \geq 2.5$	90	75	45

Source: After Ref. 4.2.

When $\beta_t = 2.5$, the distribution of exterior negative moment is the same as at an interior negative-moment section, as is reasonable from the studies of elastic moments. However, it is very unlikely that any reinforced concrete beam could reasonably have such a high torsional stiffness if it is designed

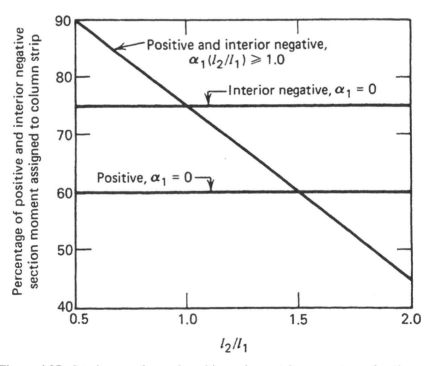

Figure 4.25 Interior negative and positive column strip moments as functions of $\alpha_1(l_2/l_1)$ and l_2/l_1.

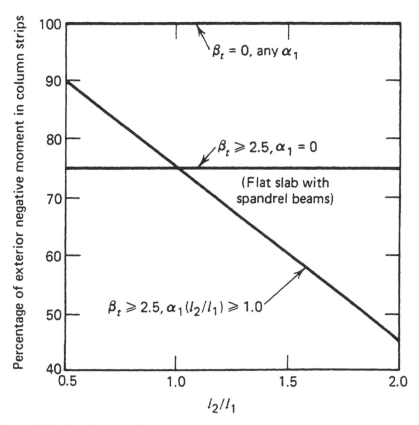

Figure 4.26 Exterior negative column strip moment as function of β_t, $\alpha_l(l_2/l_1)$, and l_2/l_1.

for only the loading from the slab plus the usual weights of curtain-wall construction, so this limit is hypothetical in most respects.

A linear variation of exterior negative moment with β_t is probably not correct, but should be accurate enough for design purposes within the actual range of β_t values likely to be encountered. In his study of slab with edge beams, Jofriet,[4.35] suggested a variation involving the term $\beta_t/(1 + \beta_t)$.

Once the column strip moments have been determined, the beam moments can be found. According to the ACI Code, if $\alpha_1(l_2/l_1) \geq 1.0$, the beam is assigned 85% of the column strip moment, and the slab in the column strip is assigned the remaining 15%. For lower values of beam stiffness, a linear reduction from the beam moment existing when $\alpha_1(l_2/l_1) = 1.0$ to zero is made as the beam stiffness parameter is reduced from 1.0 to zero, as indicated in Figs. 4.22 to 4.24.

A more accurate approximation of the beam moments could be made if the horizontal part of the design curve were given an upward slope with increasing beam stiffness. This is particularly true for the long-span beam moments when the aspect ratio is 2.0, although the horizontal curve is quite satisfactory for the short span of the same panel. Although this step could probably be taken fairly easily, it might add considerably to the complexity of the code, and such a step would have to be considered very carefully to determine the real benefits of such a change. In addition to such a change in the beam moments, a matching change would have to be made in the column strip moments.

Shear forces are required before the beam designs can be completed. The ACI Code specifies that all shear shall be assigned to the beams when $\alpha_1(l_2/l_1) \geq 1.0$. For that case, the shear forces are computed assuming that each beam supports the load on a tributary area bounded by lines drawn from each corner of the panel which make a 45° angle with the panel borders. This calculation is illustrated in Fig. 4.27 for a typical case. The reactions will be modified by the end moments of spans, and calculations of shears due to end moments should be included. For lower beam stiffnesses, the shear forces found from the tributary areas for the stiff-beam case may be reduced linearly to zero as α_1 goes to zero.

The load distributions shown in Fig. 4.27 are very good approximations to the loadings applied to very stiff beams, according to the results of elastic analyses. When $\alpha_1(l_2/l_1) = 1.0$, the total load and reactions are correct, but the loads are uniformly distributed along the lengths of the members instead of being distributed as in Fig. 4.27. Thus, the reactions are the same, but because of the different distributions of load, the shear force reduces unnecessarily slowly as one moves away from the support, perhaps resulting in the use of some extra stirrups. Whether or not the provisions for shear in the case of relatively flexible beams are reasonable depends to some extent on the actual details of column and beam widths, since it would appear necessary that even quite flexible beams carry all the shear if they are of the same width as the columns. However, this could be viewed as a transition between the quite different shear stresses assigned to the concrete in beams (typically, $2\sqrt{f_c'}$ lb/in² [$\sqrt{f_c'}/6$ N/mm²]) and to slabs ($4\sqrt{f_c'}$ lb/in² [$\sqrt{f_c'}/3$ N/mm²]) if without shear reinforcement at slab–column connections. The code correctly warns the designer to be sure to account for the entire load when working with the shears and reactions.

Moments and shears due to the weight of the beam stem and any loads applied directly to the beams must be added to those caused by the slab loading. This is an approximation to the real situation, since a load applied to a beam will always produce some slab forces unless the beam is rigid, but it is a reasonable approximation and will produce appreciable errors only when the beam relative stiffness is quite low.

Example 4.3. The design example shown in Fig. 4.6 will be continued to determine the distribution of the moments to the various sections of the slab. Section moments already found from the equivalent frame method will be

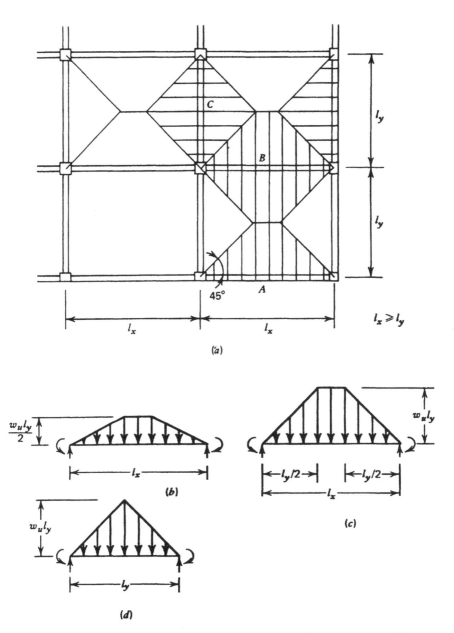

Figure 4.27 Determination of loadings for computation of beam shears in slabs having stiff beams: (a) tributary loading areas for computation of beam shears; (b) loading on beam A; (c) Loading on beam B; (d) loading on beam C.

considered; those from the direct design method could be handled in exactly the same way. The section design moments are shown in the upper left quadrant of Fig. 4.28.

Two additional stiffness parameters are needed. The spandrel beam flexural stiffness ratio is, from Eq. 3.1,

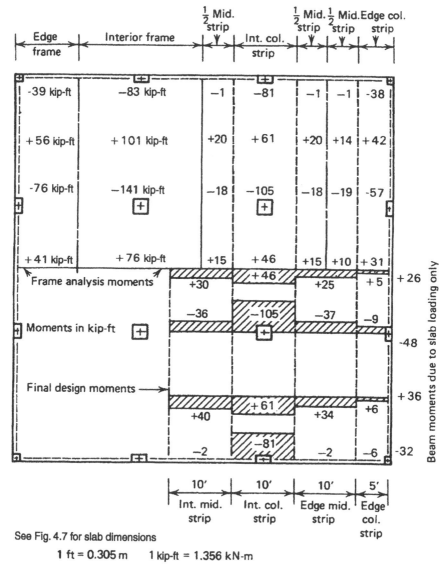

See Fig. 4.7 for slab dimensions

1 ft = 0.305 m   1 kip-ft = 1.356 kN-m

Figure 4.28 Final design moments for slab determined using the equivalent frame analysis method.

$$\alpha_1 = \frac{EI_{\text{beam}}}{EI_{\text{slab}}}$$

The beam section is shown in Fig. 4.12a, and $I_{\text{beam}} = 8082$ in⁴ (3.364×10^9 mm⁴). A slab of 120 in. (3.05 mm) width is considered, and $I_{\text{slab}} = 4219$ in⁴ (1.756×10^9 mm⁴). Assuming that the same concrete is used in slab and beam, $\alpha_1 = 8082/4219 = 1.92$. This is a reasonably stiff beam.

The spandrel beam torsional stiffness, for the same cross section, is, from Eq. 3.4,

$$\beta_t = \frac{E_{cb}C}{2E_{cs}I_{\text{slab}}}$$

where E_{cb} and E_{cs} are the Young's modulus values for the beam and slab concretes, respectively. The torsional constant is $C = 3647$ in⁴ (1.518×10^9 mm⁴) as earlier, and I_{slab} is for a width of slab equal to the span of the beam, 240 in. (6.10 m). For this case, $\beta_t = 3647/(2 \times 8438) = 0.22$. With these two stiffness parameters, the design moment distributions can be found using either Tables 4.2 to 4.4 or Figs. 4.25 and 4.26.

Consider first the interior frame. There is no interior beam (i.e., $\alpha_1 = 0$). Figure 4.25 indicates that 75% of the interior negative moment is column strip moment, with the remainder in the middle strips. Of the positive moments, 60% is in the column strips. At the exterior negative-moment section, Fig. 4.26 may be used to determine the distribution. When $\beta_t = 0$, 100% goes to the column strip, and 75% is so assigned when $\beta_t = 2.5$, since $\alpha_1 = 0$. For this slab $\beta_t = 0.22$, and the linear interpolation can be expressed as

$$100 - (100 - 75)\frac{\beta_t}{2.5} = 100 - 2.2 = 97.8\%$$

of the exterior negative moment assigned to the column strip. The resulting interior frame moments are shown in the upper right quadrant of Fig. 4.28. The middle strip has been divided into two half-strips, one on each side of the column strip. Half the middle strip moment is assigned to each.

The edge frame moments are handled in the same manner. Since $\alpha_1(l_2/l_1) = 1.95 \times 20/20 > 1.0$, 75% of the moment is assigned to the column strip (slab plus beam) at both positive and interior negative sections. At the corner of the structure, 97.8% of the exterior negative moment is assigned to the column strip, found from the type of interpolation used above. Note that for a square panel, the magnitude of α_1 does not affect the distribution of exterior negative moment. The resulting moments are also shown in the upper right quadrant of Fig. 4.28.

Because of the influence of the edge beam on the moment distributions, the moments contributed to the two halves of the edge middle strip by the edge and interior frames are different. Rather than design this middle strip as two substrips, the total moment may be assumed to be uniformly distributed over the width of the strip.

The spandrel beam moments are 85% of the edge column strip moments, because $\alpha_1(l_2/l_1) > 1.0$. The slab in the column strip is designed for the remaining 15% of the moment. Minimum reinforcement will generally govern the slab in the column strip when the beams are stiff. The final design moments for the slab and beams are shown in the lower right section of Fig. 4.28, both graphically and numerically.

The design moments may be modified, at the discretion of the designer, before the selection of steel. According to Sec. 13.6.7 of the *direct design method* of the 1995 Code, any moment may be increased or decreased by 10% as long as compensating changes at other sections are made so that the total static design moment for each panel is not reduced. Moments may be decreased to decrease steel congestion in some areas. Column strip positive moments may be decreased and middle strip moments increased to obtain a more nearly uniform spacing of slab bars. Column strip negative moments may be increased to take advantage of the extra depth of a drop panel and thus to decrease the total amount of steel.

Curiously, the same freedom of 10% redistribution of moments does not exist for the *equivalent frame method* within the limits of the code language. There is no reason why this small amount of redistribution should not be as acceptable with one method as with the other. The 10% redistribution allowed in the code was an arbitrary value, and in some cases much greater moment changes would be satisfactory. However, a reduction of more than 10% in the interior negative column strip moment in a beamless slab could cause the initiation of yielding of reinforcement at loads near the service load level, as discussed below. On the other hand, assuming the positive moments to be the same in both column and middle strips of the same beamless slab would undoubtedly produce an acceptable structure, even though this would entail a 25% increase in the middle strip moment and a 16.7% decrease in the column strip moment. This is outside the scope of the code and could be done only if justified to the satisfaction of the building official concerned.

Section 8.4 of the ACI Code also allows a redistribution of design moments by 10 to 20% in some cases, depending on the reinforcement ratio. Some designers apparently have interpreted the code to endorse using both redistribution clauses additively for slabs, but this clearly was not intended during the preparation of the slab provisions. Any liberalization of the 10% redistribution would probably have to be accompanied by definite restrictions on the amount of moment that could be distributed away from the areas around the columns of beamless slabs. The 1995 (and 1971) ACI Code allows the interior negative moments in the column strips of beamless slabs to be uniformly distributed over the full width of the strip. This is a change from both earlier codes (1956[4.6] and 1963[4.5]) and from the standard practice of some design firms, as shown by the slabs described in Refs. 4.4 and 4.9. The 1963 Code had some not-very-restrictive requirements on bar spacings, where at least

25% of the column strip steel was required to be within the width $c_2 + 2d$, where c_2 is width of column and d is the effective depth of the slab, and half the column strip steel was required to cross the drop panel if one existed. The first limitation required a steel concentration only for quite small columns. In the example slab of Fig. 4.28, $c_2 + 2d$ is slightly greater than 0.3 of the column strip width, so this limitation would not have controlled even if the provision had been retained in the code. The requirement that half the steel cross the drop panel appears superfluous on first examination, since the drop panel width is normally about two-thirds of the column strip width. However, the drop panel depth is taken into account when the reinforcement is selected, since it affects the internal lever arm, and if the moment is assumed to be uniformly distributed and an extra-thick drop panel used, the computed reinforcement requirement for the area over the drop panel might become relatively small. The requirement simply set a limit on how far this trend could be pursued, although the economics of the use of reinforcement would seem to encourage the concentration of steel over the drop panel in order to gain the benefit of the increased effective depth of the section.

The 1956 ACI Code[4.6] had a potentially more restrictive requirement on reinforcement distribution, but it was made a function of the shear stress at a critical section surrounding the column. For the maximum allowable shear stress to be used, at least half the column strip steel had to be within the width $c_2 + 2d$. For the slabs described in Refs. 4.4 and 4.9, the steel spacing over the column was half that in the outer areas of the column strip, apparently because of this requirement.

By the time the 1963 Code[4.5] was prepared, enough beamless slab shear specimens had been tested to demonstrate rather conclusively that concentration of the reinforcement over the column did not increase the shear strength, and the requirement was consequently dropped. However, concentration of reinforcement near the columns may be necessary when large unbalanced moments are to be transferred between the column and slab. A complete review of the shear strength data for slabs up to 1961 was published by Moe.[4.36] All recent information on transfer of shear, and shear and moment, between slabs and columns is presented and examined in Chapter 10 of this book. It also became apparent that the spacing requirements in the 1963 Code were only controlling in the case of a very few slabs, and these requirements were dropped from the later codes.

It may reasonably be argued that concentration of reinforcement over or near the columns would be helpful from the serviceability point of view. The elastic moments around columns, especially near the corners of the columns, are extremely high (theoretically infinite just at the corner), and consequently the first bars would be expected to reach yield stress at a relatively low load level when the steel is uniformly distributed. This, in turn, could cause relatively large crack widths in the immediate vicinity of the column.

Jofriet[4.37] studied this problem using a finite element analysis that took into account the effects of cracking on stress distributions and recommended that the column strip be subdivided into inner and outer sections. The inner strip, half the column strip width, was recommended to be assigned two-thirds of the total strip moment, with the remaining one-third in the outer sections. Following this recommendation, the interior negative column strip moment of 105 kip-ft (146 kN-m) acting over a 10-ft (3.05-m) width would be divided into 70 kip-ft (95 kN-m) for the central 5-ft (1.52-m) width plus 17.5 kip-ft (24 kN-m) for each 2.5-ft (0.76-m) quarter strip width lying outside the central portion.

Moe's tests[4.36] confirm that this degree of steel concentration can be assumed to increase the load or moment at first yielding of steel. On the basis of very few tests of simply supported square slabs loaded through short column stubs, it appears that a 10 to 15% increase in load at first yield might be expected. In these simple specimens, concentration of reinforcement near the columns decreased the deflections slightly, but this increase in stiffness would probably not be detectable in a more complex multipanel beamless slab floor. In addition, yielding of the first few bars did not affect the trends of the load–deflection curves, and in some cases not the load–steel strain curves either. Unfortunately crack widths were not reported.

Construction problems can be cited as an argument against excessive concentration of the steel over the column. If the columns are heavily reinforced, it may become difficult to obtain adequate compaction of the concrete in the slab near the column, and this difficulty is compounded if the horizontal slab (and beam) reinforcement is also closely spaced in the two orthogonal directions. High-quality concrete is probably more important to the shear strength than is the steel distribution, and concrete should not be jeopardized for small steel spacings. A spacing of slab bars of less than perhaps 4 in. (100 mm) center to center should be adopted only cautiously and after determination that construction and inspection procedures will ensure adequate concrete compaction.

The question of whether or not to concentrate some of the reinforcement over the column is a typical example of the conflict between design approaches based on the elastic and plastic analysis methods. Elastic theory seems to require concentration; plastic theory indicates that it does not matter as far as strength is concerned. Test results are not particularly conclusive in resolving this question, as they indicate only minor benefits for reasonable concentrations, which must be weighed against the potential difficulties introduced on the job site by close bar spacings.

According to the code, if a structure that has been analyzed by the equivalent frame method has relative beam stiffnesses in the two directions that meet the restrictions imposed on the direct design method, the distributions of moments across the various design strips may be those given for the direct design method. The panel aspect ratio limitation of 2.0 imposed on the direct design method is not mentioned, although it appears that this limit should also be imposed.

If a structure analyzed by the equivalent frame method does not meet the required relative beam stiffnesses, the designer is left without guidance from the code concerning the distributions of the moments. However, some information can usually be gained from the trends of the material, given in the code, in Chapter 3 of this book, and in some of the references cited in that chapter. The total section moments are known from the frame analysis, which greatly simplifies the task. From the point of view of strength, the most important factor is to supply a total moment capacity that is at least equal to the required. The lateral distribution of the reinforcement is of consequence as far as serviceability is concerned, but the total quantity is of much more significance since it affects safety.

4.5 DIRECT DESIGN METHOD

4.5.1 Negative–Positive Distribution of Moments

The basic negative–positive distribution of moments adopted for an interior panel in the ACI Code direct design method is a negative moment of $0.65M_0$ and a positive moment of $0.35M_0$, where M_0 is the static moment. For the end spans, the distribution of the static moment is controlled by the relative stiffness of the exterior equivalent column, that is, by the stiffness of the edge column as reduced by the influence of the torsional flexibility of the spandrel beam. The following recommendations are from the 1971 and 1977 ACI Codes. Later editions of the code replaced these equations with tabulated values of the three moments for several arbitrary exterior support conditions.

The distribution of end span moments is essentially that resulting from a single cycle of distribution of the fixed-end negative moments at the edge of the structure plus the carryover to the first interior negative–moment section. These moments are expressed as

$$\text{interior negative moment} = M_0 \left[0.75 - \frac{0.10}{1 + (1/\alpha_{ec})} \right] \tag{4.9}$$

$$\text{positive design moment} = M_0 \left[0.63 - \frac{0.28}{1 + (1/\alpha_{ec})} \right] \tag{4.10}$$

$$\text{exterior negative moment} = M_0 \frac{0.65}{1 + (1/\alpha_{ec})} \tag{4.11}$$

where α_{ec} is the equivalent relative stiffness of the exterior column, K_{ec}/Σ $(K_b + K_s)$, K_{ec} the flexural stiffness of an equivalent column (moment per unit rotation), K_b the flexural stiffness of beam (moment per unit rotation), and K_s the flexural stiffness of slab (moment per unit rotation). The computation of the flexural stiffness of the equivalent column was described in Section 3.1 and was discussed further in Section 4.3.

The derivation of the exterior negative-moment expression will help in understanding Eqs. 4.9 to 4.11. To a first approximation, the total moment in the exterior (equivalent) columns above and below a joint, which equals the exterior negative slab system moment, is the fixed-end moment (FEM) times the moment distribution factor to the equivalent columns. This may be written as

$$\Sigma \text{ equivalent column moment } = \text{ exterior negative moment}$$

$$= \frac{K_{ec}}{K_{ec} + \Sigma(K_b + K_s)} \text{ FEM}$$

Dividing both top and bottom of the term by K_{ec} gives

$$\text{exterior negative moment } = \text{ FEM } \frac{1}{1 + [\Sigma(K_b + K_s)/K_{ec}]}$$

In this expression, $\Sigma(K_b + K_s)/K_{ec} = 1/\alpha_{ec}$, which with the FEM put equal to $0.65M_0$ leads directly to Eq. 4.11.

The moment distributions are shown graphically in Fig. 4.29. If the negative moments on the two sides of a support are not equal, the sections on both sides of the support are to be designed for the larger moment unless an analysis, taking into account the relative stiffness of both spans and the supports, is made to determine the distribution of the unbalanced moment. The critical section for negative moment is taken as the line through the face of the column and extending perpendicular to the span being considered, as shown in Fig. 3.1 and as line A–A in Fig. 4.1b.

Example 4.4. The design moments for panels I and III of the slab shown in Fig. 4.2 will be computed to illustrate the process. For panel I, the positive and negative moments are simply $0.35M_0$ and $0.65M_0$, respectively, giving the numerical values of 71 and 132 kip-ft (96 and 179 kN-m) for the $M_0 = 203$ kip-ft (275 kN-m) determined previously. For panel III, the moment computations require the determination of α_{ee} for the edge column. The calculations for K_{ec} and K_s are given in detail in Section 4.3.

The Commentary to the 1977 Code,[4.17] Sec. 13.6.3.3, gave some simplifications that may be used when finding α_{ec} for use with the direct design method. Although these are reasonable simplifications, they should be used carefully and only after the designer is satisfied that they are appropriate for the type of structure being considered. In Section 4.3 it was found that for the example slab of Fig. 4.2, $K_{ec} = 263E$ (kip-in./rad). Also, $K_s = 145E$ (kip-in./rad), where E has units of kips/in.², computed considering a nonprismatic slab section to take into account the column thickness. For small values of c/l, taking the stiffness as $4EI/l$, as in a prismatic section, is also satisfactory, giving $K_s = 140E$ for this particular slab. Thus, $\alpha_{ec} = K_{ec}/\Sigma(K_s + K_b) = 263E/145E = 1.81$. Note that $K_b = 0$ for panel III since there is no interior beam in the l_1 direction.

$M_0 = 211$ kip-ft (286 kN-m) for the end span, as determined previously, and the design moments are computed using Eqs. 4.5 to 4.7.

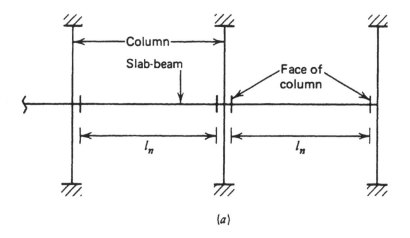

(a)

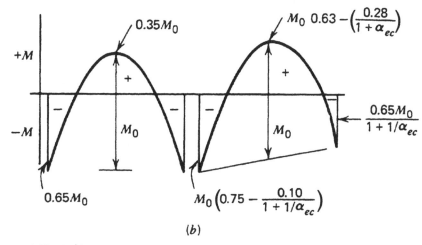

(b)

Figure 4.29 Positive–negative distribution of moments according to direct design method: (a) basic structure; (b) moment diagrams for end and interior spans.

$$\text{interior negative moment} = 211 \left[0.75 - \frac{0.10}{1 + (1/1.81)} \right]$$

$$= 145 \text{ kip-ft } (197 \text{ kN-m})$$

$$\text{positive moment} = 211 \left[0.63 - \frac{0.28}{1 + (1/1.81)} \right]$$

$$= 95 \text{ kip-ft } (129 \text{ kN-m})$$

$$\text{exterior negative moment} = 211 \left[\frac{0.65}{1 + (1/1.81)} \right]$$

$$= 88 \text{ kip-ft } (119 \text{ kN-m})$$

The end span interior negative moment is greater than the interior span negative moment and therefore controls at the interior column line unless an analysis is made to distribute the difference between it and the interior span negative moment. The difference is not large in this case.

If an analysis is made to distribute the unbalanced moment at the interior supports, the interior span positive moment may be reduced by the increase in the interior span negative moment. The end span positive moment must be increased by half the reduction in the end span interior negative moment so that the full static moment capacity, M_0, is maintained.

If the unbalanced moment is not distributed, some benefit of the higher interior span negative moment can still be obtained. The 1995 Code, Sec. 13.6.7, allows an arbitrary 10% change in any design moment as long as the total static design moment for the panel considered does not fall below M_0. Consequently, the interior span positive moment could be reduced from 71 kip-ft to 64 kip-ft, a change of 7 kip-ft, since the negative-moment capacity will be $145 - 132 = 13$ kip-ft more than the computed amount of 132 kip-ft because the end span interior negative moment, 145 kip-ft, controls. From the point of view of limit analysis, the interior span positive moment could be reduced by the entire 13 kip-ft, but this is a greater change than is allowed by the code unless supporting calculations demonstrate that plastic hinging can occur as required and that the serviceability is adequate.

A design negative moment of $0.65\,M_0$ for interior spans was selected for use by the 1971 Code for a number of reasons, some of which have more recently been shown to be not completely correct. In an elastic interior panel with column width/span ratio $c/l = 0$, the negative moment is $\frac{2}{3}M_0$, which agrees well enough with the design value selected. However, real slabs have $c/l > 0$, and Figs. 3.15 and 3.16, plus the work reported by Jofriet,[4.35] indicate a significant reduction in the percentage of M_0 that should be assigned to the negative-moment regions as c/l increases. This reduction was overlooked in the formulation of the design procedure at least partially because the first proposals for the 1971 Code used the Nichols equation, Eq. 4.3, for the computation of the static moment. As noted in Chapter 3, if M_0 is expressed in terms of Eq. 4.3, the negative moment/positive moment ratio is indeed about 2 and is insensitive to the value of c/l. When the code proposal was changed to Eq. 4.1, the clear span concept, the influence of this change on the moment distribution was not realized.

The earliest proposals for the code had incorporated a recommendation that the negative moment be $0.60M_0$ and the positive $0.40M_0$, but these values were not related to the influence of c/l. They had been picked for two reasons: (1) cracking begins first in the more highly stressed negative-moment regions, resulting in some changes in relative stiffnesses and a small redistribution of moment from negative to positive moments at service load levels; and (2) the positive-moment sections are considerably more sensitive to increases in moments due to partial loadings. The positive design moment was increased from

$0.33M_0$ to $0.40M_0$ in compensation for these effects. The distribution used in the code was finally adopted because of concern about high steel stresses near the columns in beamless slabs if the lower negative moment was used. As good average values, negative and positive moments of $0.60M_0$ and $0.40M_0$, respectively, appear to be better than those used in the 1971 Code. Making the fraction of M_0 assigned to either section a linear function of c/l would be a relatively simple additional step.

The use of the stiffness of the equivalent column, K_{ec}, rather than the combined stiffness of the actual columns, ΣK_c, is an unfortunate complication of what is essentially a quite simple method of obtaining design moments. However, the use of ΣK_c instead of K_{ec} in slabs with either no beams or only spandrel beams results in exterior negative column strip moments which are considerably too large.[4.38] The correction of this problem required either some empirical reduction of ΣK_c or of the exterior negative moment, or the use of a more nearly rational analysis, and consequently, the equivalent column concept was added to the direct design method. When there are reasonably stiff beams perpendicular to the edge of the structure, there will not be large differences between moments computed using ΣK_c or K_{ec}. However, final proportioning should always be done using the moments computed using the equivalent column concept.

The distribution of the total section bending moments found at the negative- and positive-moment regions of the slab system to the column and middle strips of the panels, and to the beams (if any), is described in Section 4.4.

The ACI Codes after 1977 omitted the use of the stiffness calculations of Eqs. 4.9 to 4.11 in the determination of the end-span moments and replaced the calculation with a table having five arbitrary classifications of support conditions. The classifications are:

1. Exterior edge unrestrained
2. Slab with beams between all supports

For slabs without beams between interior supports

3. Without edge beams
4. With edge beams
5. Exterior edge fully restrained

This classification is too simple to be rationally useful, since there are no limitations on relative sizes or lengths of members, nor is there a distinction between top stories, which have columns only below the slab, and intermediate stores with columns both above and below the story being considered. The use of this table cannot be recommended, as the potential errors can be large.

For panel III of Fig. 4.2, which has an edge beam but no interior beams, the table gives the exterior negative moment as $0.30M_0$. This leads to exterior negative moment = $0.030 \times 211 = 63$ kip-ft (85 kN-m), as opposed to the

88 kip-ft (119 kN-m) found using the relative stiffness calculation. Assuming that the 88 kip-ft moment is reasonably accurate, the 63 kip-ft moment is low enough to suggest that serious crack width problems might arise at service load. Other cases may be found which suggest that yielding might occur at the service load, an unacceptable design situation. The moment from the equivalent frame method analysis, Fig. 4.13, is somewhat higher than the 88 kip-ft found from the earlier direct design method.

The commentary to the current ACI Code has retained a reference to the 1977 Code, with a statement that the earlier approach is acceptable.

4.5.2   Requirements for Column Moments and Stiffnesses

The parts of the direct design method that were presented in Section 4.5.1 did not give design moments for interior supports (although exterior supports were treated) and did not account for the possible effects of pattern loadings. The interior column moments are covered in the 1995 Code, Sec. 13.6.9. The effects of pattern loads, which can be limited by adequate column stiffnesses, are no longer considered in the direct design method. The maximum live load was reduced from three times the dead load to twice the dead load, and this significantly reduced the potential changes in positive moments which can be attributed to pattern loads.

Moments in Interior Supports. The 1995 Code, Sec. 13.6.9, instructs the designer to proportion the interior supporting columns and walls for the moments caused by slab loadings and then gives an equation for the interior column moments to be considered unless a more general analysis is used to find the moments. The equation in the 1995 Code is a simplification from that in earlier editions. The earlier equation is necessary if the present equation is to be understood as other than completely empirical.

The equation from the 1971 and 1977 Codes for the total column moment at any particular interior joint was

$$M = \frac{0.08[(w_d + 0.5w_l)l_2(l_n)^2 - w_d'l_2'(l_n')^2]}{1 + (1/\alpha_{ec})} \tag{4.12}$$

where w_d is the design dead load (including the load factor) and w_l = design live load (including the load factor). The primed values, w_d' l_2', and l_n' are for the shorter of the two spans intersecting at the column being considered. The relative column stiffness, α_{ec}, is for the equivalent column described earlier.

The equation is the first cycle distribution to the columns of the approximate fixed-end moments resulting from loading both spans with their design dead loads and the longer of the two spans with 0.5 of the design live load. The 0.08 factor is approximately the $\frac{1}{12}$ factor used in the equation for the fixed-end moment of a uniformly loaded prismatic beam. The $1/[1 + (1/\alpha_{ec})]$

term is the combined moment distribution factor for the columns. The single-cycle moment distribution is less satisfactory for an interior connection than for a edge connection, and the use of α_{ec} introduced a major complication to the calculation.

The moment M as computed must be divided between the columns above and below the slab in proportion to their stiffnesses, as is done in the equivalent frame method (see Section 4.3). The moment and thrust combination found in the columns may or may not control the design of the column section. The thrust at full live load, together with the minimum eccentricity, could be a more critical loading combination than the loading used for Eq. 4.12.

The 1995 Code contains a simplified version of Eq. 4.12. As can be seen in the following equation, the 0.08 factor has been replaced by 0.07, and the column stiffness parameter α_{ec} has been eliminated. Thus, the equation is transformed to

$$M = 0.07[(w_d + 0.5w_l)l_2(l_n)^2 - w_d'l_2'(l_n')^2] \tag{4.13}$$

The interior column moments found from Eq. 4.13 will generally be larger than from Eq. 4.12, but both will be less than those from an equivalent frame analysis of the same structure, at least as long as the live load is large enough to require consideration of pattern loadings. This occurs because the direct design method considers 0.5 of the design live load and the equivalent frame method 0.75 of the design live load as the appropriate partial loading intensities. This inconsistency should be corrected.

An additional clarification is needed for the column moment problem. The moment computed using Eq. 4.12 or 4.13 clearly acts in one span direction only, because of the way the long and short, and loaded and unloaded, spans are specified. A bending moment should also be calculated assuming the short span to be loaded and the long span subjected to dead load only, to determine whether this loading causes controlling moments with opposite signs.

Effects of Pattern Loadings. The direct design method cannot consider pattern loading cases except through indirect processes. The 1977 Code gave the designer two options: (1) provide some certain minimum column stiffness so as to partially isolate one span from the effects of loading on the next span, or (2) increase the positive design moments to take into account the effects of pattern loadings. The intent was to produce a structure in which the average stress in the positive-moment reinforcement could not be increased to more than 133% of the basic (full live load on all spans) service load stress by the effects of partial loadings.

The 1995 Code no longer contains these limitations. The 1995 Code limits the live load to twice the dead load, and this reduction from the three times the dead load made these provisions pointless. The effects of pattern loadings were discussed extensively in Chapter 3, where it was shown that the most important variables influencing the maximum moments due to pattern loadings were the ratio of dead load to live load, β_a, the panel aspect ratio, l_2/l_1, and the beam relative stiffness, in addition to the column relative stiffness ratio, α_c.

Example 4.5. The column moments will be computed for the interior column of the slab of Fig. 4.2 considered earlier. The column moment then becomes, from Eq. 4.13,

$$M = 0.07 \{1.4 \times 0.094 + [(1.7 \times 0.070)/2]\}20$$
$$\times 18.33^2 - 1.4 \times 0.094 \times 20 \times 18^2$$
$$= 30.2 \text{ kip-ft } (41.0 \text{ kN-m})$$

This moment is to be resisted by the columns above and below the slab in direct proportion to their stiffnesses (i.e., $0.5M$ to each column in this case). For the nine-panel slab of Fig. 4.2, this moment acts in both the x- and y-directions simultaneously.

This moment is somewhat larger than the $M = 17.0$ kip-ft (before correction to the top and bottom of the slab) obtained using the equivalent frame method considering the case of all panels loaded (see Fig. 4.13). The equivalent frame analysis gives $M = 45.8$ kip-ft for the pattern loading with three-fourths of the factored live load on the end spans and factored dead load on all spans. This moment is considerably larger than that from the direct design method. Moments and stiffnesses for edge columns, considering bending parallel to the edge of the structure, are computed in the same way.

REFERENCES

4.1   *Building Code Requirements for Reinforced Concrete,* (ACI 318-71), American Concrete Institute, Detroit, 1971, 78 pp.

4.2   *Building Code Requirements for Structural Concrete,* (ACI 318-95) and *Commentary* ACI 318R-95, American Concrete Institute, Farmington Hills, Mich., 1995, 371 pp.

4.3   M. A. Sozen and C. P. Siess, "Investigation of Multi-panel Reinforced Concrete Floor Slabs: Design Methods—Their Evolution and Comparison," *Proc. ACI,* Vol. 60, August 1963, pp. 999–1028.

4.4   S. A. Guralnick, and R. W. LaFraugh, "Laboratory Study of a 45-Foot Square Flat Plate Structure," *Proc. ACI,* Vol. 60, September 1963, pp. 1107–1186.

4.5   *Building Code Requirements for Reinforced Concrete,* ACI 318-63, American Concrete Institute, Detroit, 1963, 144 pp.

4.6   *Building Code Requirements for Reinforced Concrete* ACI 318-56, American Concrete Institute, Detroit, 1956, 73 pp.

4.7   J. R. Nichols, "Statical Limitations upon the Steel Requirement in Reinforced Concrete Flat Slab Floors," *Trans. ASCE,* Vol. 77, 1914, pp. 1670–1681, "Discussion," pp. 1682–1736.

4.8   S. H. Simmonds, "Flat Slabs Supported on Columns Elongated in Plan," *Proc. ACI,* Vol. 67, December 1970, pp. 967–975.

4.9   D. S. Hatcher, M. A. Sozen, and C. P. Siess, "Test of a Reinforced Concrete Flat Plate," *J. Struct. Div., ASCE,* Vol. 91, No. ST5, October 1965, pp. 205–231.

4.10 D. S. Hatcher, M. A. Sozen, and C. P. Siess, "Test of a Reinforced Concrete Flat Slab," *J. Struct. Div., ASCE,* Vol. 95, No. ST6, June 1969, pp. 1051–1072.

4.11 J. O. Jirsa, M. A. Sozen, and C. P. Siess, "Test of a Flat Slab Reinforced with Welded Wire Fabric," *J. Struct. Div., ASCE,* Vol. 92, No. ST3, June 1966, pp. 199–224.

4.12 W. L. Gamble, "Moments in Beam Supported Slabs," *Proc. ACI,* Vol. 69, March 1972, pp. 149–157.

4.13 W. L. Gamble, M. A. Sozen, and C. P. Siess, "Tests of a Two-Way Reinforced Concrete Floor Slab," *J. Struct. Div., ASCE,* Vol. 95, No. ST6, June 1969, pp. 1073–1096.

4.14 M. D. Vanderbilt, M. A. Sozen, and C. P. Siess, "Tests of a Modified Reinforced Concrete Two-Way Slab," *J. Struct. Div., ASCE,* Vol. 95, No. ST6, June 1969, pp. 1097–1116.

4.15 M. P. Van Buren, "Staggered Columns in Flat Plates," *J. Struct. Div., ASCE,* Vol. 97, No. ST6, June 1971, pp. 1791–1797.

4.16 W. G. Corley and J. O. Jirsa, "Equivalent Frame Analysis for Slab Design," *Proc. ACI,* Vol. 67, No. 11 November 1970, pp. 875–884.

4.17 *Commentary on Building Code Requirements for Reinforced Concrete,* ACI 318-77, American Concrete Institute, Detroit, 1977, 132 pp.

4.18 S. H. Simmonds, and J. Misic, "Design Factors for the Equivalent-Frame Methods," *Proc. ACI,* Vol. 68, No. 11, November 1971, pp. 825–831.

4.19 H. Cross, *The Column Analogy,* Bulletin 215, University of Illinois Engineering Experiment Station, Urbana, Ill. 1930, 68 pp. Reprinted in *Arches, Continuous Frames, Columns, and Conduits,* Selected Papers of Hardy Cross, University of Illinois Press, Urbana, Ill., 1963, 265 pp.

4.20 *Handbook of Frame Constants,* Portland Cement Association, Skokie, Ill., 1958, 33 pp.

4.21 H. Cross and N. D. Morgan, *Continuous Frames of Reinforced Concrete,* Wiley, New York, 1932, 343 pp.

4.22 E. S. Hoffman, D. P. Gustason, and A. J. Gouwens, *Structural Design Guide to the ACI Building Code,* Chapman & Hall, New York, 1998, 437 pp.

4.23 P. F. Rice, "Practical Approach to Two-Way Slab Design," *J. Struct. Div., ASCE,* Vol. 99, No. STl, January 1973, pp. 131–143.

4.24 *CRSI Handbook,* 8th ed., Concrete Reinforcing Steel Institute, Schaumberg, Ill., 1996.

4.25 A. Zweig, "Design Aids for the Direct Design Method of Flat Slabs," *Proc. ACI,* Vol. 70, April 1973, pp. 285–299.

4.26 W. L. Gamble, Plane-Frame Analysis Applied to Slabs, *The Design of Two-Way Slabs,* ACI SP-183, American Concrete Institute, Farmington Hills, Mich., 1999, pp. 119–129.

4.27 M. T. Cano and R. Klingner, "Comparison of Analysis Procedures for Two-Way Slabs," *ACI Struct. J.,* Vol. 85, No. 6, November–December 1988, pp. 597–608.

4.28 M. D. Vanderbilt, "Equivalent Frame Analysis for Lateral Loads," *J. Struct. Div., Proc. ASCE,* Vol. 105, No. ST10, October 1979, pp. 1981–1998.

4.29 F. R. Khan and J. A. Sbarounis, "Interaction of Shear Walls and Frames," *J. Struct. Div., ASCE,* Vol. 90, No. ST3, June 1964, pp. 285–335.

4.30 D. A. Pecknold, "Slab Effective Width for Equivalent Frame Analysis," *Proc. ACI,* Vol. 72, No. 4, April, 1975, pp. 135–137.

4.31 F. H. Allen and P. LeP. Darvall, "Lateral Load Equivalent Frame," *Proc. ACI,* Vol. 74, No. 7, July 1977, pp. 294–299.

4.32 D. G. Morrison, "Response of Reinforced Concrete Plate-Column Connections to Dynamic and Static Horizontal Loads," Ph.D. thesis, University of Illinois at Urbana–Champaign, 1981, 249 pp. plus appendices. Also issued as *Civil Engineering Studies,* Structural Research Series 490, Department of Civil Engineering, University of Illinois Urbana, Ill.

4.33 M. D. Vanderbilt, and W. G. Corley, "Frame Analysis of Concrete Buildings," *Concrete International,* Vol. 5, No. 12, December 1983, pp. 33–43.

4.34 C. D. Goode, "Reinforced Concrete Beams Subjected to a Sustained Torque," *Struct. Eng.,* Vol. 53, No. 5, May 1975, pp. 215–220.

4.35 J. C. Jofriet, "Analysis and Design of Concrete Flat Plates," Ph.D. thesis, University of Waterloo, 1971, 466 pp.

4.36 J. Moe, *Shearing Strength of Reinforced Concrete Slabs and Footings under Concentrated Loads,* Bulletin D47, Research and Development Laboratories, Bulletin D47, Portland Cement Association, Skokie, Ill., April 1961, 135 pp.

4.37 J. C. Jofriet, "Flexural Cracking of Concrete Flat Plates," *Proc. ACI,* Vol. 70, December 1973, pp. 805–809.

4.38 W. L. Gamble, Discussion of "Proposed Revisions of ACI 318-63: Building Code Requirements for Reinforced Concrete," *Proc. ACI,* Vol. 67, September 1970, p. 700.

5 General Lower Bound Limit Analysis and Design

5.1 INTRODUCTION

Classical plastic theory for plates is founded substantially on work conducted by Prager and Hodge[5.1] and others at Brown University. Limit analysis based on such theory indicates that exact solutions for plates are not always possible. In general, the calculated ultimate (collapse) load lies between two limits, lower and upper bounds, depending on the approach adopted. A rigorous solution of a particular plate would attempt to make the ultimate loads given by the two bounds converge, and thus obtain the exact solution.

For a lower bound solution a distribution of moments for the whole plate is found such that:

1. The equilibrium conditions are satisfied at all points in the plate.
2. The yield criterion defining the strength of the plate elements is not exceeded anywhere in the plate.
3. The boundary conditions are complied with.

The ultimate load of the plate is calculated from the equilibrium equation and the distribution of moments. For a given plate the ultimate load so calculated is either too low or correct. Prager and Hodge[5.1] give proofs of the general lower bound principles. A rigid–perfectly plastic plate is assumed. A summary of the fundamental principles has been given by Crawford.[5.2]

The lower bound method was referred to as *equilibrium theory* by Hillerborg[5.3] and suggested by him in 1956 as a method of design for reinforced concrete slabs. The design method, based on the foregoing principles, is stated as follows: "If a distribution of moments can be found which satisfies the plate equilibrium equation and boundary conditions for a given external load, and if the plate is at every point able to carry these moments, then the given external load will represent a lower limit of the carrying capacity of the plate." Hillerborg's object was to present an ultimate load method of design for slabs that would be simple to apply and give results on the safe side.

It is evident that lower bound solutions give information on possible safe distributions of moments and shears in the slab and loads on the supporting structure. Ways of minimizing reinforcement in slab systems can therefore be found. There are theoretically an infinite number of possible lower bound solutions for a given plate, each based on a distribution of moments that satisfies the requirements. In the strict limit analysis of a given slab one seeks the distribution of moments that gives the highest, and therefore most nearly correct, ultimate load. It is to be noted that the elastic theory distribution of moments fitting within the strength of the slab sections is, in fact, a possible lower bound solution because it satisfies the equilibrium conditions and the boundary conditions. The elastic theory moments are proportional to the curvatures at the slab sections, and therefore only one distribution of elastic theory moments is possible for a given moment capacity. However, in plastic theory any number of moment distributions are possible, because in the plastic range the moments do not depend on the curvature. Wood[5.4] has shown that slabs reinforced to follow the elastic theory moments give surprisingly good minimum steel solutions which will also satisfy serviceability requirements well because the steel stresses at service load will be nearly uniform and not show locally high values.

If a distribution of moments is chosen that is different from the elastic theory distribution, moment redistribution must occur before the ultimate load is reached. In fact, even elastic theory designs will require some moment redistribution unless the moments used to design the reinforcement are based on the complex distribution of stiffnesses present in the slab after cracking of concrete has occurred at the highly stressed regions. Therefore, limit analysis solutions can only be applied to reinforced concrete slabs with reasonably ductile sections.

The moment–curvature relationship for an element of a typical reinforced slab is illustrated in Fig. 5.1. The curve has an approximately trilinear shape, consisting of an initial linear portion up to first cracking of the concrete, then a linear portion up to first yielding of the tension steel, followed by a region where the moment remains practically constant near the ultimate moment of

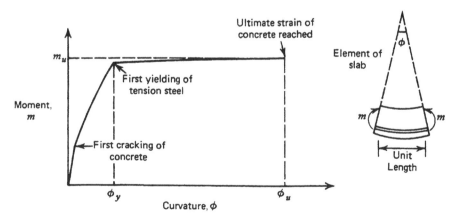

Figure 5.1   Moment–curvature relationship of reinforced concrete element.

resistance until the concrete reaches ultimate strain. For singly reinforced slabs with a concrete compressive cylinder strength of 3000 lb/in² (20.7 N/mm²) and a steel yield strength of 40,000 lb/in² (276 N/mm²), it can be shown[5.5] that for tension steel ratios of 0.02, 0.015, 0.01, and 0.005 the ratio of ultimate curvature to first yield curvature, ϕ_u/ϕ_y, is approximately 4, 6, 10, and 23, respectively, if the ultimate compressive strain of the concrete is 0.004. The presence of compression steel may increase these ϕ_u/ϕ_y values, depending on the depths of tension and compression steel in the section and on the tension steel ratio. Therefore, lightly reinforced concrete slabs can be extremely ductile. Limit analysis of slabs assumes that the sections have sufficient ductility (i.e., the moment–curvature curve has a sufficiently long horizontal branch after first steel yield) to enable the chosen distribution of moments to be reached.

5.2 GOVERNING EQUATIONS FOR GENERAL LOWER BOUND LIMIT ANALYSIS

The equilibrium equation, moment transformation equation, boundary conditions, and yield criterion required for lower bound solutions are discussed in the following sections.

5.2.1 Equilibrium Equation

Consider the equilibrium of a small element with sides dx and dy taken from the slab. The external load acting on the element is w per unit area, as in Fig. 5.2a. The shear forces per unit width, V_x and V_y, the bending moments per unit width, m_x and m_y, and the torsional moments per unit width, m_{xy} and m_{yx}, acting on the faces of the element in the x- and y- directions are shown in Fig. 5.2b and c. The actions on opposite sides of the element are different because in the general case the actions vary throughout the slab. It is to be noted that because complementary shear stresses are equal but opposite, $m_{xy} = m_{yx}$. Positive directions of all actions are as shown in Fig. 5.2.

For equilibrium of the vertical forces:

$$\left(V_x + \frac{\partial V_x}{\partial x}\,dx\right)dy + \left(V_y + \frac{\partial V_y}{\partial_y}\,dy\right)dx - V_x\,dy - V_y\,dx + w\,dx\,dy = 0$$

$$\frac{\partial V_x}{\partial x} + \frac{\partial V_y}{\partial y} = -w \tag{5.1}$$

For equilibrium of moments about a y-direction axis passing through the center of the element (higher-order terms ignored):

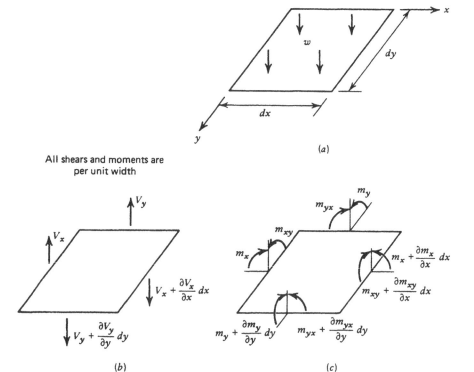

Figure 5.2   Shears and moments acting on small element of slab: (*a*) element and external loads; (*b*) shear forces; (*c*) bending and torsional moments.

$$\left(2V_x + \frac{\partial V_x}{\partial x}\,dx\right) dy\,\frac{dx}{2} - \frac{\partial m_x}{\partial x}\,dx\,dy - \frac{\partial m_{xy}}{\partial y}\,dx\,dy = 0$$

$$\frac{\partial m_x}{\partial x} + \frac{\partial m_{xy}}{\partial y} = V_x \tag{5.2}$$

Similarly, for equilibrium of moments about an *x*-direction axis passing through the center of the element:

$$\frac{\partial m_y}{\partial y} + \frac{\partial m_{xy}}{\partial x} = V_y \tag{5.3}$$

Substituting Eqs. 5.2 and 5.3 into Eq. 5.1 to eliminate V_x and V_y gives

$$\frac{\partial^2 m_x}{\partial x^2} + 2\frac{\partial^2 m_{xy}}{\partial x\,\partial y} + \frac{\partial^2 m_y}{\partial y^2} = -w \tag{5.4}$$

Equation 5.4 is the equilibrium equation for the slab. The equilibrium equation applies whether the slab is in the elastic or plastic range. The equation was derived in Chapter 2 (see Eq. 2.5), but it has been rederived here to emphasize that it is based purely on equilibrium requirements. Every solution of Eq. 5.4 that satisfies the boundary conditions and the yield criterion is a possible configuration of moments as far as lower bound analysis is concerned. Thus, as discussed previously, there are a great number of possible lower bound solutions, of which the elastic theory solution is one. The determination of moment configurations that satisfy Eq. 5.4 and the other two lower bound requirements is difficult because some measure of experience and even intuition is necessary. To obtain possible lower bound solutions using Eq. 5.4, the load w on the right-hand side can be arbitrarily apportioned between the terms

$$-\frac{\partial^2 m_x}{\partial x^2}, \quad -2\frac{\partial^2 m_{xy}}{\partial x\,\partial y}, \quad \text{and} \quad -\frac{\partial^2 m_y}{\partial y^2}$$

That is, the load can be carried by any combination of slab bending and/or twisting in the two directions. This concept gives a good physical insight into the method.

5.2.2 Transformation of Moments to Different Axes

In lower bound solutions it is often necessary to transfer moments found for an x–y system of axes to a general n–t system. The equations can be obtained by summing the components of m_x, m_y, and m_{xy} in the n and t directions to find the bending moments m_n and m_t and torsional moment m_{nt} in those directions. Figure 5.3 illustrates the moments in vector form using the right-hand screw rule. The angle α between the x- and n-axes is measured clockwise from the x-axis. Equilibrium of the moments acting on the element in the n-direction requires that

$$m_n = m_x \cos\alpha \cos\alpha + m_y \sin\alpha \sin\alpha + 2m_{xy} \sin\alpha \cos\alpha$$

$$m_n = m_x \cos^2\alpha + m_y \sin^2\alpha + m_{xy} \sin 2\alpha \tag{5.5}$$

Similarly,

$$m_t = m_x \sin^2\alpha + m_y \cos^2\alpha - m_{xy} \sin 2\alpha \tag{5.6}$$

$$m_{nt} = (m_x - m_y)\frac{\sin 2\alpha}{2} - m_{xy} \cos 2\alpha \tag{5.7}$$

Note that if bending moments m_n and m_t are principal moments (the maximum and minimum moments), the torsional moment $m_{nt} = 0$, which gives

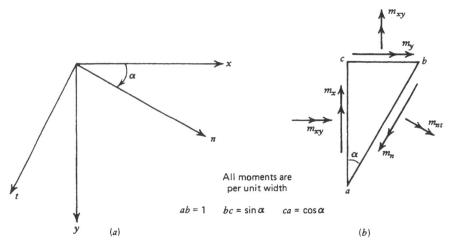

Figure 5.3 Transformation of moments from $x–y$ to $n–t$ systems of axes: (a) axes; (b) bending and torsional moments.

$$\tan 2\alpha = \frac{2m_{xy}}{m_x - m_y} \tag{5.8}$$

To find the principal moments, α from Eq. 5.8 is substituted into Eqs. 5.5 and 5.6.

5.2.3 Boundary Conditions

Boundary conditions were discussed in detail in Section 2.2.5. In summary, the boundary conditions can be expressed as follows:

at a simply supported edge:

y-direction: $m_x = 0$
x-direction: $m_y = 0$

at a free (unsupported) edge:

y-direction: $m_x = 0$, $R_x = 0$
x-direction: $m_y = 0$, $R_y = 0$

At a supported edge the reactions are given by the equations derived in Section 2.2.6 and Eqs. 5.2 and 5.3 as

$$R_x = V_x + \frac{\partial m_{xy}}{\partial y} = \frac{\partial m_x}{\partial x} + 2\frac{\partial m_{xy}}{\partial y} \tag{5.9}$$

at a y-direction edge, and

$$R_y = V_y + \frac{\partial m_{xy}}{\partial x} = \frac{\partial m_y}{\partial_y} + 2\frac{\partial m_{xy}}{\partial x} \tag{5.10}$$

at an x-direction edge. The corner force is

$$R_0 = 2m_{xy} \tag{5.11}$$

where m_{xy} is the torsional moment in the corner.

5.2.4 Yield Criterion

The yield criterion defines the strength of a given slab element subjected to a general moment field. Thus, in the case of slab reinforcement in the x- and y-directions, the yield criterion relates the ultimate moments of resistance per unit width, m_{ux} and m_{uy}, of the slab element in the x- and y-directions to the applied moments per unit width, m_x, m_y, and m_{xy} due to the external loading when the element yields. In the general case the resisting moments in the x- and y-directions will be unequal because the areas of reinforcing steel and the effective depths may be different in those directions. Hence, in general, $m_{ux} \neq m_{uy}$.

Johansen's Yield Criterion. The most commonly used yield criterion is that due to Johansen.[5.6] A line in the plane of the slab about which plastic rotation occurs, and across which the reinforcing bars are yielding, is referred to as a *yield line.* It is assumed that the reinforcing bars in both directions crossing the yield line reach the yield strength. The ultimate moment of resistance about a yield line, which is at some general angle to the reinforcement, is assumed to be due to the components of the ultimate resisting moments m_{ux} and m_{uy} in the directions of the reinforcement. Figure 5.4a shows a slab element subjected to general applied moments with a yield line in direction t at angle α to the y-axis. The actual yield line is assumed to be replaced by a "stepped" yield line consisting of a series of small steps, at right angles to the directions of the reinforcement. Figure 5.4b shows a small triangular element abc at the yield line under the action of the ultimate resisting moments. The ultimate normal resisting moment acting in the n-direction along the yield line is found from consideration of the equilibrium of the element by taking moments of the components of m_{ux} and m_{uy} about ab. This gives

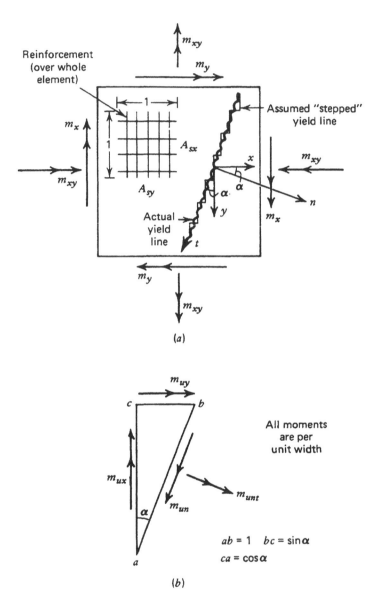

Figure 5.4 Yield criterion for reinforced concrete slab: (*a*) element of slab with general applied moment field and yield line; (*b*) equilibrium of small element of slab at the yield line under ultimate resisting moments.

$$m_{un}ab = m_{ux}ac \cos \alpha + m_{uy}bc \sin \alpha$$

$$m_{un} = m_{ux} \cos^2\alpha + m_{uy} \sin^2\alpha \tag{5.12}$$

It is also evident that equilibrium of the element requires a torsional moment m_{unt} to exist along the yield line in the general case. This torsional moment is found by taking moments in the direction of ab.

$$m_{unt}ab = m_{ux}ac \sin \alpha - m_{uy}bc \cos \alpha$$

$$m_{unt} = (m_{ux} - m_{uy}) \sin \alpha \cos \alpha \tag{5.13}$$

Note that if $m_{ux} = m_{uy} = m_u$, then $m_{un} = m_u(\cos^2\alpha + \sin^2\alpha) = m_u$ and $m_{unt} = 0$. This case of isotropic reinforcement therefore produces equal ultimate normal resisting moments and zero ultimate torsional moments in all directions.

The yield criterion assumes that the element reaches its strength when the normal moment m_n due to the moments m_x, m_y, and m_{xy} from the applied loading becomes equal to the ultimate normal resisting moment m_{un}. That is, with reference to Eq. 5.5, the normal moment yield criterion requires for the element that

$$m_x \cos^2\alpha + m_y \sin^2\alpha + m_{xy} \sin 2\alpha \leq m_{ux} \cos^2\alpha + m_{uy} \sin^2\alpha \quad (5.14)$$

Discussion of Johansen's Yield Criterion. There are a number of aspects concerning the assumptions made or implied in the derivation of Eqs. 5.12 and 5.13 which have caused controversy and led to research attempts to find a better yield criterion. One aspect is that the derivation assumes that the ultimate bending moments m_{ux} and m_{uy} acting in the directions of the reinforcing steel are unaccompanied by any torsional moments in those directions. That is, the ultimate resisting moments m_{ux} and m_{uy} are assumed to be principal moments, and hence no torsion moments exist on sides bc and ac of the element in Fig. 5.4b. This rather intuitive assumption leads to simplification in the derivation of the equations, and test results discussed later appear to show that such torsions, if they do exist, have a negligible effect on the yield criterion. Also, the derivation shows the presence of a torsional moment m_{unt} along the yield line, but this twist is not generally considered, since the yield criterion (Eq. 5.14) is based only on the normal moment. This aspect is discussed further later.

Another aspect is that summing the components of m_{ux} and m_{uy} to find m_{un}, rather than determining m_{un} directly from the components of bar forces in the n-direction leads to slight errors in the internal lever arm of the ultimate moments of resistance. This may be seen as follows. Let A_{sx} and A_{sy} be the areas of tension steel per unit width due to bars in directions x and y, respectively; d_x and d_y be the effective depths of bars in directions x and y, respectively; f_y be the steel yield strength; and f'_c be the concrete compressive strength. Then the ultimate resisting moments per unit width in the x- and y-directions may be written according to standard flexural strength theory[5.5] as

$$m_{ux} = A_{sx}f_y\left(d_x - 0.59A_{sx}\frac{f_y}{f'_c}\right) \tag{5.15}$$

$$m_{uy} = A_{sy}f_y\left(d_y - 0.59A_{sy}\frac{f_y}{f'_c}\right) \tag{5.16}$$

which on substituting into Eq. 5.12 gives

$$m_{un} = A_{sx}f_y\left(d_x - 0.59A_{sx}\frac{f_y}{f'_c}\right)\cos^2\alpha + A_{sy}f_y\left(d_y - 0.59A_{sy}\frac{f_y}{f'_c}\right)\sin^2\alpha$$

$$\tag{5.17}$$

A more accurate approach to find m_{un} would be to determine the resultant bar force T_n per unit width in direction n and to use this force to calculate the depth of concrete in compression along the yield line. Hence,

$$T_n = A_{sx}f_y\cos^2\alpha + A_{sy}f_y\sin^2\alpha = 0.85f'_c a_n \tag{5.18}$$

where a_n is the depth of the equivalent concrete compressive rectangular stress block with mean stress $0.85\,f'_c$ for the concrete force in direction n. The centroid of concrete compression is $0.5a$ from the compression edge of the section, and on this basis

$$m_{un} = A_{sx}f_y\cos^2\alpha\left[d_x - \frac{0.59}{f'_c}(A_{sx}f_y\cos^2\alpha + A_{sy}f_y\sin^2\alpha)\right]$$

$$+ A_{sy}f_y\sin^2\alpha\left[d_y - \frac{0.59}{f'_c}(A_{sx}f_y\cos^2\alpha + A_{sy}f_y\sin^2\alpha)\right] \tag{5.19}$$

Comparison of Eqs. 5.17 and 5.19 shows that the difference in the m_{un} values arises from the difference in the lever arms. However, although Eq. 5.17 is seen to be on the unsafe side, the difference between the m_{un} values is very small in the case of practical slabs. For example, for reinforcement ratios A_{sx}/d_x and A_{sy}/d_y in each direction of less than 0.01, and where $f_y = 40,000$ lb/in² (276 N/mm²) and $f'_c = 3000$ lb/in² (20.7 N/mm²), the maximum difference between the m_{un} values given by Eqs. 5.17 and 5.19 is less than 2%.[5.7]

A further aspect is that the uniaxial compressive strength of the concrete is used in Eqs. 5.17 and 5.19, whereas much of the compressed concrete in the slab is actually in a state of biaxial compression. For example, where the principal moments of the slab are both positive or both negative, the concrete compression zones will be subjected to biaxial compression. Kupfer et al.[5.8] have shown that for equal biaxial compressive stresses the compressive strength of concrete is increased by approximately 16%. This increase in concrete strength in the slab would reduce the depth of compressed concrete by about 16%, but the resulting increase in the internal lever arm of the resisting moment would be much smaller. For example, for a steel ratio of 0.01, and with $f_y = 40,000$ lb/in^2 (276 N/mm^2) and $f_c' = 3000$ lb/in^2 (20.7 N/mm^2), the depth of the equivalent concrete compressive rectangular stress block is $a = 0.157d$, where d is the effective depth of the steel, giving an internal lever arm of $d - 0.157d/2 = 0.922d$. A 16% increase in concrete strength would reduce a to 0.132d, giving an internal arm of 0.934d, resulting in only a 1% increase in moment capacity. Thus, the effect of biaxial compression is small and merely compensates for the small error due to summing the components of m_{ux} and m_{uy} discussed in the preceding paragraph. A more serious case could be where the biaxial concrete stresses are tension in one direction and compression in the other direction. Such a case can exist in some areas of the slab, for example on part of the column lines of a flat plate floor. The investigation by Kupfer et al.[5.8] showed that the compressive strength of concrete can be reduced significantly by the presence of transverse tension. Note, however, that once the concrete cracks in a slab, the transverse tension is relieved. Again, in practical cases the internal lever arm may not be affected significantly by this stress case. Lenkei[5.9] has found in tests that when significant twisting is present along a yield line, decreases in m_{un} of between 3 and 21% were measured, apparently due to the principal concrete stresses of opposite sign in the concrete due to torsion. However, Sozen et al.[5.10,5.11] have reported test results that show no decrease in m_{un} due to torsion at the yield line.

Another aspect is the phenomenon known as *kinking*, which was first discussed by Wood.[5.4] Johansen's yield criterion assumes that the reinforcing bars retain their original direction when crossing the crack in the concrete at the yield line (see Fig. 5.5a). Wood has pointed out that because the crack has finite width, reinforcement that is inclined to the yield line may be dragged across the crack at right angles and kinked, as in Fig. 5.5b. Kinking would cause a reorientation of the bar forces and therefore a change in the ultimate resisting moment at the yield line. For the case of maximum possible kinking shown in Fig. 5.5b, in which the bars are to be bent at right angles to the cracks, m_{ux} now acts as $m_{ux} \cos \alpha$ about the yield line; the $\cos \alpha$ term arises because m_{ux} is spread along side ab rather than side ac of the triangular element. Similarly, m_{uy} now acts as $m_{uy} \sin \alpha$ about the yield line. Hence, for equilibrium of moments about ab,

$$m_{un} = m_{ux} \cos \alpha + m_{uy} \sin \alpha \qquad (5.20)$$

Kinking can cause the ultimate resisting moment at the yield line to increase

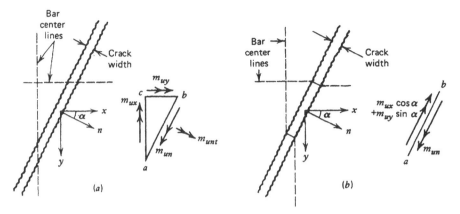

Figure 5.5   Kinking of reinforcing bars at cracks: (*a*) no kinking; (*b*) full kinking.

significantly. For example, if $m_{ux} = m_{uy}$ and $\alpha = 45°$, m_{un} from Eq. 5.20 is 41% greater than m_{un} from Eq. 5.12. However, full kinking obviously cannot occur, because the high bearing stresses created in the concrete when the bars tend to bend means that the bars remain at least partially straight. Also, the crack width is generally much smaller than the bar diameter, and hence the bar straightness will not be greatly affected unless the bar diameters are very small and the crack widths large. Kwiecinski[5.12] has developed a theory to take partial kinking into account in isotropically reinforced ($m_{ux} = m_{uy}$) slabs and concluded from tests[5.13] on slabs with bars at various inclinations to the direction of bending that up to 18% increase in ultimate moment strength occurred due to partial kinking. Wood[5.4] has also reported tests where increases in strength of that order were reported. However, much more extensive test results reported by Lenschow and Sozen,[5.10] Cardenas and Sozen,[5.11] Lenkei,[5.9] and Jain and Kennedy,[5.7] show that the effects of kinking are not significant. Hence, kinking should be ignored in strength calculations.

It should also be pointed out that Johansen's yield criterion is for the case where in-plane (membrane) forces do not exist in the slab. It is well known that if in-plane compressive forces are present, the ultimate resisting moment may be enhanced considerably. This effect is obvious from interaction diagrams commonly derived for columns and may be extremely significant in lightly reinforced slabs, as shown in Fig. 5.6. Also apparent is the reduction in resisting moment due to the presence of in-plane axial tension.

For a given set of applied moments m_x, m_y, and m_{xy}, if the yield line is free to form in any direction, it will form in the direction that results in minimum strength. Lenschow and Sozen[5.10] have found analytically that at such a yield line of minimum resistance not only was the applied normal bending moment m_n equal to the ultimate normal resisting moment m_{un} but also that the applied torsional moment m_{nt} was equal to the resisting torsional moment m_{unt} from Eq. 5.13. Intuitively, one might expect the yield line to

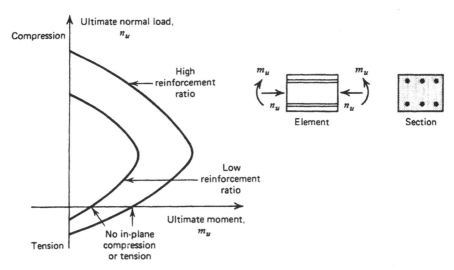

Figure 5.6 Interaction diagram showing combination of moment and axial load at the flexural strength of the section.

form perpendicular to the direction of the applied maximum principal moment. (The determination of the principal moments for a general m_x, m_y, and m_{xy} moment field was discussed in Section 5.2.2.) However, Lenschow and Sozen have shown analytically that this is not the case for nonisotropically reinforced slabs. This finding explains the presence of torsional moments along the yield lines. In the slabs tested by Lenschow and Sozen with $m_{ux} \neq m_{uy}$ and various orientations of steel direction and applied principal moments, deviations of the yield line of up to 20° from the perpendicular to the direction of applied maximum principal moment were measured and confirmed by the analysis. The analysis was also confirmed by the tests conducted by Cardenas and Sozen.[5.11]

In summary, despite considerable criticism and controversy, the ultimate normal resisting moment, Eq. 5.12, used in Johansen's yield criterion, does not appear to require improvement. The weight of experimental evidence, for example,[5.7,5.9–5.11] indicates that the equation is sufficiently accurate for general use when in-plane forces in the slab are not significant.

5.3 ANALYSIS OF SLABS BY GENERAL LOWER BOUND METHOD

Apart from elastic theory moment solutions, only a few general lower bound solutions for slabs with given loading, boundary conditions, and reinforcing steel exist. Wood[5.4] presents some solutions, and solutions for special cases are occasionally published in the literature. The paucity of lower bound solutions is unfortunate because lower bound solutions furnish information on the distribution of bending and torsional moments throughout the slab and the reactions on the supporting system, and ways of minimizing the reinforcement, including bar cutoff points, can be found.

Lower bound solutions are generally found by trial and error using mathematical functions for bending and torsional moments which give likely looking distributions which are checked to ensure that the equilibrium equation, Eq. 5.4, is satisfied, that the yield criterion is not violated, and that the boundary conditions are complied with. Some examples of lower bound analysis will be given below to illustrate the procedure.

Example 5.1. The square slab shown in Fig. 5.7 is simply supported at all edges and is reinforced by bars parallel to the edges. The ultimate positive resisting moments per unit width in the x- and y-directions are equal, $m_{ux} = m_{uy} = m_u$. For a valid distribution of moments, calculate the ultimate uniformly distributed load per unit area w_u, the negative moment strength required, if any, and the reactions at the edges.

SOLUTION. For the moment distribution at ultimate load, try:

$$m_x = m_u\left(1 - \frac{4x^2}{l^2}\right)$$

$$m_y = m_u\left(1 - \frac{4y^2}{l^2}\right)$$

$$m_{xy} = -4m_u\frac{xy}{l^2}$$

Check the boundary conditions:

When $x = \pm l/2$, $m_x = 0$, as required.
When $y = \pm l/2$, $m_y = 0$, as required.

Check the equilibrium equation:

$$\frac{\partial^2 m_x}{\partial x^2} = \frac{\partial}{\partial x^2}\left[m_u\left(1 - \frac{4x^2}{l^2}\right)\right] = -\frac{8m_u}{l^2}$$

Similarly,

$$\frac{\partial^2 m_{xy}}{\partial x\,\partial_y} = -\frac{4m_u}{l^2} \quad \text{and} \quad \frac{\partial^2 m_y}{\partial y^2} = -\frac{8m_u}{l^2}$$

Therefore, from Eq. 5.4,

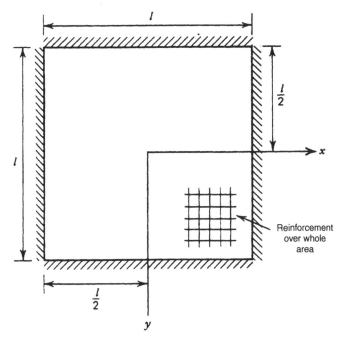

Figure 5.7 Slab of Example 5.1.

$$-\frac{24m_{ux}}{l^2} = -w_u$$

$$w_u = \frac{24m_x}{l^2}$$

Hence, a uniform load is carried as required (i.e., load is not a function of x or y).

Check the yield criterion: From Eqs. 5.5 and 5.12, the applied normal moments and ultimate normal resisting moments are

$$m_n = m_u \left(1 - \frac{4x^2}{l^2}\right) \cos^2\alpha + m_u \left(1 - \frac{4y^2}{l^2}\right) \sin^2\alpha - 4m_u \frac{xy}{l^2} \sin 2\alpha$$

$$m_{un} = m_u \cos^2\alpha + m_u \sin^2\alpha = m_u$$

It is required that $m_n \le m_{un}$ everywhere in the slab.

At midspan, $x = y = 0$, so $m_n = m_u$, as required.
At corners, $x = y = \pm l/2$, so $m_n = -m_u \sin 2\alpha$.

That is, when $\alpha = \pm 45°$, $m_n = \pm m_u$. This indicates, as expected, strong twisting in the corners, which requires isotropic top steel to provide $-m_n$ capacity as well as the existing $+m_u$ capacity, if corner failure is to be prevented. The extent of negative-moment steel in the corners can be calculated by finding the regions of the slab where the condition $m_n \leq m_{un}$ only requires positive-moment capacity (i.e., where the torsional moments become small compared to the positive bending moments).

Reactions: The reaction along the edge where $x = l/2$ may be found from Eq. 5.9:

$$R_x = \left(-\frac{8m_u x}{l^2} - \frac{8m_u x}{l^2} \right)_{x=l/2} = \frac{8m_u}{l}$$

and since $m_u = w_u l^2 / 24$,

$$R_x = -\frac{w_u l}{3}$$

This reaction acts upward on the slab, and because it is not a function of y, it is uniform along the beam. Therefore, a beam, simply supported between the slab corners, with an ultimate moment capacity of $(w_u l/3)l^2/8 = w_u l^3/24$, is required to support the slab at each edge. Note that the total upward reaction on all four beams is $4(w_u l/3)l = (4/3)w_u l^2$, which is $w_u l^2/3$ more than the total uniformly distributed load applied to the slab. This is because "hold-down" forces are necessary in the corners to prevent the slab from lifting there. For equilibrium, each downward corner force must be $R_0 = w_u l^2/12$. R_0 could also be found using Eq. 5.11 as

$$R_0 = 2 \left(\frac{4w_u l^2}{24} \right) \left(\frac{l^2}{4l^2} \right) = \frac{w_u l^2}{12}$$

Note: The ultimate load of the slab is $w_u = 24m_u/l^2$, provided that a isotropic negative-moment bending strength of $-m_u$ exists in the corners as well as an isotropic positive bending strength of $+m_u$ over the whole slab. The ultimate uniform load of $24m_u/l^2$ is, in fact, identical to that obtained for the slab by the upper bound (yield line) method of Chapter 7, again noting the presence of the negative-moment reinforcement at corners of the slab.

Example 5.2. The rectangular slab shown in Fig. 5.8 is simply supported at all edges and reinforced by bars parallel to the edges. The ultimate positive resisting moments per unit width in the x- and y-directions are m_{ux} and m_{uy}, respectively. For a valid distribution of moments, calculate the ultimate uni-

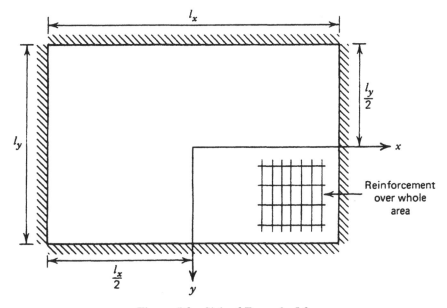

Figure 5.8 Slab of Example 5.2.

formly distributed load per unit area w_u and the negative-moment strength required.

SOLUTION. For the moment distribution at ultimate load, try

$$m_x = 1 - \frac{4x^2}{l_x^2}$$

$$m_y = 1 - \frac{4y^2}{l_y^2}$$

$$m_{xy} = -4m_{xy}\frac{xy}{l_x l_y}$$

where $l_x \geq l_y$ and $m_{ux} \leq m_{uy}$.
Check the boundary conditions:

When $x = \pm l_x/2$, $m_x = 0$, as required.
When $y = \pm l_y/2$, $m_y = 0$, as required.

Check the equilibrium equation:

$$\frac{\partial^2 m_x}{\partial x^2} = \frac{\partial}{\partial x^2}\left[m_{ux}\left(1 - \frac{4x^2}{l_x^2}\right)\right] = -\frac{8m_{ux}}{l_x^2}$$

Similarly,

$$\frac{\partial^2 m_{xy}}{\partial x\,\partial_y} = -\frac{4m_{ux}}{l_x l_y} \quad \text{and} \quad \frac{\partial^2 m_y}{\partial y^2} = -\frac{8m_{uy}}{l_y^2}$$

Therefore, from Eq. 5.4,

$$-\frac{8m_{ux}}{l_x^2} - \frac{8m_{ux}}{l_x l_y} - \frac{8m_{uy}}{l_y^2} = -w_u$$

$$w_u = \frac{8m_{ux}}{l_x^2}\left(1 + \frac{l_x}{l_y} + \frac{m_{uy}}{m_{ux}}\frac{l_x^2}{l_y^2}\right)$$

Hence, uniform load is carried as required, and the solution converges to the previous one (Example 5.1) when $l_x = l_y$ and $m_{ux} = m_{uy}$.

Check the yield criterion: From Eqs. 5.5 and 5.12,

$$m_n = m_{ux}\left(1 - \frac{4x^2}{l_x^2}\right)\cos^2\alpha + m_{uy}\left(1 - \frac{4y^2}{l_y^2}\right)\sin^2\alpha - 4m_{ux}\frac{xy}{l_x l_y}\sin 2\alpha$$

$$m_{un} = m_{ux}\cos^2\alpha + m_{uy}\sin^2\alpha$$

It is required that $m_n \le m_{un}$ everywhere in the slab.

At midspan, $x = y = 0$, so $m_n = m_{ux}\cos^2\alpha + m_{uy}\sin^2\alpha$, as required.
At corners, $x = l_x/2$, $y = l_y/2$, so $m_n = -m_{ux}\sin 2\alpha$.

That is, when $\alpha = \pm45°$, $m_n = \pm m_{ux}$. Hence, isotropic top steel is required in the corners to provide $-m_{ux}$ capacity to avoid corner failure. The extent of negative-moment corner steel can be calculated from the yield criterion equations.

Note: The ultimate load of the slab w_u found above is approximately 10% less than the upper bound (yield line) solution. The reason is that the yield criterion is only reached at $x = y = 0$ (disregarding the corner requirements). Kemp[5.14] has shown that a w_u which is within $1\frac{1}{2}\%$ of the upper bound solution can be obtained if the m_{xy} expression used above is multiplied by $\sqrt{m_{uy}/m_{ux}}$.

A number of other lower bound solutions may be found in the book by Wood.[5.4] In addition, it should be remembered that all elastic theory solutions for slab systems can be regarded as lower bound solutions because they satisfy the equilibrium equation and the boundary conditions. The ultimate load for such solutions is computed by finding the load at which the yield criterion is reached.

5.4 DESIGN OF REINFORCEMENT FOR SLABS IN ACCORDANCE WITH A PREDETERMINED FIELD OF MOMENTS

5.4.1 General Approach

So far in this chapter the analysis of slabs with given reinforcement, loading, and boundary conditions has been considered. In design the procedure is to determine the design moment field m_x, m_y, and m_{xy} and to provide reinforcement to carry those moments. The 1995 ACI Code[5.15] states that "a slab system may be designed by any procedure satisfying conditions of equilibrium and geometrical compatibility if shown that the design strength at every section is at least equal to the required strength . . . , and that all serviceability conditions, including the specified limits on deflections, are met." There is no doubt that use of the elastic theory distribution of moments (discussed in Chapters 2 and 3) would satisfy the code, provided that the minimum thickness or deflection requirements of the code are complied with, since the strength and serviceability requirements would then be met. Crack widths at service load would not be a problem, since at the service load the steel stresses would be well within the elastic range, because the reinforcement would match the actual distribution of moments well. This applies to other lower bound solutions as well, provided that the designer does not deviate significantly from the elastic theory ratios of positive to negative moments and x- to y-direction moments. For an explicit demonstration of serviceability in cases where significant deviations from the elastic theory moments occur, the designer may have to resort to the elastic theory moments to calculate steel stresses at service load to check that crack widths are not too large (see Chapter 9) and to calculate deflections to check that they are not excessive (see Chapter 9), or to rely on the results of laboratory or field tests. In any case the usual deflection checks are required.

To determine the slab reinforcement by the general lower bound approach, the first step would be to determine the required factored (ultimate) load of the slab system. For gravity loads, according to the 1995 ACI Code,[5.15] the factored load is

$$U = 1.4D + 1.7L \tag{5.21}$$

where D is the service dead load and L the service live load. Then the design (ultimate) moments m_x, m_y, and m_{xy} in the slab produced by the factored load can be determined from the moment distribution given by elastic theory or by a general lower bound solution that satisfies equilibrium and the boundary conditions. Reinforcement is then provided for these moments. If the ultimate resisting moment per unit width in a particular direction is to be m_u, the design equation for steel in that direction is [5.5]

$$m_u = \phi A_s f_y \left(d - 0.59 A_s \frac{f_y}{f_c'} \right)$$

where ϕ is the strength reduction factor, taken by the ACI Code as 0.9; A_s the tension steel area per unit width; f_y the steel yield strength; d the effective depth of the tension steel; and f_c' the concrete compressive cylinder strength. Steel ratios A_s/d less than $0.5\rho_b$ should be used to ensure reasonably ductile sections, where ρ_b is the balanced failure steel ratio.[5.5] The effect of compression steel on strength is negligible and hence may be neglected. The minimum amount of steel placed in the slab in the direction of the spans should not be less than that required for shrinkage and temperature reinforcement. According to the ACI Code, this minimum amount is either 0.002 of the gross concrete area if grade 40 ($f_y = 276$ N/mm²) or grade 50 ($f_y = 345$ N/mm²) deformed bars are used, or 0.0018 where grade 60 ($f_y = 414$ N/mm²) deformed bars or welded wire fabric are used, or $0.0018 \times 60{,}000/f_y$ but not less than 0.0014 where reinforcement with $f_y > 60{,}000$ lb/in² (414 N/mm²) is used.

Generally, reinforcing bars are placed at right angles in the x- and y-directions because it is impracticable for the bars to follow the directions of the principal moments over the slab. Determination of the ultimate resisting moments required for a general design moment field m_x, m_y, and m_{xy} presents a problem if torsional moment is present. Generally, designers have ignored the torsional moment m_{xy} because of lack of a method to account for it, but clearly this is unsafe, particularly where twists are high, such as in the corner regions of slabs. The ultimate resisting moments required for a general design moment field including torsion are considered below using a method developed by Wood[5.16] based on early work by Hillerborg[5.17] for reinforcing bars placed in two directions at right angles.

5.4.2 Reinforcement Arranged at Right Angles

The normal moment yield criterion (see Section 5.2.4) requires that at any point in the slab the normal moment per unit width m_n due to the design moments m_x, m_y, and m_{xy} should not exceed the ultimate normal resisting moment per unit width m_{un} in that direction. This criterion should hold in all directions, since yield lines may occur in any direction. The reinforcement in the derivation below will be assumed to be in the directions of the x- and y-axes, as in Fig. 5.9.

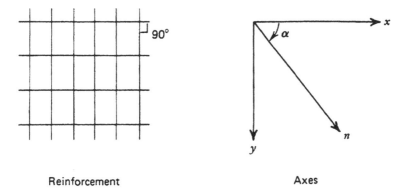

Reinforcement   Axes

Figure 5.9   Orthogonal reinforcement.

Positive-Moment Fields. Reference to Eqs. 5.12 and 5.5 shows that the requirement $m_{un} \geq m_n$ may be expressed as

$$m_{ux} \cos^2\alpha + m_{uy} \sin^2\alpha \geq m_x \cos^2\alpha + m_y \sin^2\alpha + m_{xy} \sin 2\alpha \quad (5.22)$$

which should apply for all values of α. Dividing through by $\cos^2\alpha$ gives

$$m_{ux} + m_{uy} \tan^2\alpha - m_x - m_y \tan^2\alpha - 2m_{xy} \tan\alpha \geq 0 \quad (5.23)$$

If the left-hand side is denoted by $f(\tan \alpha)$, then $f(\tan \alpha)$ is related to the excess of m_{un} over m_n. When this excess is a minimum,

$$\frac{df(\tan \alpha)}{d\alpha} = \frac{df(\tan \alpha)}{d \tan \alpha} \frac{d\tan \alpha}{d\alpha} = \frac{df(\tan \alpha)}{d \tan \alpha} \sec^2\alpha = 0$$

Since $\sec^2\alpha$ cannot be zero, then $df(\tan \alpha)/d \tan \alpha = 0$, giving

$$2m_{uy} \tan \alpha - 2m_y \tan \alpha - 2m_{xy} = 0$$

$$m_{uy} = m_y + \frac{m_{xy}}{\tan \alpha} \qquad (5.24)$$

Also, if $f(\tan \alpha)$ is to represent minimum excess moment, then

$$\frac{d^2 f(\tan \alpha)}{d(\tan \alpha)^2} = 2m_{uy} - 2m_u \geq 0$$

from which $m_{uy} \geq m_u$. Hence, in Eq. 5.24,

$$\frac{m_{uy}}{\tan \alpha} \geq 0 \qquad (5.25)$$

Therefore, if m_{xy} is positive, $\tan \alpha$ must be positive, and vice versa. Substituting m_{uy} from Eq. 5.24 into Eq. 5.23 gives

$$m_{ux} + \left(m_y + \frac{m_{xy}}{\tan \alpha}\right) \tan^2\alpha - m_x - m_y \tan^2\alpha - 2m_{xy} \tan \alpha \geq 0$$

$$m_{ux} \geq m_x + m_{xy} \tan \alpha$$

The minimum requirement is

$$m_{ux} = m_x + m_{xy} \tan \alpha \qquad (5.26)$$

where from the result of Eq. 5.25 if m_{xy} is positive, $\tan \alpha$ must be positive, and vice versa. Hence, Eqs. 5.24 and 5.26 can be rewritten as

$$m_{ux} = m_x + K|m_{xy}| \qquad (5.27)$$

$$m_{uy} = m_y + \frac{1}{K}|m_{xy}| \qquad (5.28)$$

in which $K = \tan \alpha$ is always positive. Apparently, any value of $K = |\tan \alpha|$ can be assumed, since no restriction other than its sign emerges. Thus, a value of K to suit the reinforcement arrangement can be chosen, or alternatively, the values giving minimum steel volume used. If the internal lever arms are the same in both directions, the total volume of reinforcing steel at any point in the slab is proportional to

$$m_{ux} + m_{uy} = m_x + m_y + m_{xy}\left(K + \frac{1}{K}\right)$$

so that for minimum steel volume,

$$\frac{d}{dk}(m_{ux} + m_{uy}) = m_{xy}\left(1 - \frac{1}{K^2}\right) = 0 \qquad \text{whence } K = 1$$

Hence, the most effective arrangement of steel is when

$$m_{ux} = m_x + |m_{xy}| \qquad (5.29)$$

$$m_{uy} = m_y + |m_{xy}| \qquad (5.30)$$

Equations 5.29 and 5.30 allow the bottom reinforcement of the slab to be designed. For example, if at a point in the slab $m_x = 3$, $m_y = 1$, and $m_{xy} = 1$ (moments per unit width in some system of units), reinforcement is required for $m_{ux} = 3 + 1 = 4$ and $m_{uy} = 1 + 1 = 2$.

Negative-Moment Fields. Following through the same reasoning as above, it may be shown that when the moments are negative, the equations become

$$m_{ux} = m_x - K|m_{xy}| \qquad (5.31)$$

$$m_{uy} = m_y - \frac{1}{K}|m_{xy}| \qquad (5.32)$$

in which m_x and m_y are negative and K may have a different value from that used in Eqs. 5.27 and 5.28. Again, the most efficient reinforcement arrangement is when $K = 1$, giving

$$m_{ux} = m_x - |m_{xy}| \qquad (5.33)$$

$$m_{uy} = m_y - |m_{xy}| \qquad (5.34)$$

Equations 5.33 and 5.34 allow the top steel of the slab to be designed. For example, if at a point in the slab $m_x = -3$, m_y, $= -1$, $m_{xy} = 1$ (moments per unit width in some system), reinforcement is required for $m_{ux} = -3 - 1 = -4$ and $m_{uy} = -1 - 1 = -2$.

Note that torsional moments, if large enough relative to the bending moments, may require reinforcement in both the bottom and the top of the slab because m_{xy} occurs in both Eqs. 5.29 and 5.30 and Eqs. 5.33 and 5.34. For example, in a pure torsion field of $m_{xy} = 2$, bottom steel would need to be placed for $m_{ux} = m_{uy} = 2$ and top steel for $m_{ux} = m_{uy} = -2$.

Mixed (Positive and Negative)-Moment Fields. Awkward cases arise when the principal moments are of mixed signs. For example, if $m_x = 2$, $m_y = -1$ and $m_{xy} = -1$, Eqs. 5.29 and 5.30 require that $m_{ux} = 2 + 1 = 3$ and $m_{uy} = -1 + 1 = 0$, whereas Eqs. 5.33 and 5.34 require that $m_{ux} = 2 - 1 = 1$ and $m_{uy} = -1 - 1 = -2$. Clearly, the results are in conflict. This arises because Johansen's yield criterion, $m_{un} = m_{ux} \cos^2\alpha + m_{uy} \sin^2\alpha$, was not intended for use with moments m_{ux} and m_{uy} of different signs. For this case, Wood[5.16] suggests putting either m_{ux} or m_{uy} equal to zero and determining the other moment value. Such an approach is valid because any positive value for K may be used in the equations.

Beginning with Eq. 5.23, let $m_{uy} = 0$. Then, from Eq. 5.24,

$$\tan \alpha = -\frac{m_{xy}}{m_y} = K$$

where either m_{xy} is positive and m_y is negative, or vice versa, so that K is now defined. Then Eq. 5.27 becomes

$$m_{ux} = m_x + \left| \frac{m_{xy}^2}{m_y} \right| \tag{5.35}$$

with $m_{uy} = 0$. Similar equations can be written putting the other ultimate moment values equal to zero.

Rules for Placing Reinforcement.   Design rules based on the foregoing equations, derived by Hillerborg[5.17] and Wood,[5.16] can now be stated. At a point in a moment field where the moments are m_x, m_y, and m_{xy}, the reinforcement should be provided in the slab in the x- and y-directions so that the ultimate resisting moments are as follows.

Bottom Reinforcement.   Generally,

$$m_{ux} = m_x + |m_{xy}|$$

and

$$m_{uy} = m_y + |m_{xy}|$$

If either m_{ux} or m_{uy} is found to be negative, the negative value of moment is changed to zero and the other moment is given as follows: either

$$m_{ux} = m_x + \left| \frac{m_{xy}^2}{m_y} \right| \qquad \text{with } m_{uy} = 0$$

or

$$m_{uy} = m_y + \left| \frac{m_{xy}^2}{m_x} \right| \qquad \text{with } m_{ux} = 0$$

If negative m_{ux} or m_{uy} still occurs, no bottom reinforcement is required. If both m_{ux} and m_{uy} are negative, no bottom reinforcement is required.

Top Reinforcement. Generally,

$$m_{ux} = m_x - |m_{xy}|$$

and

$$m_{uy} = m_y - |m_{xy}|$$

If either m_{ux} or m_{uy} is found to be positive, the positive value of moment is changed to zero and the other moment is given as follows: either

$$m_{ux} = m_x - \left| \frac{m_{xy}^2}{m_y} \right| \qquad \text{with } m_{uy} = 0$$

or

$$m_{uy} = m_y - \left| \frac{m_{xy}^2}{m_x} \right| \qquad \text{with } m_{ux} = 0$$

If positive m_{ux} or m_{uy} still occurs, no top reinforcement is required. If both m_{ux} and m_{uy} are positive, no top reinforcement is required.

Example. For $m_x = 1$, $m_y = 2$ and $m_{xy} = 4$ (moments per unit width in some system). This is an example of strong torsional moments, as at slab corners.

Bottom reinforcement: $m_{ux} = 1 + 4 = 5$ $m_{uy} = 2 + 4 = 6$

Top reinforcement: $m_{ux} = 1 - 4 = -3$ $m_{uy} = 2 - 4 = -2$

Example. For $m_x = 2$, $m_y = -1$ and $m_{xy} = 1$ (moments per unit width in some system).

Bottom reinforcement: $m_{ux} = 2 + 1 = 3$ $m_{uy} = -1 + 1 = 0$

Top reinforcement: $m_{ux} = 2 - 1 = 1$ $m_{uy} = -1 - 1 = -2$

The positive value of m_{ux} for top reinforcement is not permitted, so use $m_{ux} = 0$ and $m_{uy} = -1 - \frac{1}{2} = -1.5$ for that reinforcement. Therefore, no steel is required for strength in the top in the x-direction or in the bottom in the y-direction.

Example. For $m_x = 2$, $m_y = 3$ and $m_{xy} = 1$ (moments per unit width in some system).

Bottom reinforcement: $m_{ux} = 2 + 1 = 3$ $m_{uy} = 3 + 1 = 4$

Top reinforcement: $m_{ux} = 2 - 1 = 1$ $m_{uy} = 3 - 1 = 2$

Both m_{ux} and m_{uy} for top reinforcement are positive, and therefore no top steel is required (the torsional moment is not strong enough to overcome the positive bending moments). Equation 5.22 may be used to find the reinforcement required for each ultimate resisting moment.

In skew slabs it is often convenient to place the reinforcement parallel to the edges of the slab so that the. angle between the steel directions is not 90°. The determination of the required ultimate resisting moments for skew reinforcement to carry a general design moment field m_x, m_y, and m_{xy}, has been determined by Armer.[5.18] The procedure for determining the equations for the ultimate resisting moments in the directions of the bars is analogous to that for bars at right angles, but the derived expressions are lengthy. Reference can be made to Armer's work[5.18] for the design equations. Note that the sign of the torsional moment makes a difference to the ultimate resisting moments required in the case of skew reinforcement, whereas it did not in the case of orthogonal reinforcement. Hence, correct interpretation of signs is essential.

5.5   COMMENT ON GENERAL LOWER BOUND LIMIT DESIGN

It is evident that the determination of design moment fields m_x, m_y, and m_{xy}, by the general lower bound limit design method is difficult and only a few solutions exist other than those obtained by elastic theory. However, the elastic theory moment fields are valid lower bound solutions, and limit design is helpful in indicating how Johansen's yield criterion can be used to determine reinforcement for general moment fields, including torsional moments. Such a design based on the elastic moments will always satisfy strength (safety) requirements, and if the elastic theory moments are used, crack widths will not be a problem, although the normal deflection checks will be necessary. It is evident that the method of section design for a general field of moments could well be included in computer programs for elastic theory moment determination, enabling the computer output to take the form of the required ultimate resisting moments in the x- and y-directions, including the effects of torsion, to allow the direct selection of reinforcement.

REFERENCES

5.1   W. Prager and P. O. Hodge, Jr., *Theory of Perfectly Plastic Solids,* Wiley, New York, 1951.

5.2   R. E. Crawford, "Limit Design of Reinforced Concrete Slabs," Ph.D. thesis, University of Illinois at Urbana–Champaign, 1962, 163 pp.

5.3   A. Hillerborg, "Jämviktsteori för armerade betongplattor," *Betong,* Vol. 41, No. 4, 1956, pp. 171–182.

5.4   R. H. Wood, *Plastic and Elastic Design of Slabs and Plates,* Thames and Hudson, London, 1961, 344 pp.

5.5   R. Park and T. Paulay, *Reinforced Concrete Structures,* Wiley, New York, 1975, 769 pp.

5.6   K. W. Johansen, *Brudlinieteorier,* Jul. Giellerups Forlag, Copenhagen, 1943, 191 pp. (*Yield-Line Theory,* translated by Cement and Concrete Association, London, 1962, 181 pp.)

5.7   S. C. Jain and J. B. Kennedy, "Yield Criterion for Reinforced Concrete Slabs," *J. Struct. Div., ASCE,* Vol. 100, No. ST3, March 1974, pp. 631–644.

5.8   H. Kupfer, H. K. Hilsdorf, and H. Rüsch, "Behavior of Concrete Under Biaxial Stress," *Proc. ACI,* Vol. 66, August 1969, pp. 656–666.

5.9   P. Lenkei, discussion of Ref. 5.10, *Proc. ACI,* Vol. 64, November 1967, pp. 786–789.

5.10   R. Lenschow and M. A. Sozen, "A Yield Criterion for Reinforced Concrete Slabs," *Proc. ACI,* Vol. 64, May 1967, pp. 266–273.

5.11   A. E. Cardenas and M. A. Sozen, "Flexural Yield Capacity of Slabs," *Proc. ACI,* Vol. 70, February 1973, pp. 124–126.

5.12   M. W. Kwiecinski, "Yield Criterion for Initially Isotropic Reinforced Slab," *Mag. Concr. Res.,* Vol. 17, No. 51, June 1965, pp. 97–100.

5.13   M. W. Kwiecinski, "Some Tests on the Yield Criteria for a Reinforced Concrete Slab," *Mag. Concr. Res.,* Vol. 17, No. 52, September 1965, pp. 135–138.

5.14   K. O. Kemp, "A Lower Bound Solution to the Collapse of an Orthotropically Reinforced Slab on Simple Supports," *Mag. Concr. Res.,* Vol. 14, No. 41, July 1962, pp. 79–84.

5.15   *Building Code Requirements for Structural Concrete,* ACI 318-95 and *Commentary,* ACI 318R-95, American Concrete Institute, Farmington Hills, Mich., 1995, 371 pp.

5.16   R. H. Wood, "The Reinforcement of Slabs in Accordance with a Pre-determined Field of Moments," *Concrete,* Vol. 2, No. 2, February 1968, pp. 69–76.

5.17   A. Hillerborg, "Reinforcement of Slabs and Shells Designed According to the Theory of Elasticity," *Betong,* Vol. 38, No. 2, 1953, pp. 101–109. (Translated by Building Research Station, Watford, Hertfordshire, England, 1962. Library Communication 1081.)

5.18   G. S. T. Armer, discussion of Ref. 5.16, *Concrete,* Vol. 2, No. 8, August 1968, pp. 319–320.

6 Design by the Strip Method and Other Equilibrium Methods

6.1 INTRODUCTION

As has been discussed in Chapter 5, the lower bound design method for reinforced concrete slabs suggested by Hillerborg[6.1] in 1956 can be stated as follows: "If a distribution of moments can be found which satisfies the plate equilibrium equation and boundary conditions for a given external load, and if the plate is at every point able to carry these moments, then the given external load will represent a lower limit of the carrying capacity of the plate."

Hillerborg[6.1,6.2] has simplified the general lower bound method for slab design by eliminating the necessity to consider torsional moments when deriving the design moments. If no external load is required to be carried by torsion, the designer can treat the slab as if composed of systems of strips, generally in two directions at right angles, which enables the design bending moments to be calculated by simple statics involving the equilibrium of the strips. This leads to a very attractive design procedure termed the *simple strip method*. A later publication by Hillerborg[6.3] introduced the *advanced strip method,* which features rectangular corner-supported elements, as well as triangular and rectangular edge-supported elements, for use in the design of beamless slabs supported directly on columns and slabs with reentrant corners and openings. The design moments are found by consideration of the equilibrium of the elements. The strip method has also been treated in publications by Crawford,[6.4] Wood and Armer,[6.5–6.7] Kemp,[6.8,6.9] Shukla,[6.10] and others. In particular, Wood and Armer[6.5] have made the simple strip method much more powerful by introducing the concept of *strong bands* to enable beamless slabs supported on columns, and slabs with reentrant corners and openings, to be treated. Hillerborg[6.11] later wrote a book that brought together a great deal of the material. More recently, Hillerborg has written a handbook[6.12] that contains many example slab solutions, considering a wide range of panel shapes, support conditions, and loading conditions. Wiesinger[6.13] has also described an equilibrium method for flat plate floors with regular or irregular column layouts that divides the loading on the plate into triangular segments and, disregarding torsion, calculates the design bending moments using a form of the equilibrium equation.

The strip method has been used in Sweden since around 1960 and is becoming widely recognized elsewhere as an alternative limit design method to the yield line method described in Chapter 7. A feature of the method, in common with yield line theory, is that the distribution of moments between positive and negative moment sections, and between the sections in the two spanning directions, is left to the designer. This freedom is almost embarrassing to designers, and if used unwisely could lead to slabs which, although satisfying strength (safety) requirements, may be unserviceable due to wide cracking or excessive deflections at service load. Also, such freedom may result in extensive moment redistribution being necessary to develop the design (ultimate) load, thus requiring extremely ductile sections which may be available only in very lightly reinforced slabs. Hence, the designer when using the method needs to allocate moments in proportions reasonably close to the distribution given by elastic theory. However, this requirement is not difficult for a designer who has a good feel for the elastic moments, and some guidance is given by rules in codes. For example, the 1997 British code of practice for the structural use of concrete[6.14] explicitly recognizes the yield line or strip methods "provided the ratio between support and span moments is similar to those obtained by the use of elastic theory." This is significantly more restrictive than a predecessor version which recommended that the ratio between negative and positive moments be between 1.0 and 1.5. Generally, solutions for design moments quite close to the elastic theory moments can be obtained, thus ensuring that the steel stresses remain well within the elastic range at service loads and that the redistribution of moments required to reach the design (ultimate) load is not excessive. Some ductility will be required at slab sections, but this can be achieved in design by limiting the steel ratios to ensure that sections are not over-reinforced.

In the following sections we describe the application of the strip and segment equilibrium methods to a range of slabs. In the figures the boundary conditions of the slabs are illustrated using the convention shown in Fig. 6.1.

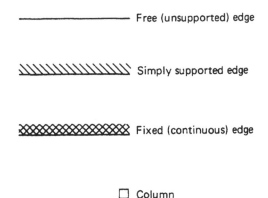

Figure 6.1 Convention for illustrating slab support conditions.

6.2 SIMPLE STRIP METHOD

6.2.1 Strip Action

The equilibrium equation for an element of a slab, as derived in Section 5.2.1, is

$$\frac{\partial^2 m_x}{\partial x^2} + 2\frac{\partial^2 m_{xy}}{\partial x\,\partial y} + \frac{\partial^2 m_y}{\partial y^2} = -w \tag{6.1}$$

where x and y are rectangular axes in the plane of the slab, m_x and m_y the bending moments per unit width in the x- and y-directions, m_{xy} is the torsional moment per unit width in the x- and y-directions, and w the uniformly distributed load per unit area acting on the element. According to the lower bound theory, any combination of m_x, m_y, and m_{xy} that satisfies Eq. 6.1 at all points in the slab and the boundary conditions when the ultimate load is applied is a valid design solution provided that reinforcement can be placed to carry these moments. Thus, the external load w can be apportioned arbitrarily between the terms $\partial^2 m_x/\partial x^2$, $\partial^2 m_{xy}/\partial x\,\partial y$, and $\partial^2 m_y/\partial y^2$.

Hillerborg chooses as his solution to Eq. 6.1 the condition $m_{xy} = 0$ and carries the load entirely by the $\partial^2 m_x/\partial x^2$ and $\partial^2 m_y/\partial y^2$ terms. This means that load is carried entirely by bending in the x- and y-directions, and hence that the slab can be visualized as being composed of two systems of strips running in the x- and y-directions. It is evident that Eq. 6.1 can be replaced by two equations that represent twistless strip action:

$$\frac{\partial^2 m_x}{\partial x^2} = -\gamma w \tag{6.2}$$

$$\frac{\partial^2 m_y}{\partial y^2} = -(1 - \gamma)w \tag{6.3}$$

where γ is a factor chosen by the designer. Normally, γ has a value between 0 and 1. The value of γ chosen can fluctuate throughout the slab without affecting its validity. Note that if $\gamma = 1$, all the load is carried by bending of the x-direction strips, and if $\gamma = 0$, all the load is carried by bending of the y-direction strips.

The use of Eqs. 6.2 and 6.3 to find possible design moment fields will be illustrated for the case of a square, simply supported slab carrying a uniformly distributed ultimate load per unit area of w_u. The slab and the equivalent system of strips is shown in Fig. 6.2. Of the many solutions possible, three that have been considered by Hillerborg[6.11] are presented.

Solution 1, shown in Fig. 6.3, is obtained by putting $\gamma = 0.5$ over the entire area of the slab. That is, one-half of the load is allocated uniformly to the strips in each direction, as indicated by the load dispersion arrows in Fig.

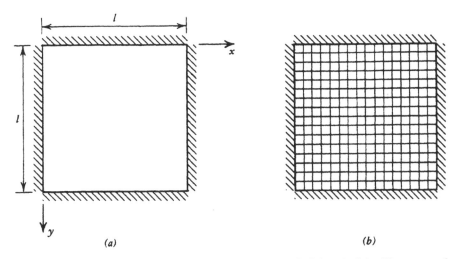

Figure 6.2 Uniformly loaded square simply supported slab: (*a*) slab; (*b*) system of twistless strips.

6.3. The resulting x-direction moments, obtained by simple statics for uniform load per unit area $w_u/2$ on the strips, are shown in the figure. The distribution of y-direction moments is similar to the x-direction moments. Thus, the max-

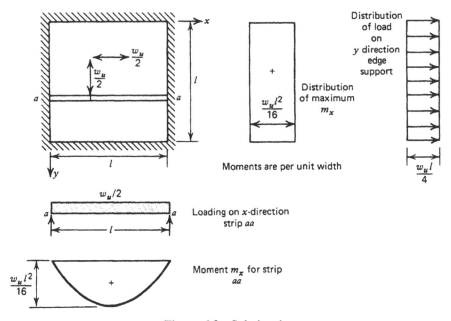

Figure 6.3 Solution 1.

imum moment per unit width in each direction is $w_u l^2 / 16$ and has a constant value at the midspan sections across the slab. The distribution of the loading acting on the edge support of the slab is also shown. This loading is simply the end reactions of the strips and acts on the supporting beam or wall.

Solution 2, shown in Fig. 6.4, is obtained by giving γ values that depend on the region of the slab. The slab is divided into three regions, marked I, II, and III, corresponding to the slab corners, middle edges, and center, respectively. The load is allocated to the strips in each direction within the regions in the manner indicated by the load-dispersion arrows in Fig. 6.4. Two basic types of strip loading exist, shown as strips aa and bb in the figure. The resulting x-direction moments can be obtained by simple statics. The maximum x-direction moments for strips aa and bb are different. The distribution of y-direction moments is similar to the x-direction moments. Thus, the maximum moment per unit width in each direction is $5w_u l^2 / 64$, constant across the middle half of the slab, with a moment per unit width of $w_u l^2 / 64$ in the edge strips. The distribution of the loading acting on the edge supports of the slab is also shown.

Solution 3, shown in Fig. 6.5, is obtained by giving γ values of either 0 or 1, depending on the region of the slab. The load is divided into triangular regions by diagonal lines and is transferred to the nearest support, as indicated by the load-dispersion arrows in Fig. 6.5. Each strip therefore carries a uniform load per unit area of w_u over the end regions. The resulting x-direction moments can be obtained by simple statics. The maximum x-direction moment is a function of y and rises sharply to a peak at the slab center. The distribution of y-direction moments is similar to the x-direction moments. The maximum moment per unit width is $w_u l^2 / 8$. The distribution of the loading acting on the edge supports of the slab is also shown. The triangular shape of the edge load is similar to that assumed by many designers.

The foregoing three solutions illustrate two features of the simple strip method. The first is the ease with which the moments in the slab and the loads on the supporting system can be obtained by the use of simple statics. The second is the variety of moment and load distributions possible depending on the assumed manner of load dispersion. For many years designers have used strip action intuitively to approximate the moments in slabs of awkward shape or boundary conditions. It is of interest to note that such an approach has the full formal backing of lower bound limit design.

It is of interest to look at the relative economy, from the point of view of the reinforcing steel requirements, of the three solutions obtained in Figs. 6.3 to 6.5. The area of steel per unit width is proportional to the moment per unit width. Suppose that all the slab bars run the full length l of the slabs and that all the bars have the same effective depth. Then the value of steel in the slab is proportional to the area of the diagram showing the distribution of maximum m_x. For solutions 1, 2, and 3, these areas are in the ratio 1:0.75:0.67, respectively, indicating the relative economies. Solution 3 is seen to be the most economical, but note that the difference between solution 3 and solutions

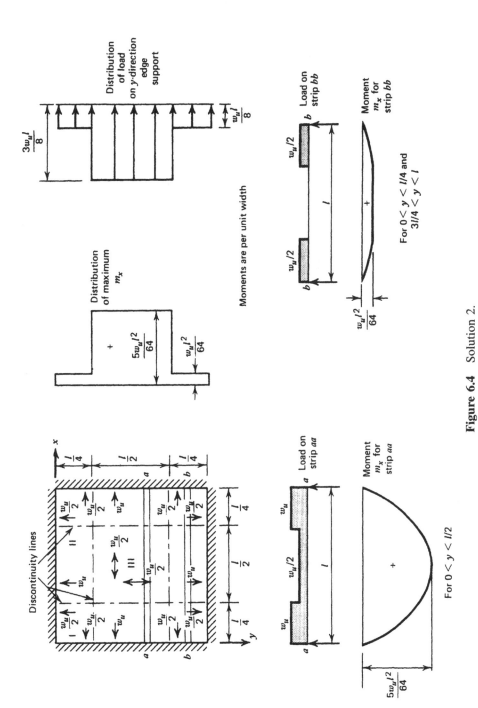

Figure 6.4 Solution 2.

237

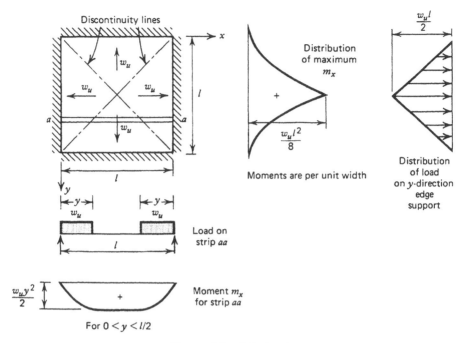

Figure 6.5 Solution 3.

2 and 1 will be less if some bars are cut smaller than length l, as is allowed by the reduction in moment at the ends of the strips. Also, in solution 3 the moment varies continuously across the slab, requiring (theoretically) a continuously variable bar spacing, which is obviously impracticable. Thus, in practice, for solution 3 the bars would need to be placed in several uniform bands to cope with the distribution of moments.

The lines on the slabs that indicate the region of different load dispersion will be referred to as *discontinuity lines*. Solution 1 uses the simplest possible load dispersion assumption. Solutions 2 and 3 offer the alternatives of discontinuity lines originating from either the slab corners or the slab sides. These two possibilities are discussed further in the next two sections.

6.2.2 Discontinuity Lines Originating from Slab Corners

Determination of Moments. In his early work Hillerborg[6.1] favored the approach whereby the strips were usually laid out so as to carry the load to the nearest support. Moments throughout the slab were found for the strips so loaded. Strictly, the discontinuity lines can enter a slab corner at any angle, but the angles are best selected on the basis of the moments giving economy of reinforcing steel and reasonable accord with the elastic moment distribution. The following rules for right-angle corners (see Fig. 6.6) were suggested by Hillerborg:

1. Where two simply supported, or two fixed edges, meet, the discontinuity line should bisect the corner angle.
2. Where a simply supported and a fixed edge meet, the discontinuity line should make an angle with the fixed edge about 1.5 to 2 times the size of the angle with the simply supported edge.

The second rule comes from the observation that a fixed edge attracts moment, and therefore more load should be transferred to it than to the simply supported edge. Another way of looking at it is that the discontinuity lines in fact define where the maximum positive moments are reached in the strips, and it is obvious that a fixed edge causes the positive moment zone to move away from the edge. The foregoing two rules will not lead to slabs with minimum steel volume but give solutions that are reasonably close to minimum steel volume.

As demonstrated previously, the moments in the strips in the two directions can be found by statics. When a fixed edge or edges exist, elastic theory for fixed-end beams can be used to determine the moments, but some moment redistribution would be allowable between the positive- and negative-moment sections. Some examples of the positioning of discontinuity lines for uniformly loaded slabs using this approach are shown in Figs. 6.7 to 6.9. Strips for the determination of moments throughout the slabs in the x- and y-directions are also illustrated. It is of interest to note that Hillerborg[6.11] comments that for a rectangular slab with all edges simply supported (Fig. 6.7), the most economical load dispersion, for straight discontinuity lines originating from the corners, occurs when the angle to the long edge is $\theta = \tan^{-1}(l_x/l_y)$, where $l_x > l_y$. Figure 6.8 illustrates a slab with edges either simply supported or fixed, and the resulting unsymmetrical shape of the regions divided by the discontinuity lines. Figure 6.9 shows a nonrectangular

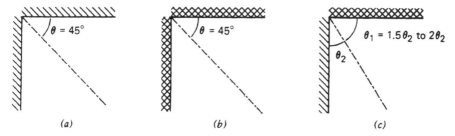

Figure 6.6 Position of discontinuity lines at right-angle corners. (*a*) both edges simply supported; (*b*) both edges fixed; (*c*) one edge simply supported and one edge fixed.

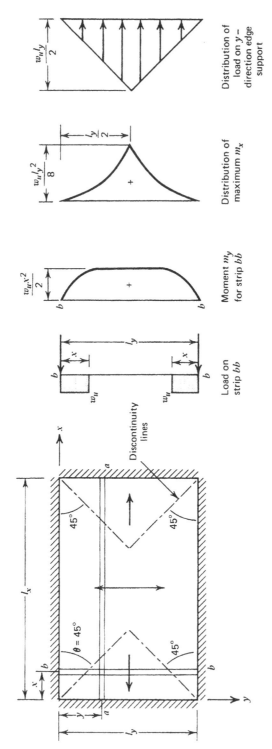

Figure 6.7 Bending moments and edge reactions for a uniformly loaded rectangular slab with all edges simply supported.

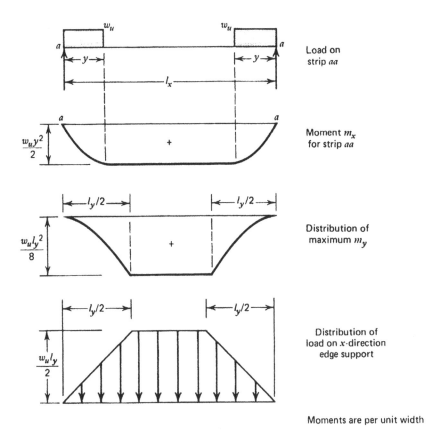

Load on
strip aa

Moment m_x
for strip aa

Distribution of
maximum m_y

Distribution of
load on x-direction
edge support

Moments are per unit width

Figure 6.7 *(Continued)*

slab. In such a case the discontinuity lines can be taken to approximately bisect the corner angles if the edge conditions are similar or follow approximately the rule of Fig. 6.6c if the edge conditions are dissimilar.

The moment diagrams shown in Fig. 6.8 are in general not those that would be calculated for elastic beams fixed at one end and simply supported at the other end and subjected to the loadings shown. The moment diagrams have in this case intentionally been drawn with constant moment in the central unloaded region. This is within the freedom allowed the designer by the strip method and is useful because it greatly simplifies the eventual selection of reinforcement. If both ends of a strip that is nonsymmetrically loaded are fixed, the moment diagram can also be adjusted to give constant moment (or zero shear) within the unloaded length. In this case Hillerborg suggests that the negative moments at the ends of the spans be 1.5 to 2.5 times the span positive moment, which is the same range as the expected elastic moments. The moment diagrams shown in Fig. 6.9 cannot be made to have constant moment regions, because they are statically determinate and are not sym-

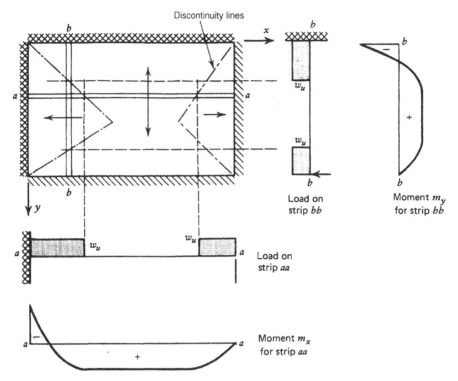

Figure 6.8   Bending moments in a uniformly loaded slab with fixed and simply supported edges.

metrically loaded. The maximum moment is not between the loaded portions, but rather is in the loaded region with the greater length.

Banding of Reinforcement. A problem that arises when reinforcement is being designed for the bending moments obtained from the strip method with discontinuity lines originating from the corners is that over a large proportion of rectangular slabs, and throughout nonrectangular slabs, the bending moments can change rapidly and theoretically require continuously variable bar spacing. Reinforcement to follow such a distribution of moments is obviously impracticable. In such cases Hillerborg suggests placing the reinforcement uniformly in bands of reasonable width, with the design moment for each band taken as the average maximum moment for the strips in the band. A number of bands can be taken across the slab side by side. Design on the basis of such bands is strictly not in accordance with lower bound theory because at the ultimate load the theoretical moments will exceed the ultimate moments of resistance over a part of each band. However, once yielding occurs, it is reasonable to expect the moments to redistribute themselves; also,

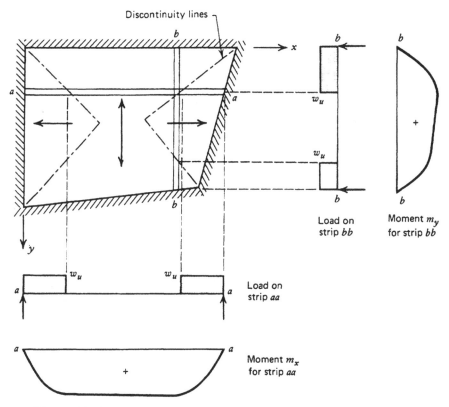

Figure 6.9 Bending moments in a uniformly loaded nonrectangular slab.

the total available ultimate moment of resistance across a band is equal to the required value.

Each band is composed of a number of strips, and in general the end region between the slab edge and the discontinuity line has the trapezoidal shape *abdc* shown in Fig. 6.10 for part of a rectangular slab. Let the width of the band be t and let the discontinuity lines intersect the band symmetrically. The ultimate uniform load per unit area is w_u. Then for a strip at distance z from the edge *ab* of the band,

$$m_{\max} = w_u \frac{(\beta l)^2}{2}$$

where $\beta l = l_1 + (l_2 - l_1)z/t$ and l_1, l_2 are the smaller and longer distances on edge of band from the slab edge to the discontinuity line, respectively. Hence, the average maximum moment per unit width at *bd* on the band is

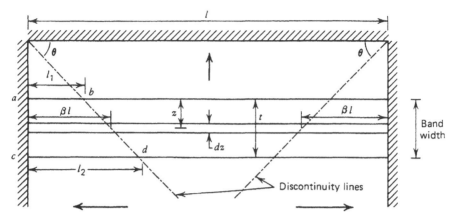

Figure 6.10   Band in a rectangular slab.

$$m_{\text{av max}} = \int_0^t \frac{m_{\max} \, dz}{t}$$

$$= \int_0^t \frac{w_u}{2t} \left[l_1 + (l_2 - l_1) \frac{z}{t} \right]^2 dz$$

$$= \frac{w_u}{2} \left(\frac{l_1 + l_2}{2} \right)^2 \left[1.333 - \frac{1.333}{(l_1/l_2) + 2 + (l_2/l_1)} \right]$$

$$= \text{maximum moment at midstrip per unit width} \times K \qquad (6.4)$$

where

$$K = 1.333 - \frac{1.333}{(l_1/l_2) + 2 + (l_2/l_1)} \qquad (6.5)$$

Values of K given by Eq. 6.5 are shown plotted in Fig. 6.11 and vary from $K = 1.333$ for a triangular shape ($l_1 = 0$) to $K = 1.0$ for a rectangular shape ($l_1 = l_2$).

Equation 6.5 can also be used to find the average maximum moment in a band of a uniformly loaded nonrectangular slab. For example, Fig. 6.12 shows a band crossing a slab of more general shape. This band can best be analyzed by first isolating the narrow strip marked at the center of the band. The loading on the strip is shown in Fig. 6.12b. If the loaded lengths, $\beta_1 l$ and $\beta_2 l$, are the same, Eq. 6.4 can be used to determine the average maximum moment. If they are not the same, the moment diagram will be shaped like that in Fig. 6.12c and the maximum moment in the strip must be found. This maximum moment is then multiplied by the K factor given by Eq. 6.5 to find the average maximum moment for the band, using the l_1 and l_2 values for the end with

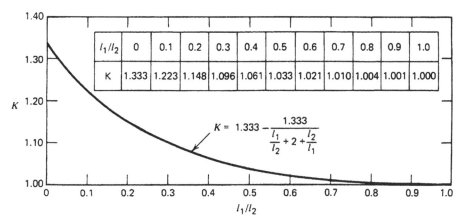

Figure 6.11 K values for average moments in bands.

the greatest loaded length when finding K. Direct use of Eq. 6.4 when β_1 and β_2 are different will result in a maximum moment that is too small, with the error increasing as $\beta_2 l$ diminishes relative to $\beta_1 l$. This is because the maximum moment is within the loaded length $\beta_1 l$, but using the moment at the edge of the loaded length will be satisfactory unless $\beta_2 l$ becomes too small.

As an example, consider the uniformly loaded square simply supported slab shown in Fig. 6.13. Consider the slab to be divided into four bands in each direction of width $l/4$. There are two types of band, marked 1 and 2 in the figure. The average maximum moments in each band for use in design can be calculated as follows:

Band	l_1	l_2	$\dfrac{l_1}{l_2}$	K	Average Maximum Design Moment per Unit Width
1	0	$0.25l$	0	1.33	$\dfrac{w_u}{2}\left(\dfrac{0.25l}{2}\right)^2 \times 1.33 = 0.0104 w_u l^2$
2	$0.25l$	$0.5l$	0.5	1.04	$\dfrac{w_u}{2}\left(\dfrac{0.25l + 0.5l}{2}\right)^2 \times 1.04 = 0.0731 w_u l^2$

The reinforcement placed for these moments can be carried through to the supports. The distribution of these average maximum moments is shown in Fig. 6.13.

As another example, consider the uniformly loaded rectangular simply supported slab shown in Fig. 6.7. In the x-direction consider the slab to be divided into four bands of width $l_y/4$ as in Fig. 6.14. In the y-direction, consider the slab to be divided into four bands of width $l_y/4$, two at each end, and a central

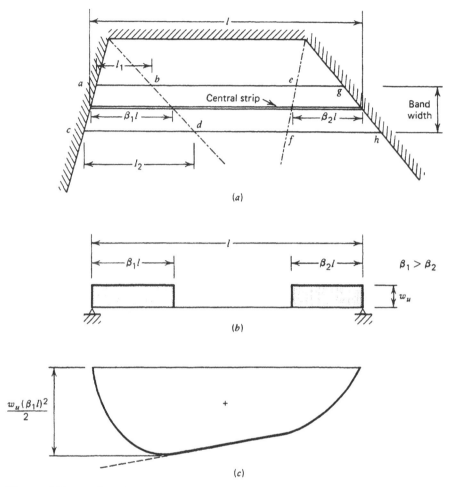

Figure 6.12 (*a*) Band in nonrectangular slab; (*b*) loading on strip at center of band; (*c*) moment diagram for strip at center of band.

band of width $l_x - l_y$. The average maximum moments in each band for use in design can be calculated as follows:

Band	l_1	l_2	$\dfrac{l_1}{l_2}$	K	Average Maximum Design Moment per Unit Width
1	0	$0.25l_y$	0	1.33	$\dfrac{w_u}{2}\left(\dfrac{0.25l_y}{2}\right)^2 \times 1.33 = 0.0104w_u l^2$
2	$0.25l_y$	$0.5l_y$	0.5	1.04	$\dfrac{w_u}{2}\left(\dfrac{0.75l_y}{2}\right)^2 \times 1.04 = 0.0731w_u l^2$
3	$0.5l_y$	$0.5l_y$	1.0	1.00	$\dfrac{w_u}{2}\left(\dfrac{l_y}{2}\right)^2 \times 1.00 = 0.125w_u l^2$

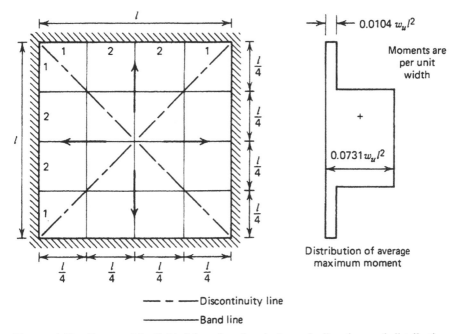

Figure 6.13   Square slab divided into four bands in each direction and distribution of average maximum moments.

The distributions of these moments are shown in Fig. 6.14.

6.2.3   Discontinuity Lines Originating from Slab Sides

There is no reason why the discontinuity lines should originate from the corners or be straight. Wood and Armer[6.5] have pointed out that rather than complicating the calculations by using triangular and trapezoidal shapes for the loaded regions of bands, the discontinuity lines could be drawn to cross each band at right angles and thus allow direct determination of the maximum design moment in the bands without any averaging. In addition to simplifying the calculations, the solution is now exact and in accordance with strict lower bound theory. Such a procedure is illustrated in Fig. 6.15.

A similar approach has been adopted by Hillerborg.[6.11] An example of a simply supported rectangular slab with ultimate uniformly distributed load is shown in Fig. 6.16. For convenience it is reasonable to take four bands across the slab with the edge bands each of width equal to the short span/4 (i.e., $l/4$). The two middle bands taken together, of width $l_y/2$ for the long span and of width $l_x - l_y/2$ for the short span, are referred to as the middle strips in the ACI Code.[6.15] Similarly, the bands at the edge, of width $l_y/4$, are referred to as the column strips by that code. Moments in the x- and y-directions for the slab of Fig. 6.16 resulting from the loading allocated to the strips are shown derived in the figure. The distributed loading acting on the x- and y-

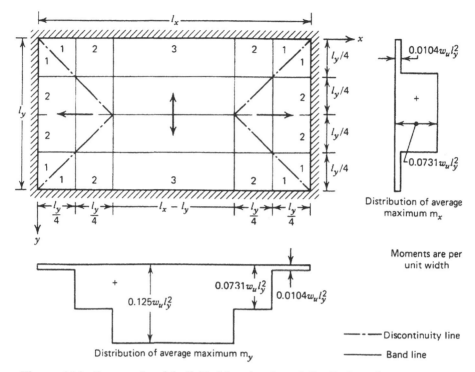

Figure 6.14 Rectangular slab divided into bands and distribution of average maximum moments.

direction edge supports is also shown in the figure. It is evident that calculation of the design moments by this direct procedure is simpler than using the averaging procedure described previously.

It is of interest to compare the steel volumes required for the moments of the uniformly loaded rectangular simply supported slab obtained as in Fig. 6.16 by the direct procedure without averaging and as in Fig. 6.14 by the averaging procedure described previously. Let each reinforcing bar run the full length of its direction and the internal lever arm be the same for the moments in the two directions. Then the volume of steel in the slab is proportional to l_x times the area of the diagram of distribution of maximum m_x plus l_y times the area of the diagram of distribution of maximum m_y. Such calculations show that the ratio of steel volume for the slab obtained by the direct procedure to the steel volume for the slab obtained by the averaging procedure is 1.12, 1.05, 1.01, 0.98, and 0.97 for l_x/l_y = 1.0, 1.25, 1.5, 1.75, and 2.0, respectively. Thus, for square slabs there is some justification for using the solution shown in Fig. 6.14 (gives Fig. 6.13), but for most rectangular slabs the more direct solution shown in Fig. 6.16 is reasonable. Minimum reinforcement requirements will often reduce the differences between the two cases once final designs are prepared.

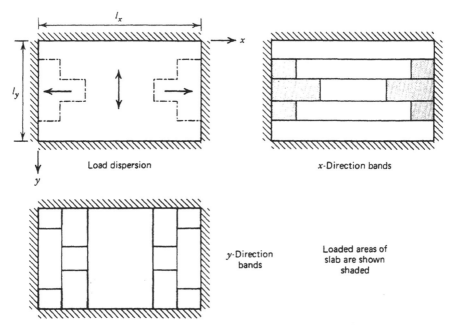

Figure 6.15 Load discontinuity lines giving uniform moments in bands.

An example of a rectangular slab with edges either simply supported or fixed is shown in Fig. 6.17. The discontinuity lines are positioned to take into account the greater load-carrying capacity of fixed edges than of simply supported edges, due to the cantilever behavior of the element of the slab between the fixed edge and the line of contraflexure. The extent of the region between the discontinuity lines is defined by a coefficient β, where β is chosen to give a reasonable distribution of moments. For this example Hillerborg[6.11] suggests taking the middle strip in the x-direction to be of width $l_y/2$ and the middle strip in the y-direction to be of width $l_x - l_y/2$, as for the case of all edges simply supported (Fig. 6.16). It is convenient if the regions of the strips carrying no load are constant-moment zones (i.e., are free of shear). This enables the positive-moment reinforcement to be fully utilized along much of the strip. The moments can be readily calculated assuming zero shear in unloaded regions of strips. Referring to the x-direction middle strip, the right-hand-end reaction is then $\beta w_u l_y/2$, and the maximum positive moment per unit width is therefore $w_u\beta^2 l_y^2/8$. Similarly, this positive moment plus the negative moment at the left-hand support must be equal to $w_u(1 - \beta)^2 l_y^2/8$, and therefore the negative moment per unit width is $w_u(1 - 2\beta)l_y^2/8$. The moments may be found similarly for the x- and y-direction edge strips. For the y-direction middle strip, if the distance from the right-hand end to the section of maximum positive moment is taken as βl_y, the maximum positive moment per unit width must be $w_u\beta^2 l_y^2/2$. The left-hand length of $(1 - \beta)l_y/2$ is

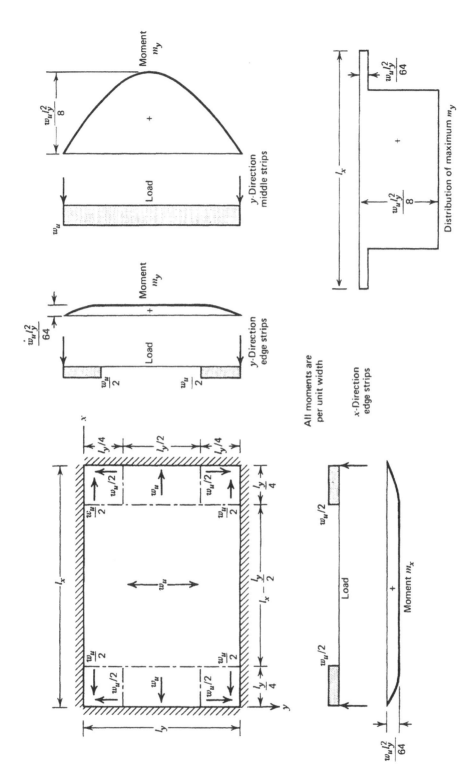

Figure 6.16 Bending moments and edge reactions for a uniformly loaded slab with all edges simply supported.

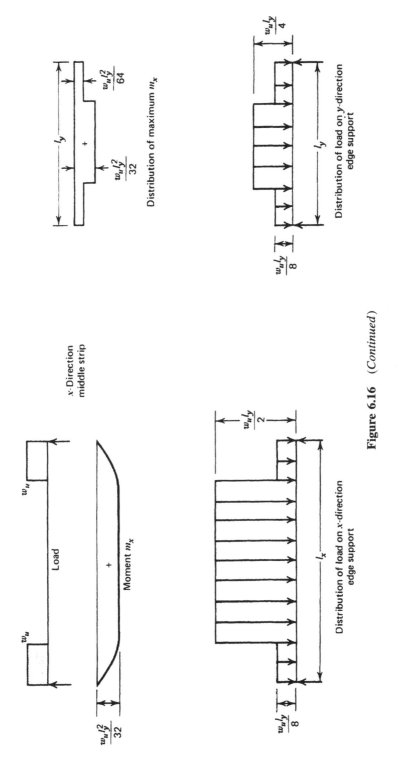

Figure 6.16 (*Continued*)

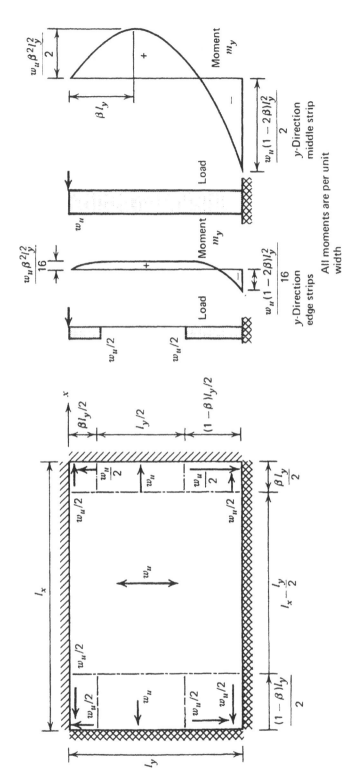

Figure 6.17 Bending moments for a uniformly loaded slab with two adjacent edges fixed and other edges simply supported.

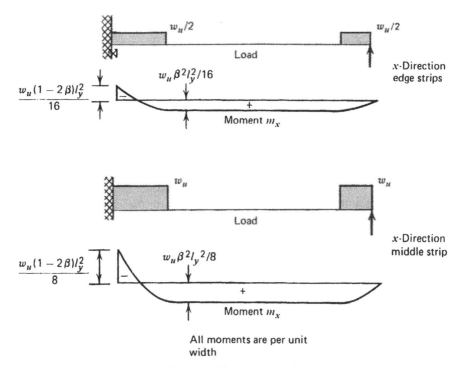

All moments are per unit
width

Figure 6.17 (*Continued*)

analogous to the left-hand length of $(1 - \beta)l_y/2$ of the other strips, and hence the negative moment per unit width is $w_u(1 - 2\beta)l_y^2/2$. The moments are shown in Fig. 6.17. The value of β selected will depend on the desired ratio of maximum negative to maximum positive moments. From the moments in Fig. 6.17 it is evident that the ratio of maximum negative to positive moments is $(1 - 2\beta)/\beta^2$, giving values of 2.0, 1.5, and 1.0 for values of β of 0.366, 0.387, and 0.414, respectively. It would appear that values of β should normally be in the range 0.36 to 0.40. The value of this moment ratio is surprisingly sensitive to the value of β. It is evident that the elastic bending moment distribution has not been used for the fixed-end strip moments, and hence moment redistribution is assumed to take place throughout the slab.

The technique of defining the unequal lengths of the loaded end regions, to obtain a constant positive-moment zone, by a β value that is a function of the desired ratio of the maximum negative to positive moments, as in the example above, is necessary only when the support conditions at each end of the strip are different. If both ends are fixed, or both simply supported as in Fig. 6.16, a value of $\beta = 0.5$ can be adopted. It is to be noted that the ratio of maximum negative to positive moments for the both-ends-fixed case is not a function of β but rather is determined arbitrarily by the designer.

6.2.4 Strong Bands

The simple strip method cannot deal with slabs with openings, reentrant corners, and beamless slabs with column supports without use of strong bands to help distribute the load to the supports. The approach using strong bands was suggested by Wood and Armer[6.5] and adds considerable power to the method. A strong band is a strip of slab of reasonable width that contains a concentration of reinforcement and hence acts as a beam within the slab. Such a slab strip can be thickened if necessary to allow the reinforcing steel in it to better carry the required design moment. The approach using strong bands is also useful for some slabs with free (unsupported) edges to help distribute the moments. The design procedure for these cases is discussed below.

Slabs containing openings can be dealt with by the provision of strong bands around the opening. A rectangular slab, simply supported around all edges, carrying uniformly distributed loading, is illustrated in Fig. 6.18. The strong bands around the opening are marked *aa*, *bb*, *cc*, and *dd*. The discontinuity lines are shown. The slab strips transfer load in the directions shown between two edge supports or between one edge support and a strong band. Strong bands *aa* and *bb* transfer their loading to strong bands *cc* and *dd*,

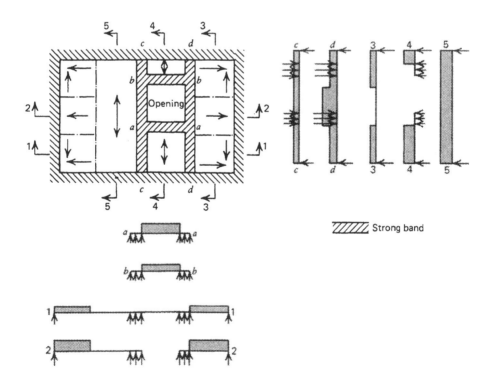

Figure 6.18 Simply supported uniformly loaded slab with opening.

which in turn transfer their total loading to the edge supports. Hence, as previously, the moments throughout the slab in the two directions, and the edge reactions, can be determined.

Similarly, uniformly loaded slabs with reentrant corners, or beamless slabs with column supports, can be treated using strong bands, as shown in Figs 6.19 and 6.20. The slab strips span between the edge supports or strong bands. The strong bands are designed to transfer the loading into the supports. Figure 6.21 shows a uniformly loaded slab with one edge free (unsupported). Although not essential, a strong band along the free edge assists in a convenient design solution, and also stiffens the edge.

Note that in general for a solution using strong bands a great number of possible arrangements of load transfer are possible. Each design case must be treated on its merits and the designer should have a feel for the elastic theory distribution of moments to ensure that a reasonable distribution of moments results.

6.2.5 Skewed and Triangular Slabs

The two systems of strips need not cross at right angles. Skewed slabs may be designed using skewed strips parallel to the edges. An example of a uniformly loaded skewed slab with all edges simply supported is shown in Fig. 6.22a. An example of a uniformly loaded one-way slab with one free edge skewed is shown in Fig. 6.22b, which illustrates a possible method of load dispersal for that case.

Triangular slabs with one edge free and uniformly loaded are shown in Fig. 6.23. A simple one-way spanning system may be adopted whether edges are fixed or simply supported (Fig. 6.23a). A strong band along the free edge is a useful alternative, particularly where loads are heavy (Fig. 6.23b). If the two edges are fixed, a further alternative is to select the contraflexure lines and to use "crooked" strips which logically run perpendicular to the supports to the contraflexure lines, and then parallel to the free edge between the

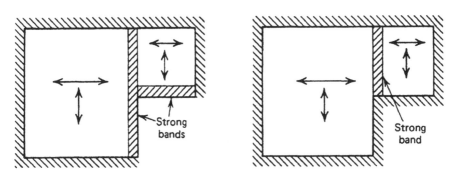

Figure 6.19   Uniformly loaded slabs with reentrant corners.

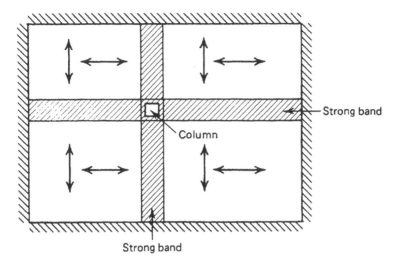

Figure 6.20   Uniformly loaded slab with a column support.

contraflexure lines (Fig. 6.23c). The middle region between the lines of contraflexure can be designed as a simply supported strip carrying uniform loading, and the remaining regions can be designed as cantilevers carrying uniformly distributed loads and concentrated loads at the ends.

6.2.6   Comparison with the Yield Line Theory Ultimate Load

The yield line theory method for calculating the ultimate load of a given slab system is an upper bound approach based on postulated collapse mechanisms.

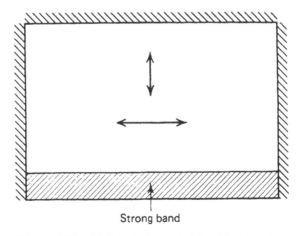

Figure 6.21   Uniformly loaded slab with free edge.

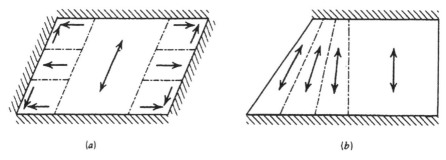

Figure 6.22 Uniformly loaded skewed slabs.

Wood and Armer[6.5] made some examination of slabs designed according to the strip method and concluded that when the reinforcement is placed precisely in accordance with the strip moments, the ultimate load given by yield line theory is identical to the ultimate load used in the strip design. For a uniformly loaded simply supported slab designed by the strip method, the yield line solution in fact found an unlimited number of simultaneous collapse modes, indicating all-over yielding of the slab somewhat like a "plastic ham-

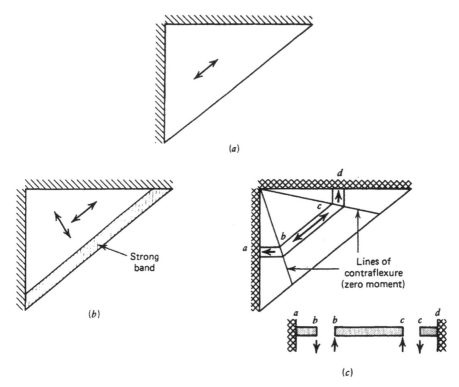

Figure 6.23 Uniformly loaded triangular slabs supported at two edges.

mock." Wood and Armer concluded that Hillerborg's method provides an exact solution. This result indicates that a strip method design utilizes reinforcement efficiently and economically. A later study by Fernando and Kemp[6.16] found that for some moment fields, the yield line theory ultimate load is higher than that used in the strip design, indicating that the strip method is always safe but not necessarily exact. However, it was found difficult to determine a practical example of a slab designed by the strip method for which there was not at least one collapse mechanism giving the same ultimate load as used in the strip design.

6.2.7 Design Applications

The preceding sections have illustrated how the simple strip method may be used to determine the moments throughout the slab for a given ultimate load. For gravity loading, according to the 1995 ACI Code,[6.15] the factored (ultimate) load is

$$U = 1.4D + 1.7L \tag{6.6}$$

where D is the service dead load and L is the service live load. Reinforcement is provided in the directions of the strips for the design (ultimate) moments. If the ultimate resisting moment per unit width in a particular direction is to be m_u, the design equation for steel in that direction is[6.17]

$$m_u = \phi A_s f_y \left(d - 0.59 A_s \frac{f_y}{f_c'} \right) \tag{6.7}$$

where ϕ the strength reduction factor, taken by the ACI Code as 0.9; A_s the tension steel area per unit width; f_y the steel yield strength; d the effective depth to the tension steel; and f_c' the concrete compressive cylinder strength. Steel ratios A_s/d less than $0.5\rho_b$ should be used to ensure reasonably ductile sections, where ρ_b is the balanced failure steel ratio.[6.17] The effect of compression steel on strength is negligible and may be neglected. According to the 1995 ACI Code, the bar spacing should not exceed the smaller of twice the slab thickness or 18 in. (500 mm), and the minimum amount of steel placed in the slab in the direction of the spans should not be less than that required for shrinkage and temperature reinforcement. This minimum amount is either 0.002 of the gross concrete area if grade 40 [grade 300 N/mm²] or grade 50 [grade 350 N/mm²] deformed bars are used, or 0.0018 where grade 60 [grade 420 N/mm²] deformed bars or welded wire fabric are used, or $0.0018 \times 60,000/f_y$ but not less than 0.0014 where reinforcement with $f_y > 60,000$ lb/in² [420 N/mm²] is used.

Some attempt may be made at minimizing the volume of steel in the slab by comparing the steel volumes resulting from different positions of discontinuity lines and different ratios of negative to positive design moments. The required steel area per unit width is proportional to the design moment per unit width. If the lever arm for moments in both directions is assumed to be the same, the volume of steel in the slab is proportional to the moment volume, given by

$$V_m = \iint (m_{ux} + m_{uy}) \, dx \, dy \qquad (6.8)$$

where m_{ux} and m_{uy} are the ultimate resisting moments in the x- and y-directions per unit width, respectively, and dx and dy are the lengths of the sides of the element in the x- and y-directions, respectively. Curtailment of some bottom steel at the ends of spans may be impracticable, and all bottom steel could continue the full length of the span. For example, for the slab shown in Fig. 6.13. if the bars are all of span length l, the volume of steel is proportional to

$$V_m = 2[(0.0104wl^2)(0.5l)l + (0.0731wl^2)(0.5l)l] = 0.0835w_ul^4$$

To find the actual volume of steel in the slab, V_m is divided by the internal lever arm and the steel yield strength.

The required extent of top steel at fixed or continuous edges is given by the bending moment diagrams for the strips. For a uniformly loaded strip of span l fixed at both ends, where m' and m are the negative and positive moments per unit width, it is evident that (see Fig. 6.24a)

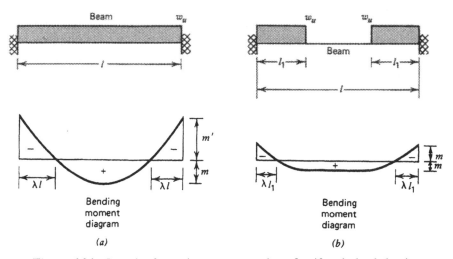

Bending
moment
diagram

(a)

Bending
moment
diagram

(b)

Figure 6.24 Length of negative moment region of uniformly loaded strips.

$$m' + m = w_u \frac{l^2}{8} \quad \text{and} \quad m = w_u \frac{l^2(1 - 2\lambda)^2}{8}$$

where λl is the length of negative moment region at each end. Therefore,

$$m' + m = \frac{m}{(1 - 2\lambda)^2}$$

$$\lambda = 0.5 \left[1 - \frac{1}{\sqrt{1 + (m'/m)}} \right]$$

(6.9)

Similarly, in strips loaded only over the end l_1 as in Fig. 6.24b, it is evident that the length of the negative-moment region at each end is λl_1, where λ is given by Eq. 6.9.

Considerable care should be used when calculating the lengths of top steel from Eq. 6.9, since the positions of the lines of contraflexure at ultimate load when the flexural strengths of the critical negative and positive moment sections have been reached may be considerably different from the positions in the service load range when the slab behavior is essentially elastic. Therefore, the ratio of moments m'/m used in Eq. 6.9 should be that giving the greatest value of λ. For example, if the value of λ used is that found from the ratio of the ultimate negative to positive moments per unit width, and if that ratio of m'/m is less than the ratio of moments given by elastic theory, the length of top bars will be too short for service load conditions and undesirable cracking could occur in the top of the slab under heavy service loads. Thus, to find λ from Eq. 6.9, the value of m'/m used should be either the elastic theory ratio of maximum negative to positive moments per unit width (e.g., 2 for fixed-end strips loaded uniformly over their entire length) or the ratio of the ultimate negative to positive moments of resistance per unit width of the slab sections, whichever is greater. Also, the top steel should extend at least a distance d or 12 bar diameters, whichever is greater, beyond the length of the negative moment region given by Eq. 6.9, where d is the effective depth of the slab.[6.15]

The strip method sets the torsional moment equal to zero and carries all the load by bending the strips. This is a valid lower bound procedure for strength; nevertheless, at the service loads, torsional moments, as indicated by elastic theory moments, will exist. Such twisting is particularly high in the corner regions of slabs simply supported on stiff beams or walls. Although the strip method does not require torsional moments to be resisted, both top and bottom steel should be present in the corner regions when one or both edges at the corner are simply supported. This reinforcement in the top and bottom should be provided for a distance of at least 0.2 of the longer span in each direction from the corner and should be capable of resisting the design maximum positive moment per unit width in the slab. The principal moments caused by torsion in the corner are negative in the direction of the slab diagonal and positive in the direction perpendicular to the diagonal. This reinforcement may be placed in the direction of these moments or in two directions at right angles parallel to the sides of the slab in the top and bottom.

Also, as has been emphasized previously, the strip moments should not be too far removed from the elastic theory distribution if serviceable slabs are to be designed. If large differences exist between the distribution of design moments and the distribution of elastic theory moments, it could mean that the cracking at service load will be excessive because low steel ratios at sections with high bending moments will lead to high steel stresses and correspondingly large crack widths. Such regions of high steel stress may also result in large deflections. Thus, it is important that the designer keep a feeling for the elastic theory distribution of moments and use it to help decide the ratios between negative and positive moments and the moments in the two directions. It is recommended that the ratios of negative to positive design moments should be between 1.0 and 2.0, and that some account should be taken of the degree of restraint at the edges. For example, if the edge is only partially restrained, a value for the ratio of less than 1.0 may be more appropriate. Ratios of design moments in the two directions should take into account the direction of maximum elastic bending moment. For example, in two-way panels supported at the edges, the greatest design moment should be in the direction of the short span. For a check of serviceability in cases where significant deviations from elastic theory moments occur, the designer may need to resort to the elastic theory moments to calculate the steel stresses at service load to allow the crack widths to be checked (see Chapter 9).

When a number of types of load act on the slab, it is evident that the moments required for each of them can be found separately and summed to find the total design moments. That is, the theory of superposition applies because no principles of lower bound design are violated. Line loads on slabs, such as those due to walls, are conveniently carried by designing strong bands in the slab under the wall.

At edges that have been considered as simply supported, care should be taken to provide top steel to control cracking due to fortuitous restraining moments. Such reinforcement could be for approximately 0.33 to 0.5 of the maximum positive moment. The thickness of the slab in most practical cases will be set by the need for adequate stiffness for deflection control. The thickness should be not less than the minimum values specified in the ACI Code[6.15] unless deflections are computed and shown to be within the limiting values allowed by the code (see Chapter 9).

The simple approach to the determination of the design moments allowed by the strip method, even for slabs with irregular shape and boundary conditions, should appeal in many cases. The examples given below demonstrate the design procedure.

Example 6.1. A rectangular interior panel of a continuous slab-and-beam floor system has clear spans of 16 ft (4.88 m) and 24 ft (7.32 m), as in Fig. 6.25. The panel carries a uniformly distributed service live load of 150 lb/ft²

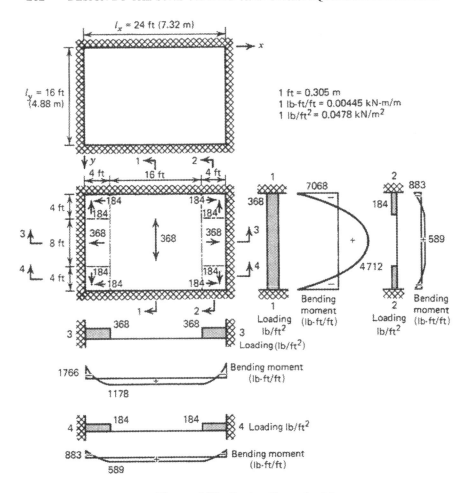

Figure 6.25 Design Example 6.1.

(7.18 kN/m^2). The concrete is of normal weight with a cylinder strength of 4000 lb/in^2 (27.6 N/mm^2), and the steel has a yield strength of 60,000 lb/in^2 (414 N/mm^2). Design a suitable panel.

SOLUTION. Assuming stiff beams, the minimum slab thickness according to ACI 318-95[6.15] is given by Eq. 9.7, which for a slab aspect ratio of $\beta = 24/16 = 1.5$ gives $h = l_n/46.4 = 24 \times 12/46.4 = 6.21$ in. Use, say, a $6\frac{1}{2}$-in.-thick slab.

Assuming that the unit weight of the concrete is 150 lb/ft^3, the service dead load is $D = (6 \ 1/2/12)150 = 81$ lb/ft^2. The service live load is $L = 150$ lb/ft^2. Therefore, the factored load according to ACI 318-95, Eq. 6.6, is

$$w_u = 1.4D + 1.7L = 1.4 \times 81 + 1.7 \times 150 = 368 \text{ lb/ft}^2$$

Figure 6.25 shows the assumed discontinuity lines for load dispersion. The edge strips are 0.25 of the short span = 4 ft wide. Four types of strips exist in the slab, shown by views 11, 22, 33, and 44. The ratio of design negative to positive moments will be assumed to be 1.5. The design moments may be found by determining the static moment for each strip and allocating 60% of it to negative moment and 40% to positive moment. The design moments for the strips in lb-ft/ft width of slab are given below.

STRIP 11. y-direction middle strip:

$$\text{static moment} = w_u \frac{l_y^2}{8} = 368 \times \frac{16^2}{8} = 11{,}780 \text{ lb-ft/ft } (= 11{,}780 \text{ lb-in./in.})$$

Therefore,

$$\text{maximum negative moment} = 0.6 \times 11{,}780 = 7068 \text{ lb-ft/ft}$$

$$\text{maximum positive moment} = 0.4 \times 11{,}780 = 4712 \text{ lb-ft/ft}$$

STRIP 22. y-direction edge strip:

$$\text{static moment} = \frac{w_u}{2} \left(\frac{l_y}{4}\right)^2 \Big/ 2 = 184 \times \frac{4^2}{2} = 1472 \text{ lb-ft/ft}$$

Therefore,

$$\text{maximum negative moment} = 883 \text{ lb-ft/ft}$$

$$\text{maximum positive moment} = 589 \text{ lb-ft/ft}$$

STRIP 33. x-direction middle strip:

$$\text{static moment} = w_u \left(\frac{l_y}{4}\right)^2 \Big/ 2 = 368 \times \frac{4^2}{2} = 2944 \text{ lb-ft/ft}$$

Therefore,

$$\text{maximum negative moment} = 1766 \text{ lb-ft/ft}$$

$$\text{maximum positive moment} = 1178 \text{ lb-ft/ft}$$

STRIP 44. x-direction edge strip:

moments are one-half of those for strip 33

Therefore,

$$\text{maximum negative moment} = 883 \text{ lb-ft/ft}$$
$$\text{maximum positive moment} = 589 \text{ lb-ft/ft}$$

The ultimate moment of resistance per unit width is given by Eq. 6.7 as

$$m_u = \phi A_s f_y \left(d - 0.59 A_s \frac{f_y}{f_c'} \right) \qquad \text{where } \phi = 0.9$$

Using No. 4 bars with $\frac{3}{4}$ in. of cover in the short span, $d = 6.5 - 1.0 = 5.5$ in. in the short span and $d = 6.5 - 1.5 = 5.0$ in. in the long span. The minimum amount of steel permitted is 0.0018 of the gross section, giving $A_s = 0.0018 \times 6.5 = 0.0117$ in.^{2}/in. width $= 0.140$ in.^{2}/ft width. Therefore, the minimum ultimate moment of resistance is

$$m_u = 0.9 \times 0.0117 \times 60,000 \left(5.0 - 0.59 \times 0.0117 \times \frac{60,000}{4000} \right)$$
$$= 3094 \text{ lb-ft/ft}$$

It is evident that minimum steel can carry the required ultimate moments in strips 22, 33, and 44. Minimum steel requires No. 3 (0.11-in.^{2}) bars on 0.11 × 12/0.14 = 9.4-in. centers. Therefore, use No. 3 bars on 9-in. centers in top and bottom of long-span strips (strips 33 and 44) and in top and bottom of short-span edge strips (strip 22).

Top steel for the short-span middle strip (strip 11) is found from

$$7068 = 0.9 \times A_s \times 60,000 \left(5.5 - 0.59 \times A_s \times \frac{60,000}{4000} \right)$$

giving $A_s = 0.025$ in.^{2}/in. $= 0.300$ in.^{2}/ft, requiring No. 4 (0.2-in.^{2}) bars on 8-in. centers. Bottom steel for the short-span middle strip (strip 11) is found from

$$4712 = 0.9 \times A_s \times 60,000 \left(5.5 - 0.59 \times A_s \times \frac{60,000}{4000} \right)$$

giving $A_s = 0.016$ in.^{2}/in. $= 0.197$ in.^{2}/ft, requiring No. 4 bars on 12-in. centers. Note that according to the code,[6.15] the bar spacing should not exceed twice the slab thickness (i.e., 13 in.), which is complied with.

The required length of top steel can be calculated from the moment diagrams or Eq. 6.9 may be used. The ratio of negative to positive moment,

m'/m, used in the design is 1.5 for the ultimate moments, but elastic theory indicates that this ratio is 2 for the short-span middle strips at service load. The greater m'/m ratio given by elastic theory governs, and substituting $m'/m = 2$ in Eq. 6.9 gives $\lambda = 0.21$. Therefore, the distance from the slab edge to the critical position of the point of contraflexure in the short-span middle strips is $0.21l_y = 0.21 \times 16 = 3.36$ ft. In all other strips this distance can be taken as $0.21l_y/2 = 1.68$ ft. Top steel should extend d or 12 bar diameters, whichever is greater, beyond the point of contraflexure.[6.15] Hence, the length of top steel from each edge is found from the distances given above plus 6 in.

Note that $0.5\rho_b = 0.0143$ for $f'_c = 4000$ lb/in² and $f_y = 60,000$ lb/in², where ρ_b is the balanced failure steel ratio.[6.15] The maximum steel ratio used is for the negative moment in the short-span middle strip, where $\rho = 0.3/(12 \times 5.5) = 0.00455$, which is less than $0.5\rho_b$ and is therefore satisfactory.

Example 6.2. A 12-ft (3.6-m) by 18-ft (5.49-m) rectangular panel at the edge of a floor system is free (unsupported) along one long side and is continuous with adjacent panels at supporting beams along the other three sides. The slab is shown in Fig. 6.26. The panel carries a uniformly distributed service live load of 100 lb/ft² (4.78 kN/m²). The concrete is of normal weight with a cylinder strength of 4000 lb/in² (27.6 N/mm²) and the steel has a yield strength of 40,000 lb/in² (276 N/mm²). Design a suitable panel.

SOLUTION. The minimum slab thickness may be computed conservatively by referring to Table 9.3. For both edges continuous and a span of 18 ft, the minimum thickness allowed is $l/28 = 18 \times 12/28 = 7.7$ in. Use, say, an 8-in.-thick slab.

Assuming that the unit weight of the concrete is 150 lb/ft³, the service load dead load is $D = (8/12)150 = 100$ lb/ft². The service live load is $L = 100$ lb/ft². Therefore, the factored load, according to ACI 318-95, Eq. 6.6, is

$$w_u = 1.4D + 1.7L = 1.4 \times 100 + 1.7 \times 100 = 310 \text{ lb/ft}^2$$

Figure 6.26 shows the panel. When one edge is unsupported, as in this case, the greater proportion of the load is carried by strips spanning between the supported sides (in the x-direction). A strong band will be designed along the free x-direction edge. The load will be assumed to be dispersed uniformly in each direction, with $0.6w_u = 0.6 \times 310 = 186$ lb/ft² assumed to be carried by fixed-end strips spanning in the x-direction and $0.4w_u = 0.4 \times 310 = 124$ lb/ft² assumed to be carried by propped cantilever strips in the y-direction. The 2-ft-wide fixed-end strong band along the x-direction free edge carries the uniform load on it plus the prop reaction from the y-direction strips. The prop reaction will be assumed to be sufficient to reduce the maximum negative moment in the y-direction strips to one-half of the free cantilever value. Note that this requires a prop reaction of two-thirds of the value obtained from elastic theory for nondeflecting supports, which is a reasonable value since

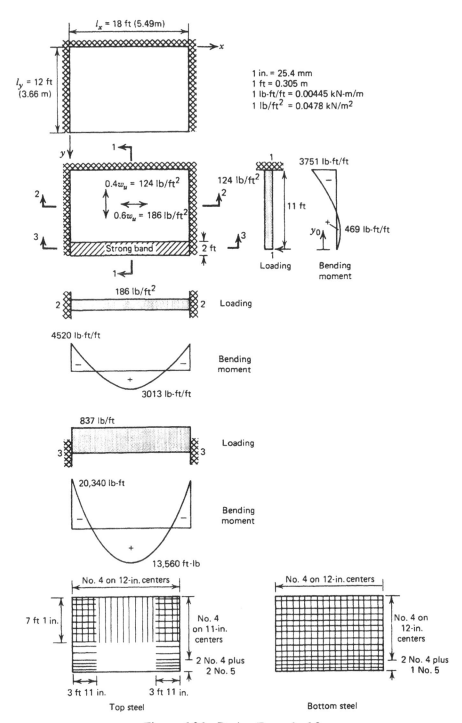

Figure 6.26   Design Example 6.2.

the strong band will actually deflect under the load. The ratio of maximum negative to positive moments in the x-direction will be assumed to be 1.5 for both the determination of the design moments and for the calculation of the length of the top steel (a value for this ratio approaching 2 may have been more realistic for calculating the length of the top steel). The design moments for the strips and the strong band are given below.

STRIP 11. y-direction strip. The cantilever will be assumed to extend to the middle of the strong band (i.e., 11 ft long). Free cantilever maximum negative moment (without prop reaction) = $124 \times 11^2/2 = 7502$ lb-ft/ft. Say that the prop reaction reduces the maximum negative moment to $0.5 \times 7502 = 3751$ lb-ft/ft; then the prop reaction is $R = 3751/11 = 341$ lb/ft. The positive moments per unit width in the strips is given by $m = 341y_0 - 124y_0^2/2$, where y_0 = distance from propped end. For maximum m, $dm/dy_0 = 0$. Therefore, $341 - 124y_0 = 0$. Therefore, $y_0 = 2.75$ ft and the maximum positive moment = $341 \times 2.75 - 124 \times 2.75^2/2 = 469$ lb-ft/ft. The extent of the positive moment is given when $m = 0 = 341y_0 - 124 y_0^2/2$. Therefore, the length of the positive-moment region is $y_0 = 341/62 = 5.5$ ft.

STRIP 22. x-direction strip:

$$\text{static moment} = 186 \times 18^2/8 = 7533 \text{ lb-ft/ft}$$

Therefore,

$$\text{maximum negative moment} = 0.6 \times 7533 = 4520 \text{ lb-ft/ft}$$

$$\text{maximum positive moment} = 0.4 \times 7533 = 3013 \text{ lb-ft/ft}$$

STRONG BAND. Free edge in the x-direction: Loading on the strong band is the prop reaction plus the remaining uniform load on it. (124 lb/ft² on a 1-ft-wide portion has already been allocated to the y-direction propped cantilever strips.)

$$\text{band load} = 341 + (310 \times 2 - 124 \times 1) = 837 \text{ lb/ft}$$

$$\text{static moment} = 837 \times 18^2/8 = 33,900 \text{ lb-ft}$$

$$\text{maximum negative moment} = 0.6 \times 33,900 = 20,340 \text{ lb-ft}$$

$$\text{maximum positive moment} = 0.4 \times 33,900 = 13,560 \text{ lb-ft}$$

From Eq. 6.9 with $m'/m = 1.5$, the x-direction negative moment extends $0.184 \times 18 = 3.31$ ft from each support. The ultimate moment of resistance per unit width is given by Eq. 6.7 as

$$m_u = \phi A_s f_y \left(d - 0.59 A_s \frac{f_y}{f_c'} \right) \qquad \text{where } \phi = 0.9$$

Using No. 4 bars with a $\frac{3}{4}$-in. clear cover in the x-direction, $d = 8 - 1 = 7$ in. In the y-direction, $d = 8 - 1.5 = 6.5$ in. The minimum amount of steel permitted is 0.002 of the gross section, giving $A_s = 0.002 \times 8 = 0.016$ in.^{2}/in. width = 0.192 in.^{2}/ft width. Therefore, the minimum ultimate moment of resistance is

$$m_u = 0.9 \times 0.016 \times 40{,}000 \left(6.5 - 0.59 \times 0.016 \times \frac{40{,}000}{4000} \right)$$

$$= 3960 \text{ lb-ft/ft}$$

It is evident that minimum steel can carry the required ultimate positive moments in both directions. Minimum steel requires No. 4 (0.2-in.^{2}) bars on $0.2 \times 12/0.192 = 12.5$-in. centers.

Therefore, use No. 4 bars on 12-in. centers in the bottom in both directions (except in the strong band parallel to free edge). Top steel for the y-direction (strip 11) is found from

$$3751 = 0.9 \times A_s \times 40{,}000 \left(6.5 - 0.59 \times A_s \times \frac{40{,}000}{4000} \right)$$

giving $A_s = 0.0163$ in.^{2}/in. $= 0.195$ in.^{2}/ft, requiring No. 4 bars on 12-in. centers.

Top steel for the x-direction (strip 22) is found from

$$4520 = 0.9 \times A_s \times 40{,}000 \left(7 - 0.59 \times A_s \times \frac{40{,}000}{4000} \right)$$

giving $A_s = 0.0182$ in.^{2}/in. $= 0.219$ in.^{2}/ft, requiring No. 4 bars on $0.2 \times 12/0.219 = 11$-in. centers.

For the x-direction 2-ft-wide strong band at the slab edge, the top steel is found from

$$20{,}340 \times 12 = 0.9 \times A_s \times 40{,}000 \left(7 - 0.59 \times A_s \times \frac{40{,}000}{24 \times 4000} \right)$$

giving $A_s = 1.00$ in.^{2}, requiring two No. 4 and two No. 5 bars (giving 1.02 in.^{2}). The bottom steel in the strong band is found from

$$13,560 \times 12 = 0.9 \times A_s \times 40,000 \left(7 - 0.59 \times A_s \times \frac{40,000}{24 \times 4000} \right)$$

giving $A_s = 0.661$ in.^{2}, requiring two No. 4 and one No. 5 bars (giving 0.71 in.^{2}).

The negative-moment steel should extend a distance d or 12 bar diameters, whichever is greater, beyond the point of contraflexure.[6.15] The extent of the negative-moment regions was calculated above. Hence, the top steel in the y-direction extends $(12 - 5.5)$ ft $+ 6.5$ in. $= 7$ ft 1 in. from the fixed edges, and the top steel in the x-direction extends 3.31 ft $+ 7$ in. $= 3$ ft 11 in. from each fixed edge.

The layout of reinforcement is shown in Fig. 6.26. Note that $0.5\rho_b = 0.0247$ for $f'_c = 4000$ lb/in^2 and $f_y = 40,000$ lb/in^2, where ρ_b is the balanced failure steel ratio.[6.17] The maximum steel ratio used is for the negative moment in the strong band, where $\rho = 1.02/(24 \times 7) = 0.0061$, which is less than $0.5\rho_b$ and therefore is satisfactory.

Example 6.3. A corner panel of a floor system is continuous with adjacent panels at supporting beams along two edges and simply supported at the other two edges except for a rectangular opening that is unsupported at its edges. Figure 6.27 shows the slab. The service loads are a uniformly distributed live load of 100 lb/ft^2 (4.78 kN/m^2) and a line load of 300 lb/ft (4.38 kN/m) positioned as shown in the figure. The concrete is of normal weight with a cylinder strength of 4000 lb/in^2 (27.6 N/mm^2) and the steel has a yield strength of 40,000 lb/in^2 (276 N/mm^2). Design a suitable panel. (The same slab panel is shown in Fig. 2.11, which includes some information on service load elastic moments.)

SOLUTION. The minimum slab thickness may be computed conservatively by referring to Table 9.3. For one edge continuous and a span of 16 ft, the minimum thickness allowed is $l/24 = 16 \times 12/24 = 8$ in. Use, say, an 8-in.-thick slab.

Assuming that the unit weight of concrete is 150 lb/ft^3, the service dead load is $D = (8/12)150 = 100$ lb/ft^2. The service live loads L are 100 lb/ft^2 uniformly distributed and a 300-lb/ft line load. Therefore, the factored loads, according to ACI 318-95, Eq. 6.6, are a uniformly distributed load

$$w_u = 1.4D + 1.7L = 1.4 \times 100 + 1.7 \times 100 = 310 \text{ lb/ft}^2$$

and a line load

$$p_u = 1.7L = 1.7 \times 300 = 510 \text{ lb/ft}$$

Figure 6.27 shows the manner in which the loads will be assumed to be

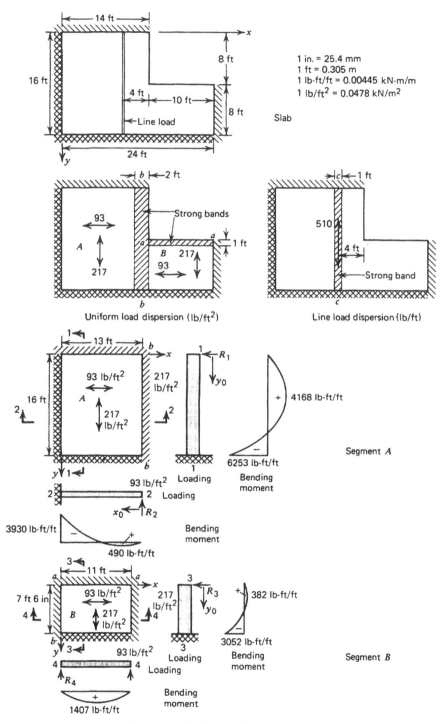

Figure 6.27   Design Example 6.3.

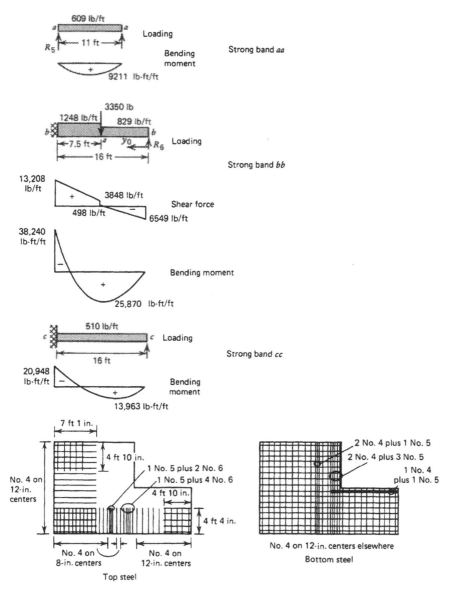

Figure 6.27 *(Continued)*

dispersed in this design. The design moments required for the uniformly distributed and line loads can be found separately and summed.

DESIGN MOMENTS FOR UNIFORM LOAD. For the uniformly distributed loading, a 2-ft-wide strong band *bb* will be designed to span the short-span direction adjacent to the opening, and a 1-ft-wide strong band *aa* will be designed to span between *bb* and the supported edge adjacent to the opening.

The panel is then divided into two rectangular segments, marked A and B in Fig. 6.27. For segment A, in order not to unduly load the strong band bb, $0.7w_u = 0.7 \times 310 = 217$ lb/ft^2 will be assumed to be carried by y-direction strips spanning between fixed and simple supports, and $0.3w_u = 0.3 \times 310 = 93$ lb/ft^2 will be assumed to be carried by x-direction strips spanning between fixed support and strong band bb. For segment B, because of the fixed edge and shorter span in the y-direction, $0.7w_u = 217$ lb/ft^2 will be assumed to be carried by the y-direction strips spanning between fixed support and strong band aa, and $0.3w_u = 93$ lb/ft^2 will be assumed to be carried by x-direction strips spanning between simple support and strong band bb. The 1-ft-wide strong band aa carries the uniform load on it plus the reaction from the y-direction strips of segment B. The 2-ft-wide strong band bb carries the uniform load on it plus the reactions from the x-direction strips of segments A and B, including a reaction from strong band aa. The reactions on the strong bands from the cantilever strips (x-direction strips of segment A and y-direction strips of segment B) will be assumed to be sufficient to reduce the maximum negative moments in those strips to one-half of the free cantilever value. This requires a reaction of two-thirds of the value obtained from elastic theory for nondeflecting supports, which is a reasonable value since the strong band will actually deflect under the load. The ratio of maximum negative to maximum positive moments in the y-direction strips of segment A and strong band bb will be taken to be 1.5 for both the determination of the design moments and for the calculation of the length of top steel (a value approaching 2 may have been more realistic for calculating the length of top steel). The design moments for the strips and the strong bands are given below.

SEGMENT A. The x-direction strips will be assumed to extend to the middle of strong band bb (i.e., 13-ft span) (see Fig. 6.27).

STRIP 11. y-direction strips: The moment per unit width in the strips is given by $m = R_1 y_0 - 217y_0^2/2$, where R_1 is the prop reaction and y_0 the distance from the propped end. For the maximum positive moment, $dm/dy_0 = 0$. Therefore, $y_0 = R_1/217$ and the maximum positive moment $= R_1^2/434$. The maximum negative moment occurs when $y = 16$ ft. Therefore, the maximum negative moment $= 16R_1 - 217 \times 16^2/2 = 16R_1 - 27,780$. If the maximum negative moment is to be 1.5 times the maximum positive moment, $16R_1 - 27,780 = -1.5R_1^2/434$. Therefore, $R_1^2 + 4629R_1 - 8.038 \times 10^6 = 0$, giving $R_1 = 1345$ lb/ft and

$$\text{maximum negative moment} = 6253 \text{ lb-ft/ft}$$

$$\text{maximum positive moment} = 4168 \text{ lb-ft/ft}$$

The extent of the positive moment is given when $m = 0 = 1345y_0 - 217y_0^2/2$, giving the length of the positive-moment region as $y_0 = 12.40$ ft.

STRIP 22. x-direction strips: The free cantilever maximum negative moment (without prop reaction) $= 93 \times 13^2/2 = 7859$ lb-ft/ft. If the prop reaction reduces the maximum negative moment to $0.5 \times 7859 = 3930$ lb-ft/ft, the prop reaction $R_2 = 3930/13 = 302$ lb/ft. The positive moments per unit width in the strips are given by $m = 302x_0 - 93x_0^2/2$, where x_0 is the distance from the propped end. For maximum m, $dm/dx_0 = 0$. Therefore, $302 - 93x_0 = 0$ and $x_0 = 3.25$ ft.

$$\text{maximum positive moment} = 302 \times 3.25 - 93 \times \frac{3.25}{2} = 490 \text{ lb-ft/ft}$$

The extent of the positive moment is given when $m = 0 = 302x_0 - 93x_0^2/2$, giving the length of the positive-moment region as $x_0 = 302/46.5 = 6.50$ ft.

SEGMENT B. The x-direction strips will be assumed to extend to the middle of strong band bb (i.e., 11-ft span) and the y-direction strips to the middle of strong band aa (i.e., 7.5-ft span) (see Fig. 6.27).

STRIP 33. y-direction strips: The free cantilever maximum negative moment (without prop reaction) $= 217 \times 7.5^2/2 = 6103$ lb-ft/ft. If the prop reaction reduces the maximum negative moment to $0.5 \times 6103 = 3052$ lb-ft/ft, the prop reaction $R_3 = 3052/7.5 = 407$ lb/ft. The positive moments per unit width in the strips are given by $m = 407y_0 - 217y_0^2/2$, where y_0 is the distance from the propped end. For maximum m, $dm/dy_0 = 0$. Therefore, $407 - 217y_0 = 0$ and $y_0 = 1.88$ ft:

$$\text{maximum positive moment} = 407 \times 1.88 - 217 \times \frac{1.88^2}{2} = 382 \text{ lb-ft/ft}$$

The extent of the positive moment is given when $m = 0 = 407y_0 - 217y_0^2/2$, giving the length of the positive-moment region as $y_0 = 407/108.5 = 3.75$ ft.

STRIP 44. x-direction strips: The maximum positive moment $= 93 \times 11^2/8 = 1407$ lb-ft/ft. The prop reaction $R_4 = 93 \times 5.5 = 512$ lb/ft.

STRONG BAND aa. Strong band aa will be assumed to extend to the middle of strong band bb (i.e., 11-ft span). The loading on the strong band is the prop reaction from the y-direction strips of segment B plus the remaining uniform load on it (217 lb/ft^2 on a 6-in-wide portion has already been allocated to the y-direction strips). Therefore,

$$\text{band load} = 407 + (310 \times 1 - 217 \times 0.5) = 609 \text{ lb/ft}$$

$$\text{maximum positive moment} = 609 \times \frac{11^2}{8} = 9211 \text{ lb-ft}$$

The reaction at the end $R_5 = 609 \times 5.5 = 3350$ lb.

STRONG BAND bb. The loading on strong band bb is the distributed prop reactions from the x-direction strips of segments A and B, plus the concentrated reaction from strong band *aa*, plus the remaining uniform load on it (93 lb/ft² on a 1-ft-wide portion has already been allocated to the *x*-direction strips of panel *A*; similarly, for panel *B*). Between the fixed support and the intersection with band *aa*, the uniform load is 302 + 512 + (310 × 2 − 93 × 2) = 1248 lb/ft. Between the simple support and the intersection with band *aa*, the uniform load is 302 + (310 × 2 − 93 × 1) = 829 lb/ft. The concentrated reaction at the intersection with band *aa* is 3350 lb. The loads on strong band *bb* are shown in Fig. 6.27. The lack of symmetry makes the calculation of moments less easy. To find the moments when the ratio of maximum negative to positive moments is 1.5, a graphical approach can be adopted whereby the static moment diagram is drawn and the fixing moment line placed by trial so as to give the required ratio between maximum moments. An analytical approach is to adjust the point of zero shear until the required ratio of moments is obtained. For example, let the distance of the point of zero shear (i.e., maximum positive bending moment) from the simple support be 7.9 ft. Then the end reaction R_6 = 829 × 7.9 = 6549 lb.

$$\text{maximum positive moment} = 6549 \times 7.9 - 829 \times \frac{7.9^2}{2}$$

$$= 25{,}870 \text{ lb-ft}$$

$$\text{maximum negative moment} = 1249 \times \frac{7.5^2}{2} + 3350 \times 7.5$$

$$+ \ 829 \times 8.5 \times 11.75 - 6549 \times 16$$

$$= 38{,}240 \text{ lb-ft}$$

The ratio of maximum negative to positive moments = 38,240/25,870 = 1.48, as required. In the left portion of the span, if y_0 is the distance from simple support,

$$m = 6549y_0 - 829 \times 8.5(y_0 - 4.25) - 3350(y_0 - 8.8)$$

$$- \ \frac{1248 \ (y_0 - 8.5)^2}{2}$$

Solution of this equation for $m = 0$ gives $y_0 = 12.54$ ft.

DESIGN MOMENTS FOR LINE LOAD. For the line load a 1-ft-wide strong band *cc* will be designed. The line load is $p_u = 510$ lb/ft. The *y*-direction strips of segment A carry 217 lb/ft², and hence the moments for the line load can be obtained by multiplying these moments by 510/217 = 2.35.

$$\text{maximum negative moment} = 2.35 \times 6253 = 14{,}695 \text{ lb-ft}$$

$$\text{maximum positive moment} = 2.35 \times 4168 = 9795 \text{ lb-ft}$$

and the length of the positive-moment region is 12.40 ft.

The line load moments are added to the uniform load moments in the strong band cc to find the total design moments as follows:

$$\text{maximum positive moment} = 9795 + 4168 = 13{,}963 \text{ lb-ft}$$

$$\text{maximum negative moment} = 14{,}695 + 6253 = 20{,}948 \text{ lb-ft}$$

DESIGN OF REINFORCEMENT. The ultimate moment of resistance per unit width is given by Eq. 6.7 as

$$m_u = \phi A_s f_y \left(d - 0.59 A_s \frac{f_y}{f'_c} \right) \qquad \text{where } \phi = 0.9$$

Using No. 4 bars with a 0.75-in. clear cover in the y-direction, $d = 8 - 1 = 7$ in. In the x-direction, $d = 8 - 1.5 = 6.5$ in. The minimum amount of steel permitted is 0.002 of the gross section, giving $A_s = 0.002 \times 8 = 0.016$ in.^{2} /in. width $= 0.192$ in.^{2}/ft width. Therefore, the minimum ultimate moment of resistance is

$$m_u = 0.9 \times 0.016 \times 40{,}000 \left(6.5 - 0.59 \times 0.016 \times \frac{40{,}000}{4000} \right)$$

$$= 3960 \text{ lb-ft/ft}$$

It is evident that minimum steel can carry the required ultimate positive and negative moments in the y-direction in segment B and can carry the required ultimate positive moments in the x-direction over the entire slab. Minimum steel requires No. 4 (0.2-in.^{2}) bars on $0.2 \times 12/0.192 = 12.5$-in. centers.

Therefore, use No. 4 bars on 12-in. centers in the top and bottom in the y-direction in segment B and in the bottom in the x-direction over the entire slab.

Bottom steel in the y-direction in segment A is found from

$$4168 = 0.9 \times A_s \times 40{,}000 \left(7 - 0.59 \times A_s \times \frac{40{,}000}{4000} \right)$$

giving $A_s = 0.0168$ in.^{2}/in., requiring No. 4 bars on $0.2/0.0168 = 12$-in. centers.

Top steel in the y-direction in segment A is found from

$$6253 = 0.9 \times A_s \times 40{,}000 \left(7 - 0.59 \times A_s \times \frac{40{,}000}{4000} \right)$$

giving A_s = 0.0254 in.^{2}/in., requiring No. 4 bars on 0.2/0.0254 = 8-in. centers.

Top steel in the x-direction in segment A is found from

$$3930 = 0.9 \times A_s \times 40{,}000 \left(6.5 - 0.59 \times A_s \times \frac{40{,}000}{4000} \right)$$

giving A_s = 0.0171 in.^{2}/in., requiring No. 4 bars on 0.2/0.0171 = 12-in. centers.

In strong band aa, bottom steel is found from

$$9211 \times 12 = 0.9 \times A_s \times 40{,}000 \left(6.5 - 0.59 \times A_s \times \frac{40{,}000}{12 \times 4000} \right)$$

giving A_s = 0.491 in.^{2}, requiring one No. 4 and one No. 5 bar (0.51 in.^{2}).

In strong band bb, top steel is found from

$$38{,}240 \times 12 = 0.9 \times A_s \times 40{,}000 \left(7 - 0.59 \times A_s \times \frac{40{,}000}{24 \times 4000} \right)$$

giving A_s = 1.96 in.^{2}, requiring one No. 5 and four No. 6 bars (2.07 in.^{2}), and bottom steel is found from

$$25{,}870 \times 12 = 0.9 \times A_s \times 40{,}000 \left(7 - 0.59 \times A_s \times \frac{40{,}000}{24 \times 4000} \right)$$

giving A_s = 1.29 in.^{2}, requiring two No. 4 and three No. 5 bars (1.33 in.^{2}).

In strong band cc, top steel is found from

$$20{,}948 \times 12 = 0.9 \times A_s \times 40{,}000 \left(7 - 0.59 \times A_s \times \frac{40{,}000}{12 \times 4000} \right)$$

giving A_s = 1.08 in.^{2}, requiring one No. 5 and two No. 6 bars (1.19 in.^{2}), and bottom steel is found from

$$13{,}963 \times 12 = 0.9 \times A_s \times 40{,}000 \left(7 - 0.59 \times A_s \times \frac{40{,}000}{12 \times 4000} \right)$$

giving A_s = 0.70 in.^{2}, requiring two No. 4 and one No. 5 bars (0.71 in.^{2}).

The negative-moment steel should extend from the fixed edges a distance d or 12 bar diameters, whichever is greater, beyond the point of contraflexure.[6.15] The extent of the positive-moment regions has been calculated above. Hence, in the x-direction, the top steel at the fixed edge extends $(13 - 6.5)$ ft $+ 6.5$ in. $= 7$ ft 1 in. into the span. In the y-direction, the distances the top steel are required to extend into the span from the fixed edge are for segment A $(16 - 12.4)$ ft $+ 7$ in. $= 4$ ft 2 in., for segment B $(7.5 - 3.75)$ ft $+ 7$ in. $= 4$ ft 4 in., for strong band bb $(16 - 12.54)$ ft $+ 9$ in. $= 4$ ft 3 in., and for strong band cc $(16 - 12.4)$ ft $+ 9$ in. $= 4$ ft 4 in. For convenience, make all y-direction top steel extend into span 4 ft 4 in.

In the slab corners adjacent to the simply supported edges, No. 4 bars on 12-in. centers for torsion should be placed both ways top and bottom if not already present. This steel extends 0.2×24 ft $= 4$ ft 10 in. from the edge. The layout of reinforcement is shown in Fig. 6.27. Note that $0.5\rho_b = 0.0247$ for $f'_c = 4000$ lb/in² and $f_y = 40,000$ lb/in², where ρ_b is the balanced steel ratio.[6.17] The maximum steel ratio used is for the negative moment in strong band bb, where $\rho = 2.01/(24 \times 7) = 0.012$, which is less than $0.5\rho_b$ and is therefore satisfactory.

It is evident that the foregoing design could be refined if necessary by a more complex dispersion of load involving discontinuity lines dividing slab up into edge and middle strips. Such refinement may be warranted in more heavily loaded floors where most design moments are well above the capacity given by minimum steel requirements.

6.3 ADVANCED STRIP METHOD

To extend the strip method to the design of beamless slabs with column supports, and to slabs with reentrant corners, Hillerborg[6.3] introduced a more complex version of the strip method in 1959. Crawford,[6.4] who made the first study of this theory in English, has referred to it as the *advanced strip method*. The advanced strip method introduces a *type 3 element* with a complex moment field in order to transfer the slab loading into columns and reentrant corners.

It has been shown in Section 6.2 that beamless slabs supported on columns, and slabs with reentrant corners, can be solved by the simple strip method by the use of strong bands between columns and supports, and between reentrant corners and supports, as suggested by Wood and Armer.[6.5] This development has meant that the simple strip method can now be used to solve all cases of practical slabs. Hence, it is debatable whether the advanced strip method should be regarded as an essential addition to general strip theory. Nevertheless, the method will be briefly described here, since it has attractions. A full treatment may be seen in Hillerborg's book.[6.11]

In the advanced strip method, the slab is divided into elements bounded by lines of zero shear force. The design moments are the bending moments found throughout the slab which are compatible with these zero shear lines and which are in equilibrium with the applied design loading.

6.3.1   Types of Slab Element

The slab can be divided into three different types of element:

Type 1: rectangular in shape and supported at one edge, dispersing load in one direction

Type 2: triangular in shape and supported at one edge, dispersing load in one direction

Type 3: rectangular in shape and supported at one corner, dispersing load in two directions

An example of a slab divided into elements is shown in Fig. 6.28. Type 1 elements occupy the edge portions of the slab away from the corners, type 2 elements occupy the corner portions of the slab, and type 3 elements occupy the portions of the slab adjacent to the column. Elastic theory for continuous strips may be used to determine the positions of the zero shear lines. At the column a knife-edge support is assumed to exist across the slab. The points of maximum positive moment in the elastic bending moment diagram for the strip locate the lines of zero shear between elements types 1 and 3. This is illustrated for section 11 of the slab in Fig. 6.28. If elastic theory is used to divide the slab up into elements, a distribution of design moments will be obtained that will ensure that the stresses in the reinforcement are well within the elastic range at the service loads, thus minimizing any serviceability problems. The design moments are found from the moment equilibrium equation obtained considering the design load and edge moments of each element.

Type 1 Elements.   A type 1 element is shown in Fig. 6.29. All the load on the element is assumed to be carried by the moments acting perpendicular to the support. Assuming zero shear at all internal boundaries of the element, for a uniformly distributed load per unit area w_u acting on the element, the moment equilibrium equation is

$$m_s + m_f = w_u \frac{a^2}{2} \qquad (6.10)$$

where m_s and m_f are the support (maximum negative) and field (maximum positive) moments per unit width, respectively, and a is the length of the element in the direction of the moments. If the exterior edge is simply supported, then m_s is put equal to zero in Eq. 6.10. The reinforcement for these moments is distributed uniformly across the width of the element. The bottom reinforcement continues into the adjacent elements, as is required by the bend-

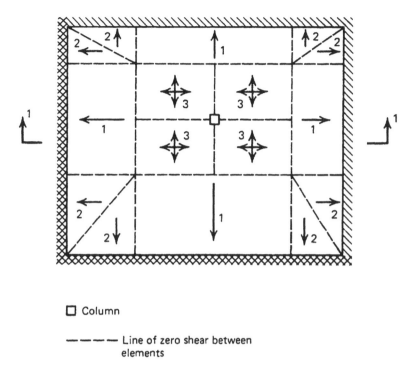

□ Column

— — — — Line of zero shear between
elements

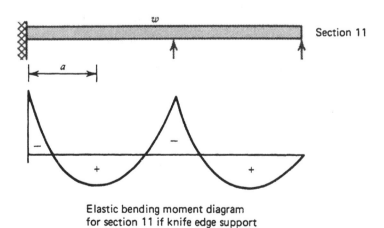

Elastic bending moment diagram
for section 11 if knife edge support
extends from column

Figure 6.28 Uniformly loaded slab with internal column divided into elements.

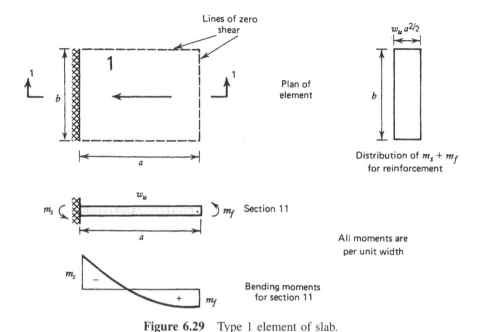

Figure 6.29   Type 1 element of slab.

ing moment distribution. Bottom reinforcement is required in the direction parallel to the supported edge and is a continuation of that in the type 2 elements in the adjacent corners dispersing load parallel to the edge.

Type 2 Elements. A type 2 element is shown in Fig. 6.30. All the load on the element is considered to be carried by the moments acting perpendicular to the support. Assuming zero shear at both internal boundaries of the element, for a uniformly distributed load per unit area w_u acting on the element, the moment equilibrium equation is

$$m_s + m_f = w_u \frac{(az/b)^2}{2} \qquad (6.11)$$

where m_s and m_f are the support (maximum negative) and field (maximum positive) moments per unit width, respectively, and az/b is the length of the element at the section, as shown in Fig. 6.30. If the exterior edge is simply supported, m_s is put equal to zero in Eq. 6.11. The maximum value for $m_s + m_f$ occurs when $z = b$. The variation of $m_s + m_f$ with z is shown in Fig. 6.30. Practical reinforcement cannot be placed to follow this variation of $m_s + m_f$ with z exactly, but the reinforcement can be placed in uniform bands to give the required total moment. If the reinforcement is placed in a single uniform band, Eq. 6.4 indicates that the average $m_s + m_f$ (per unit width) is 1.33 $\times$ mid-element moment = $1.333w_u(0.5a)^2/2 = w_u a^2/6$. Alternatively, if the re-

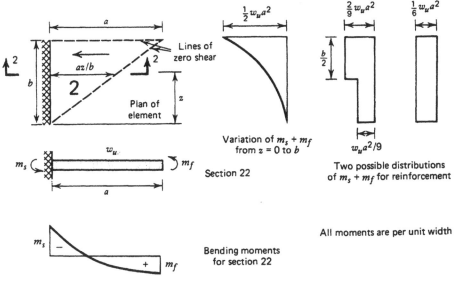

Figure 6.30   Type 2 element of slab.

inforcement is placed in two uniform bands to match more closely the actual distribution of the moments $m_s + m_f$, steel could be placed for $w_u a^2/9$ over one-half of the element width and for $2w_u a^2/9$ over the remaining half of the element width, as shown in Fig. 6.30, to give the same total moment. The bottom reinforcement continues into the adjacent elements, as is required by the bending moment distribution. It is evident that for the reinforcement perpendicular to the supported edge, the average steel area required per unit width for a type 2 element is one-third of that for the adjacent type 1 elements. The bottom reinforcement parallel to the supported edge in the element is a continuation of that in the adjacent type 2 element, dispersing load parallel to the edge.

Type 3 Elements.   A type 3 element is shown in Fig. 6.31. Since the rectangle is assumed to be bounded by lines of zero shear, all the loading on the slab must be carried by the column reaction at one corner. Furthermore, equilibrium of the element requires the *total* load on the element to be carried in *each* direction. For a uniformly distributed load per unit area w_u acting on the element, the moment equilibrium equations in the *x*- and *y*-directions are

$$m_{px} + m_{fx} = w_u \frac{a^2}{2} \qquad (6.12)$$

$$m_{py} + m_{fy} = w_u \frac{b^2}{2} \qquad (6.13)$$

where m_{px} and m_{fx} are the column (maximum negative) and field (maximum

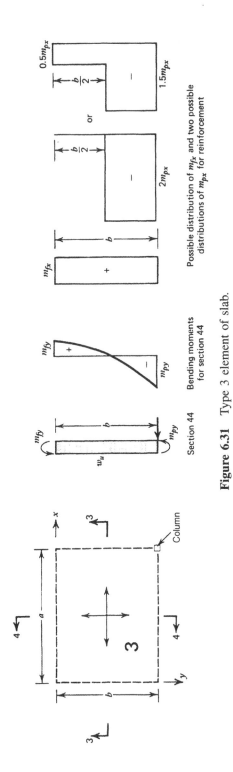

Figure 6.31 Type 3 element of slab.

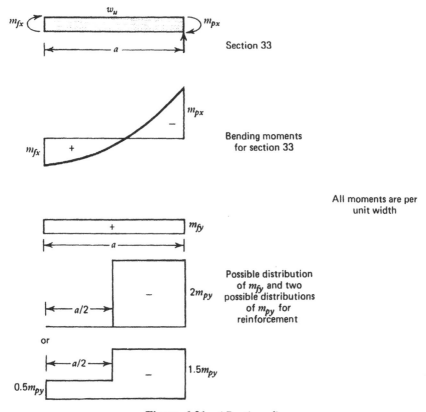

Figure 6.31 (*Continued*)

positive) moments per unit width in the x-direction, respectively; m_{py} and m_{fy} the column and field moments per unit width in the y-direction, respectively; and a and b the lengths of the element in the x- and y-directions, respectively. Hillerborg[6.3] recommended that reinforcement should be placed uniformly across the element for positive moments. For negative moments Hillerborg[6.3] recommended placing reinforcement for twice m_p in the half-element width adjacent to the column and no reinforcement in the remaining half-width (see the first possible distribution of m_{px} and m_{py} in Fig. 6.31). This distribution of negative-moment steel is to take some account of the actual elastic theory distribution of negative moments across the slab. The elastic negative moments are small away from the column and increase rapidly near the column. Hillerborg's recommended distribution satisfies statics in that the total moment is as required. Some moment redistribution as ultimate load is approached is implied by this chosen distribution of negative-moment strength. Many designers will prefer to have some top steel at the negative-moment edge in the outer region of the element to take care of possible cracking there at service load. Hillerborg's book[6.11] suggests various other distributions of moments along the element edges. A readily acceptable distribution of negative slab moments across the column line would be $1.5m_p$ in the half-element width by the column and $0.5m_p$ in the remaining outside half-element width (see the second possible distribution of m_{px} and m_{py} in Fig. 6.31).

Although the suggested distributions of moments along the edges of a type 3 element appear to be reasonable, the actual distribution of moments within the element is complex. Because the load on the element is carried by a single vertical reaction at one corner, there must be strong twists within the element. Thus, while type 1 and 2 elements preserve the concept of strip theory by being composed of twistless strips, type 3 elements involve a more general moment condition and, strictly, plate theory should be used to determine the moment field within the element for use in design. Hillerborg has analyzed the moments within a square type 3 element assuming that one-half of the uniform load is carried in each direction by two systems of strips interacting with each other and supported by part of the slab in the shape of a quadrant of a circle centered on the column. The slab portion within the quadrant is treated by polar coordinates and the moment field for different radii of quadrants found by plate theory. Rectangular elements were also considered. Full details of the procedure are given in Hillerborg's book.[6.11] The aim of this analysis is to show that the edge moments on the element are the governing design moments for the reinforcement, and hence that the reinforcement will not be subjected to higher moments within the element than at the edges. As a result of this study of the moment field within a type 3 element, Hillerborg has concluded that the design of reinforcement based on the recommended edge moments (see Fig. 6.31) will lead to reinforcement that can safely carry the moment field within the element in practical cases.

Rules for steel curtailment were also determined from this study. All the bottom steel designed for m_f must be continued through the element. Top steel may be curtailed in regions of the element well away from the column, but farther away than the strip bending moments indicate, because of the complex moment field in the element. Hillerborg[6.3] recommends that all top steel be required up to at least the midpoint of the element and gives a graph for length of top steel which indicates that if the ratio of m_p/m_f is equal to or less than 2.0, the top steel is no longer required beyond 0.57 of the element length from the column.

An alternative design method for type 3 elements has been suggested by Wood and Armer,[6.5] who produced a lower bound solution that gives the moment field within the element for uniform applied moments along the edges and uniform loading on the element. The required design ultimate resisting moments per unit width in the x- and y-directions for four equal-sized regions within the element were found, including the effect of torsional moment, for elements with various aspect ratios and negative to positive edge moment ratios. The effects of torsional moments were included using the method described in Section 5.5. The design moments were tabulated and enable the reinforcement for a type 3 element to be determined. Bottom steel extending uniformly over the whole of the element is found to be necessary. Top steel may be curtailed in the regions of the element away from the column.

6.3.2   Design Applications

The crucial first step in design by the advanced strip method is to choose the positions of the lines of zero shear. This should be carried out with the aid of the elastic distribution of strip moments, as described previously. Then the slab is divided into type 1, 2, and 3 elements with the zero shear lines as boundaries and the design moments found using Eqs. 6.10 to 6.13. It is obvious that the design moments for each adjacent element should be equal at a zero shear line. An example is given below to illustrate the procedure.

Example 6.4. A square beamless slab with sides of length $2l$, forming the roof of a tank, is simply supported around all four sides on walls and at the center by a column. The slab carries a uniformly distributed design load per unit area of w_u. Determine the design bending moments for the slab.

SOLUTION. The slab is shown in Fig. 6.32. Considering a continuous strip passing through the central part of the slab, simply supported by the walls and by a knife edge at the column, the bending moments due to w_u may be found by elastic theory and are shown in the figure. It is apparent that the points of zero shear occur at distance $0.375l$ from each end support. Hence, the type 3 elements are each $0.625l$ square and the slab can be divided into elements as shown in Fig. 6.32.

Section $A–A$ shows the x-direction moments in the types 2 and 1 elements adjacent to an x-direction boundary. The length of the loaded part of the type 2 elements at each end varies for different sections across the elements. Reinforcement can be placed for the average design moment. At mid-element the maximum positive design moment per unit width is $w_u(0.1875l)^2/2 = 0.0176w_ul^2$, and hence according to Eq. 6.4, the average maximum positive design moment per unit width is $1.33 \times 0.0176w_ul^2 = 0.0234w_ul^2$. Bottom reinforcement should be placed for this moment parallel to the walls in the types 1 and 2 elements around the slab.

Section $B–B$ shows the x-direction moments in the type 1 elements near the columns. The maximum positive moment per unit width is $0.0703w_ul^2$ constant across the elements. Bottom reinforcement should be placed for this moment in both directions without curtailment through the type 3 elements and to the walls. The maximum negative moment per unit width is $0.125w_ul^2$ and, adopting Hillerborg's suggested distribution of this moment, a design moment per unit width of $0.25w_ul^2$ allocated to the $0.313l$ width each side of the column. Top reinforcement should be placed for this moment and, using the Hillerborg rule quoted previously, may be curtailed after extending 0.57 of the element length. In addition, because the slab corners are simply supported, top and bottom steel is required for a moment per unit width of

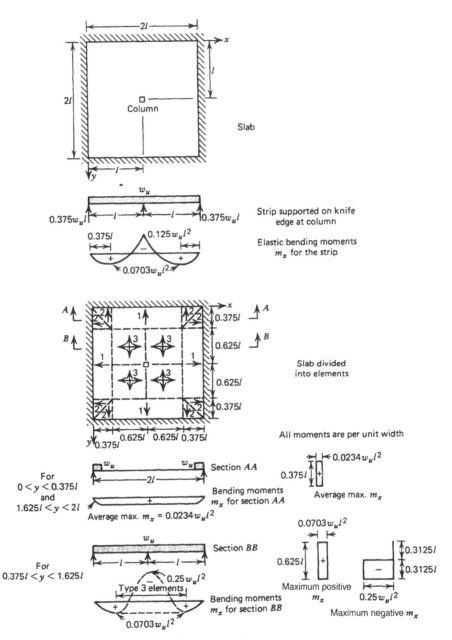

Figure 6.32   Design Example 6.4.

$0.0703w_ul^2$ in the four corners of the slab. This steel is required for torsion and should extend $0.2l$ into the span in each direction.

The division of some other types of slabs into elements for design by the advanced strip method is shown in Figs. 6.33 and 6.34. The uniformly loaded beamless multipanel floor of Fig. 6.33 is assumed to have columns or small interior walls of significant size. To take into account the effect of these on the span moments, type 1 elements are assumed to span between the column faces and the sides of the floor. A uniformly loaded L-shaped slab is shown in Fig. 6.34a. Design moments are determined from the moment equilibrium requirements of the elements, as previously. More detailed treatment of various slab types by this approach is given by Hillerborg.[6.11] It is of interest that in his book[6.11] Hillerborg prefers to call type 3 elements "corner-supported elements," and rather than type 2 elements in the corners he tends to use rectangular elements composed of strips spanning in two directions, as shown in Fig. 6.34b. Despite the somewhat violent discontinuities at times (e.g., along line aa of Fig. 6.34a and b), the resulting moment fields appear reasonable.

It is evident that the introduction of type 3 elements offers a powerful alternative to the use of strong bands in strip design theory.

6.4   SEGMENT EQUILIBRIUM METHOD

Wiesinger[6.13] has presented a segment equilibrium method involving triangular loaded areas for the design of uniformly loaded flat plates with regular or irregular column layouts. The method assumes that the perpendicular bisectors of the spans, and the lines connecting the corners of the columns with the panels centers (points of intersection of the bisectors), are lines of zero shear and form boundaries between the tributary load areas of the slab. The portions of the slab supported by the faces of the columns and equal in width to the column are assumed to span as one-way strips between columns. Torsional moments are disregarded. The sum of the negative and positive moments for the column and middle strips is found by considering the equilibrium of the various loaded areas of the slab. The distribution of the total moment between negative- and positive-moment sections of column and middle strips is carried out according to the percentages recommended in the direct design method for rectangular panels of the ACI Code[6.15] (see Chapter 4). For rectangular panels the column and middle strips are defined as in the ACI Code and for irregular panels by an equivalent scheme.

6.4.1   Flat Plates with Columns on a Rectangular Grid

Part of a flat plate floor with columns on a rectangular grid is shown in Fig. 6.35. The assumed zero shear lines are shown. The factored uniform load per

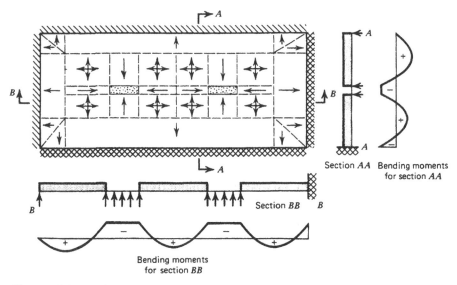

Figure 6.33   Uniformly loaded beamless slab supported on walls and two large interior columns.

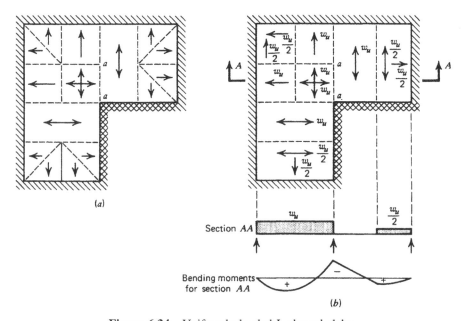

Figure 6.34   Uniformly loaded L-shaped slabs.

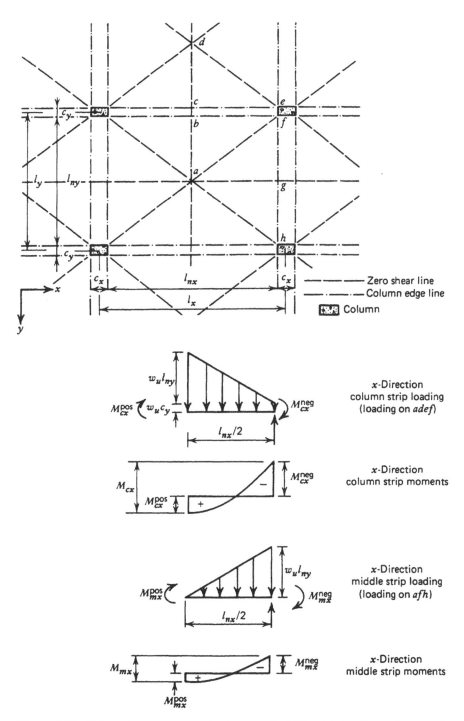

Figure 6.35   Triangular segments and column and middle strip loading and moments for uniformly loaded flat plate with regular column layout.

unit area is w_u. According to Wiesinger, the total x-direction moments acting on the column strip are found by considering the equilibrium of the loaded area $adef$ of the slab. The shears are zero at the edges of the loaded area within the panel, and taking moments about ef gives the total column strip moment as

$$M_{cx} = M_{cx}^{neg} + M_{cx}^{pos} = w_u l_{ny} \frac{l_{nx}^2}{12} + w_u c_y \frac{l_{nx}^2}{8} \qquad (6.14)$$

where M_{cx}^{neg} and M_{cx}^{pos} are the total negative and positive x-direction moments in the column strip (not moments per unit width), respectively; l_{nx} and l_{ny} the clear spans in the x- and y-directions, respectively; and c_y the width of the column in the y-direction.

Similarly, the total x-direction moments acting on the middle strip are found by considering the equilibrium of the loaded area afh of the slab. The shears are zero at the edges of the loaded area within the panel, and hence taking moments about fh gives the total middle strip moment as

$$M_{mx} = M_{mx}^{neg} + M_{mx}^{pos} = w_u l_{ny} \frac{l_{nx}^2}{24} \qquad (6.15)$$

where M_{mx}^{neg} and M_{mx}^{pos} are the total negative and positive x-direction moments in the middle strip (not moments per unit width), respectively, and l_{nx} and l_{ny} are the clear spans in the x- and y-directions, respectively. The total panel moment in the x-direction is, from Eqs. 6.14 and 6.15,

$$M_0 = M_{cx} + M_{mx} = (3c_y + 2l_{ny} + l_{ny})w_u \frac{l_{nx}^2}{24}$$

$$= w_u l_y \frac{l_{nx}^2}{8} \qquad (6.16)$$

which is the design static moment for the panel based on the clear span and is the same as in the ACI Code direct design method. Also, from Eqs. 6.14 to 6.16,

$$\frac{M_{cx}}{M_0} = \frac{2}{3}\frac{l_{ny}}{l_y} + \frac{c_y}{l_y} = \frac{2}{3} + \frac{1}{3}\frac{c_y}{l_y} \qquad (6.17)$$

$$\frac{M_{mx}}{M_0} = \frac{1}{3}\frac{l_{ny}}{l_y} = \frac{1}{3} - \frac{1}{3}\frac{c_y}{l_y} \qquad (6.18)$$

In the direct design method for flat plates of the ACI Code[6.15] (see Chapter 4), 65% of the static moment M_0 is distributed to negative moment, of which the column strip resists $0.75 \times 0.65M_0 = 0.4875M_0$ and the middle strip resists $0.25 \times 0.65M_0 = 0.1625M_0$. The remaining $0.35M_0$ is distributed to positive moment, of which the column strip resists $0.6 \times 0.35M_0 = 0.21M_0$ and the middle strip resists $0.4 \times 0.35M_0 = 0.14M_0$. Thus, according to the code, the total column strip moment is $0.4875M_0 + 0.21M_0 = 0.6975M_0$ and the total middle strip moment is $0.1625M_0 + 0.14M_0 = 0.3025M_0$. It is evident that the total column and middle strip moments given by Eqs. 6.17 and 6.18 are in excellent agreement with the ACI-recommended values, particularly as the code allows a 10% redistribution of moments from the recommended values. Therefore, the distribution of loading assumed by Wiesinger gives total column and middle strip moments of the required magnitude.

To find the negative and positive moments from his determined total strip moments M_{cx} and M_{mx}, Wiesinger suggests using the ratios of moments given by the ACI Code: namely, $M_{cx}^{neg}/M_{cx}^{pos} = 0.4875/0.21 = 2.32$ and $M_{mx}^{neg}/M_{mx}^{pos} = 0.1625/0.14 = 1.16$. The column strip and middle strip regions of the slab are defined as in the ACI Code, as shown in Fig. 6.36.

Equations 6.14 to 6.18 have only been derived for the x-direction moments. It is evident that a similar set of equations can be derived for the y-direction moments. The equations for the y-direction actions may be obtained directly from the preceding equation by reversing the subscripts, that is, by substituting subscript x for y and subscript y for x. The full load w_u on each triangular area is carried in both the x- and y-directions. For example, the load on triangle abf of Fig. 6.35 contributes to x-direction column strip moments and y-direction middle strip moments.

The method proposed by Wiesinger for establishing the equilibrium equations has similarities with Hillerborg's advanced strip method. The rectangular segment $abfg$ of the slab in Fig. 6.35 is, in fact, a Hillerborg type 3 element (see Section 6.3). The diagonal zero shear line af merely indicates the part of the load that is assumed to be carried by the column strip moments and the part that is assumed to be carried by the middle strip moments. Thus, Hillerborg considers the equilibrium of the total element $abfg$ to find the total contribution to column and middle strip moments, and then allocates the moment so found to the column and middle strips. Wiesinger, on the other hand, divides the rectangular element into two triangular elements with load on one triangle being carried by the column strip moments and load on the other triangle being carried by the middle strip moments.

There is, in fact, no justification for thinking in terms of "triangular elements." Figure 6.37a shows a Hillerborg type 3 element ($abfg$ of Fig. 6.35) supported at one corner at the column. Only the x-direction moments are shown. It is evident that the sum of the x-direction negative and positive column and middle strip moments are in equilibrium with the moment of the external loading about the support. That is, $\Sigma M_x = w_u l_y l_{nx}^2/16$, where ΣM_x is the sum of the x-direction negative and positive column and middle strip moments acting on the edges of the element. Figure 6.37b shows the element divided into two of Wiesinger's triangular elements (abf and fga of Fig.

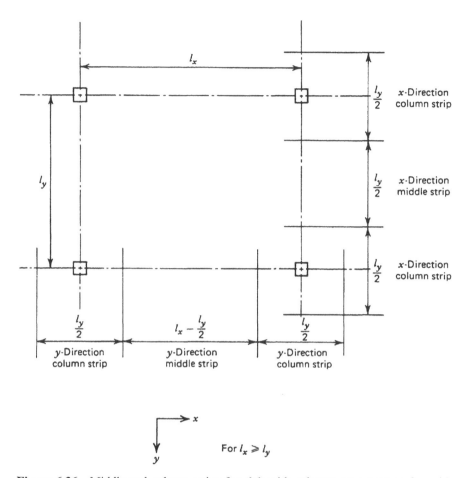

Figure 6.36   Middle and column strips for slab with columns on a rectangular grid.

6.35). Each triangular element must have both column and middle strip moments acting on it because the elements extend into those regions of the slab. Hence, the equilibrium equation for triangular element A will involve both column and middle strip moments (i.e., not just the column strip moments assumed by Wiesinger in Eq. 6.14). Also, the equilibrium equation for segment B will involve both column and middle strip moments (i.e., not just the middle strip moments assumed by Wiesinger in Eq. 6.15). Therefore, triangular elements cannot be considered as carrying either column or middle strip moments alone, and Wiesinger's equations in fact relate to rectangular elements, similar to Hillerborg's type 3 elements, with an assumed triangular loading carried by the column strip moments and the remaining triangular loading carried by the middle strip moments. Thus, Wiesinger's approach to flat plates with columns on a rectangular grid gives equations that are per-

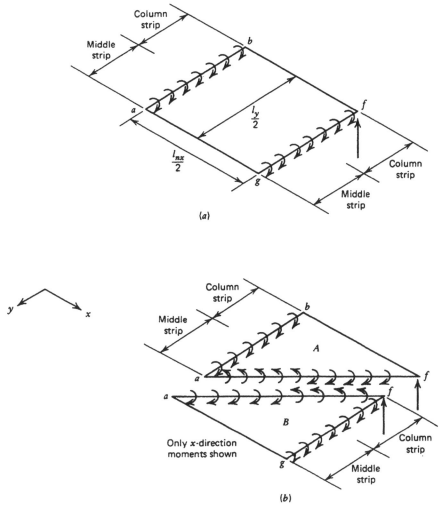

Figure 6.37 Rectangular and triangular elements: (*a*) Hillerborg's type 3 rectangular element; (*b*) Wiesinger's triangular elements.

fectly valid but relate to rectangular slab elements with triangular loading and not to triangular elements.

6.4.2 Flat Plates with Irregular Column Layouts

Wiesinger[6.13] has extended his concept of triangular "elements" to flat plates with irregular column layouts. Figure 6.38 shows a floor with columns arrayed in an irregular manner, such as may be necessary to suit a particular floor layout. In the first instance, the columns may be assumed to have zero width

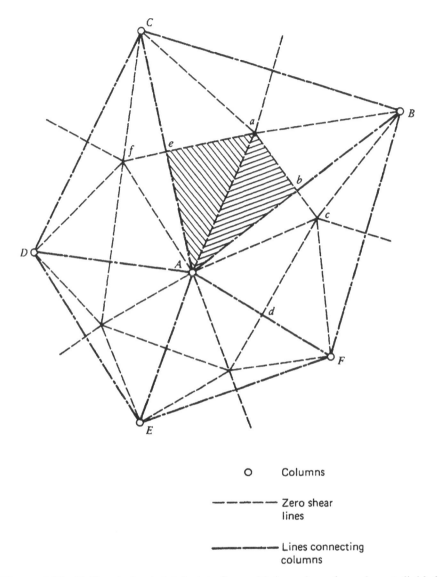

Figure 6.38 Uniformly loaded flat plate floor with irregular column layout divided into Wiesinger's triangular elements.

in order to simplify the determination of moments. The panel is imagined to be divided by lines of zero shear consisting of perpendicular bisectors of the lines between columns which intersect at a point (the slab "center") and lines joining those points with the columns.

Wiesinger recommends determining the column and middle strip moments by considering the equilibrium of triangular elements. The full load on each triangle is carried in two directions. For example, by analogy with the procedure used for rectangular panels, the load on triangle Aab in Fig. 6.38 contributes to the column strip moments in the direction AB and to the middle strip moments in direction ac. Similarly, the load on triangle Aae contributes toward the column strip moments in direction AC and to the middle strip moments in direction af. It is assumed for positive moments that the column strip for moment in direction AB is one-half of distance ac, and that the middle strip for moment in direction ac is one-half of distance AB. Similar definitions for the width of positive-moment strips apply elsewhere. Negative-moment requirements at columns are determined in each of the directions to the adjacent columns. For example, at column A negative-moment requirements will be calculated in directions AB, AC, AD, AE, and AF. To develop the required total negative-moment strength for all these moments simultaneously, orthogonal reinforcement can be placed to resist the sum of the components of the moments in the directions of the steel.

As discussed before, it is strictly not valid to think in terms of triangular elements unless all the moments acting on the element are considered. For slabs with a rectangular column grid it was apparent that it was necessary to think in terms of rectangular elements with loading on two triangles, the column strip moments being in equilibrium with the load on one triangle and the middle strip moments being in equilibrium with the load on the other triangle, in order to obtain Wiesinger's equations. In slabs with irregular column layouts, the analogy involves trapezium-shaped elements with load on two triangles. For example, consider trapezium-shaped element $Aeab$. To obtain Wiesinger's solution, the column strip moments in direction Ae on one side of column A would be found by taking moments of the load on triangle Aea about A in direction Ae, and the middle strip moments to one side of ba would be found by taking moments of the load on triangle Aba about A in the direction ba. However, the moments acting on the trapeziform element are more complex than this simple equilibrium procedure suggests. Figure 6.39 shows all the bending moments acting at the edges of the trapeziform element. It is not obvious that the column strip moments in direction Ae are in equilibrium with the loading on triangle Aea, or that the middle strip moments in direction ba are in equilibrium with the load on triangle Aba. Therefore, it is difficult to be assured of the validity of the equilibrium equations used in Wiesinger's method for irregular column layouts, although they are valid for columns on a rectangular grid.

Also, for the procedure to be a valid lower bound solution, the designer would need some assurance that the moments within the elements did not exceed the ultimate resisting moments provided on the basis of the edge moments. Another factor is that with such irregular-shaped elements, torsional moments, which have been ignored, may have a significant effect on the principal moments in the vicinity of the columns. Therefore, the validity of Wiesinger's approach for irregular column layouts is to be questioned. This is a pity because the approach intuitively appears correct and leads to a rel-

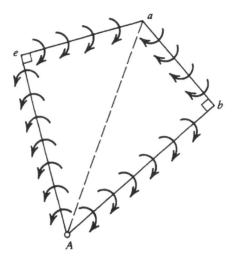

Figure 6.39   Bending moments acting on edges of trapezium shaped element.

atively simple procedure for determination of design moments in cases where the column layout is irregular.

It is to be noted that a strip design based on strong bands is a simple alternative. That is, strong bands can be placed in the slab between the columns, dividing the slab up into triangular areas such as shown in Fig. 6.40. Continuous triangular slabs can be designed supported on the strong bands and the strong bands in turn can be designed to support the triangular slabs.

6.5   COMPARISON WITH TEST RESULTS

Test results obtained from slabs designed by the strip method and other equilibrium procedures are relatively scarce. Most of the available test results are from scale models of relatively simple structures: for example, uniformly loaded square or rectangular panels with simple supports. One series of tests has been reported on slabs with internal columns designed by the simple strip method with strong bands or the advanced strip method. Some of the available experimental studies on reinforced concrete slabs are discussed below.

Taylor et al.[6.18] tested 10 uniformly loaded square slabs simply supported at all edges over 6-ft (1.83-m) spans. The span/thickness ratios were either 41, 36, or 24. The slabs were designed either by the strip method or by yield line theory (see Chapters 7 and 8). Uniformly spaced bars at right angles parallel to the supports or at 45° to the supports, bars with different spacing, and some bars curtailed before reaching the supports were the variables studied. The design ultimate load of the slabs was either 479 or 498 lb/ft² (22.9 or 23.8 kN/m²). The maximum load reached in the tests was 1.26 to 1.80

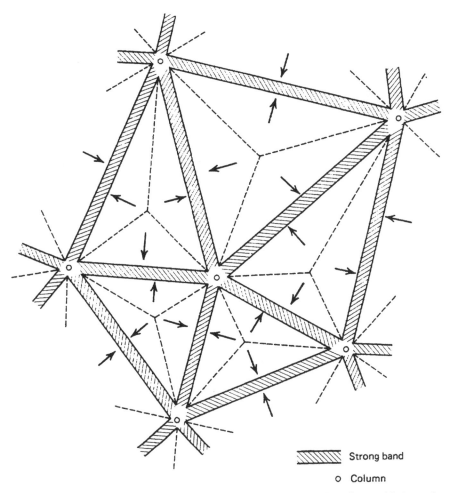

Figure 6.40 Alternative solution for uniformly loaded flat plate floor with irregular column layout.

times the design ultimate load. This large enhancement of strength was attributed to strain hardening of the reinforcement and to tensile membrane action that developed at large deflections. At large deflections load is no longer carried to the supports by ordinary bending action, as is evident from the penetration of cracking to the top surface in the central region of the slab, and the change of geometry can be considered as an increase in the internal lever arm. For the slabs with curtailed bars the ultimate load was not enhanced to the same extent. Cracking on the underside of the slabs was observed at about 311 lb/ft^2 (14.9 kN/m^2), and at that load the maximum crack widths observed were approximately 0.006 in. (0.15 mm) or rather less. Deflections reached span/360 at short-term loadings varying between 243 and 610 lb/ft^2

(11.6 to 29.2 kN/m^2). From these test results it was concluded that for the simple case of slabs tested, the strip (and yield line) design method resulted in slabs with safe predictions of ultimate loads and satisfactory behavior at service loads.

Other tests on square or rectangular slabs with simple or continuous supports, with uniform loading, have been reported by Sharpe and Clyde,[6.19] Rozvany,[6.20] and Muspratt.[6.21] The slabs were designed either by the strip method or by yield line theory. The ultimate loads measured were significantly greater than the yield line theory load, owing mainly to strain hardening of the reinforcement and to tensile membrane action, and cracking and deflections at service load were not a major problem. Some of these tests investigated various optimum reinforcement arrangements leading to minimum volume of reinforcing steel. Such solutions are attractive but pose the problem of more difficult bar placing and may also lead to areas of steel in some regions of the slab less than the minimum value required by codes.

A most interesting series of slab tests has been reported by Armer.[6.6] Seven rectangular slabs of spans 12 ft (3.66 m) by 6 ft (1.83 m) were tested under uniform loading. The slabs were $2\frac{1}{4}$ in. (57 mm) thick. The slabs were designated T1 to T7. Slabs T1, T2, and T3 were cast with integral edge beams and supported simply at the corners. Slabs T4 to T7 were not cast with edge beams and were simply supported along all edges. Slab T3 to T7 were also supported on a column at the slab center. Slab T2 had a 4.5 ft by 3.0 ft (1.37 m by 0.91 m) rectangular opening near the slab center. Slab T1 was designed by simple strip theory. Slab T2 was designed by simple strip theory with strong bands in the slab around the opening. Slabs T3 and T4 were designed by Hillerborg's advanced strip method (see Section 6.3). Slabs T5 and T6 were designed by the modified version of the advanced strip method suggested by Wood and Armer.[6.5] Slab T7 was designed by the simple strip method with strong bands within the slab passing each way over the central column. The design ultimate loads of the slabs varied between 420 and 840 lb/ft^2 (20.1 and 40.2 kN/m^2). The actual ultimate loads reached were 1.30 to 2.22 times the design ultimate loads. These very large enhancements of slab strength were evidently due to the assumption that load is carried by flexure only. In fact, during the tests membrane action within the slab was very apparent, particularly around the column. Therefore, there is no doubt as to the safety of the design method. Figures 6.41 to 6.43 show crack patterns on two of the slabs after failure. At a load of about one-half the theoretical ultimate load, the maximum deflections measured relative to the supporting beams were 0.30 in. (7.6 mm) for slab T1, 0.23 in. (5.8 mm) for slab T2, and between 0.08 and 0.18 in. (2.0 and 4.6 mm) for slabs T3 to T7. Armer concluded that the behavior at service load in terms of deflections and cracking was satisfactory. Nevertheless, it should be noted that a 0.3-in. deflection for slab T1 is short span/240 relative to the beams, and it may be that this deflection could be regarded as excessive at service load if the additional

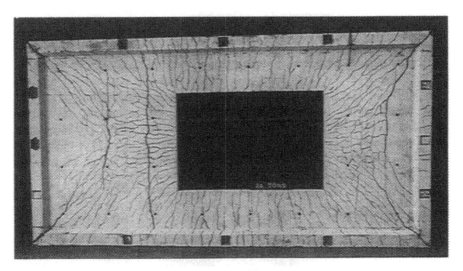

Figure 6.41 Bottom of slab T2 after failure.[6.6] (Crown copyright. Building Research Station, U.K.)

deflection due to the beams and to long-term loading is to be considered as well.

Cardenas and Kaar[6.22] have reported on a load test conducted on a full-scale flat plate floor. Although not designed by the strip method, the plate is

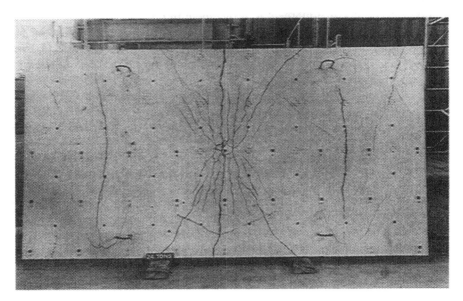

Figure 6.42 Top of slab T4 after failure.[6.6] (Crown copyright. Building Research Station, U.K.)

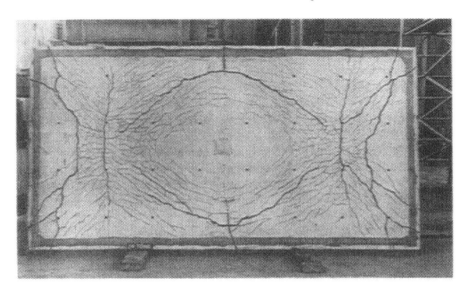

Figure 6.43   Bottom of slab T4 after failure.[6.6] (Crown copyright. Building Research Station, U.K.)

of interest in that in one-half of the test area, the total negative-moment reinforcement was concentrated in the column strips with no negative-moment steel in the middle strips, while in the other half of the test area the negative-moment reinforcement was distributed according to the requirements of ACI 318-63.[6.23] The positive-moment reinforcement in both halves of the test area was designed according to ACI 318-63.[6.23]. The slab had a thickness of $5\frac{1}{2}$ in. (140 mm) and the service live load was 50 lb/ft² (2.39 kN/m²). A live load of up to 104 lb/ft² (4.98 kN/m²) was applied during the test. At the full test load, measured crack widths in the plate were small and had no major effect on stiffness or serviceability. No significant difference in deflections on both halves of the test area were measured. Hence, the placing of all the negative-moment steel in the column strip evidently had no effect on the serviceability of the structure.

It is apparent that the various possible designs by the strip method have not been tested extensively. A range of testing remains to be conducted, particularly on slabs with free edges. Nevertheless, the available evidence is that the design ultimate loads are on the safe side and that serviceability will not be a problem if sensible arrangements of reinforcement are used.

REFERENCES

6.1   A. Hillerborg, "Jamviktsteori for armerade betongplattor," *Betong,* Vol. 41, No. 4, 1956, pp. 171–182.

6.2   A. Hillerborg, "A Plastic Theory for the Design of Reinforced Concrete Slabs," *Proceedings of the 6th Congress of the International Association for Bridge and Structural Engineering,* Stockholm, 1960.

6.3   A. Hillerborg, *Strimlemetoden for platlor pa pelare, vinkelplattor m m,* Utgiven av Svenska Riksbyggen, Stockholm, 1959. (*Strip Method for Slabs on Columns, L-Shaped Plates, etc.,* translated by F. A. Blakey, Commonwealth Scientific and Industrial Research Organisation, Melbourne, Victoria, Australia, 1964.)

6.4   R. E. Crawford, "Limit Design of Reinforced Concrete Slabs," Ph.D. thesis, University of Illinois at Urbana–Champaign, 1962, 163 pp.

6.5   R. H. Wood and G. S. T. Armer, "The Theory of the Strip Method for Design of Slabs," *Proc. Inst. Civ. Eng.,* Vol. 41, October 1968, pp. 285–311.

6.6   G. S. T. Armer, "Ultimate Load Tests of Slabs Designed by the Strip Method," *Proc. Inst. Civ. Eng.,* Vol. 41, October 1968, pp. 313–331.

6.7   G. S. T. Armer, "The Strip Method: A New Approach to the Design of Slabs," *Concrete,* Vol. 2, No. 9, September 1968, pp. 358–363.

6.8   K. O. Kemp, "Continuity Conditions in the Strip Method of Slab Design," *Proc. Inst. Civ. Eng.,* Vol. 45, February 1970, pp. 283 (supplement paper 7268 S).

6.9   K. O. Kemp, "A Strip Method of Slab Design with Concentrated Loads or Supports," *Struct. Eng.,* Vol. 49, No. 12, December 1971, pp. 543–548.

6.10   S. N. Shukla, *Handbook for Design of Slabs by Yield-Line and Strip Methods,* Structural Engineering Research Centre, Roorkee, India, 1973, 180 pp.

6.11   A. Hillerborg, *Dimensionering av armerade betongplattor enligt strimlemetoden,* Almqvist & Wiksell, Stockholm, 1974, 327 pp. (*Strip Method of Design,* Viewpoint Publications, Cement and Concrete Association, London, 1975, 256 pp.)

6.12   A. Hillerborg, *Strip Method Design Handbook,* E&FN Spon, London, 1996, 302 pp.

6.13   F. P. Wiesinger, "Design of Flat Plates with Irregular Column Layout," *Proc. ACI,* Vol. 70, February 1973, pp. 117–123.

6.14   *Structural Use of Concrete,* BS 8110, Part 1:1997, *Code of Practice for Design and Construction,* British Standards Institution, London, 1997, 121 pp.

6.15   *Building Code Requirements for Structural Concrete,* ACI 318-95, and *Commentary,* ACI 318R-95, American Concrete Institute, Farmington Hills, Mich., 1995, 371 pp.

6.16   J. S. Fernando and K. O. Kemp, "The Strip Method of Slab Design: Unique or Lower-Bound Solution?" *Mag. Concr. Res.,* Vol. 27, No. 90, March 1975, pp. 23–29.

6.17   R. Park and T. Paulay, *Reinforced Concrete Structures,* Wiley, New York, 1975, 769 pp.

6.18   R. Taylor, D. R. H. Maher, and B. Hayes, "Effect of the Arrangement of Reinforcement on the Behaviour of Reinforced Concrete Slabs," *Mag. Concr. Res.,* Vol. 18, No. 55, June 1966, pp. 85–94.

6.19   R. Sharpe and D. H. Clyde, "The Rational Design of Reinforced Concrete Slabs," *Civ. Eng. Trans., Inst. Eng., Aust.,* Vol. CE9, No. 2, October 1967, pp. 209–214.

6.20 G. I. N. Rozvany, "The Behaviour of Optimized Reinforced Concrete Slabs," *Civil Eng. Trans., Inst. Eng., Aust.,* Vol. CE9, No. 2, October 1967, pp. 283–292.

6.21 M. A. Muspratt, "Destructive Tests on Rationally Designed Slabs," *Mag. Concr. Res.,* Vol. 22, No. 70, March 1970, pp. 25–36.

6.22 A. E. Cardenas and P. H. Kaar, "Field Test of a Flat Plate Structure," *Proc. ACI,* Vol. 68, January 1971, pp. 50–59.

6.23 *Building Code Requirements for Reinforced Concrete,* ACI 318-63, American Concrete Institute, Detroit, 1963, 144 pp.

7 Basis of Yield Line Theory

7.1 INTRODUCTION

The method for the limit analysis of reinforced concrete slabs known as yield line theory was initiated by Ingerslev[7.1] and greatly extended and advanced by Johansen.[7.2,7.3] This method is an upper bound approach. The ultimate load of the slab system is estimated by postulating a collapse mechanism that is compatible with the boundary conditions. The moments at the plastic hinge lines are the ultimate moments of resistance of the sections, and the ultimate load is determined using the principle of virtual work or the equations of equilibrium. Being an upper bound approach the method gives an ultimate load for a given slab that is either correct or too high. The regions of the slab between the lines of plastic hinges are not examined to ensure that the moments there do not exceed the ultimate moments of resistance of the sections, but the ultimate moments of resistance between the lines of plastic hinges will be exceeded only if an incorrect collapse mechanism is used. Thus, all the possible collapse mechanisms of the slab must be examined to ensure that the load-carrying capacity of the slab is not overestimated. The correct collapse mechanisms in nearly all common cases are well known, however, and therefore the designer is not often faced with the uncertainty of whether further alternatives exist. It is to be noted that yield line theory assumes a flexural collapse mode, that is, that the slab has sufficient shear strength to prevent a shear failure.

The early literature on yield line theory was mainly in Danish, and in 1953 Hognestad[7.4] produced the first summary of this work in English. More recently, yield line theory has been treated extensively in publications by Wood,[7.5] Jones,[7.6] Wood and Jones,[7.7] the European Concrete Committee,[7.8,7.9] the Dutch Committee for Concrete Research,[7.10] Shukla,[7.11] and others. In this chapter we summarize the theory and enable the ultimate load of a range of slabs with known boundary conditions and type of loading to be determined.

7.2 SLAB REINFORCEMENT, SECTION BEHAVIOR, AND CONDITIONS AT ULTIMATE LOAD

7.2.1 Slab Reinforcement

Yield line theory is applicable to slabs that are reinforced uniformly. The sectional area of reinforcement per unit width is assumed to be constant across the slab, but may be different for the reinforcement in the two directions and different for reinforcement in the top and bottom of the slab. For such slabs the ultimate moment of resistance per unit width will have a constant value along any straight line in the plane of the slab. Usually, the reinforcement is placed in two directions at right angles, and this will be the main case considered. In some cases it is convenient for the angle between the reinforcement directions to be other than 90°, such as in skew slabs, and this case will also be considered later in the chapter. Yield line analysis of slabs with nonuniformly distributed reinforcement is also possible, but there are some extra problems, and the processes have not been generalized to the extent that they have been for uniformly distributed reinforcement.

7.2.2 Ductility of Slab Sections

The configuration of moments in a slab at the ultimate load depends on the flexural strength of the slab sections as well as on the loading and the boundary conditions. Significant redistribution of bending moments may be necessary to develop the collapse mechanism. This implies that the slab sections should be ductile enough to allow plastic rotation to occur at the critical sections while the plastic hinging develops throughout the slab. The available ductility is dependent on the shape of the moment–curvature relationship of the sections. Figure 7.1 illustrates a typical moment–curvature relationship for a singly reinforced concrete section. The relationship has approximately a trilinear shape, with an initial elastic portion to first cracking of the concrete, then a linear portion to first yielding of the tension steel, and finally, a nearly horizontal region in which the moment of resistance remains at close to the ultimate value until the concrete reaches its ultimate strain. A measure of the ductility of the section is the curvature ductility factor, defined as the ratio of curvature when the ultimate concrete strain is reached to curvature when the

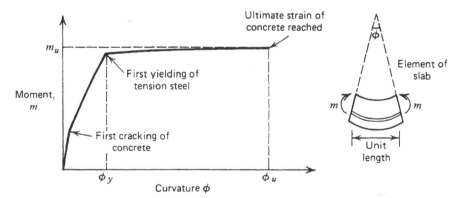

Figure 7.1   Moment–curvature relationship for a singly reinforced concrete slab section.

tension steel first yields, ϕ_u/ϕ_y. For singly reinforced slabs with a concrete compressive cylinder strength of 3000 lb/in² (20.7 N/mm²) and a steel yield strength of 40,000 lb/in² (276 N/mm²), it can be shown[7.12] that for tension steel ratios of 0.02, 0.015, 0.01, and 0.005 the curvature ductility factor is approximately 4, 6, 10, and 23, respectively, if a value of 0.004 for the ultimate concrete strain is assumed. For higher-strength reinforcement, as would be expected, lower steel ratios will be necessary to obtain curvature ductility factors in this range. The presence of compression steel in sections may increase the available ductility. It is evident that for most slabs the tension steel ratios will be low enough to ensure that the sections are reasonably ductile. It is assumed in yield line theory that sufficient ductility is always available at the critical sections to allow the slab to develop its collapse mechanism with the ultimate moments of resistance maintained at all plastic hinge sections. Flexural strength increase due to strain hardening of steel is excluded since the strain at which strain hardening begins is not stipulated in specifications for steel, and it is difficult to include it.

7.2.3 Conditions at Ultimate Load

Consider a reinforced concrete slab that is progressively loaded to failure. At low loads before cracking, the distribution of bending moments is as according to elastic plate theory. After cracking, the distribution of moments changes due to the decrease in flexural rigidity at the cracked sections. With further loading, yielding of tension steel eventually occurs at the section of maximum bending moment and the slab undergoes a large change in curvature at the sections of yielding, with the moment there remaining practically constant at the ultimate moment of resistance. A large redistribution of bending moments may occur with further loading. As the load on the slab is increased, the lines of intense cracking across which the tension steel has yielded (known as yield lines) are propagated from the point at which yielding originated until eventually the yield lines have formed in sufficient numbers to divide the slab into segments that can form a collapse mechanism. Once the mechanism is formed, the slab can carry no further load. A system of yield lines forming a collapse mechanism is known as a yield line pattern. Although yielding of tension steel begins at the section of maximum bending moment given by elastic theory, the positions of the yield lines developed by further loading are governed by the arrangement of reinforcement, the boundary conditions, and the type of loading. Figure 7.2 shows the development of the yield line pattern for a uniformly loaded simply supported rectangular slab. It is to be noted that a yield line is, in fact, an idealization for a band of intense cracking across which the tension steel has yielded. Figure 7.3 shows an actual crack pattern on the tension face of a simply supported slab at the ultimate uniformly distributed load. For the purpose of analysis, the band of intense cracking across which the tension steel has yielded is represented by a single yield line at the center of the band, and all plastic rotation is considered to occur

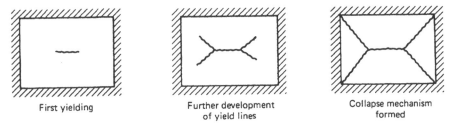

| First yielding | Further development of yield lines | Collapse mechanism formed |

Figure 7.2   Yield lines developing in a uniformly loaded simply supported slab.

along that line. For the full attainment of the ultimate load it is evident that sufficient plastic rotation must be available at the first yield lines to form, to allow the yield line pattern to be developed with the full ultimate moment of resistance at each yield line. The available curvature ductility factor of sections was discussed in Section 7.2.2 and Ref. 7.12. No simple analytical procedure has been devised to enable the required curvature ductility factor for slabs to be calculated. Instead, tests have been used to substantiate that slabs generally have the required ultimate curvature capacity to develop the full ultimate load.

7.2.4   Yield Lines as Axes of Rotation

When the collapse mechanism has developed, the plastic deformations along the yield lines are much greater than the elastic deformations of the slab segments between the yield lines, and hence in the theory it is reasonable to assume that the slab segments between the yield lines are plane, or that once

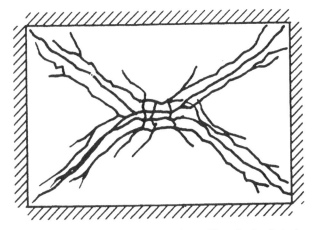

Figure 7.3   Actual crack pattern at failure of a uniformly loaded simply supported slab.

a mechanism has formed, all *additional* deformations occur as if each segment were a plane. Examination of the geometry of the deformations gives three basic rules for the determination of the pattern of yield lines:

1. To act as plastic hinges of a collapse mechanism made up of plane segments, yield lines must be straight lines forming axes of rotation for the movements of the segments.
2. The supports of the slab will act as axes of rotation. If an edge is fixed, a yield line may form along the support. An axis of rotation will pass over a column.
3. For compatibility of deformations, a yield line must pass through the intersection of the axes of rotation of the adjacent slab segments.

The foregoing rules may be used to postulate the yield line patterns for slabs. With practice, the yield line pattern for the normal range of slabs can readily be sketched.

The convention used to illustrate types of supports, axes of rotation, and yield lines is indicated in Fig. 7.4. This convention is used throughout the chapter. Figure 7.5 shows some examples of the yield line patterns for uniformly loaded slabs of various shapes and boundary conditions. The angles between the yield lines and the supports can be established by methods to be given later. Note that for each slab there may be more than one family of possible yield line patterns, any of which may be the critical pattern.

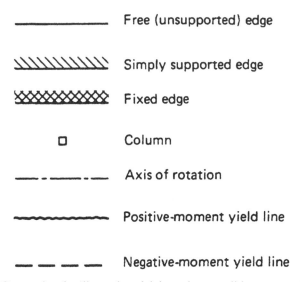

Figure 7.4 Convention for illustrating slab boundary conditions, axes of rotation, and yield lines.

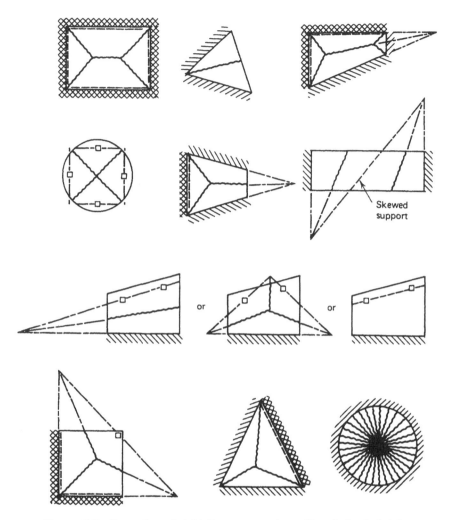

Figure 7.5   Examples of yield line patterns for uniformly loaded slabs.

7.2.5   Ultimate Moments of Resistance at Yield Lines

For a yield line that runs at right angles to the reinforcement, the ideal ultimate moment of resistance per unit width due to that reinforcement is given by[7.12]

$$m_u = A_s f_y \left(d - 0.59 A_s \frac{f_y}{f_c'} \right) \tag{7.1}$$

where A_s is the area of tension steel per unit width, f_y the yield strength of the reinforcement, d the distance from the centroid of the tension steel to the extreme concrete compression fiber, and f_c' the compressive cylinder strength of the concrete. In design the right-hand side of Eq. 7.1 is multiplied by a strength reduction factor $\phi = 0.9$ to obtain the dependable (design) strength. The effect of compression steel can be neglected in flexural strength calculations, since for typical underreinforced slab sections it makes little difference to the strength of the section.[7.12]

In the usual case of a slab reinforced by bars at right angles in the x- and y-directions, the ultimate moments per unit width in the x- and y-directions will generally be unequal, because the areas of steel and the effective depths of the steel will be different in those directions. Also, it is often necessary to determine the ultimate moment per unit width along a yield line that is at an angle of other than 90° to the x- and y-axes. In this general case, torsional moments will exist along the yield lines as well as the ultimate (bending) moments. The ultimate bending moment per unit width m_{un} and torsional moment per unit width m_{unt} acting at a general yield line may be found from Johansen's yield criterion. This criterion assumes that the actual yield line can be replaced by a "stepped" line consisting of a series of small steps in the x- and y-directions, that the torsional moments acting in the x- and y-direction are zero, that the strength of the section is not influenced by kinking of reinforcing bars across the crack or by biaxial stress conditions in the concrete compression zone, that the tension steel in both directions crossing the yield line has reached yield strength, and that the internal lever arms for the x- and y-direction ultimate moments of resistance are not affected when bending occurs in a general direction. The foregoing assumptions are discussed in detail in Section 5.2.4. Tests on slabs have shown that despite its simplicity, Johansen's yield criterion is accurate.

Figure 7.6 shows a yield line crossing reinforcement at a general angle. The reinforcement is placed in the x- and y-directions at right angles and the yield line is inclined at angle α to the y-axis. The equivalent stepped yield line is also shown in the figure. The ultimate resisting moments per unit width in the x- and y-directions are m_{ux} and m_{uy}, respectively. These moments can be found using Eq. 7.1. The components of m_{ux} and m_{uy} contributing to the ultimate moment of resistance per unit width m_{un} and torsional moment per unit width m_{unt} acting at the yield line may be found by considering the equilibrium of the small triangular element taken from the yield line. The moments acting on the element are shown in the figure in vector notation.

Taking moments about side ab of the element shows that for equilibrium, the ultimate moment of resistance per unit width acting normal to the yield line is

$$m_{un}ab = m_{ux}ac \cos \alpha + m_{uy}bc \sin \alpha$$

$$m_{un} = m_{ux} \cos^2\alpha + m_{uy} \sin^2\alpha \tag{7.2}$$

Similarly, taking moments about an axis perpendicular to ab shows that the torsional moment per unit width acting along the yield line is

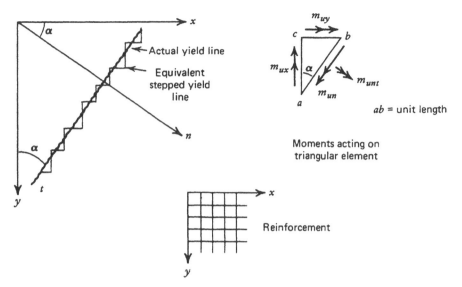

Figure 7.6   Yield line at general angle to orthogonal reinforcement.

$$m_{unt} = (m_{ux} - m_{uy}) \sin \alpha \cos \alpha \tag{7.3}$$

If $m_{ux} = m_{uy}$, then from Eq. 7.2, $m_{un} = m_{ux} \cos^2\alpha + m_{uy} \sin^2\alpha = m_{ux} = m_{uy}$, and from Eq. 7.3, $m_{unt} = 0$. Thus, for this case the ultimate moments of resistance per unit width are equal in all directions and the torsional moment at the yield line is zero. Such a slab is said to be *isotropic* or *isotropically reinforced*. When $m_{ux} \neq m_{uy}$, it is evident that the ultimate moment of resistance per unit width is dependent on the direction of the yield line and that there will be a torsional moment at the yield line. Such a slab is said to be *orthotropic* or *orthotropically reinforced*.

7.2.6   Determination of the Ultimate Load

The first step in any yield line solution is to postulate the yield line pattern using the rules given in Section 7.2.4. The patterns in general will contain unknown dimensions that locate the positions of the yield lines, and there may be more than one family of yield lines for a particular slab, as illustrated in Fig. 7.5. The designer should ensure that all possible yield line patterns are envisaged, since the correct pattern is the one giving the lowest ultimate load, and if the critical pattern is missed, the calculated ultimate load will be unsafe.

The ultimate load may be found from the yield line patterns using either the principle of virtual work or the equations of equilibrium. Both approaches are discussed in the following sections. Each approach has advantages for some situations. In general, the virtual work method is easier in principle but more difficult to manipulate algebraically.

7.3 ANALYSIS BY PRINCIPLE OF VIRTUAL WORK

7.3.1 Virtual Work Equation

Suppose that a rigid body is in equilibrium under the action of a system of forces. If the body is given a small arbitrary displacement, the sum of the work done by the forces (force times its corresponding displacement) will be zero because the resultant force is zero. Hence, the *principle of virtual work* may be stated as: If a rigid body that is in statical equilibrium under a system of forces is given a virtual displacement, the sum of the virtual work done by the forces is zero.

The virtual displacement is a small arbitrary displacement and the virtual work is the work resulting from the displacement.

To analyze a slab by the virtual work method, a yield line pattern is postulated for the slab at the ultimate load. The segments of the yield line pattern may be regarded as rigid bodies because the slab deformation with further deflection occurs only at the yield lines. The segments of the slab are in equilibrium under external loading and the bending and torsional moments and shears along the yield lines. A convenient point within the slab is chosen and given a small displacement δ in the direction of the load. Then the resulting displacements at all points of the slab, $\delta(x,y)$, and the rotations of the slab segments about the yield lines, may be found in terms of δ and the dimensions of the slab segments. Work will be done by the external loads and by the internal actions along the yield lines. The work done by a uniformly distributed ultimate load per unit area w_u is

$$\iint w_u \delta(x,y)\ dx\ dy = \Sigma\ W_u \Delta \qquad (7.4)$$

where W_u is the total load on a segment of the yield line pattern and Δ is the downward movement of its centroid. Work for all segments is summed. The reactions at the supports will not contribute to the work, as they do not undergo displacement. The work done by the internal actions at the yield lines will be due only to the bending moments, because the work done by the torsional moments and the shear forces is zero when summed over the whole slab. This follows because the actions on each side of the yield line are equal and opposite, as shown in Fig. 7.7, and for any displacement of the yield line pattern there is no relative movement between the sides of the yield line corresponding to the torsional moments and shear forces. However, there is relative movement corresponding to the bending moments, since there is a relative rotation between the two sides of the yield line. Thus, the work done

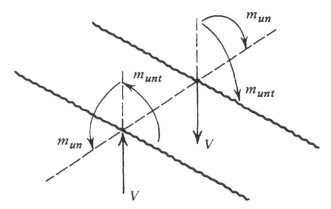

Figure 7.7 Actions at a yield line.

at the yield lines is due only to the ultimate (bending) moments. The work done by the ultimate moment of resistance per unit width m_{un} at a yield line of length l_0, where the relative rotation about the yield line of the two segments is θ_n, is $-m_{un}\theta_n l_0$. The work done here is negative because the bending moments will be acting in the direction opposite to the rotation if the slab is given a displacement in the direction of the loading. The total work done by the ultimate moments of resistance, as given by summing the work done along all the yield lines, is $-\Sigma\, m_{un}\theta_n l_0$. Therefore, the virtual work equation may be written as

$$0 = \Sigma\, W_u\Delta - \Sigma\, m_{un}\theta_n l_0$$

or

$$\Sigma\, W_u\Delta = \Sigma\, m_{un}\theta_n l_0 \qquad (7.5)$$

When applied to a particular slab, the displacement term cancels from the equation and the ultimate load is given in terms of the slab dimensions and the ultimate moments of resistance per unit width. The term $\Sigma\, W_u\Delta$ will be referred to as the external work done, and the term $\Sigma\, m_{un}\theta_n l_0$ will be referred to as the internal work done.

Example 7.1. Determine the ultimate uniformly distributed load per unit area w_u of a square simply supported slab. The slab is isotropically reinforced with ultimate positive moments of resistance per unit width m_u in the directions of both spans.

SOLUTION. The ultimate moment of resistance per unit width has a constant value m_u in all directions. A postulated yield line pattern for the slab is shown in Fig. 7.8. Because of symmetry, the junction point of the yield lines must be at the slab center. Hence, the yield line pattern has no unknown dimensions. Let the center of the slab be given a small downward displacement δ. The length of each diagonal yield line is $\sqrt{2}\, l$, where l is the span of the slab. The total rotation of the segments about each yield line is $2 \times [\delta/(l\sqrt{2})] = 2\sqrt{2}\, \delta/l$. Hence, for each of the two diagonal yield lines, the terms in the internal work done equation are $m_{un} = m_u$, $\theta_n = 2\sqrt{2}\, \delta/l$, and $l_0 = \sqrt{2}\, l$. Therefore, internal work done is

$$\Sigma\, m_{un}\theta_n l_0 = 2\left(m_u \times \frac{2\sqrt{2}\,\delta}{l} \times \sqrt{2}\, l\right) = 8m_u\delta$$

and external work done is

$$\Sigma\, W_u\Delta = 4\left(\frac{w_u l^2}{4} \times \frac{\delta}{3}\right) = w_u l^2 \frac{\delta}{3}$$

which is the total load on the triangular segments multiplied by the downward

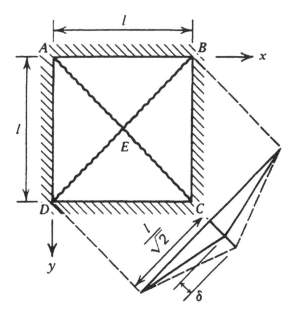

Figure 7.8 Uniformly loaded square slab of Example 7.1.

displacement of their centroids. Therefore, the virtual work equation, Eq. 7.5, may be written as

$$w_u l^2 \frac{\delta}{3} = 8 m_u \delta$$

from which the ultimate uniformly distributed load per unit area is

$$w_u = 24 \frac{m_u}{l^2} \tag{7.6}$$

Corner effects can cause a more complicated arrangement of yield lines in the corner regions of the slab than has been assumed above, and may result in a slightly lower ultimate load. Corner effects are discussed later.

7.3.2 Components of Internal Work Done

Since most slabs are rectangular with steel placed parallel to the edges in the x- and y-directions, and because the ultimate moments per unit width in the x- and y-directions are known, it is usually easier to deal separately with the x- and y-direction components of the internal work done by the ultimate moments, $\Sigma \, m_{un} \theta_n \, l_0$. For a yield line inclined at angle α to the y-axis (see Fig.7.9), with the segments of the slab undergoing a relative rotation θ_n about the yield line, reference to Eq. 7.2 shows that the internal work done is

$$\Sigma \, m_{un} \theta_n l_0 = \Sigma (m_{ux} \cos^2\alpha + m_{uy} \sin^2\alpha)\theta_n l_0$$

$$= \Sigma \, m_{ux} \theta_n \cos \alpha \; y_0 + \Sigma \, m_{uy} \theta_n \sin \alpha \; x_0$$

$$= \Sigma \, m_{ux} \theta_x y_0 + \Sigma \, m_{uy} \theta_y x_0 \tag{7.7}$$

where θ_x and θ_y are the components of θ_n in the x- and y-directions, respectively, and x_0 and y_0 are the projected lengths of the yield line in the x- and y-directions. It is to be noted that the rotation about the yield line is the sum of the rotations of the slab segments each side of the yield line.

The virtual work equation, Eq. 7.5, can now be written as

$$\Sigma \, W_u \Delta = \Sigma \, m_{ux} \theta_x y_0 + \Sigma \, m_{uy} \theta_y x_0 \tag{7.8}$$

Use of the virtual work equation in this form means that it is not necessary to find the ultimate moment of resistance normal to the yield line.

7.3.3 Minimum-Load Principle

In most cases a yield line pattern cannot be drawn without unknown dimensions locating the yield line positions. Figure 7.10 illustrates some examples

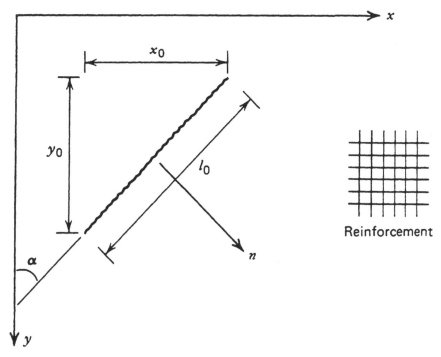

Figure 7.9   Yield line inclined to directions of orthogonal reinforcement.

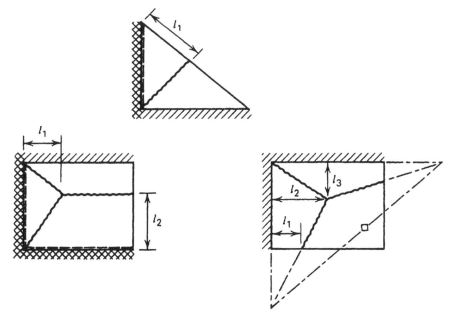

Figure 7.10   Yield line patterns with unknown dimensions.

in which the unknown dimensions are shown as l_1, l_2, l_3, and so on. In these cases the unknown dimensions will be included in the virtual work equation. The equation for ultimate load is in the form $W_u = f(l_1, l_2, l_3, \ldots)$. Since an upper bound approach is used, the values for l_1, l_2, l_3, $\ldots$ required are those values that give the minimum value for W_u and may be found by solving simultaneously the equations

$$\frac{\partial W_u}{\partial l_1} = 0, \quad \frac{\partial W_u}{\partial l_2} = 0, \quad \frac{\partial W_u}{\partial l_3} = 0, \quad \ldots \qquad (7.9)$$

The values of l_1, l_2, l_3, $\ldots$ obtained from the solutions of Eq. 7.9 are substituted back into the ultimate load equation to obtain the minimum ultimate load.

Example 7.2. Determine the ultimate uniformly distributed load per unit area w_u of a rectangular simply supported slab. The slab is orthotropically reinforced with ultimate positive moments of resistance per unit width m_{ux} and m_{uy} in the directions of the long and short spans, respectively.

SOLUTION. A postulated yield line pattern for the slab is shown in Fig. 7.11. Because of symmetry, the yield line *EF* lies on the slab centerline and l_1 is the only unknown dimension. Let the center of the slab be given a small downward displacement δ.

The internal work done can be found from Eq. 7.7 as below.

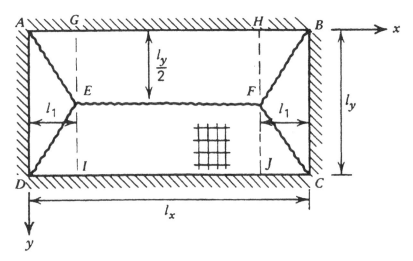

Figure 7.11   Uniformly loaded rectangular slab of Example 7.2.

Segment	Components of Rotation		Components of Work	
	θ_x	θ_y	$m_{ux}\theta_x y_0$	$m_{uy}\theta_y x_0$
ADE	$\dfrac{\delta}{l_1}$	0	$m_{ux}\dfrac{\delta}{l_1}l_y$	0
ABFE	0	$\dfrac{\delta}{0.5l_y}$	0	$m_{uy}\dfrac{2\delta}{l_y}l_x$
BCF	$\dfrac{\delta}{l_1}$	0	$m_{ux}\dfrac{\delta}{l_1}l_y$	0
DCFE	0	$\dfrac{\delta}{0.5l_y}$	0	$m_{uy}\dfrac{2\delta}{l_y}l_x$

Therefore,

$$\Sigma\, m_{un}\theta_n l_0 = 2m_{ux}\frac{l_y}{l_1}\delta + 4m_{uy}\frac{l_x}{l_y}\delta$$

The external work done is, from Eq. 7.4,

$$\Sigma\, W_u\Delta = w_u 2l_y l_1 \frac{\delta}{3} + w_u l_y(l_x - 2l_1)\frac{\delta}{2}$$

$$= w_u l_y(3l_x - 2l_1)\frac{\delta}{6}$$

The first term on the right-hand side of the external work done equation is from the two end regions of the slab, AGID and BHJC, each of area $l_y l_1$, which are each composed of three triangles with centroids moving down by $\delta/3$; the second term is from the remaining central region of area l_y $(l_x - 2l_1)$, composed of two rectangles with centroids moving down by $\delta/2$.

Note that the external work done can be thought of as being the ultimate load per unit area multiplied by the volume swept out by the segments when being displaced. From the virtual work equation, Eq. 7.5,

$$w_u l_y(3l_x - 2l_1)\frac{\delta}{6} = 2m_{ux}l_y\frac{\delta}{l_1} + 4m_{uy}l_x\frac{\delta}{l_y}$$

$$w_u = \frac{12}{l_y^2(3l_x/l_y - 2l_1/l_y)}\left(\frac{m_{ux}l_y}{l_1} + \frac{2m_{uy}l_x}{l_y}\right)$$

For minimum w_u from Eq. 7.9,

$$0 = \frac{dw_u}{dl_1} = \left(\frac{3l_x}{l_y} - \frac{2l_1}{l_y}\right)\left(-\frac{m_{ux}l_y}{l_1^2}\right) - \left(\frac{m_{ux}l_y}{l_1} + \frac{2m_{uy}l_x}{l_y}\right)\left(-\frac{2}{l_y}\right)$$

Therefore,

$$\left(\frac{l_1}{l_y}\right)^2 + \frac{m_{ux}}{m_{uy}}\frac{l_y}{l_x}\left(\frac{l_1}{l_y}\right) - \frac{3}{4}\frac{m_{ux}}{m_{uy}} = 0$$

$$\frac{l_1}{l_y} = \frac{1}{2}\left\{\left[\left(\frac{l_y}{l_x}\frac{m_{ux}}{m_{uy}}\right)^2 + 3\frac{m_{ux}}{m_{uy}}\right]^{1/2} - \frac{l_y}{l_x}\frac{m_{ux}}{m_{uy}}\right\}$$

Note that for the yield line pattern to be valid, l_1 cannot exceed $0.5l_x$. On substituting the expression for l_1/l_y into the equation for w_u, the ultimate uniform load per unit area is given as

$$w_u = \frac{24m_{uy}}{l_y^2\{[3 + (m_{ux}/m_{uy})(l_y/l_x)^2]^{1/2} - (l_y/l_x)(m_{ux}/m_{uy})^{1/2}\}^2} \qquad (7.10)$$

Corner effects can cause a more complicated arrangement of yield lines in the corner regions of the slab than has been assumed above and will be discussed later.

In some cases, especially if a single slab with known values of l_x, l_y, m_{ux}, and m_{uy} is to be solved, a trial-and-error solution rather than the general solution involving the derivative of the load with respect to the unknown variable(s) may be more convenient. In a trial-and-error solution, successive values of the variable dimension(s), l_1 in the example above, are selected until the one corresponding to the minimum load w_u has been found. The nature of the problem is such that the load is not particularly sensitive to the value of the variable if values close to the correct one are selected. Optimization procedures in spreadsheet and math-solving computer programs provide very convenient means of finding the critical values of the variables and the minimum load once the virtual work equation has been written.

Example 7.2 illustrates that the advantage of using a procedure that sums the x- and y-direction components of internal work is that in the usual case of slabs with edges in the x- and y-directions, either the x- or y-direction component of work is zero. Thus, the calculation is simplified. With familiarity with the method, the terms involving the components of internal work done can be summed without the necessity of a table.

Example 7.3. Determine the ultimate uniformly distributed load per unit area w_u of a triangular slab simply supported at two sides, with the third side free. The slab is isotropically reinforced with ultimate positive moments of resistance per unit width m_u.

SOLUTION. The slab and postulated yield line pattern are shown in Fig. 7.12. The solution will apply to slabs with various values for the side ratio k and internal angle β.

For given values of k and β, there is one unknown dimension in the yield line pattern. The unknown can be defined by the angle ϕ. The choice of whether to define the unknown by a length dimension such as y_0 or x_0 or an angle such as ϕ is a matter of convenience. Let the point D of the slab edge be given a downward displacement δ.

Internal work done:

Segment *ABD:* $\theta_x = \dfrac{\delta}{aD} = \dfrac{\delta}{y_0 \tan \phi}$

$\theta_y = 0$

Segment *CBD:* $\theta_x = \dfrac{\delta}{cD} = \dfrac{\delta}{y_0 \tan \beta - y_0 \tan \phi}$

$\theta_y = \dfrac{\delta}{bD} = \dfrac{\delta}{y_0 - y_0 \tan \phi \cot \beta}$

Therefore,

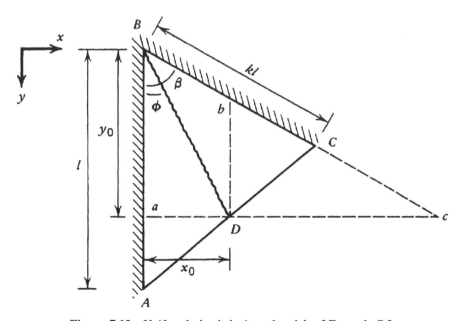

Figure 7.12   Uniformly loaded triangular slab of Example 7.3.

$$\Sigma \, m_{un}\theta_n l_0 = \Sigma \, m_{ux}\theta_x y_0 + \Sigma \, m_{uy}\theta_y x_0$$

$$= m_u \frac{\delta}{y_0 \tan \phi} y_0 + m_u \frac{\delta}{y_0(\tan \beta - \tan \phi)} y_0$$

$$+ m_u \frac{\delta}{y_0(1 - \tan \phi \tan \beta)} y_0 \tan \phi$$

$$= m_u\delta\left(\cot \phi + \frac{1}{\tan \beta - \tan \phi} + \frac{\tan \phi}{1 - \tan \phi \tan \beta}\right)$$

$$= m_u\delta\left(\cot \phi + \frac{1 + \tan \beta \tan \phi}{\tan \beta - \tan \phi}\right)$$

$$= m_u\delta[\cot \phi + \cot(\beta - \phi)]$$

External work done:

$$\Sigma \, W_u\Delta = w_u \times \text{area of segments} \times \frac{\delta}{3}$$

$$= w_u \frac{lkl \sin \beta}{2} \frac{\delta}{3}$$

$$= w_u\delta kl^2 \frac{\sin \beta}{6}$$

From the virtual work equation, Eq. 7.5,

$$w_u\delta kl^2 \frac{\sin \beta}{6} = m_u\delta[\cot \phi + \cot(\beta - \phi)]$$

Therefore,

$$w_u = \frac{6m_u[\cot \phi + \cot(\beta - \phi)]}{kl^2 \sin \beta}$$

which on expansion gives

$$w_u = \frac{6m_u}{kl^2 \sin \phi \sin(\beta - \phi)}$$

For minimum $w_u = dw/d\phi = 0$, which gives

$$\cos \phi \sin(\beta - \phi) - \sin \phi \cos(\beta - \phi) = 0$$

Therefore,

$$\tan \phi = \tan(\beta - \phi)$$

$$\phi = \frac{\beta}{2}$$

Thus, minimum load is when the corner angle between the supported edges is bisected by the yield line. On substituting for ϕ', the ultimate load per unit area is given as

$$w_u = \frac{6m_u}{kl^2 \sin^2(\beta/2)} \tag{7.11}$$

as obtained by Jones.[7.6] Again, possible more complex arrangements of yield lines in the corner of the slab have been neglected but are discussed later.

7.4 ANALYSIS BY EQUATIONS OF EQUILIBRIUM

7.4.1 Equilibrium Equations

An alternative method for determining the ultimate load of a slab from the yield line pattern is to use the equations of equilibrium. In this method the equilibrium of each individual segment of the yield line pattern, under the action of its bending and torsional moments, shear forces, and external loads, is considered. Generally, the equilibrium equations are written by taking moments of the actions about suitable axes. Sufficient equilibrium equations need to be written to be solved simultaneously to enable the unknown dimensions that define the yield line pattern to be eliminated and to find the ultimate load. No differentiation process is required and in most cases the algebraic manipulation necessary to obtain a solution is less than for the virtual work method.

In the virtual work method, the magnitudes and distributions of the shear forces and torsional moments that act at the yield lines were not required to be known because they do no work when summed over the entire slab when the yield line pattern is given a small displacement. For an equilibrium solution, however, because the segments are considered separately, all the actions at the yield lines need to be known before a solution can be obtained. The torsional moment at yield lines given by Johansen's yield criterion was derived previously (see Eq. 7.3). The magnitudes of the shear forces at yield lines are considered in the next sections.

The derivation of the shear forces acting at yield lines is the most difficult aspect of yield line theory and has caused controversy. It has been shown[7.13,7.14] that some of Johansen's original theorems for the determination of shear forces at yield lines have limitations, but it appears that these have now been resolved. It has also become apparent in recent years that the equilibrium method is actually the virtual work method presented in another form.[7.7]

The equilibrium method described in this chapter should not be confused with the equilibrium method used in the lower bound solutions of Chapters 5 and 6. Both approaches use the equations of equilibrium, but in yield line theory no check is made that the moments within the segments between the yield lines do not exceed the ultimate moments of resistance of the sections, and therefore this yield line theory method is not a lower bound solution.

7.4.2 Statical Equivalents of Shear Forces Along a Yield Line

Figure 7.13a shows a typical yield line pattern of a slab, and Fig. 7.13b shows the bending and torsional moments, and the shear forces, acting at the three internal edges of a segment of the yield line pattern. A vector notation is used for the moments: The moment acts in a clockwise direction when looking along the arrow (right-hand rule). The ultimate bending and torsional moments at the yield lines are assumed to be given by Eqs. 7.2 and 7.3.

It is convenient to replace the actual shear forces acting at each straight length of yield line by two statically equivalent forces, one at each end of the length of yield line. This is permissible since any system of coplanar forces can be replaced by two statically equivalent forces. Thus, for the yield line abcd of Fig. 7.13b, the shear forces acting along ab are replaced by single forces q_{ab} at a and q_{ba} at b. Similarly, the shear forces along lines bc and cd can be replaced by statically equivalent forces at the ends of these lines. The magnitude and direction of these forces is unknown. The forces are shown in Fig. 7.13c and will be taken as positive if acting upward. The total force acting on a segment at a point of intersection of two adjacent straight portions of a yield line is known as a *nodal force*. The nodal force acting at b of segment A will be referred to as Q_{Ab}, and in Fig. 7.13c,

$$Q_{Ab} = q_{ba} + q_{bc} \tag{7.12}$$

Similarly, the nodal force at b in segment B is Q_{Bb}, and at b in segment C is Q_{Cb}, and so on, as in Fig. 7.13d.

Equilibrium requires the shear forces acting on each side of a yield line to be equal and opposite. Therefore, if the statically equivalent force q_{ba} for yield line ba of segment A acts upward, the statically equivalent force q_{ba} for yield line ba of segment B must act downward. The equal and opposite argument also applies to the statically equivalent forces q_{bc} of segments A and C and the statically equivalent forces q_{bf} of segments B and C. It is evident that if all six statically equivalent forces at intersection point b of yield lines

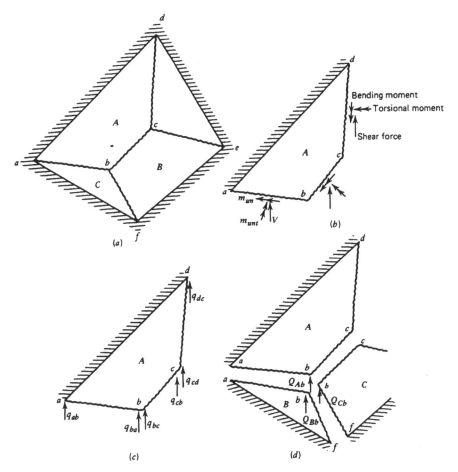

Figure 7.13 (*a*) Slab of general shape with yield line pattern; (*b*) actions at yield lines of segment *A*; (*c*) statically equivalent shear forces; (*d*) intersection of yield lines.

ba, *bc*, and *bf* of Fig. 7.13*d* are added, the sum must be zero, as is required by the equilibrium of the slab in any case. This argument can be extended to the junction of any number of yield lines. Hence, it may be stated that "at the junction of any number of yield lines, the sum of the nodal forces is zero." This statement can be written as

$$Q_{Ab} + Q_{Bb} + Q_{Cb} + \cdots = 0 \qquad (7.13)$$

7.4.3 Magnitude of Nodal Forces

To calculate the magnitude of the nodal force between two yield lines, the derivation and method given by Jones[7.6] will be closely followed. The method is basically that due to Johansen,[7.2] but it has been extended by Jones to give a more general equation.

The term *mesh* is used below to imply reinforcement bars placed in two directions. Consider the equilibrium of a small triangular element between two yield lines, such as element A', as shown in Fig. 7.14, where ab and ac are yield lines but ce is not a yield line. The angle between the yield lines is ϕ and is measured in an anticlockwise direction from ac. The angle ace is $\delta\phi$, and as $\delta\phi \to 0$ it is evident that the moments that exist along ac are the same as those along ec. The slab is orthotropic. The moments at the yield lines are taken in the positive directions. The moments at the first yield line are governed by reinforcement mesh 1 and at the second yield line by mesh 2. Mesh 1 is under the first yield line and mesh 2 is under the second yield line. Meshes 1 and 2 may be different and may be inclined at different angles. At point a the statical equivalent of the shear forces along ac is q_{ac} and that due to the short length ae of yield line ab is q_{ae}, both acting upward, and therefore $Q_{A'a} = q_{ae} + q_{ac}$. It is to be noted that $Q_{A'a}$ is not the total nodal force at point a between yield lines ab and ac, since it only contains forces for a part of yield line ab.

Consider the effect of the bending and torsional moments along lines ec and ca. If the lines are considered to be stepped in the direction of mesh 1, it can be seen in Fig. 7.15 that the resultant moment along path eca is the same as that along path efa. Also, the same resultant moment is obtained by going from e to a directly, while associating that path with the reinforcement of mesh 1. Hence, the total effect of bending and torsional moments along lines ec and ca is a bending moment $(m_{un})_1$ and a torsional moment $(m_{unt})_1$

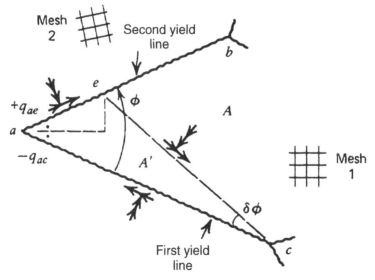

Figure 7.14   Equivalent shear forces at junction of two yield lines.

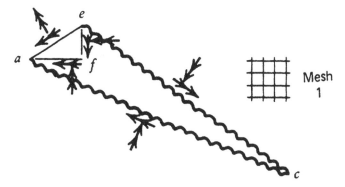

Figure 7.15 Equivalent moments along *ac* and *ce*.

acting along path *ea*. The subscript 1 outside the parentheses implies that these ultimate resisting moments are governed by mesh 1. Note that $(m_{un})_1$ acts perpendicular to, and $(m_{unt})_1$ acts in the direction of, the second yield line, *ba*. The moments acting along path *ae* due to the actual moments along *ab* are $(m_{un})_2$ and $(m_{unt})_2$ acting in the direction *ae*. These ultimate resisting moments are governed by mesh 2. Thus, the effect of the moments along lines *ae*, *ec*, and *ac* of element A' is a bending moment $(m_{un})_2 - (m_{un})_1$ and a torsional moment $(m_{unt})_2 - (m_{unt})_1$ along path *ae*. Note that the subscripts 1 and 2 imply that the ultimate resisting moments in the path are due to meshes 1 and 2, respectively. These moments are shown in Fig. 7.16.

Taking moments about the axis *ec* for the equilibrium of triangle A' gives

$$(q_{ec} + q_{ac})ae \sin(\phi + \delta\phi) = [(m_{un})_2 - (m_{un})_1]ae \cos(\phi + \delta\phi)$$

$$+ [(m_{unt})_2 - (m_{unt})_1]ae \sin(\phi + \delta\phi)$$

When $\delta\phi \to 0$, and on dividing through by $ae \sin \phi$, the expression becomes

$$Q_{A'a} = q_{ae} + q_{ac} = [(m_{un})_2 - (m_{un})_1] \cot \phi + [(m_{unt})_2 - (m_{unt})_1] \quad (7.14)$$

where $(m_{un})_1$ and $(m_{unt})_1$ are the ultimate bending and torsional resisting moments due to mesh 1 in the path of the second yield line, and $(m_{un})_2$ and $(m_{unt})_2$ are the ultimate bending and torsional resisting moments due to mesh 2 in the path of the second yield line. Note that the bending moment acts perpendicular to the path and the torsional moment acts along the path.

As a final step, consider three yield lines meeting at a point and let their ultimate resisting moments be governed by three orthotropic meshes, 1, 2, and 3. These yield lines are shown in Fig. 7.17. Consider initially the equilibrium of element A' bounded by *ae*, *ec*, and *ac*. The *first* and *second* yield lines bounding the element are *ac* and *ad*, and their ultimate resisting mo-

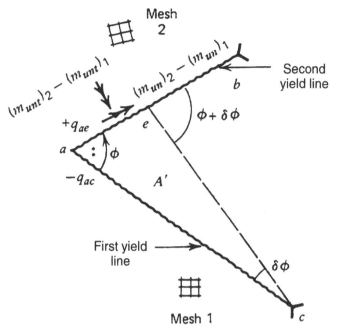

Figure 7.16   Equivalent moments along *ae*.

ments are determined by meshes 1 and 3, respectively. Thus, for this element, Eq. 7.14 becomes

$$Q_{A'a} = q_{ae} + q_{ac} = [(m_{un})_3 - (m_{un})_1] \cot \phi_{13} + [(m_{unt})_3 - (m_{unt})_1] \quad (7.15)$$

where all the moments are along the path of line *ad* and the subscripts outside the parentheses indicate the governing mesh. The angle ϕ_{13} is between lines *ac* and *ad*.

Consider now the equilibrium of element *B'*, which is bounded by lines *ae*, *eb*, and *ab*. The *first* and *second* yield lines for this element are *ab* and *ad*, and their ultimate resisting moments are determined by meshes 2 and 3, respectively. For this element, Eq. 7.14 becomes

$$Q_{B'a} = q_{ae} + q_{ab} = [(m_{un})_3 - (m_{un})_2] \cot \phi_{23} + [(m_{unt})_3 - (m_{unt})_2]$$

where all the moments are along the path of line *ad* and the subscripts outside the parentheses indicate the governing mesh. The angle ϕ_{23} is between lines *ab* and *ad*. Now

$$Q_{A'a} - Q_{B'a} = (q_{ae} + q_{ac}) - (q_{ae} + q_{ab}) = q_{ac} - q_{ab} \quad (7.16)$$

The statically equivalent shear force $-q_{ab}$ in Eq. 7.16 is that acting above

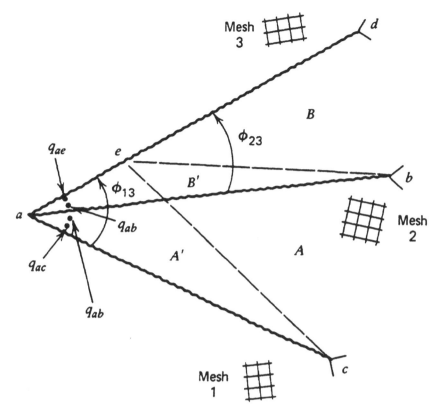

Figure 7.17 Nodal forces at the junction of three yield lines.

line *ab* in Fig. 7.17. Its equal but opposite statically equivalent shear force acting below line *ab* will therefore be $+q_{ab}$. But the total nodal force Q_{Aa} acting between full yield lines *ab* and *ac* is $q_{ab} + q_{ac}$, and hence Eq. 7.16 gives Q_{Aa}. Therefore, substituting Eqs. 7.14 and 7.15 into Eq. 7.16 gives

$$Q_{Aa} = [(m_{un})_3 - (m_{un})_1] \cot \phi_{13} - [(m_{un})_3 - (m_{un})_2] \cot \phi_{23}$$
$$+ [(m_{unt})_2 - (m_{unt})_1] \qquad (7.17)$$

Rather than Q_{Aa}, it is preferable to call the nodal force of Eq. 7.17 Q_{12} (see Fig. 7.18), because it is the nodal force between yield lines 1 and 2, which surround the element. Therefore,

$$Q_{12} = [(m_{un})_3 - (m_{un})_1] \cot \phi_{13} - [(m_{un})_3 - (m_{un})_2] \cot \phi_{23}$$
$$+ [(m_{unt})_2 - (m_{unt})_1] \qquad (7.18)$$

where Q_{12} is the nodal force between yield lines 1 and 2; $(m_{un})_1$ the ultimate

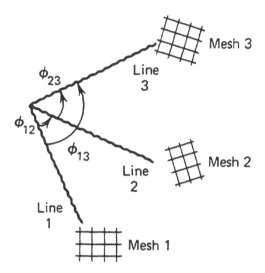

Figure 7.18 Numbering of yield lines.

bending moment per unit width due to mesh 1, which is under yield line 1, along the path of yield line 3; $(m_{un})_2$ the ultimate bending moment per unit width due to mesh 2, which is under yield line 2, along the path of yield line 3; $(m_{un})_3$ the ultimate bending moment per unit width due to mesh 3, which is under yield line 3 along the path of yield line 3; $(m_{unt})_1$ the torsional moment per unit width due to mesh 1, which is under yield line 1, along the path of yield line 3; $(m_{unt})_2$ the torsional moment per unit width due to mesh 2, which is under yield line 2, along the path of yield line 3; ϕ_{13} the angle between yield lines 1 and 3 measured counterclockwise from yield line 1; and ϕ_{23} the angle between yield lines 2 and 3 measured counterclockwise from yield line 2. Refer to Fig. 7.18. The expression "along the path of yield line 3" means that the bending moment vector points along yield line 3 and therefore that the bending moment acts perpendicular to yield line 3 and the torsional moment acts along yield line 3. When Eq. 7.18 is used to find a nodal force, the yield lines are numbered in a counterclockwise direction with lines 1 and 2 surrounding the nodal force required. Equation 7.18 may appear cumbersome for constant use, but fortunately, its application to two commonly found cases gives simple results, as discussed below.

Case 1: Junction of Three Yield Lines All Governed by the Same Mesh. If all three reinforcement meshes are the same, then in Eq. 7.18

$$(m_{un})_1 = (m_{un})_2 = (m_{un})_3$$

$$(m_{unt})_1 = (m_{unt})_2$$

Therefore,

$$Q_{12} = 0$$

Similarly, it may be shown that the other nodal forces at the junction are zero. This is true whether the moment at the yield lines are all negative or all positive, or whether the mesh is isotropic or orthotropic. Thus, it may be stated that: At the junction of three yield lines governed by the same mesh, each of the nodal forces is zero if the yield lines are either all negative-moment or all positive-moment lines.

Case 2: Intersection of a Yield Line and a Free Edge. Figure 7.19 shows a yield line intersecting a free (unsupported) edge. Lines 2 and 3 are the free edge and therefore may be considered as yield lines of zero strength. Hence, the moments with subscripts 2 and 3 outside the parentheses in Eq. 7.18 are zero. Therefore, from Eq. 7.18,

$$Q_{12} = -(m_{un})_1 \cot \phi_{13} - (m_{unt})_1$$

But

$$\cot \phi_{13} = \cot(\pi + \beta) = \cot \beta$$

Therefore,

$$Q_{12} = -(m_{un})_1 \cot \beta - (m_{unt})_1 \tag{7.19}$$

where $(m_{un})_1$ and $(m_{unt})_1$ are the ultimate bending and torsional moments along the path of line 3, that is, bending in the direction perpendicular to the free edge and torsion in the direction of the free edge.

Also, since the sum of the nodal forces at the intersection is zero, $Q_{13} = -Q_{12}$. According to the sign convention, a negative value for the nodal force indicates a downward force and a positive value indicates an upward force when a positive-moment yield line intersects the edge.

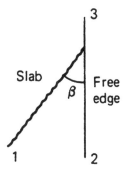

Figure 7.19   Yield line intersecting a free edge.

For the usual case of reinforcement placed parallel and perpendicular to the free edge, $\alpha = 0$ or $90°$ in Eq. 7.3 and therefore $(m_{unt})_1 = 0$. If the ultimate moment of resistance per unit width acting perpendicular to the free edge is m_{ue}, then $(m_{un})_1 = m_{ue}$ and from Eq. 7.19 the nodal force at the free edge in Fig. 7.20 is

$$Q_e = \pm m_{ue} \cot \beta \qquad (7.20)$$

For a positive-moment yield line, the nodal force is downward in the acute angle and upward in the obtuse angle. For a negative-moment yield line, the directions of the nodal forces are reversed. Note that when $\beta = 90°$, the nodal forces Q_e become zero.

Restrictions. Wood and Jones[7.7] have more recently pointed out that Eq. 7.18 cannot be used to determine the nodal forces in all situations. It has been realized for some time that certain "breakdown" cases existed, but it is only in recent years (see, e.g., Refs. 7.13 to 7.17) that these have been clarified. The exceptions are:

1. Equation 7.18 cannot be used to determine the nodal forces at a junction that is a fixed point (i.e., an anchored point) in a yield line pattern. For the equation to be valid, the junction point must be free to move to the position that gives the minimum load to cause collapse. For example, for the uniformly loaded slab with the opening shown in Fig. 7.21a, the equation may be used to find the nodal forces at A, because A may be any point on the edge of the opening. (Part of the ultimate load determination for the slab would be finding the position of A that gave the least load.) However, the equation cannot be used to find the nodal

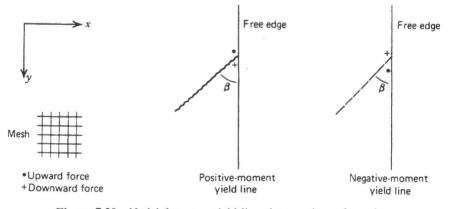

Figure 7.20   Nodal forces at yield lines intersecting a free edge.

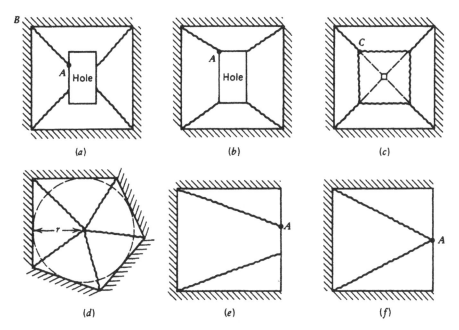

Figure 7.21 Uniformly loaded slabs.

forces at A in Fig. 7.21b, since there the yield line has been assumed to run into the corner of the opening and hence is anchored there.

2. Equation 7.18 cannot be used to determine the nodal force between a yield line and a supported boundary. For example, the nodal forces acting at corner B of the slab in Fig. 7.21a cannot be determined from the equation.

3. Equation 7.18 cannot be used at nodes where more than three yield lines meet. For example, the equation cannot be used to determine the nodal forces at C of the uniformly loaded slab with an internal column shown in Fig. 7.21c. However, for multisided isotropic slabs with edges that are either all simply supported or all fixed, and where any number of positive moment yield lines meet at the center of the inscribed circle as shown in Fig. 7.21d, the nodal forces at the center may be shown to be zero.

4. Similarly, Eq. 7.18 gives the correct nodal force at A for the uniformly loaded slab shown in Fig. 7.21e, but does not give the correct nodal force at A in Fig. 7.21f, since it is basically a case where four yield lines meet.

The breakdown cases and special theory to cover these exceptions are discussed at length by Wood and Jones.[7.7] The nodal forces in these cases may be found by comparing the virtual work and equilibrium solutions and determining what the nodal forces should be for agreement.

7.4.4 Method of Solution by the Equilibrium Equations

To solve a slab by the equilibrium method, the steps are:

1. Postulate a yield line pattern.
2. Calculate the values for the required nodal forces.
3. Write the equilibrium equations by taking moments about axes of rotation and, if necessary, resolving the vertical forces of each slab segment. The number of equilibrium equations required is one more than the number of unknown dimensions necessary to define the positions of yield lines in the yield line pattern. When taking moments for the equilibrium equation, it is best to consider the yield line to be stepped in the directions of the reinforcement. If moments are taken in the directions of the reinforcement, the torsional moments do not have to be considered, since they are zero in those directions (see Section 7.2.5) according to Johansen's yield criterion.
4. Solve the equilibrium equations simultaneously to determine the unknown dimensions and hence to find the ultimate load.

The following examples illustrate the procedure.

Example 7.4. Determine the ultimate uniformly distributed load per unit area w_u of a square, simply supported slab. The slab is isotropically reinforced with ultimate positive moments of resistance per unit width m_u in the directions of both spans.

SOLUTION. The ultimate moment of resistance per unit width has a constant value m_u in all directions. A postulated yield line pattern is shown in Fig. 7.22. Because of symmetry the junction point of the yield lines must be at the slab center. Hence, the yield line pattern has no unknown dimensions and only one equilibrium equation is required. The yield lines are all governed by the same reinforcement mesh, and are of the same sign, and hence the nodal forces at e are all zero because e is at the center of the inscribed circle. For segment A of the yield line pattern, taking moments about the support axis gives

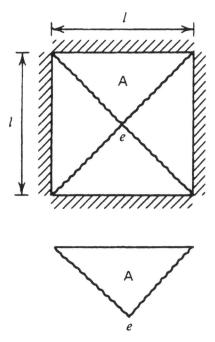

Figure 7.22   Uniformly loaded square slab of Example 7.4.

$$m_u l = \frac{w_u l^2}{4} \frac{l}{6}$$

Therefore,

$$w_u = 24 \frac{m_u}{l^2} \tag{7.21}$$

Again, the possible more complex arrangement of yield lines in the corners of the slab have been neglected.

Example 7.5.   Determine the ultimate uniformly distributed load per unit area w_u of a rectangular slab that is fixed at two adjacent edges and is free at the other edges. The slab is orthotropically reinforced in the top and the bottom and the reinforcing bars run parallel to the slab edges.

SOLUTION. The slab and a postulated yield line pattern (there may be others) are shown in Fig. 7.23. The ultimate negative and positive moments of resistance per unit width are m'_{ux} and m_{ux}, respectively, in the x-direction, and m'_{uy} and m_{uy}, respectively, in the y-direction. The yield line pattern shown in

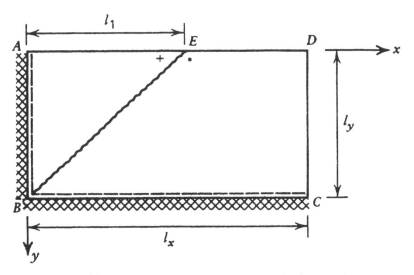

Figure 7.23   Uniformly loaded rectangular slab of Example 7.5.

Fig. 7.23 has one unknown dimension l_1, and therefore two equilibrium equations are required for solution. The nodal forces at the free edge at E, according to Eq. 7.20, are $\pm m_{uy} l_1 / l_y$ acting downward in the acute angle and upward in the obtuse angle. The two equilibrium equations are obtained by taking moments about the supports for segments ABE and $CBED$.

For segment ABE, taking moments about AB gives

$$(m'_{ux} + m_{ux})l_y - m_{uy} \frac{l_1}{l_y} l_1 - \frac{w_u l_1 l_y}{2} \frac{l_1}{3} = 0$$

Therefore,

$$w_u = \frac{6(m'_{ux} + m_{ux})}{l_1^2} - \frac{6 m_{uy}}{l_y^2} \qquad (7.22)$$

The second term in the equilibrium equation above is due to the downward nodal force at E.

For segment $CBED$, taking moments about CB gives

$$m'_{uy} l_x + m_{uy} l_1 + m_{uy} \frac{l_1}{l_y} l_y - \frac{w_u l_1 l_y}{2} \frac{l_y}{3} - w_u (l_x - l_1) l_y \frac{l_y}{2} = 0$$

Therefore,

$$w_u = \frac{6(m'_{uy}l_x + 2m_{uy}l_1)}{l_y^2(3l_x - 2l_1)} \tag{7.23}$$

Equating Eqs. 7.22 and 7.23 gives

$$\frac{6(m'_{ux} + m_{ux})}{l_1^2} - \frac{6m_{uy}}{l_y^2} = \frac{6(m'_{uy}l_x + 2m_{uy}l_1)}{l_y^2(3l_x - 2l_1)}$$

Therefore,

$$0 = \left(\frac{l_1}{l_x}\right)^2 + 2\frac{l_1}{l_x}\left(\frac{l_y}{l_x}\right)^2 \frac{m'_{ux} + m_{ux}}{m'_{uy} + 3m_{uy}} - 3\left(\frac{l_y}{l_x}\right)^2 \frac{m'_{ux} + m_{ux}}{m'_{uy} + 3m_{uy}}$$

$$\frac{l_1}{l_x} = \frac{\sqrt{1 + 3K_1} - 1}{K_1} \tag{7.24}$$

where

$$K_1 = \left(\frac{l_x}{l_y}\right)^2 \left(\frac{m'_{uy} + 3m_{uy}}{m'_{ux} + m_{ux}}\right) \tag{7.25}$$

The ultimate uniformly distributed load per unit area w_u is found by substituting l_1 from Eqs. 7.24 and 7.25 into either Eq. 7.22 or 7.23. Note that for the yield line pattern to be valid, l_1 cannot exceed l_x.

A trial-and-error solution is also possible. For this case, a value of l_1 is assumed, and the loads w_u of the two segments are computed using Eqs. 7.22 and 7.23. If they are not equal, the value of l_1 is adjusted in the direction required to reduce the higher load and increase the lower load, and a new trial is completed. When the loads are equal, or within some small tolerance, the correct values of l_1 and w_u have been found. This process may be particularly useful when there are several variable dimensions in the pattern, as the relative loads give some guidance to the adjustments needed to reach the correct mechanism, where the loads on all segments are the same. This process is discussed further in Section 7.16. A simple program in a spreadsheet can greatly facilitate the trial-and-error process because each trial can be done so easily.

Example 7.6. Determine the ultimate uniformly distributed load per unit area w_u of a regular n-sided slab fixed around the edges. The slab is isotropically reinforced in the top and bottom with ultimate negative and positive moments of resistance per unit width m'_u and m_u, respectively.

SOLUTION. Let the inscribed radius of the slab be r and the length of each side be l. The postulated yield line pattern is shown in Fig. 7.24. The nodal forces at the slab center are zero and there are no unknown dimensions in the yield line pattern. Consider the equilibrium of segment ABC by taking moments about the support line AR.

$$(m_u' + m_u)l = \frac{w_u lr}{2}\frac{r}{3}$$

Therefore,

$$w_u = \frac{6(m_u' + m_u)}{r^2} \tag{7.26}$$

This solution can be applied to a range of cases of regular sided slabs.
For $n = 3$, a triangular slab (Fig. 7.25a),

$$r = 0.5l \tan 30° = \frac{0.5l}{\sqrt{3}}$$

Therefore,

$$w_u = \frac{72(m_u' + m_u)}{l^2} \tag{7.27}$$

For $n = 4$, a square slab (Fig. 7.25b),

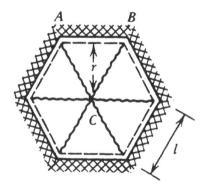

Figure 7.24 Uniformly loaded regular sided slab of Example 7.6.

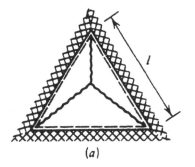

(a)

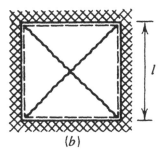

(b)

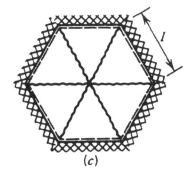

(c)

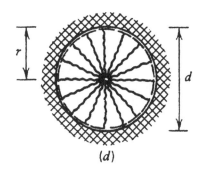

(d)

Figure 7.25 Special cases of uniformly loaded regular sided slabs of Example 7.6.

$$r = \frac{l}{2}$$

Therefore,

$$w_u = \frac{24(m'_u + m_u)}{l^2} \tag{7.28}$$

For $n = 6$, a hexagonal slab (Fig. 7.25c),

$$r = 0.5l \tan 60° = \frac{\sqrt{3}\, l}{2}$$

Therefore,

$$w_u = \frac{8(m'_u + m_u)}{l^2} \qquad (7.29)$$

For $n \to \infty$, a circular slab (Fig. 7.25d), the radius is r. An infinitive number of radial yield lines form and $l = 0$. Therefore,

$$w_u = \frac{6(m'_u + m_u)}{r^2} \qquad (7.30)$$

Note that the conical-shaped yield line pattern of Fig. 7.25d is also a possible yield line pattern for slabs of other regular shapes. Figure 7.26a shows it forming within a square slab. For the square slab $r = l/2$, and hence from Eq. 7.30,

$$w_u = \frac{24(m'_u + m_u}{l^2} \qquad (7.31)$$

Equations 7.28 and 7.31 are the same, and hence the yield line patterns of Figs. 7.25b and 7.26a give the same uniformly distributed ultimate load. Similarly, the conical yield line pattern of Fig. 7.25d could form within a hexagonal slab and give the same ultimate uniform load as that given by Eq. 7.29.

A good approximation for the ultimate load of a uniformly loaded isotropic irregular slab such as that shown in Fig. 7.26b may be obtained from the yield line pattern shown in that figure with the junction point of the yield

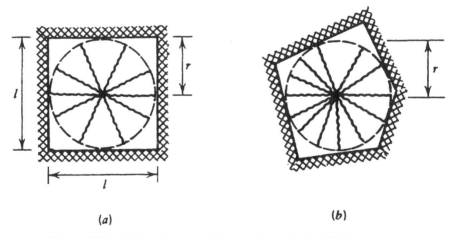

(a)   (b)

Figure 7.26   Uniformly loaded slabs with conical yield line patterns.

lines at the center of the inscribed circle. The ultimate uniform load as given by the equilibrium of any segment is

$$w_u = \frac{6(m_u' + m_u)}{r^2} \tag{7.32}$$

This is not an exact solution, because for minimum load the junction point may be off-center in this unsymmetrical case. but the approximation is quite accurate.

7.5   CONCENTRATED LOADS

7.5.1   Types of Yield Line Patterns

When concentrated loads act, the flexural failure modes are likely to involve concentrations of yield lines around the loaded area. In general, for concentrated loads, yield line patterns involving curved negative-moment lines with radial positive-moment lines are liable to be more critical than patterns involving large triangular segments between yield lines.

Example 7.7. Determine the ultimate central concentrated load P_u of a square slab fixed around the edges. The slab is isotropically reinforced in the top and bottom with ultimate negative and positive moments per unit width m_u' and m_u, respectively.

SOLUTION. The ultimate load given by two possible yield line patterns will be compared. Figure 7.27a shows a yield line pattern with four large triangular segments between the yield lines. The nodal forces induced by shear forces along the yield lines are zero at the slab center, but by symmetry the external load causes a load $P_u/4$ to act at the center of the slab on each segment. Consider the equilibrium of segment ABC by taking moments about support line AB.

$$(m_u' + m_u)l = \frac{P_u}{4}\frac{l}{2}$$

$$P_u = 8(m_u' + m_u) \tag{7.33}$$

Figure 7.27b shows a yield line pattern with curved (circular) negative-moment yield lines and n radial positive-moment yield lines, where $n \to \infty$. The curved negative-moment yield line is really composed of a series of short straights between the radial lines. By symmetry, each segment has a load

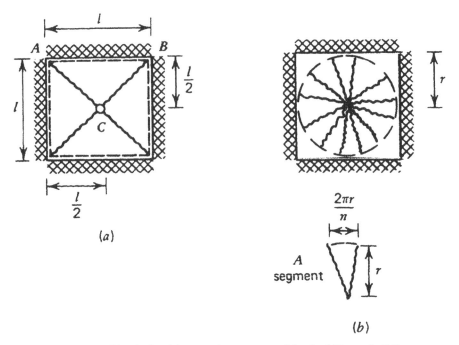

Figure 7.27   Slab with central concentrated load of Example 7.7.

P_u/n acting at the center of the slab. Consider the equilibrium of a segment by taking moments about its support line.

$$(m_u' + m_u)\frac{2\pi r}{n} = \frac{P_u}{n} r$$

$$P_u = 2\pi(m_u' + m_u) \tag{7.34}$$

The ultimate load given by Eq. 7.34 is smaller than that of Eq. 7.33, and hence Fig. 7.27b shows a more nearly correct collapse mode. This indicates that for slabs with concentrated loads, the collapse mechanisms involving curved yield negative-moment lines are more critical than the straight-line mechanisms involving large triangular segments. Note also that in Eq. 7.34 the radius of the failure cone r has disappeared from the expression for ultimate load. Therefore, the failure cone could have any radius that lies within the slab, and hence the ultimate concentrated load is the same for any position of the concentrated load and for any shape of slab with fixed edges.

However, in practice a uniform load will also be present on the slab. If an ultimate uniformly distributed load per unit area w_u acts with a central ultimate concentrated load P_u on a regular-sided fixed-edge slab, a conical collapse mode of the type shown in Fig. 7.27b occurs, and the ultimate load equation is found by combining Eqs. 7.26 and 7.34 to give

$$\frac{P_u}{2\pi} + \frac{w_u r^2}{6} = m'_u + m_u \qquad (7.35)$$

where r, the radius of the cone, must be taken as large as possible (i.e., extending to the slab edge) for the ultimate load to be as small as possible.

7.5.2 Circular Fans

Circular fans, comprising part or the whole of a full failure cone, are liable to occur anywhere where major concentrated loads are present. Slabs with circular fans are most easily solved by the virtual work method. It is convenient to derive an expression for the internal work done by the fan for use in such analyses.

Consider an isotropic slab with ultimate negative and positive moments of resistance per unit width m'_u and m_u, respectively. Let the circular fan shown in Fig. 7.28 be a portion of a yield line pattern. Consider the internal work done by the ultimate moments of the shaded segment if the center of the fan is given a downward displacement δ and the segment rotates about the negative-moment yield line as axis. The rotation of the segment is δ/r and the internal work done by the moments of the segment is

$$m_{un}\theta_n l_0 = (m'_u + m_u)\frac{\delta}{r}r\,d\phi = (m'_u + m_u)\delta\,d\phi$$

Therefore, for the entire fan, if ϕ is the angle subtended by the fan,

$$m_{un}\theta_n l_0 = \int_0^\phi (m'_u + m_u)\delta\,d\phi = (m'_u + m_u)\delta\phi \qquad (7.36)$$

This expression for the internal work done by the fan can be incorporated in solutions where fans are present.

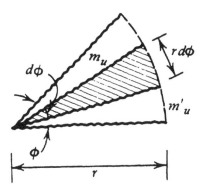

Figure 7.28 Part of a circular fan.

Example 7.8. A rectangular slab bridge is simply supported at two opposite edges and is free at the remaining two edges. The slab is isotropically reinforced in the top and in the bottom with ultimate negative and positive moments per unit width m'_u and m_u, respectively. Determine the ultimate concentrated load P_u acting alone anywhere on the transverse centerline at midspan, neglecting the self-weight of the slab.

SOLUTION. Figure 7.29a shows the slab. There are a number of possible yield line patterns, the critical pattern depending on the aspect ratio of the slab and the position of the load on the transverse centerline. The virtual work method will be used to calculate the ultimate loads of the various yield line patterns.

MODE 1 (Fig. 7.29b). This mode involves the slab failing as a wide beam. Let the load undergo a small downward displacement δ. The virtual work equation is

$$P_u\delta = 2m_u \frac{\delta}{0.5l} b$$

$$P_u = \frac{4m_u b}{l} \tag{7.37}$$

MODE 2a (Fig. 7.29c). This mode is confined to the interior of the slab and is composed of triangular-shaped segments. The dimension a of the yield line pattern is unknown. Let the load undergo a small displacement δ. Work is done by the moments at the four triangular segments. The virtual work equation is

$$P_u\delta = 2m_u \frac{\delta}{0.5l} 2a + 2(m'_u + m_u)\frac{\delta}{a} l$$

$$P_u = \frac{8m_u a}{l^2} + \frac{2(m'_u + m_u)l}{a}$$

For minimum P_u,

$$0 = \frac{dP_u}{da} = \frac{8m_u}{l} - \frac{2(m'_u + m_u)l}{a^2}$$

Therefore,

$$a = 0.5l \sqrt{\frac{m'_u + m_u}{m_u}}$$

which on substitution into the equation for P_u gives

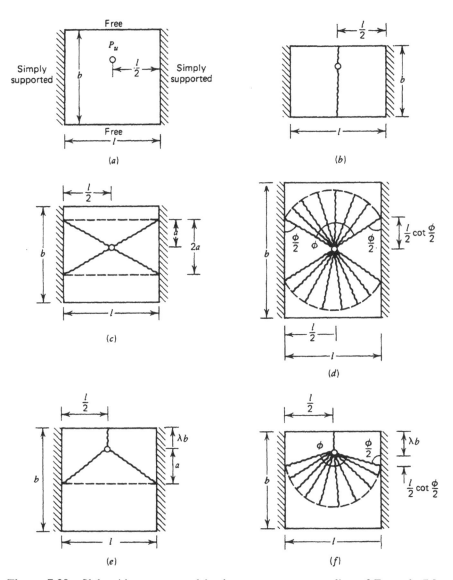

Figure 7.29 Slab with concentrated load on transverse centerline of Example 7.8: (a) slab; (b) mode 1; (c) mode 2a; (d) mode 2b; (e) mode 3a; (f) mode 3b.

$$P_u = 8\sqrt{(m'_u + m_u)m_u} \tag{7.38}$$

MODE 2b (Fig. 7.29d). This mode is also confined to the interior of the slab and is composed of two circular fans. The included angle ϕ of each fan is unknown. Let the load undergo a small downward displacement δ. Work is done by the moments at the two triangular segments in between the fans and at the two fans (given by Eq. 7.36). The virtual work equation is

$$P_u \delta = 2m_u \frac{\delta}{0.5l} l \cot \frac{\phi}{2} + 2(m'_u + m_u)\delta\phi$$

$$P_u = 4m_u \cot \frac{\phi}{2} + 2(m'_u + m_u)\phi \tag{7.39}$$

For minimum P_u,

$$0 = \frac{dP_u}{d\phi} = -2m_u \csc^2 \frac{\phi}{2} + 2(m'_u + m_u)$$

Therefore,

$$\csc^2 \frac{\phi}{2} = \frac{m'_u + m_u}{m_u} = 1 + \frac{m'_u}{m_u}$$

$$\cot \frac{\phi}{2} = \sqrt{\frac{m'_u}{m_u}} \tag{7.40}$$

Comparison of modes 2a and 2b: Let $m'_u = m_u$. Then from Eq. 7.40, $\cot(\phi/2) = 1$ and $\phi = \pi/2$ rad.

For mode 2a, Eq. 7.38: $P_u = 8\sqrt{2m_u^2} = 11.31m_u$

For mode 2b, Eq. 7.39: $P_u = 4m_u + \dfrac{4m_u\pi}{2} = 10.28m_u$

Let $m'_u = 0$. Then from Eq. 7.40, $\cot(\phi/2) = 0$ and $\phi = \pi$ rad. That is, the two fans join to form a complete circle.

For mode 2a, Eq. 7.38: $P_u = 8\sqrt{m_u^2} = 8.00m_u$

For mode 2b, Eq. 7.39: $P_u = 2m_u\pi = 6.28m_u$

These and other comparisons show that mode 2b is always more critical than mode 2a.

MODE 3a (Fig. 7.29e). This mode has a single yield line extending to one free edge and triangular-shaped yield lines confined to the interior of the slab at the other side of the load. The dimension a of the yield line pattern is unknown. The coefficient λ defines the position of the load. Let the load undergo a small downward displacement δ. Work is done by the moments at the three segments. The virtual work equation is

$$P_u\delta = 2m_u \frac{\delta}{0.5l}(a + \lambda b) + (m_u' + m_u)\frac{\delta}{a}l$$

$$P_u = \frac{4m_u(a + \lambda b)}{l} + \frac{(m_u' + m_u)l}{a}$$

For minimum P_u,

$$0 = \frac{dP_u}{da} = \frac{4m_u}{l} - \frac{(m_u' + m_u)l}{a^2}$$

Therefore,

$$a = 0.5l\sqrt{\frac{m_u' + m_u}{m_u}}$$

which on substitution into the equation for P_u gives

$$P_u = 4\sqrt{(m_u' + m_u)m_u} + \frac{4m_u\lambda b}{l} \qquad (7.41)$$

MODE 3b (Fig. 7.29f). This mode has a single yield line extending to the free edge and a circular fan on the other side of the load. The included angle ϕ of the fan is unknown. Let the load undergo a small downward displacement δ. Work is done by the moments at the two segments and at the fan (given by Eq. 7.36). The virtual work equation is

$$P_u\delta = 2m_u\frac{\delta}{0.5l}\left(\lambda b + \frac{l}{2}\cot\frac{\phi}{2}\right) + (m_u' + m_u)\delta\phi$$

$$P_u = \frac{4m_u\lambda b}{l} + 2m_u\cot\frac{\phi}{2} + (m_u' + m_u)\phi \qquad (7.42)$$

For minimum P_u,

$$0 = \frac{dP_u}{d\phi} = -m_u\csc^2\frac{\phi}{2} + (m_u' + m_u)$$

Therefore,

$$\csc^2 \frac{\phi}{2} = \frac{m'_u + m_u}{m_u} = 1 + \frac{m'_u}{m_u}$$

$$\cot \frac{\phi}{2} = \sqrt{\frac{m'_u}{m_u}} \qquad (7.43)$$

Comparison of modes 3a and 3b: Let $m'_u = m_u$. Then from Eq. 7.43, $\cot(\phi/2) = 1$ and $\phi = \pi/2$ rad.

For mode 3a, Eq. 7.41: $P_u = 5.66m_u + \dfrac{4m_u\lambda b}{l}$

For mode 3b, Eq. 7.42: $P_u = 5.14m_u + \dfrac{4m_u\lambda b}{l}$

Let $m'_u = 0$. Then from Eq. 7.43, $\cot(\phi/2) = 0$ and $\phi = \pi$ rad.

For mode 3a, Eq. 7.41: $P_u = 4.00m_u + \dfrac{4m_u\lambda b}{l}$

For mode 3b, Eq. 7.42: $P_u = 3.14m_u + \dfrac{4m_u\lambda b}{l}$

These, and other comparisons, show that mode 3b is always more critical than mode 3a.

When analyzing a particular slab, modes 1, 2b, and 3b need to be examined to determine which gives the smallest load to cause collapse. Usually, it will be found that mode 1 will govern if the slab is narrow, mode 2b will govern if the slab is wide with the load near the center, and mode 3b will govern if the slab is wide with the load near a free edge.

7.6 SUPERPOSITION OF MOMENT STRENGTHS FOR COMBINED LOADING CASES

It is sometimes necessary to determine the ultimate load for a slab that may be subjected to various combinations of loading: for example, combinations of concentrated and uniform loading. Although strictly, the principle of superposition is only applicable to linear elastic structures, it may be shown that distributing the moment strength of the slab between the yield line patterns for the separate load cases gives a result that is on the safe side even though in yield line theory the slab is behaving nonlinearly.

To illustrate this, consider a slab with moment strength $m_{u1} + m_{u2}$ subjected to two types of loading. Let m_{u1} carry ultimate type 1 load W_{u1} applied alone to the slab, where W_{u1} is calculated from the exact yield line pattern for that load type. For all other yield line patterns the ultimate type 1 load carried by m_{u1} will exceed W_{u1}. Let m_{u2} carry ultimate type 2 load W_{u2} applied alone to the slab, where W_{u2} is calculated from the exact yield line pattern for that load type. For all other yield line patterns the ultimate type 2 load carried by m_{u2} will exceed W_{u2}. Now, if load types 1 and 2 are applied simultaneously to the slab, the exact yield line pattern for the combined loading case may be either one of the exact patterns given above for separate loading or some other pattern. Say that the ultimate load given by moment strengths m_{u1} and m_{u2} for this combined loading case are $\mathcal{W}_{u1}$ and $\mathcal{W}_{u2}$, respectively. Then the correct ultimate load for slab strength $m_{u1} + m_{u2}$ for this combined loading case is $\mathcal{W}_{u1} + \mathcal{W}_{u2}$. But since $\mathcal{W}_{u1} \geq \mathcal{W}_{u1}$ and $\mathcal{W}_{u2} \geq \mathcal{W}_{u2}$, then $\mathcal{W}_{u1} + \mathcal{W}_{u2} \geq \mathcal{W}_{u1} + \mathcal{W}_{u2}$. Hence, the ultimate load found by distributing the slab moment strength between the yield line patterns for each load applied separately will always be equal to or less than the correct ultimate load. This reasoning is valid for any number of loads.

Thus, the distribution of the slab moment strength between the patterns gives a safe estimate of the ultimate load. This estimate can be very conservative if the separate yield line patterns are significantly different.

Example 7.9. A square slab is simply supported at two opposite sides over a 15-ft (4.57-m) span and is free at the other two sides. The slab is isotropically reinforced with an ultimate positive moment of resistance of 20,000 lb-ft/ft width (89 kN-m/m) and has no top steel. The slab carries a uniformly distributed load of 300 lb/ft² (14.35 kN/m²). Calculate the ultimate concentrated load that, acting at the center of a free edge of the slab with the uniformly distributed load, would cause a flexural failure.

APPROXIMATE SOLUTION. For the ultimate uniformly distributed load of 300 lb/ft² on the slab, the yield line pattern is that of mode A in Fig. 7.30. For this pattern the required ultimate moment of resistance per unit width is

$$m_{u1} = w_u \frac{l^2}{8} = 300 \times \frac{15^2}{8} = 8438 \text{ lb-ft/ft width}$$

Hence, ultimate moment capacity remaining to carry the concentrated load is

$$m_{u2} = 20,000 - 8438 = 11,562 \text{ lb-ft/ft width}$$

For the concentrated load at the center of the free edge, the yield line pattern is given by either mode A or mode B of Fig. 7.30. For mode A the ultimate concentrated load, according to Eq. 7.37, is

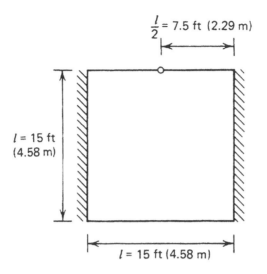

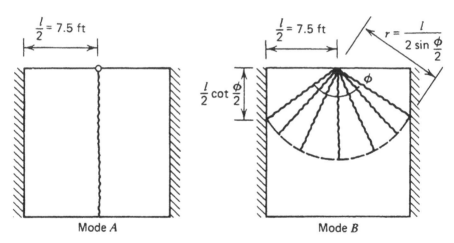

Figure 7.30   Slab with concentrated and uniform loads of Example 7.9.

$$P_u = \frac{4m_{u2}b}{l} = 4 \times 11{,}562 = 46{,}250 \text{ lb}$$

For mode B, the ultimate concentrated load is, according to Eqs. 7.42 and 7.43 with $m'_u = 0$ and $\lambda = 0$, given when $\phi = \pi$ rad and is

$$P_u = \pi m_{u2} = \pi \times 11{,}562 = 36{,}320 \text{ lb}$$

It is evident that mode B governs and the ultimate concentrated load calculated by this approach is 36,320 lb.

Exact Solution. If the uniformly distributed load and the concentrated load act simultaneously, the yield line pattern that develops is either mode *A* or mode *B* of Fig. 7.30.

For mode *A*, let the center of the free edge undergo a small downward displacement δ. The virtual work equation is

$$P_u\delta + w_u l^2 \frac{\delta}{2} = 2m_u l \frac{2\delta}{l}$$

Therefore,

$$P_u + w_u \frac{l^2}{2} = 4m_u$$

$$P_u = (4 \times 20{,}000) - \left(300 \times \frac{15^2}{2}\right)$$

$$= 46{,}250 \text{ lb}$$

For mode *B*, let the center of the free edge undergo a small downward displacement δ. The virtual work equation, assuming that $\phi = \pi$ rad and $m_u' = 0$, is

$$P_u\delta + w_u \frac{\pi}{2}\left(\frac{l}{2}\right)^2 \frac{\delta}{3} = m_u\delta\pi$$

Therefore,

$$P_u + 300 \times \pi \times \frac{15^2}{24} = 20{,}000 \times \pi$$

$$P_u = 54{,}000 \text{ lb}$$

Hence, mode *A* governs and the ultimate concentrated load calculated by this approach is 46,250 lb. This load is 27% higher than that given by the approximate solution, indicating that the approximate solution is safe but can be very conservative. Hence, for better accuracy it is best to calculate the ultimate load from the yield line pattern that actually develops when the loads are applied simultaneously. Nevertheless, the approximate approach may be useful for a relatively quick design check.

7.7 CORNER EFFECTS

So far it has been assumed that a yield line forming in the corner of a slab enters directly into the corner, as in Fig. 7.31*a*. However, it is evident from

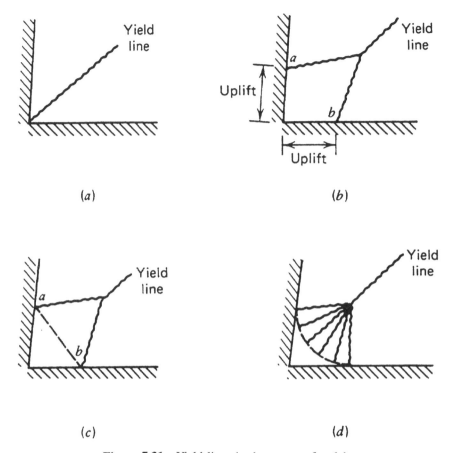

Figure 7.31   Yield lines in the corner of a slab.

the elastic theory for slabs that there are strong torsional moments in the corner regions and that if a corner of a simply supported slab is not held down, it will tend to lift off the support. The analogy in yield line theory is the phenomenon referred to by Johansen as *corner levers*. If the corners of a simply supported slab are not held down, the yield line tends to fork before reaching the corner, as shown in Fig. 7.31*b*, and the triangular portion of the corner region tends to rotate about *ab* as axis and lift off the support. Alternatively, if the corner is held down and no top steel is provided, the slab tends to crack across line *ab*, as shown in Fig. 7.31*c*. Line *ab* is then a yield line of zero strength. If some top steel is provided and the corner is held down, the corner yield line pattern of Fig. 7.31*c* may still form with *ab* as a yield line with some negative moment strength. If sufficient top steel is provided and the corner is held down, the corner yield line of Fig. 7.31*a* will develop. If the corner yield line patterns of Fig. 7.31*b* or *c* form, the ultimate load of the slab will be lower than for the pattern with a single line entering the corner. The reduction is greater when circular fans, illustrated in Fig. 7.31*d*, rather than triangular segments (Fig. 7.31*c*), form in the corners.

Although when top steel is absent, or present only in small quantities, corner levers lead to a reduced ultimate load, they have been neglected in the analysis presented so far because the analysis including them becomes much more complex, and in many cases the error introduced by ignoring them is small. The example given below of a square slab illustrates the complexities of the analysis including them and the order of the difference made to the ultimate load.

Example 7.10. Determine the ultimate uniformly distributed load per unit area w_u of a square simply supported slab. The slab is isotropically reinforced with ultimate positive moments of resistance per unit width m_u in the directions of both spans. The slab has no top steel and is held down at the corners.

SOLUTION. Assume that the yield line pattern is composed of corner levers in the form of fans that extend to the slab center as shown in Fig. 7.32. The unknown dimensions of each fan are defined by the included angle ϕ. The radius of each fan is $l/[2 \sin(\pi/4 + \phi/2)]$. Since there is no top steel, the

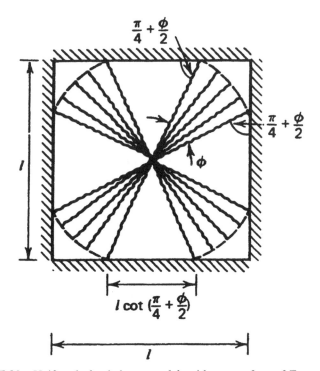

Figure 7.32 Uniformly loaded square slab with corner fans of Example 7.10.

negative-moment yield line across each fan has zero strength. Consider the solution by the virtual work method. Let the center of the slab be given a small downward displacement δ. The work done by the ultimate moments is due to ultimate moments at the four fans and the four triangular segments in between. The internal work done by each fan is given by Eq. 7.36. The total internal work done is

$$\Sigma\, m_{um}\theta_n l_0 = 4\left[m_u\delta\phi + m_u \frac{\delta}{0.5l}\, l \cot\left(\frac{\pi}{4} + \frac{\phi}{2}\right)\right]$$

$$= 4m_u\delta\phi + 8m_u\delta \cot\left(\frac{\pi}{4} + \frac{\phi}{2}\right)$$

The work done by the uniform loading is equal to the load on the four fans plus the four triangles multiplied by $\delta/3$. The external work done is

$$\Sigma\, W_u\Delta = 4\left\{ \frac{\phi}{2}\left[\frac{l}{2\,\sin(\pi/4 + \phi/2)}\right]^2 \frac{w_u\delta}{3} + \frac{l^2}{4}\cot\left(\frac{\pi}{4} + \frac{\phi}{2}\right)\frac{w_u\delta}{3}\right\}$$

$$= \frac{w_u\delta\phi l^2}{6\,\sin^2(\pi/4 + \phi/2)} + \frac{w_u\delta l^2}{3}\cot\left(\frac{\pi}{4} + \frac{\phi}{2}\right)$$

Hence, from the work equation, Eq. 7.5, the ultimate uniform load is

$$w_u = \frac{24 m_u}{l^2}\,\frac{\phi + 2\cot(\pi/4 + \phi/2)}{[\phi/\sin^2(\pi/4 + \phi/2)] + 2\cot(\pi/4 + \phi/2)} \tag{7.44}$$

Note that when $\phi = 0$, $w_u = 24m_u/l^2$, as obtained by Eqs. 7.6 and 7.21.

The value of ϕ giving the minimum value of w_u in Eq. 7.44 is found from $dw_u/d\phi = 0$. The value of ϕ so found is 30° and gives

$$w_u = 21.7\,\frac{m_u}{l^2} \tag{7.45}$$

This value for the ultimate load is 9.6% less than the ultimate load given by the straight-line yield line patterns leading to Eqs. 7.6 and 7.21. It is of interest to note that if the fan boundaries were hyperbolic rather than circular, the ultimate load is further reduced to $21.4m_u/l^2$, according to Wood and Jones.[7.7] However, circular fans give adequate accuracy. If the slab had had top steel in the corners so that $m_u' = m_u$, it could be shown that the development of corner fans would not cause the ultimate load to be less than that found from the simple straight-line yield line patterns.

Extensive fans can also form in the corners of slabs with fixed edges. For example, if the slab of Example 7.10 shown in Fig. 7.32 had fixed edges and top steel with an ultimate negative moment of resistance per unit width m_u'

in the direction of the spans, negative-moment yield lines of strength m'_u will form along the edges between the fans and the negative-moment yield lines across the fans will have strength m'_u. The ultimate uniform load for this fixed-edge case is given by Eq. 7.44, with $m'_u + m_u$ substituted for m_u, and therefore the corner fans cause a reduction of 9.6% in the fixed-edge case as well.

When the corners of slabs are acute angles, the reduction in strength due to corner fans is even larger. Hence, triangular slabs may be particularly affected. In triangular slabs the presence of a free edge opposite an acute angle corner has been shown by Wood[7.5] to lead to very significant strength reductions when a corner is considered. For example, Fig. 7.33 shows the case of a uniformly loaded triangular isotropic slab simply supported at two edges and free at the remaining edge. The angle between the supported edges is 45° and the slab has no top steel. For the yield line patterns of Fig. 7.33a with the yield line entering straight into the corner, the ultimate uniform load is calculated to be $w_u = 40.97 m_u/a^2$. However, the ultimate uniform load for the yield line pattern of Fig. 7.33b with the corner fan is found[7.5] to be $w_u = 30.1 m_u/a^2$. Thus, the fan-shaped mechanism causes a 26.5% reduction in the ultimate load in this case. Reductions in the ultimate load of slabs of various shapes, boundary conditions, and loading caused by corner levers are discussed more extensively in the text by Wood.[7.5]

It is evident that the exact solution of slabs considering corner effects in detail could require a large amount of effort. This exact solution is too difficult to be called a design procedure, and the time spent on it can be out of proportion to the relative importance of corner effects. It is considered that the best procedure is to calculate the ultimate load on the basis of the simple straight-line mechanisms (i.e., those yield line patterns in which the yield lines run straight into the corners), and then to subtract from the ultimate load

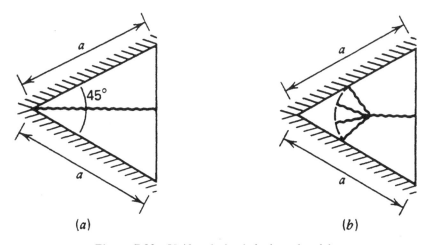

Figure 7.33 Uniformly loaded triangular slab.

of these simple "basic" mechanisms to allow for the effect of corner levers. Wood[7.5] has suggested the percentage reductions of collapse load per corner shown in Table 7.1. The table includes an allowance for the strength of the concrete in tension. The trend is that acute angled corners, absence of top steel in corners, presence of restrained edges, and point loads all contribute toward a reduction in the ultimate load due to corner effects. In applying the reductions given in Table 7.1, Wood[7.5] recommends that the following rules be observed:

1. The "basic" mechanism is the worst straight-line type of mechanism that can be found.
2. Only those corners to which a yield line runs in the basic mechanism should be considered.
3. No extremely acute angled corners should be treated using the basic mechanism. The table does not apply for acute angles of less than 40°.
4. With heavy point loads the reductions allow for the formation of circular fans in the corners, but the possibility of forming a complete conical yield line pattern around the point load should be watched.

Example 7.11. A four-sided slab is simply supported around all edges and is uniformly loaded. The slab has one obtuse, one right, and two acute angled corners. In the basic collapse mechanism, yield lines run into each corner. Assess the reduction in the ultimate load of the slab due to corner levers if there is no top steel.

SOLUTION. From Table 7.1, the reduction in the basic ultimate load is as follows:

$$
\begin{array}{ll}
\text{1 obtuse angle} & 0\% \\
\text{1 right angle} & 2 \\
\text{2 acute angles } 2 \times 5 = & \underline{10} \\
& 12\%
\end{array}
$$

The view of many research workers is that the reduction in ultimate strength due to corner levers can be neglected in practice in most cases because tests on slabs have invariably shown yield line theory to be conservative, due primarily to the neglect of membrane action in the theory. Some tests results are discussed in Section 7.17.

7.8   AFFINITY THEOREM

Many of the yield line solutions for slabs have been obtained for the case of isotropic reinforcement. Yet for most slabs the use of orthotropic reinforce-

TABLE 7.1 Suggested Percentage Reduction of Ultimate Load, per Corner, Below the Basic Ultimate Load

Type of Slab	Type of Loading	Kind of Corner Angle[a]	Reduction per Corner (%)		
			No Restraining Moments on Edges		Restrained Edges, Top Reinforced in Corners
			Top[b] Reinforced in Corners	No Top Reinforcement	
All sides supported	Distributed	Obtuse	0	0	0
		Right	0	−1	−2
		Acute	−2	−5	−5
One long free edge in square or rectangular slab	Distributed	Obtuse	0	−1	−1
		Right	−1	−3	−3
		Acute	−3	−7	−7
Free edge cutting across corner	Distributed	Obtuse	−2	−8	−4
		Right	−5	−12	−8
		Acute	−10	−20	−18
All sides supported	Point loads predominant[c]	Obtuse	0	−2	−2
		Right	0	−4	−5
		Acute	−4	−12	−12

Source: After Ref. 7.5.

[a]Reentrant corners are outside the scope of this table.
[b]Top steel area per unit width at least equal to bottom steel area per unit width.
[c]With long slabs, or free edges, very heavy point loads require special treatment.

ment is desirable on the grounds of economy. Orthotropic reinforcement also enables the elastic theory distribution of moments to be followed more closely. Johansen has derived a theorem that enables solutions for orthotropic slabs to be obtained from solutions for isotropic slabs. To determine the equivalent isotropic slab, the lengths of the sides of the orthotropic slab and the loading are altered by ratios that depend on the ratios of the ultimate moments of resistance per unit width in the two directions.

The affinity theorem may be derived with reference to segment ABC of Fig. 7.34, which is part of the yield line pattern of an orthotropic slab. To obtain a variety of edge conditions, AB is a positive-moment yield line, BC is a negative-moment yield line, and AC is a free edge. The reinforcing bars for both positive and negative moments are placed in the x- and y-directions. The negative and positive ultimate moments of resistance per unit width are m'_{ux} and m_{ux}, respectively, in the x-direction, and m'_{uy} and m_{uy} respectively, in the y-direction. Also, $\mu = m'_{uy}/m'_{ux} = m_{uy}/m_{ux}$ where μ is known as the *coefficient of orthotropy*. The slab carries a uniformly distributed ultimate load per unit area w_u. Let the segment undergo a small rotation about the axis RR so that the point O moves down by δ. Then the internal work done by the ultimate moments is, from Eq. 7.7 for the segment,

$$\Sigma m_{nn}\theta_n l_0 = (m_{ux}y_1 + m'_{ux}y_2)\frac{\delta}{h_x} + (m_{uy}x_1 + m'_{uy}x_2)\frac{\delta}{h_y}$$

$$= (m_{ux}y_1 + m'_{ux}y_2)\frac{\delta}{h_x} + (m'_{ux}x_1 + m'_{ux}x_2)\frac{\mu\delta}{h_y}$$

The work done by the loading is given by Eq. 7.4. Therefore, considering all the segments of the yield line pattern of the slab, the virtual work equation may be written as

$$\Sigma\left[(m_{ux}y_1 + m'_{ux}y_2)\frac{\delta}{h_x} + (m_{ux}x_1 + m'_{ux}x_2)\frac{\mu\delta}{h_y}\right] = \iint w_u\delta(x,y)\,dx\,dy$$

$$(7.46)$$

where $\delta(x,y)$ is the displacement at point (x,y).

Now consider an isotropic slab with negative and positive ultimate moments of resistance per unit width m'_{ux} and m_{ux}, respectively, in each direction. Let the ultimate uniform load per unit area be w_u and the x-direction dimensions be the same as for the orthotropic slab, but let the y-direction dimensions be k times those of the orthotropic slab. The yield line pattern will remain of the same form, and the deflection of the corresponding points of the slabs will remain equal if the corresponding point O moves down by the same amount δ. Figure 7.35 shows the segment of the isotropic slab. For rotation about the RR axis causing the point O to move down by δ, the virtual work equation is obtained by the same method as previously. When all the segments

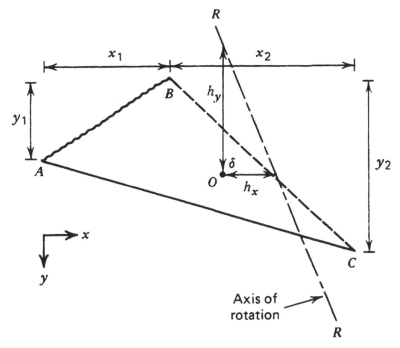

Figure 7.34 Segment of the yield line pattern of the orthotropic slab.

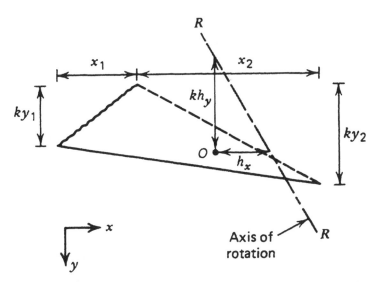

Figure 7.35 Segment of the yield line pattern of the isotropic slab.

of the yield line pattern of the isotropic slab are considered, the internal work done may be written as

$$\sum \left[(m_{ux}ky_1 + m'_{ux}ky_2) \frac{\delta}{h_x} + (m_{ux}x_1 + m'_{ux}x_2) \frac{\delta}{kh_y} \right] = \int\int w_u \delta(x,ky) \, dx \, k \, dy$$

On dividing through by k, the equation becomes

$$\sum \left[(m_{ux}y_1 + m'_{ux}y_2) \frac{\delta}{h_x} + (m_{ux}x_1 + m'_{ux}x_2) \frac{\delta}{k^2h_y} \right] = \int\int w_u \delta(x,y) \, dx \, dy$$

(7.47)

Comparison of Eqs. 7.46 and 7.47 shows that they are similar if $k = 1/\sqrt{\mu}$. This means that the dimensions of the slab are unchanged in the direction of unchanged moment, but are changed in the direction of changed moment, and also that the ultimate uniform load per unit area is unchanged. The equivalent isotropic slab is known as the *affine slab*.

A concentrated load may be considered as a uniformly distributed load over a small area. On the orthotropic slab an ultimate concentrated load may be written as $P_u = w_u \, dx \, dy$. On the affine isotropic slab, dy becomes $k \, dy = dy/\sqrt{\mu}$, and therefore the ultimate concentrated load becomes $P_u = w_u \, dx \, dy/\sqrt{\mu}$. Since the ultimate uniform load per unit area remains the same on both slabs, an ultimate concentrated load P_u on the orthotropic slab becomes $P_u/\sqrt{\mu}$ on the affine isotropic slab.

Hence, the affinity theorem may be stated as follows: An orthotropic slab with negative and positive ultimate moments of resistance per unit width m'_{ux} and m_{ux}, respectively, in the x-direction and m'_{uy} and m_{uy}, respectively, in the y-direction, where $\mu = m'_{uy}/m'_{ux} = m_{uy}/m_{ux}$, may be analyzed as an isotropic slab with negative and positive ultimate moments of resistance per unit width m'_{ux} and m_{ux} in both directions if:

	Orthotropic slab	$\rightarrow$	Isotropic slab
Dimensions:	l_x	remains	l_x
	l_y	becomes	$\dfrac{l_y}{\sqrt{\mu}}$
Loading:	w_u	remains	w_u
	P_u	becomes	$\dfrac{P_u}{\sqrt{\mu}}$

It is observed in both the case of the uniform load and that of the concentrated load that the total load on the affine isotropic slab is obtained by dividing the total load on the orthotropic slab by $\sqrt{\mu}$. It follows that in the case of a line load, the total ultimate line load on the affine isotropic slab is obtained by dividing the total ultimate line load on the orthotropic slab by $\sqrt{\mu}$.

Thus, by making use of the affinity theorem, all solutions derived for isotropic slabs can be readily rewritten for orthotropic slabs.

Example 7.12. A triangular slab is simply supported at two edges adjacent to a right-angle corner. The remaining edge is free. Given that if the slab is isotropically reinforced with ultimate positive moments of resistance per unit width m_u, the ultimate uniformly distributed load per unit area is

$$w_u = \frac{12m_u}{l_x l_y}$$

calculate the ultimate uniform load for the slab if orthotropically reinforced.

SOLUTION. The ultimate load equation for the isotropic case given above was obtained from Eq. 7.11. The slab is shown in Fig. 7.36. If the slab is orthotropic with ultimate positive moments of resistance per unit width m_{ux} and m_{uy} in the x- and y-directions, respectively, and $\mu = m_{uy}/m_{ux}$, for the affine isotropic slab m_u becomes m_{ux}, l_x remains l_x, l_y becomes $l_y/\sqrt{\mu}$, and w_u remains w_u. Hence, for an orthotropic slab of the dimensions shown in Fig. 7.36,

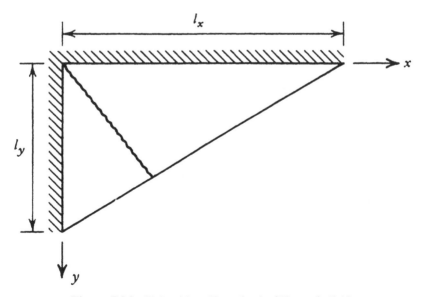

Figure 7.36   Slab with uniform load of Example 7.12.

$$w_w = \frac{12m_{ux}}{l_x l_y / \sqrt{\mu}} = \frac{12\sqrt{m_{ux} m_{uy}}}{l_x l_y} \qquad (7.48)$$

Example 7.13. A slab is isotropically reinforced with negative and positive ultimate moments of resistance per unit width m_u' and m_u, respectively. Given that the ultimate concentrated load for a conical collapse mechanism is

$$P_u = 2\pi(m_u' + m_u)$$

calculate the ultimate concentrated load for the slab if orthotropically reinforced, and sketch the shape of the yield line pattern.

SOLUTION. The ultimate load equation for the isotropic case given above was obtained from Eq. 7.34. If the slab is orthotropic with negative and positive ultimate moments of resistance per unit with m_{ux}' and m_{ux}, respectively, in the x-direction, and m_{uy}' and m_{uy}, respectively, in the y-direction, and $\mu = m_{uy}'/m_{ux}' = m_{uy}/m_{ux}$, for the affine isotropic slab m_u' becomes m_{ux}', m_u becomes m_{ux}, l_x remains l_x, l_y becomes $l_y/\sqrt{\mu}$ and P_u becomes $P_u/\sqrt{\mu}$. Thus, the circular fan of the isotropic slab becomes an elliptical fan in the orthotropic slab, as shown in Fig. 7.37. The ultimate concentrated load of the orthotropic slab becomes

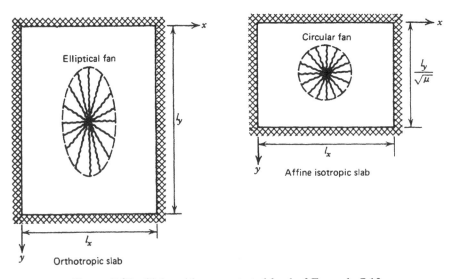

Figure 7.37   Slabs with concentrated load of Example 7.13.

$$\frac{P_u}{\sqrt{\mu}} = 2\pi(m'_{ux} + m_{ux})$$

$$P_u = 2\pi \left(m'_{ux} \sqrt{\frac{m'_{uy}}{m'_{ux}}} + m_{ux} \sqrt{\frac{m_{uy}}{m_{ux}}} \right)$$

$$= 2\pi(\sqrt{m'_{ux}m'_{uy}} + \sqrt{m_{ux}m_{uy}}) \tag{7.49}$$

The complexities of the direct analysis of orthotropic slabs with elliptical fans under concentrated loads are such that the affinity theorem provides an obvious method for obtaining solutions for orthotropic slabs with concentrated loads.

Example 7.14. A cantilever slab is orthotropically reinforced with negative and positive ultimate moments of resistance per unit width m'_{ux} and m_{ux}, respectively, in the x-direction, and m'_{uy} and m_{uy}, respectively, in the y-direction. Use the affinity theorem to calculate the ultimate concentrated load applied at one free corner.

SOLUTION. Consider the slab to be isotropically reinforced with negative and positive ultimate moments of resistance per unit width m'_u and m_u, respectively. There are three possible modes of failure, as illustrated in Fig. 7.38.

MODE 1 (Fig. 7.38a). The angle ϕ is 45° by symmetry. The equation for equilibrium of moments for the triangular segment, taken about the yield line, is

$$\frac{P_u l}{\sqrt{2}} = m'_u \sqrt{2}\, l$$

$$P_u = 2m'_u \tag{7.50}$$

Hence, at the ultimate load, l can have any value between 0 and l_x if $l_x < l_y$.

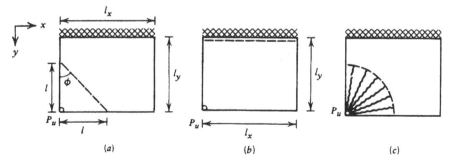

Figure 7.38 Isotropic slabs with concentrated load of Example 7.14: (a) mode 1; (b) mode 2; (c) mode 3.

MODE 2 (Fig. 7.38*b*). The equation for equilibrium of moments for the segment, taken about the yield line, is

$$P_u l_y = m'_u l_x$$

$$P_u = \frac{m'_u l_x}{l_y} \tag{7.51}$$

MODE 3 (Fig. 7.38*c*). For this fan mode, using the virtual work method, let the load be displaced downward by δ. The internal work done by the fan is given by Eq. 7.36. The virtual work equation is

$$P_u \delta = \frac{(m'_u + m_u)\delta\pi}{2}$$

$$P_u = \frac{(m'_u + m_u)\pi}{2} \tag{7.52}$$

The radius of the fan can have any value within the slab. Note that if no bottom steel is present, $m_u = 0$ in Eq. 7.52, and that equation is the only one affected by the presence of bottom steel. If there is no bottom steel, Eq. 7.52 is more critical than Eq. 7.50, and it is evident that $P_u = m'_u l_x / l_y$ when $l_x / l_y < 1.57$ and $P_u = 1.57 m_u$ when $l_x / l_y \geq 1.57$.

Now consider the slab to be orthotropically reinforced as required and to be of dimensions l_x by l_y with a corner load P_u. For the affine isotropic slab m'_u becomes m'_{ux}, m_u becomes m_{ux}, l_x remains l_x, l_y becomes $l_y / \sqrt{\mu}$, and P_u becomes $P_u / \sqrt{\mu}$. Hence, for the orthotropic slab the ultimate load equations 7.50 to 7.52 become:

Mode 1:
$$\frac{P_u}{\sqrt{\mu}} = 2m'_x$$

$$P_u = 2\sqrt{m'_{ux} m'_{uy}} \tag{7.53}$$

Mode 2:
$$\frac{P_u}{\sqrt{\mu}} = \frac{m'_{ux} l_x}{l_y / \sqrt{\mu}}$$

$$P_u = \frac{m'_{ux} l_x}{l_y} \tag{7.54}$$

Mode 3:
$$\frac{P_u}{\sqrt{\mu}} = \frac{(m'_{ux} + m_{ux})\pi}{2}$$

$$P_u = \frac{(\sqrt{m'_{ux} m'_{uy}} + \sqrt{m_{ux} m_{uy}})\pi}{2} \tag{7.55}$$

7.9 GENERAL CASES FOR UNIFORMLY LOADED RECTANGULAR SLABS

7.9.1 Ultimate Moments of Resistance of the Slabs

General cases of uniformly loaded rectangular slabs with combinations of fixed, simply supported, and free edges will be considered. The slabs will be considered to be orthotropically reinforced with bars in the directions parallel to the edges. The ultimate positive moments of resistance per unit width are m_{ux} and m_{uy} in the x- and y-directions, respectively, where $\mu = m_{uy}/m_{ux}$ is the coefficient of orthotropy. The ultimate negative moments of resistance per unit width are defined in terms of the ultimate positive moments in those directions. The ultimate negative moments of resistance per unit width in the x-direction are $i_1 m_{ux}$ or $i_3 m_{ux}$ at the slab edges, and the ultimate negative moments of resistance per unit width in the y-direction are $i_2 m_{uy}$ or $i_4 m_{uy}$ at the slab edges. The case of simple supports can be obtained by putting the value at that support equal to zero.

7.9.2 Uniformly Loaded Orthotropic Rectangular Slabs with All Edges Supported

The slab and the yield line pattern are shown in Fig. 7.39. All edges are assumed to be fixed and the ultimate negative moments of resistance per unit width are defined in terms of the positive-moment capacities as shown in the

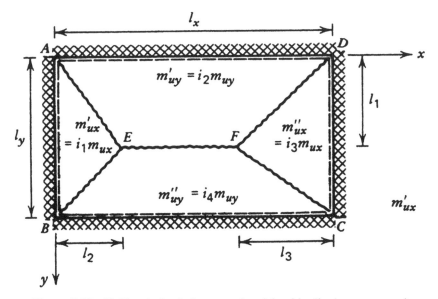

Figure 7.39 Uniformly loaded rectangular slab with all edges supported.

figure. The nodal forces at the two interior junctions of the yield lines are zero. Three unknown dimensions define the location of the yield lines, and therefore four equilibrium equations are required. These are obtained by considering the equilibrium of each segment by taking moments about the supports as axes.

Segment ABE:
$$(1 + i_1)m_{ux}l_y = \frac{w_u l_y l_2^2}{6} \tag{i}$$

Segment DAEF:
$$(1 + i_2)m_{uy}l_x = \frac{w_u l_2 l_1^2}{6} + \frac{w_u l_3 l_1^2}{6} + \frac{w_u(l_x - l_2 - l_3)l_1^2}{2} \tag{ii}$$

Segment CDF:
$$(1 + i_3)m_{ux}l_y = \frac{w_y l_y l_3^2}{6} \tag{iii}$$

Segment BCFE:
$$(1 + i_4)m_{uy}l_x = \frac{w_u l_2(l_y - l_1)^2}{6} + \frac{w_u l_3(l_y - l_1)^2}{6}$$
$$+ \frac{w_u(l_x - l_2 - l_3)(l_y - l_1)^2}{2} \tag{iv}$$

From Eq. i/Eq. iii:
$$\frac{1 + i_1}{1 + i_3} = \left(\frac{l_2}{l_3}\right)^2 \qquad l_2 = l_3 \sqrt{\frac{1 + i_1}{1 + i_3}} \tag{7.56}$$

From Eq. ii/Eq. iv:
$$\frac{1 + i_2}{1 + i_4} = \left(\frac{l_1}{l_y - l_1}\right)^2 \qquad l_1 = \frac{l_y\sqrt{1 + i_2}}{\sqrt{1 + i_2} + \sqrt{1 + I_4}} \tag{7.57}$$

Let
$$X = \sqrt{1 + i_1} + \sqrt{1 + i_3}$$
$$Y = \sqrt{1 + i_2} + \sqrt{1 + i_4}$$

On substituting l_1 from Eq. 7.57 into Eq. ii, we find that

$$(1 + i_2)m_{uy}l_x = \frac{w_u}{6}(3l_x - 2l_2 - 2l_3)\frac{l_y^2(1 + i_2)}{Y^2} \tag{v}$$

On substituting l_2 from Eq. 7.56 into Eq. v, we find that

$$6m_{uy}l_xY^2 = w_ul_y^2\left(3l_x - \frac{2l_3X}{\sqrt{1+i_3}}\right) \qquad \text{(vi)}$$

Eliminating w_u between Eqs. vi and iii gives

$$\frac{6(1+i_3)m_{ux}}{l_3^2} = \frac{6m_{uy}l_xY^2}{l_y^2[3l_x - (2l_3X/\sqrt{1+i_3})]}$$

Therefore,

$$\left(\frac{l_3}{l_y}\right)^2 \frac{1}{1+l_3} + \frac{l_3}{l_y\sqrt{1+i_3}} \frac{2Xl_y}{\mu Y^2 l_x} - \frac{3}{\mu Y^2} = 0$$

$$\frac{l_3}{l_y\sqrt{1+i_3}} = \frac{l_y}{\mu Yl_y}\left\{\left[\left(\frac{X}{Y}\right)^2 + 3\mu\left(\frac{l_x}{l_y}\right)^2\right]^{1/2} - \frac{X}{Y}\right\} \qquad (7.58)$$

Now from Eq. iii we find that

$$w_u = \frac{6(1+i_3)m_{ux}}{l_3^2}$$

and on substituting for l_3 from Eq. 7.58, we obtain

$$w_u = \frac{6m_{uy}\mu Y^2}{l_y^2(l_y/l_x)^2\{[(X/Y)^2 + 3\mu(l_x/l_y)^2]^{1/2} - (X/Y)\}^2} \qquad (7.59)$$

where

$$\mu = \frac{m_{uy}}{m_{ux}}$$

$$X = \sqrt{1+i_1} + \sqrt{1+i_3}$$

$$Y = \sqrt{1+i_2} + \sqrt{1+i_4}$$

Equation 7.59 also applies to simply supported edges by setting the i value at the appropriate edge equal to zero. Hence, the equation is applicable to uniformly loaded rectangular slabs with any combination of fixed and simply supported edges.

7.9.3   Uniformly Loaded Orthotropic Rectangular Slabs with Three Edges Supported and One Edge Free

The slab and the two alternative yield line patterns are shown in Fig. 7.40. All supported edges are assumed to be fixed and the ultimate negative moments of resistance per unit width are defined in terms of the positive-moment capacities as shown in the figure. The negative-moment strengths of the two y-direction edges are assumed to be equal.

Mode 1   (Fig. 7.40a). The nodal forces at the free edge at E and F act in the directions shown, where a dot indicates an upward force and a cross indicates a downward force. Equation 7.20 gives the nodal forces at E and F as $\pm m_{uy} l_1 / l_y$. The yield line pattern has one unknown dimension, and therefore two equilibrium equations are required. These are obtained by considering the equilibrium of each segment by taking moments about the supports as axes.

Segment *ABE:*
$$(1 + i_1)m_{ux}l_y - \frac{m_{ux}l_1^2}{l_y} = \frac{w_u l_y l_1^2}{6} \tag{i}$$

Segment *BCFE:*
$$2m_{uy}l_1 + i_2 m_{uy}l_x + \frac{2m_{uy}l_1 l_y}{l_y} = \frac{2w_u l_1 l_y^2}{6} + \frac{w_u(l_x - 2l_1)l_y^2}{2} \tag{ii}$$

From Eq. i/Eq. ii:

$$\frac{l_y^2(1 + i_1) - \mu l_1^2}{\mu(l_x i_2 + 4l_1)} = \frac{l_1^2}{3l_x - 4l_1}$$

Therefore,

$$\left(\frac{l_1}{l_x}\right)^2 \frac{3 + i_2}{1 + i_1} \left(\frac{l_x}{l_y}\right)^2 \mu + 4\frac{l_1}{l_x} - 3 = 0 \tag{iii}$$

Solution of Eq. iii gives

$$\frac{l_1}{l_x} = \frac{\sqrt{4 + 3K_2} - 2}{K_2} \tag{7.60}$$

where

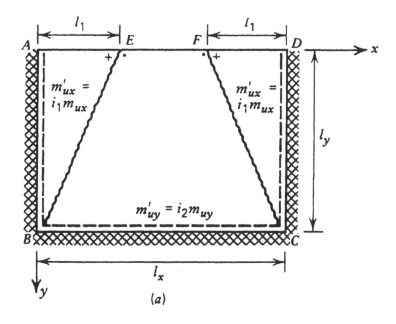

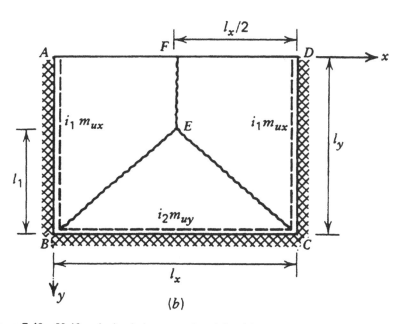

Figure 7.40 Uniformly loaded rectangular slab with three edges supported and one edge free: (*a*) mode 1; (*b*) mode 2.

$$K_2 = \mu\left(\frac{l_x}{l_y}\right)^2 \frac{3 + i_2}{1 + i_1}$$

and from Eq. ii,

$$w_u = \frac{6m_{uy}[i_2 + (4l_1/l_x)]}{l_y^2[3 - (4l_1/l_x)]} \tag{7.61}$$

Substitution of Eq. 7.60 into Eq. 7.61 gives the ultimate uniform load for this yield line pattern unless $l_1 = l_x/2$, in which case the nodal force calculation fails.

Mode 2 (Fig. 7.40*b*). The nodal forces at the free edge are zero. The yield line pattern has one unknown dimension, and therefore two equilibrium equations are required. These are obtained by considering the equilibrium of each segment by taking moments about the supports as axes.

Segment *ABEF*: $(1 + i_1)m_{ux}l_y = \dfrac{w_u l_1 l_x^2}{24} + \dfrac{w_u(l_y - l_1)l_x^2}{8}$ (i)

Segment *BCE*: $(1 + i_2)m_{uy}l_x = \dfrac{w_u l_x l_1^2}{6}$ (ii)

From Eq. ii/Eq. i:

$$\frac{(1 + i_2)\mu}{1 + i_1} = \frac{4l_1^2}{l_x^2[3 - (2l_1/l_y)]}$$

Therefore,

$$\left(\frac{l_1}{l_y}\right)^2 \frac{1 + i_1}{1 + i_2} \left(\frac{l_y}{l_x}\right)^2 \frac{4}{\mu} + 2\frac{l_1}{l_y} - 3 = 0 \tag{iii}$$

Solution of Eq. iii gives

$$\frac{l_1}{l_y} = \frac{\sqrt{1 + 3K_3} - 1}{K_3} \tag{7.62}$$

where

$$K_3 = \frac{4}{\mu}\left(\frac{l_y}{l_x}\right)^2 \frac{1 + i_1}{1 + i_2}$$

and from Eq. ii,

$$w_u = \frac{6(1 + i_2)m_{uy}}{l_1^2} \tag{7.63}$$

Substitution of Eq. 7.62 into Eq. 7.63 gives the ultimate uniform load for this yield line pattern.

The governing alternative collapse mode is the one giving the lowest ultimate load. The case of simple supports can be obtained by setting the appropriate i values equal to zero. The mode 1 breakdown case of $l_1 = l_x/2$ does not occur because, in general, this case does not represent the transition point from mode 1 to mode 2 as the slab properties change. In general, the transition occurs for slab properties leading to solutions giving the same load for mode 1 with $l_1 < l_x/2$ and for mode 2 with $l_1 < l_y$.

Demsky and Hatcher[7.18] conducted an extensive study of slabs supported on three sides and free at the fourth. They considered many combinations of l_x/l_y and moment capacities. They also varied the load linearly in the y-direction in addition to the uniformly distributed load case. They investigated corner effects as well and reported a maximum reduction in load capacity of about 9%.

7.9.4 Uniformly Loaded Orthotropic Rectangular Slabs with Two Adjacent Edges Supported and the Other Edges Free

The slab and alternative yield line patterns are shown in Fig. 7.41. Both supported edges are assumed to be fixed and the ultimate negative moments of resistance per unit width are defined in terms of the positive-moment capacities as shown in the figure.

Mode 1 (Fig. 7.41a). The nodal forces at the free edge at E act in the directions shown, where a dot indicates an upward force and a cross indicates a downward force. Equation 7.20 gives the nodal forces as $\pm m_{uy}l_1/l_y$. The yield line pattern has one unknown dimension. The solution for this case has been derived previously (see Example 7.5). From Eqs. 7.24, 7.25, and 7.22,

$$\frac{l_1}{l_x} = \frac{\sqrt{1 + 3K_1} - 1}{K_1} \tag{7.64}$$

where

$$K_1 = \mu \left(\frac{l_x}{l_y}\right)^2 \frac{3 + i_2}{1 + i_1}$$

$$w_w = \frac{6(1 + i_1)m_{ux}}{l_1^2} - \frac{6m_{uy}}{l_y^2} \tag{7.65}$$

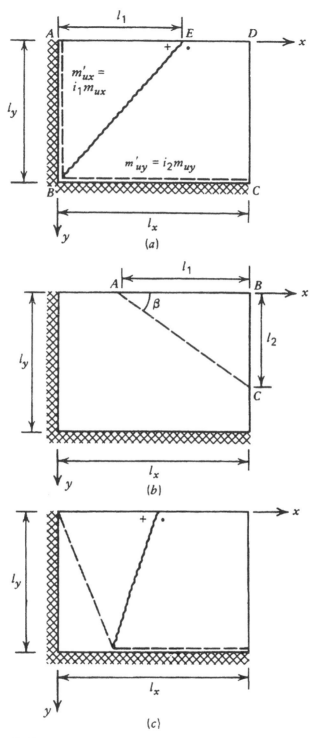

Figure 7.41 Uniformly loaded rectangular slab with two edges supported and two edges free: (*a*) mode 1; (*b*) mode 2; (*c*) mode 3.

Substitution of Eq. 7.64 into Eq. 7.65 gives the ultimate uniform load for this yield line pattern.

Mode 2   (Fig. 7.41*b*). The yield line pattern has two unknown dimensions. Consider the equilibrium of segment *ABC* by taking moments about *AC*.

$$i_1 m_{ux} l_2 \sin \beta + i_2 m_{uy} l_1 \cos \beta = \frac{w_u l_1 l_2}{2} \frac{l_1 \sin \beta}{3}$$

Therefore,

$$i_1 m_{ux} l_2^2 + i_2 m_{uy} l_1^2 = \frac{w_u l_1^2 l_2^2}{6}$$

$$w_u = 6 \left(\frac{i_1 m_{ux}}{l_1^2} + \frac{i_2 m_{uy}}{l_2^2} \right) \tag{i}$$

Inspection shows that for minimum w_u both l_1 and l_2 should be a maximum. Therefore, $l_1 = l_x$ and $l_2 = l_y$ and the yield line pattern runs across the diagonal of the slab. From Eq. i,

$$w_u = 6 \left(\frac{i_1 m_{ux}}{l_x^2} + \frac{i_2 m_{uy}}{l_y^2} \right) \tag{7.66}$$

Mode 3   (Fig. 7.41*c*). The third alternative yield line pattern is critical only when l_x is much greater than l_y and when it does govern it, will not give a load much different from that of mode 1 and has a lower limit at the load capacity of a cantilever slab.

   The governing alternative collapse mechanism of modes 1 and 2 is the one giving the lowest ultimate load. The case of simple supports may be obtained by setting the appropriate *i* value equal to zero.

7.10 COMPOSITE BEAM–SLAB COLLAPSE MECHANISMS

In the derivation of the ultimate load equations for slabs, it has been assumed that the supporting system is strong enough to support the ultimate load of the slab. It is evident that floors with relatively weak beams could fail by alternative collapse mechanisms which involve both plastic hinges in the beams and yield lines in the slab, as discussed by Wood,[7.5, 7.7,7.19] Park,[7.20,7.21] and Gamble et al.[7.41] Such composite beam–slab collapse mechanisms can be analyzed taking into account the flexural strength of the beams. A method of design for the beams based on composite beam–slab mechanisms is discussed in Chapter 8.

To illustrate composite beam–slab collapse mechanisms, consider a uniformly loaded rectangular slab resting on beams which in turn are simply supported at their ends, as shown in Fig. 7.42. Three possible collapse modes are also illustrated in the figure. Modes 1 and 2 involve yield lines across the slab centerline and plastic hinges in the beams, and mode 3 involves yield lines forming only within the slab. Mode 1 develops when the x-direction beams are relatively weak, mode 2 develops when the y-direction beams are relatively weak, and mode 3 involves yield lines forming only within the slab

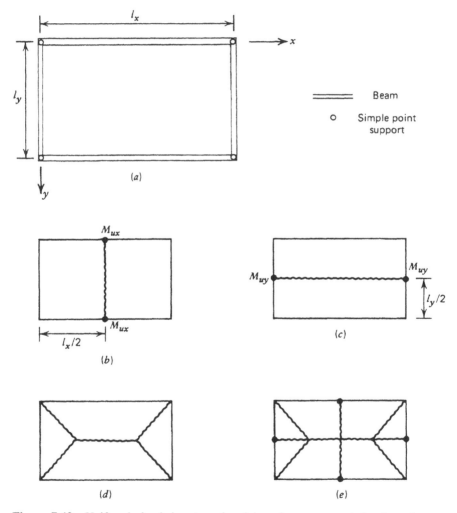

Figure 7.42   Uniformly loaded rectangular slab on beams supported only at the corners: (*a*) slab on beams; (*b*) mode 1; (*c*) mode 2; (*d*) mode 3; (*e*) combined modes, 1, 2, and 3.

when all beams are relatively strong. Let M_{ux} and M_{uy} be the ultimate moments of resistance of the x- and y-direction beams, respectively; m_{ux} and m_{uy} be the ultimate moment of resistance per unit width of the slab in the x- and y-directions, respectively; and w_u be the ultimate load per unit area.

For mode 1 (Fig. 7.42b), equilibrium of moments of a slab segment about the support as axis requires that

$$2M_{ux} + m_{ux}l_y = \frac{w_u l_y l_x^2}{8}$$

Therefore,

$$w_u = \frac{16M_{ux}}{l_y l_x^2} + \frac{8m_{ux}}{l_x^2} \tag{7.67}$$

For mode 2 (Fig. 7.42c), equilibrium of moments of a slab segment about the supports as axis requires that

$$2M_{uy} + m_{uy}l_x = \frac{w_u l_x l_y^2}{8}$$

Therefore,

$$w_u = \frac{16M_{uy}}{l_x l_y^2} + \frac{8m_{uy}}{l_y^2} \tag{7.68}$$

For mode 3 (Fig. 7.42d), Eq. 7.10 gives as the ultimate uniform load,

$$w_u = \frac{24m_{uy}}{l_y^2 \{[3 + (m_{ux}/m_{uy})(l_y/l_x)^2]^{1/2} - (l_y/l_x)(m_{ux}/m_{uy})^{1/2}\}^2} \tag{7.69}$$

The mode that occurs is the mode with the smallest ultimate load. It is possible that the relative strengths of the beams and the slab are such that all three modes have the same ultimate load. In that case the combined mode shown in Fig. 7.42e would occur.

Should the beam steel be curtailed near the supports, it is also necessary to investigate the collapse mechanisms shown in Fig. 7.43. One of these may be critical if the strength of the beams near the supports is significantly less than the midspan strength. This may also be the case if slab steel is partially curtailed short of the supports, and alternative slab mechanisms should be considered in that case.

Composite beam–slab collapse mechanisms can also occur in continuous floor systems. For example, Fig. 7.44 shows a uniformly loaded floor slab

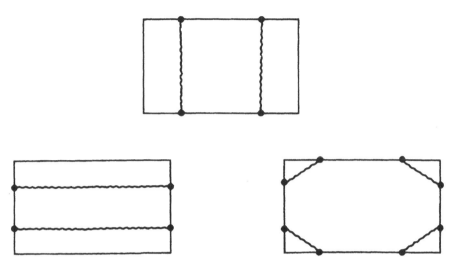

Figure 7.43   Alternative yield line patterns if beam steel is curtailed.

consisting of four rectangular panels supported by and continuous with beams at all edges. The beams are supported on columns at the panel corners. Collapse modes 1 and 2 (Fig. 7.44*b* and *c)* involve yield lines across the panels and plastic hinges in the beams. In the figure it has been assumed that at the edge of the floor the beams have a smaller ultimate negative moment of resistance than the moment capacities of the columns, and thus plastic hinges form in the beams there. Also in the figure it has been assumed that the torsional strength and/or stiffness of the edge beams is insufficient to allow the panels to develop negative-moment yield lines along the exterior edges. Collapse mode 3 (Fig. 7.44*d*) is the case where the beams are relatively strong. Again, each collapse mode would need to be examined to find the minimum load to cause collapse. For modes 1 and 2 the equilibrium equation can be written as previously, by summing the flexural strengths of the beams and slab across the floor and equating the total moment of resistance to the moment caused by the loading.

Two additional collapse modes, which are combinations of modes involving slabs and beams, are shown in Fig. 7.44*e* and *f*. These modes both include formation of plastic hinges in interior beams but not in the edge beams, and their formation is, of course, strongly dependent on the relative strengths of the various beams. The mechanism in Fig. 7.44*f* is one that could reasonably be expected to form in a slab structure that had edge beams but no interior beams. If the edge beams had adequate torsional strength and stiffness, negative-moment yield lines would also form in the slab along the edge beams, but this is unlikely in structures of normal proportions.

Where the slab and beams act compositely, as for in situ concrete construction or for steel beams with shear connectors, some of the concrete slab will

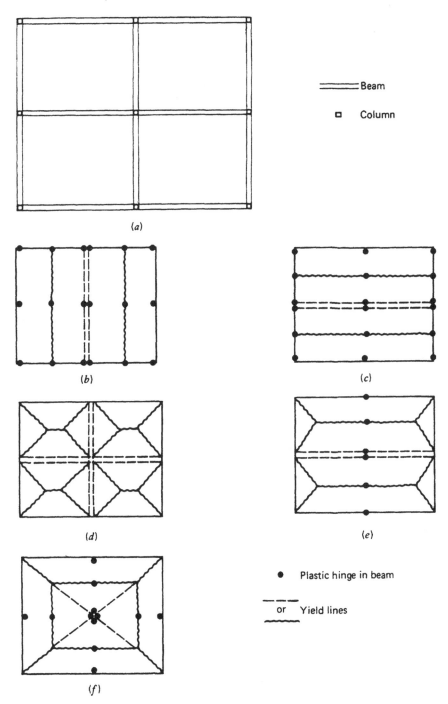

Figure 7.44 Uniformly loaded multipanel floor: (*a*) floor; (*b*) mode 1; (*c*) mode 2; (*d*) mode 3; (*e*) combined mode 4; (*f*) combined mode 5.

act as part of the beam to form T or L beams. The question then arises as to what the "slab" is and what the "beam" is. In fact, the strength of the floor is generally not very sensitive to the width of the slab assumed to be effective with the beam, because the difference made to the depth of concrete in compression is not particularly great if the members are not heavily reinforced. As a good approximation the definition of a beam given in Chapter 13 of the ACI Code[7.22] may be used: "A beam includes that portion of slab on each side of the beam extending a distance equal to the projection of the beam above or below the slab, whichever is greater, but not greater than four times the slab thickness." This code recommendation is based on test results obtained from floor systems at the University of Illinois and elsewhere. For the typical section through a floor shown in Fig. 7.45, the width of the compression flange for positive bending moment is set by this recommendation. For negative bending moments the reinforcement within that width can be included in the calculation of the ultimate moment of resistance of the beam.

It is also obvious that because of variations of neutral-axis depth and the geometry of deformations, considerable membrane (in-plane) forces may be induced in continuous floor systems when beams are present. At small deflections of the panels, stiff beams may prevent the edges of panels from spreading horizontally and result in the development of in-plane compressive membrane forces in the slab that enhance the flexural strength of the slab sections. At large deflections the center of the panel may develop full-depth

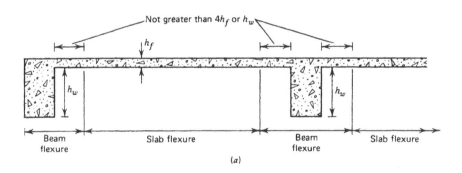

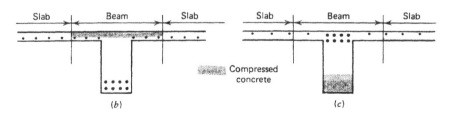

Figure 7.45   Beam-and-slab action of floor: (a) effective widths; (b) assumed positive moment sections; (c) assumed negative moment sections.

cracking but be capable of carrying large loads by tensile membrane action supported by compression in the edge regions of the slab and the beams. Thus, the simple bending action generally assumed in slab theory is conservative in many situations, and unexpected reserves of strength due to membrane action can be available, particularly if the beams are stiff and strong. Membrane action is referred to in the discussion of the test results reported in Section 7.17 and is treated in more detail in Chapter 12.

7.11   BEAMLESS FLOORS

Beamless (flat slab and flat plate) floors can also be analyzed by yield line theory. Two types of yield line pattern need to be investigated for uniformly loaded floors, one involving general folding of the floor and the other involving local slab collapse around the column. If beams are present around the exterior of the floor, other types of yield line pattern also require attention.

7.11.1   Folding Yield Line Patterns

Figure 7.46 shows a uniformly loaded flat slab floor with regular column layout and folding types of collapse mechanisms involving yield lines crossing the floor in each direction. The negative-moment yield lines are considered to be at the faces of the supports, as they must be if the yield lines are straight.

For interior bays let the negative and positive ultimate moments of resistance per unit width of the slab in the x-direction be m''_{ux} and m'_{ux} at opposite column lines and m_{ux}, respectively, and in the y-direction be m''_{uy} and m'_{uy} at opposite column lines and m_{uy}, respectively. Let l_{nx} and l_{ny} be the clear spans in the x- and y-directions, respectively, and the ultimate load per unit width area be w_u. Then the ultimate load equations are, assuming that the positive-moment yield line is at midspan, for mode 1 (Fig. 7.46b):

$$\frac{1}{2}\,(m''_{ux} + m'_{ux}) + m_{ux} = \frac{w_u l^2_{nx}}{8} \tag{7.70}$$

and for mode 2 (Fig. 7.46c):

$$\frac{1}{2}\,(m''_{uy} + m'_{uy}) + m_{uy} = \frac{w_u l^2_{ny}}{8} \tag{7.71}$$

If $m''_{ux} \neq m'_{ux}$, the positive-moment yield line will not actually be at midspan, but these equations still give good approximations of the correct load capacity.

For exterior bays, ultimate load equations can also be written taking into account the negative-moment strength of the slab–column junction and determining the position of the positive-moment yield line that gives the smallest

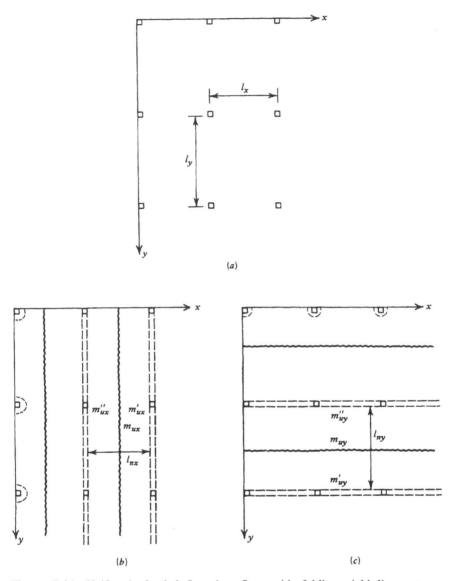

Figure 7.46 Uniformly loaded flat plate floor with folding yield line patterns: (a) floor; (b) mode 1; (c) mode 2.

ultimate load. When the steel is not uniformly spread but is placed in column and middle strips as is usual with more steel in the column strips than in the middle strips to more evenly match the elastic bending moments, the ultimate moments of resistance per unit width to be used in the equations are the average values across the slab.

7.11.2 Local Yield Line Patterns at Columns

Wood[7.5] has shown that another possible type of collapse mode involves circular fans around columns. Gesund and Dikshit[7.23] and the European Concrete Committee[7.9] also give equations for the strength of fan mechanisms around columns.

Columns as Point Supports. A simple approach to the problem is to assume that the columns have negligible width. Figure 7.47 shows circular fans occurring around point supports in a flat plate floor. The reinforcement is assumed to be uniformly spread and the negative and positive ultimate moments of resistance per unit width in the two directions are m'_u and m_u in the region of the fans. In this case of isotropic reinforcement, the fans are circular.

Had the slab been reinforced orthotropically, the fans would have had an elliptical shape with the major axis in the direction of strongest moment.

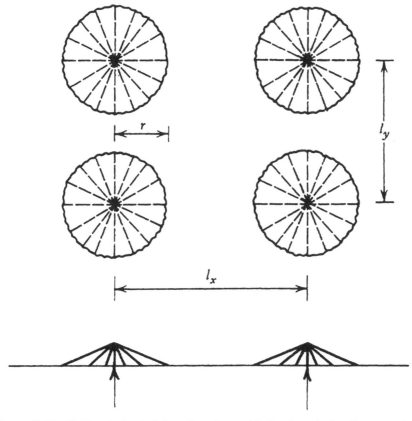

Figure 7.47 Uniformly loaded flat plate floor with local conical collapse mode at point supports.

Solutions for orthotropic reinforcement can be obtained from the isotropic solution using the affinity theorem of Section 7.8. Consider the floor to be given a small downward displacement δ at the center of each panel. External work done by the loading on one panel is given by

$$w_u(l_x l_y - \pi r^2)\delta + \frac{w_u \pi r^2 \cdot 2\delta}{3}$$

where l_x and l_y are the panel centerline dimensions and r is the radius of the failure cone. The internal work done by one failure cone, from Eq. 7.36, is

$$(m'_u + m_u)\delta \cdot 2\pi$$

Therefore, from the virtual work equation,

$$w_u(l_x l_y - \pi r^2)\delta + \frac{w_u \pi r^2 \cdot 2\delta}{3} = (m'_u + m_u)\delta \cdot 2\pi$$

Therefore,

$$w_u = \frac{2\pi(m'_u + m_u)}{l_x l_y - \frac{1}{3}\pi r^2} \tag{7.72}$$

If $l_x = l_y = l$, Eq. 7.72 may be written as

$$\frac{w_u l^2}{m'_u + m_u} = \frac{2\pi}{1 - (\pi/3)(r/l)^2} \tag{7.73}$$

which has the following values:

r/l	0.5	0.45	0.4	0.3	0.2	0.1	0
$w_u l^2/(m'_u + m_u)$	8.51	8.00	7.55	6.94	6.56	6.35	6.28

It can be seen that for uniformly spread reinforcement over the entire slab, the ultimate load given by Eq. 7.73 is smaller than that given by the folding mechanisms of Fig. 7.46 when $r/l < 0.45$. Hence, the reinforcement in the slab in the region of the column would have to be increased if the folding mechanisms of Fig. 7.46 are to govern. The reinforcement distribution normally used by designers will ensure that the folding mechanisms are more critical than the conical mechanisms, however, because of the greater concentration of reinforcement in the column strip than in the middle strip. The foregoing analysis for square panels shows that the average $m'_u + m_u$ in the column strip need only be $8.00/6.28 = 1.27$ times the average $m'_u + m_u$ for the entire panel width to prevent the formation of the conical collapse mode.

Rectangular Columns. A more rigorous approach would consider the effect of the column shape on the ultimate load. Figure 7.48 shows the collapse mode occurring around a rectangular column in a flat plate floor. The yield line pattern consists of fans centering on the column corners with rectangular segments between the fans. For isotropic reinforcement with negative and positive ultimate moments of resistance per unit width m'_u and m_u, respectively, and an ultimate uniformly distributed load per unit area w_u, if the floor is given a downward displacement δ at the center of each panel, the external work done by the loading on one panel is

$$w_u(l_x l_y - c_1 c_2 - \pi r^2 - 2c_1 r - 2c_2 r)\delta + 2w_u(c_1 r + c_2 r)\frac{\delta}{2} + w_u \pi r^2 \frac{2\delta}{3}$$

$$= w_u(l_x l_y - c_1 c_2 - c_1 r - c_2 r - \frac{\pi}{3}r^2)\delta$$

where l_x and l_y are the panel centerline dimensions and the other notation is

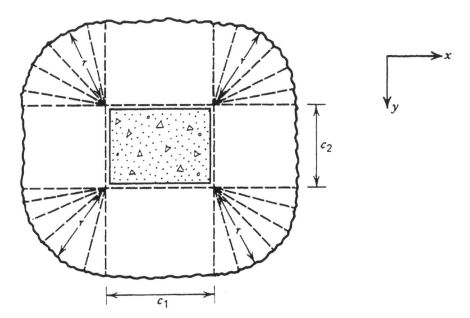

Figure 7.48 Uniformly loaded flat plate floor with local collapse mode around rectangular column.

shown in Fig. 7.48. The internal work done by one failure mechanism, utilizing Eq. 7.36, is

$$(m_u' + m_u)2\pi\delta + 2(m_u' + m_u)(c_1 + c_2) \frac{\delta}{r}$$

Therefore, from the virtual work equation,

$$w_u = \frac{2(m_u' + m_u)\{\pi + [(c_1 + c_2)/r]\}}{l_x l_y - c_1 c_2 - c_1 r - c_2 r - (\pi/3)r^2} \tag{7.74}$$

If $l_x = l_y = l$ and $c_1 = c_2 = c$, Eq. 7.74 may be written as

$$\frac{w_u l^2}{m_u' + m_u} = \frac{2[\pi + (2c/r)]}{1 - (c/l)^2 - (2cr/l^2) - (\pi/3)(r/l)^2} \tag{7.75}$$

To investigate the effect of a large column, let $c = 0.1l$; then Eq. 7.75 has the following values:

r/l	0.45	0.35	0.25	0.15	0.05
$w_u l^2/(m_u' + m_u)$	10.43	9.38	9.01	9.56	14.61

(These r/l values are each 0.05 less than those used in the preceding case because the column one-half size is $0.05l$.) For this particular case the clear span is $l_n = l - 0.1l = 0.9l$, and therefore from Eqs. 7.70 and 7.71 for the folding collapse mechanism, $m_u' + m_u = w_u l_n^2/8 = w_u l^2/9.88$. Thus, again for uniformly spread reinforcement over the entire slab, the local mechanism at the column will govern. The reinforcement in the vicinity of the columns will need to be increased if the folding collapse mechanism is to be critical. The value of $w_u l^2(m_u' + m_u)$ given by Eq. 7.75 when $c = 0.1l$ is a minimum when $r/l = 0.25$. Thus, for the case of square panels and $c = 0.1l$, the average $m_u' + m_u$ in the column strip need only be $9.88/9.01 = 1.10$ times the average $m_u' + m_u$ for the entire panel width to prevent the formation of a local collapse mechanism, a situation that will exist in all practical designs.

Circular Columns. It is also of interest to apply the more rigorous approach to circular columns. Figure 7.49 shows the collapse mode occurring around a circular column in a flat plate floor.

The yield line pattern consists of an annular-shaped fan. For isotropic reinforcement with negative and positive ultimate moments of resistance per unit width m_u' and m_u, respectively, and an ultimate uniformly distributed load

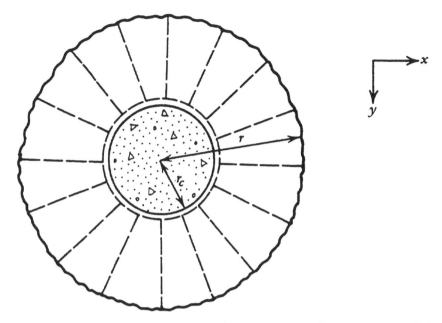

Figure 7.49 Uniformly loaded flat plate floor with local collapse mode around circular column.

per unit area w_u, if the floor is given a downward displacement δ at the center of each panel, the external work done by the loading on one panel is

$$w_u(l_x l_y - \pi r^2)\delta + w_u \int_0^{2\pi} \frac{\delta}{2}(r - r_c)r_c\,d\phi + w_u \int_0^{2\pi} \frac{2\delta}{3}\frac{(r - r_c)^2}{2}\,d\phi$$

$$= w_u(l_x l_y - \pi r^2)\delta + \frac{w_u \delta \pi (r - r_c)(2r + r_c)}{3}$$

where l_x and l_y are the panel centerline dimensions and the other notation is as in Fig. 7.49. The internal work done by the fan is

$$\int_0^{2\pi} (m_u' + m_u)\frac{\delta}{r - r_c}r\,d\phi = (m_u' + m_u)\frac{\delta r 2\pi}{r - r_c}$$

Therefore, from the virtual work equation,

$$w_u = \frac{2\pi r(m_u' + m_u)}{(r - r_c)(l_x l_y - \pi r^2) + (\pi/3)(r - r_c)^2\,(2r + r_c)} \tag{7.76}$$

If $l_x = l_y = l$, Eq. 7.76 may be written as

$$\frac{w_u l^2}{m_u' + m_u} = \frac{2\pi}{(1 - r_c/r)[1 - \pi(r/l)^2] + (\pi/3)(r/l - r_c/l)^2 (2 + r_c/r)} \quad (7.77)$$

To investigate the effect of a large column, let $r_c = 0.05l$; then Eq. 7.77 has the following values:

r/l	0.5	0.4	0.3	0.29	0.2	0.1
$w_u l^2/(m_u' + m_u)$	9.84	8.88	8.50	8.49	8.87	12.80

For this particular case, if the clear span is taken as $0.9l$, Eqs. 7.70 and 7.71 for the folding collapse mechanism give $m_u' + m_u = w_u l^2/9.88$. The value of $w_u l^2/(m_u' + m_u)$ from Eq. 7.77 for the local collapse mechanism when $r_c = 0.05l$ is a minimum when $r/l = 0.29$. Thus, for the case of square panels and $r_c = 0.05l$, the average $m_u' + m_u$ in the column strip needs to be $9.88/8.49 = 1.16$ times the average $m_u' + m_u$ for the entire panel width to prevent the formation of a local collapse mechanism.

7.11.3 Unbalanced Moment Transfer at Slab–Column Connections

The yield line patterns considered in previous sections for flat plate floors are for symmetrical floors with symmetrical gravity loading where the unbalanced bending moment to be transferred between the slab and the interior columns is insignificant. For symmetrical floors with gravity loading applied unsymmetrically (e.g., live load on some panels only), or for unsymmetrical floors, unbalanced bending moment will need to be transferred between slab and interior columns. Also, exterior columns will be subjected to unbalanced bending moment in all floors. Further, when a floor is subjected to substantial horizontal loading due to wind or earthquake, a large unbalanced bending moment may need to be transferred between slab and all columns. It is evident that unbalanced bending moment will affect the local yield line pattern. A local yield line pattern can be found which indicates the greatest unbalanced bending moment that can be transferred at a slab–column connection by flexure of the slab.[7.24] In addition to the yield line patterns considered in Section 7.11.2, there are two additional yield line patterns that could occur, the critical pattern depending on the ratio of unbalanced bending moment to vertical shear force to be transferred by the connection.

Local Yield Line Pattern When Gravity Load Is Small. Figure 7.50 shows an interior slab–column connection for a rectangular column with ultimate unbalanced bending moment M_u applied, causing a yield line pattern to develop in the slab consisting of fans centered on the column corners with

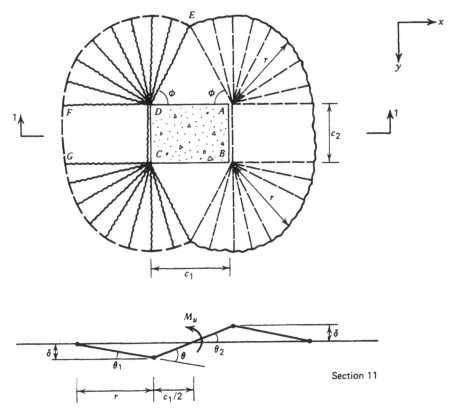

Figure 7.50 Local collapse mechanism at slab–column connection due to transfer of unbalanced bending moment when gravity load is small.

rectangular or triangular elements in between. The angle ϕ defines the unknown dimension of the yield line pattern. The radius of the fans r can be expressed in terms of the column dimension in the direction of bending c_1 and ϕ as $r = c_1/(2 \cos \phi)$. If the column rotates by angle θ_2, causing upward displacement δ at negative-moment yield line AB and downward displacement δ at positive-moment yield line CD, then with reference to the section shown in Fig. 7.50,

$$\theta_2 = \frac{2\delta}{c_1} \qquad \theta_1 = \frac{\delta}{r} = \frac{2\delta \cos \phi}{c_1}$$

$$\theta = \theta_1 + \theta_2 = \frac{2\delta(1 + \cos \phi)}{c_1}$$

For isotropic reinforcement with negative and positive ultimate moments of resistance per unit width m'_u and m_u, respectively, the internal work done by the moments at the yield lines may be obtained by considering the four fans (e.g., *FDE*), two rectangles (e.g., *CDFG*), and two triangles (e.g., *ADE*). The internal work done is

$$4(m'_u + m_u)\delta(\pi - \phi) + (m'_u + m_u)(\theta + \theta_1)c_2 + 2(m'_u + m_u)\theta_2 \times 0.5c_1 \tan \phi$$

$$= 2(m'_u + m_u) \left[2(\pi - \phi) + (1 + 2 \cos \phi) \frac{c_2}{c_1} + \tan \phi \right] \delta$$

The external work done by the uniform load on the mechanism sums to zero, since it is pushed up on one half of the pattern and pushed down on the other half by the same amount. The external work done by the unbalanced moment M_u is

$$M_u\theta_2 = \frac{2M_u\delta}{c_1}$$

Therefore, from the virtual work equation,

$$2M_u \frac{\delta}{c_1} = 2(m'_u + m_u) \left[2(\pi - \phi) + (1 + 2 \cos \phi) \frac{c_2}{c_1} + \tan \phi \right] \delta$$

$$M_u = (m'_u + m_u)[2c_1(\pi - \phi) + c_2(1 + 2 \cos \phi) + c_1 \tan \phi] \qquad (7.78)$$

The value of ϕ to be used in Eq. 7.78 is the value giving minimum M_u. For a square column, $c_1 = c_2 = c$, a value $\phi = 1.03$ rad (59°) may be shown to result in the minimum value of M_u, giving

$$M_u = 7.92(m'_u + m_u) \qquad (7.79)$$

Local Yield Line Pattern When Gravity Load Is Significant. Gesund[7.23] has derived a further local yield line pattern for the case where the gravity load is significant. This yield line pattern may be regarded as the result of combining the yield line patterns shown in Figs. 7.48 and 7.50. The fans on the left-hand side of each pattern, being of opposite moment sign in the two figures, tend to cancel each other out, and the yield line pattern shown in Fig. 7.51 results. The two fan perimeters are assumed to be quarter-circles. If the column rotation causes an upward displacement $\delta/2$ at negative moment yield line *AB* and downward displacement $\delta/2$ at positive moment yield line *CD*, then with reference to Fig. 7.51 it may be observed that the column moment M_u undergoes rotation δ/c_1 and the column load V_u rises by $\delta/2$ (or the essentially equal slab load drops by $\delta/2$).

For isotropic reinforcement with negative and positive ultimate moments of resistance per unit width m'_u and m_u, respectively, the internal work done

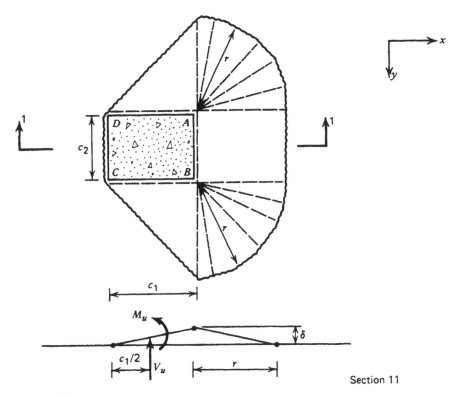

Figure 7.51 Local collapse mechanism at slab–column connection due to transfer of unbalanced bending moment when gravity load is significant.

by the moments at the yield lines may be obtained by considering the two fans, one rectangle and two triangles. The internal work done is

$$2(m'_u + m_u)\delta \cdot 0.5\pi + (m'_u + m_u)\frac{\delta}{r}c_2$$

$$+ 2(m'_u + m_u)\frac{\delta}{c_1}r + 2(m'_u + m_u)\frac{\delta}{r}c_1 + (m'_u + m_u)\frac{\delta}{c_1}c_2$$

$$= (m'_u + m_u)\delta\left(\pi + \frac{c_2}{r} + \frac{2r}{c_1} + \frac{2c_1}{r}\right) + (m'_u + m_u)\delta\frac{c_2}{c_1}$$

The external work done is

$$M_u\frac{\delta}{c_1} + V_u\frac{\delta}{2}$$

Therefore, from the virtual work equation,

$$M_u \frac{\delta}{c_1} + V_u \frac{\delta}{2} = (m'_u + m_u)\delta \left(\pi + \frac{c_2}{r} + \frac{2r}{c_1} + \frac{2c_1}{r} \right) + (m'_u + m_u)\delta \frac{c_2}{c_1}$$

$$M_u = c_1(m'_u + m_u) \left(\pi + \frac{c_2}{r} + \frac{2r}{c_1} + \frac{2c_1}{r} \right) + (m'_u + m_u)c_2 - \frac{V_u c_1}{2} \quad (7.80)$$

The value of r to be used in Eq. 7.80 is the value giving minimum M_u. For a square column ($c_1 = c_2 = c$) a value of $r = \sqrt{1.5}c$ may be shown to result in close to the minimum value of M_u, giving

$$M_u = 9.04(m'_u + m_u)c - 0.5V_u c \quad (7.81)$$

Critical Local Yield Line Pattern. It is evident that if vertical shear force V_u and unbalanced bending moment M_u are to be transferred at a slab–column connection, and if a local yield line pattern governs the strength of the slab adjacent to the column, the possible yield line patterns that could occur are as follows: (1) when $V_u = 0$ and only M_u is to be transferred, the antisymmetrical arrangement of fans shown in Fig. 7.50 will form; (2) when $M_u = 0$ and only V_u is to be transferred, the symmetrical arrangement of fans shown in Fig. 7.48 will form; and (3) when both V_u and M_u are to be transferred, either of the patterns shown in Figs. 7.50 and 7.48 or the alternative pattern shown in Fig. 7.51 could form.

As an illustration of the possible alternatives, an interaction diagram for M_u versus V_u, calculated on the basis of the aforementioned three yield line patterns, is shown in Fig. 7.52 for the common case where the column is square and the ultimate negative moment of resistance per unit width of the slab is twice the ultimate positive moment of resistance per unit width (i.e., for $m'_u = 2m_u$). The shear force V_u for the region CD is for the case where the columns are on a square grid and the ratio of length of column side to slab span between column centers is 0.1. The interaction diagram in Fig. 7.52 illustrates that when $m'_u = 2m_u$, Eq. 7.79 gives a good approximation for M_u unless the V_u / m'_u ratio for the slab is high.

7.11.4   Flat Slab Floors with Exterior Beams

In many flat slab (and flat plate) floors a beam may be present at the exterior edge of the floor to stiffen and strengthen the edge. The strength of the edge beam may be included in the strength of the general folding mechanism by the method discussed in Section 7.10. A folding mechanism is shown in Fig. 7.53a, with plastic hinges forming in the beam. A negative-moment yield line is shown forming in the slab along the edge beam parallel to the fold, but this requires a stiff and strong edge beam and column. The total ultimate negative moment along the edge of one panel may need to be replaced by the column flexural strength or by the greatest moment that can be transferred

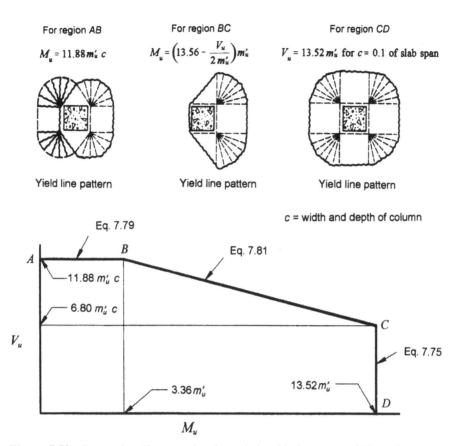

For region *AB*

$$M_u = 11.88\,m_u'\,c$$

For region *BC*

$$M_u = \left(13.56 - \frac{V_u}{2\,m_u'}\right)m_u'$$

For region *CD*

$$V_u = 13.52\,m_u' \text{ for } c = 0.1 \text{ of slab span}$$

Yield line pattern   Yield line pattern   Yield line pattern

c = width and depth of column

Figure 7.52 Interaction diagram, showing relationship between unbalanced moment strength M_u and vertical shear force V_u transferred at slab–column connection.

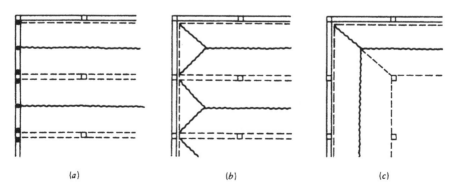

(a)   (b)   (c)

Figure 7.53 Uniformly loaded flat plate with edge beams.

between the slab and the column. Figure 7.53*b* shows the case of strong beams where the yield line pattern is confined to the panels. These two yield line patterns would also need to be checked for occurrence in the other direction. Figure 7.53*c* shows a further yield line pattern that could occur with strong edge beams.

7.11.5 Shear Strength of Slab–Column Connections

It should be emphasized that the yield line patterns given above for flat plate floors consider flexural failures only. It is important for the designer to check that shear failures at slab–column connections do not precede a flexural failure. The shear strength of slabs at the column junctions is considered in detail in Chapter 10. The shear strength of the slab is often critical and can govern the column size and slab thickness necessary at the column.

7.12 UNIFORMLY LOADED RECTANGULAR SLABS WITH OPENINGS

Reinforced concrete slabs often contain openings of considerable size for ducts, pipes, and other services. The openings tend to attract yield lines, since they represent regions of zero flexural strength in the slab. On the other hand, the slab is not required to carry load over the area of the opening and hence the total load to be carried is smaller than for a slab without an opening. Lash and Banerjee[7.25] and Zaslavsky[7.26] have used yield line theory to produce ultimate load equations for uniformly loaded simply supported slabs with rectangular openings, and Johansen[7.3] and Wood and Jones[7.7] have produced some equations for openings in various positions. Islam and Park[7.27] have produced yield line theory equations and design charts for uniformly loaded orthotropic rectangular slabs with either fixed or simply supported edges and with rectangular openings of various sizes and positions.

The effect of an opening will be illustrated with reference to the square isotropic uniformly loaded slab with a central square opening shown in Fig. 7.54. The size of the opening is defined by the value of k. The slab will be considered to be fixed around the outside edge. At the ultimate load the yield line pattern shown in the figure will develop. Because of symmetry, the positive-moment yield lines will run into the corners of the opening.

Let the negative and positive ultimate moments of resistance per unit width be m'_u and m_u, respectively, and the ultimate uniformly distributed load per unit area be w_u acting over all parts of the slab except the opening. If the edge of the opening undergoes a small downward displacement δ, the work terms for the virtual work equation may be written as follows: External work done is

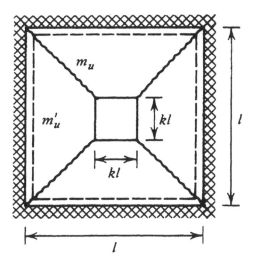

Figure 7.54   Uniformly loaded square slab with central opening.

$$4\left[kl\left(\frac{l}{2} - \frac{kl}{2}\right)\frac{\delta}{2} + \left(\frac{l}{2} - \frac{kl}{2}\right)^2\frac{\delta}{3}\right]w_u = \frac{1}{3}w_ul^2\delta(1 - k)(1 + 2k) \qquad \text{(i)}$$

Internal work done is

$$4[m_u(1 - kl) + m_u'l]\frac{\delta}{(l/2) - (kl/2)} = 8m_u\delta\left(1 + \frac{m_u'}{m_u}\frac{1}{1 - k}\right) \qquad \text{(ii)}$$

Hence, from the virtual work equation,

$$w_u = \frac{24m_u\{1 + (m_u'/m_u)[1/(1 - k)]\}}{l^2(1 - k)(1 + 2k)} \qquad (7.82)$$

Note that if $k = 0$ (i.e., no opening), $w_u = w_0$, where

$$w_0 = \frac{24m_u}{l^2}\left(1 + \frac{m_u'}{m_u}\right) \qquad (7.83)$$

and hence

$$\frac{w_u}{w_0} = \frac{1}{(1 - k)(1 + 2k)}\frac{1 + (m_u'/m_u)[1/(1 - k)]}{1 + (m_u'/m_u)} \qquad (7.84)$$

Values of w_u/w_0 from Eq. 7.84 for various values of k and m_u'/m_u are:

m'_u/m_u	\multicolumn{9}{c}{k}
	0	0.1	0.2	0.3	0.4	0.5	0.6	0.7
0	1.00	0.93	0.89	0.89	0.93	1.00	1.14	1.39
1.0	1.00	0.98	1.01	1.08	1.24	1.50	1.99	3.01
2.0	1.00	0.99	1.04	1.15	1.34	1.67	2.28	3.55

Thus, for simply supported square slabs, the opening causes a maximum reduction in the ultimate load per unit area of about 11% when $k = 0.2$ to 0.3, and when $k = 0.5$ or greater, there is no reduction in the load per unit area carried. For fixed-edge slabs, the reduction is less than for simply supported slabs since the yield lines around the edges are unaffected and usually the slab with an opening will have a greater load capacity per unit area than will the slab without an opening.

For uniformly loaded rectangular slabs, there are a large number of possible yield line patterns, depending on the opening position. Possible yield line patterns for rectangular openings at the slab center, corner, center of short side, and center of long side are shown in Figs. 7.55 to 7.58. The size of the opening is defined by the k value, and for convenience the opening is assumed to have the same aspect ratio as the slab. The ultimate negative moments of resistance per unit width are assumed to be identical at opposite edges of the slab. For a central opening there are three possible yield line patterns, each with one unknown dimension defined by a β value. For a corner opening there are four possible yield line patterns, each with three unknown dimensions. Mode B1 is likely to govern for small openings and the other modes for larger openings. For an opening at the center of the short or long side, there are three possible yield line patterns for each case, each with two unknown dimensions. Modes C1 and D1 will govern if the opening is of small size and the other modes if the opening is large.

Zaslavsky[7.26] has prepared design charts that relate the ultimate uniform load to the ultimate moments of isotropic rectangular slabs with central openings, simply supported outside edges, and various l_x/l_y and k values. The values for ultimate uniform load are the minimum given by the three yield line patterns. When these ultimate loads per unit area are compared with those of slabs without openings, it is found that the largest reduction in the load-carrying capacity is about 12% and occurs when $k = 0.2$ to 0.3. Slabs with openings with $k > 0.5$ are found to have a larger ultimate load per unit area than slabs without an opening.

Islam and Park[7.27] have considered all four cases of openings shown in Figs. 7.55 to 7.58 for uniformly loaded orthotropic slabs with all outside edges either simply supported or fixed. The full ultimate load equations for each mode of each opening type may be seen elsewhere.[7.27] The ultimate load equation for each mode can be lengthy and each equation can contain up to

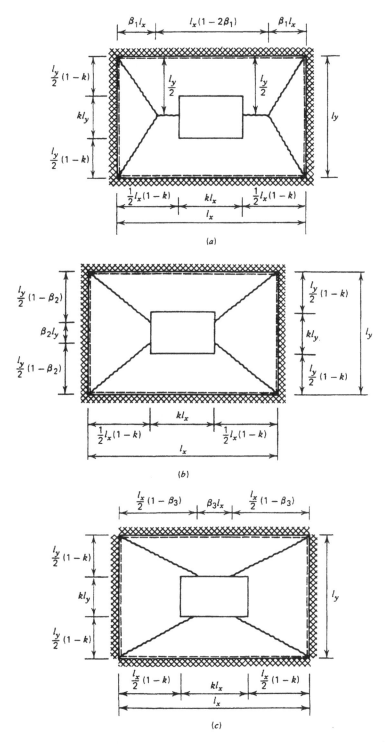

Figure 7.55 Uniformly loaded rectangular slab with central opening: (*a*) mode A1; (*b*) mode A2; (*c*) mode A3.

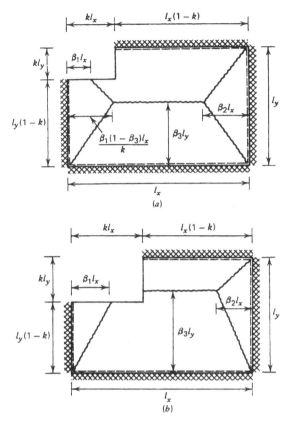

Figure 7.56 Uniformly loaded rectangular slab with corner opening: (*a*) mode B1;
(*b*) mode B2; (*c*) mode B3; (*d*) mode B4.

three of the unknown terms β_1, β_2, and β_3, which define the positions of the
yield lines. Those values of β_1, β_2, and β_3 that make the ultimate uniformly
distributed load per unit area w_u a minimum are sought in each case. The
required values of these unknown terms may be found by solving simulta-
neously the equations $\partial w_u / \partial \beta_1 = 0$, $\partial w_u / \partial \beta_2 = 0$, and $\partial w_u / \partial \beta_3 = 0$ for
each expression for w_u, but the resulting simultaneous equations are nonlinear
and the algebraic work involved is very lengthy. Also, for some yield line
patterns (e.g., modes A2 and C2), the minimum ultimate uniform load given
by the equations may not occur when the magnitude of the unknown dimension lies within the allowable range of variation. Because of these difficulties,
a numerical procedure was adopted. Computer programs were written to calculate the values of the ultimate uniform load for each mode for various
values of the dimensions β_1, β_2, and β_3 within their allowable ranges, thus
enabling the minimum value of w_u for each yield line pattern considered to
be determined. In this analysis the values of β_1, β_2, and β_3 were varied at

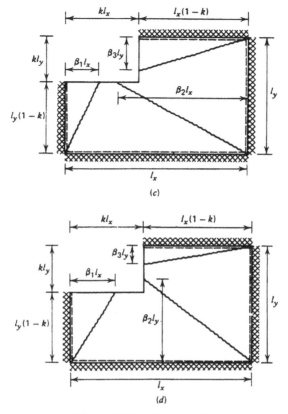

Figure 7.56 (*Continued*).

increments of either 0.01 or 0.02 and all possible combinations of values investigated. Having found the minimum w_u for each mode, the actual w_u for the particular type of opening is the lowest w_u given by the modes for that opening. Charts were plotted of the actual w_u for each type of loading versus the ratio of opening size to span k for various ratios of slab spans in the two directions l_x/l_y, ultimate positive resisting moments per unit width in the two directions m_{ux}/m_{uy}, and ultimate negative and positive resisting moments per unit width m'_{ux}/m_{ux} ($= m'_{uy}/m_{uy}$).

The findings[7.27] with regard to the variations of the ultimate uniform load follow the same trends as discussed previously. Openings in two-way slabs tend to attract yield lines to them, since they represent regions of zero flexural strength in the slab. However, the slab is not required to carry load over the area of the opening; and therefore the total load to be carried is smaller than for a slab without an opening. Hence, the ultimate load per unit area of a slab with an opening (assumed to be acting only over that part of the slab where there is no opening) may be either smaller or greater than that of a slab

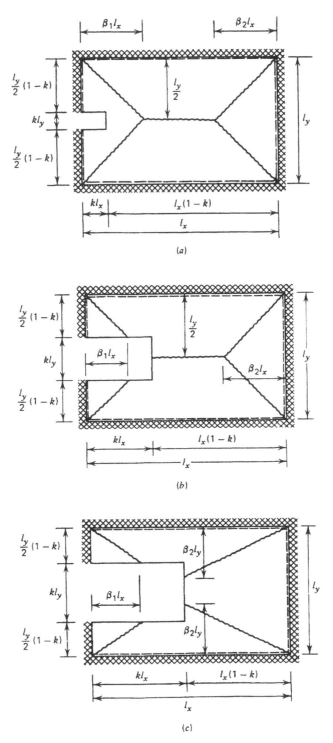

Figure 7.57 Uniformly loaded rectangular slab with opening at the middle of a short side: (*a*) mode C1; (*b*) mode C2; (*c*) mode C3.

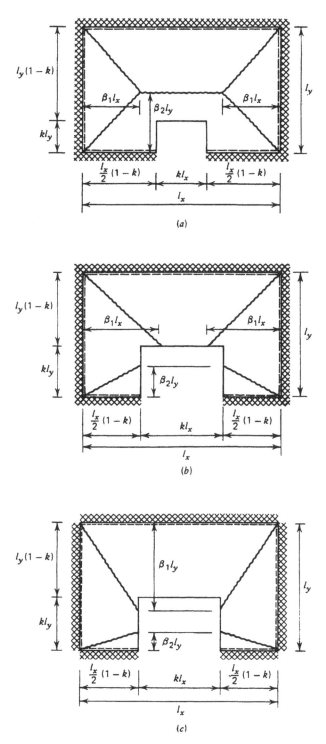

Figure 7.58 Uniformly loaded rectangular slab with opening at the middle of a long side: (*a*) mode D1; (*b*) mode D2; (*c*) mode D3.

without an opening. The effect of opening size on the ultimate load per unit area is greatest for slabs with simply supported edges, because for this case openings cause a greater reduction in the total length of yield lines than in fixed-edge slabs.

For slabs with a central opening the largest reduction in the ultimate load per unit area is about 13%, occurring when $k = 0.2$ to 0.3 for simply supported edges; when $k = 0.4$ to 0.6, the ultimate load per unit area becomes greater than that for the case without an opening. For fixed-edge slabs with a central opening, when the ratio of ultimate negative to positive moments per unit width is 1 or 2, the reduction in ultimate load is much smaller. At large k values, the ultimate load per unit area of the slab increases very significantly.

A corner opening in simply supported slabs causes a reduction in the ultimate load per unit area of up to 21% when $k = 0.3$ to 0.5. A reduction in the ultimate load of up to 10% occurs even when the ratio of ultimate negative to positive moments per unit width is 2. Slabs with a corner opening do not show such a significant increase in ultimate load carried per unit area as the opening size increases, because of the substantial reduction in total length of yield line accompanying a large opening.

An opening placed at the center of a short side of a simply supported slab causes a maximum reduction in the ultimate load per unit area of about 11% when $k = 0.3$ to 0.4. Fixed-edge slabs do not show any reduction in strength, and the ultimate load per unit area increases significantly with large k values. When the opening is at the center of a long side of a simply supported slab, the maximum reduction in ultimate load per unit area is about 22% when $k = 0.5$. The reduction in strength becomes less significant when the edges of the slab are fixed.

These considerations lead to a possible simple design approach or a means of checking the residual strength of a two-way slab if an opening is cut in it to accommodate new services. The approach is to determine the ultimate load per unit area of the slab as if the opening were not present and then modify it by the percentages indicated above for the various cases. For example, the ultimate load per unit area of a simply supported slab with a central rectangular opening will never be reduced by more than 13% regardless of the opening size. Alternatively, use can be made of the design charts[7.27] to obtain a more exact estimate of strength. (Note that the directions of the x- and y-axes in this reference are the reverse of those used in this book.) Although only rectangular openings have been considered, it is evident that the design charts will also give a very good indication of the effect of openings of other shapes.

Cases where the applied loading includes both distributed loads over the surface of the slab plus line loads at the edge of the hole do not appear to have been investigated. Such loadings may be important where an access hole is normally closed by a hatch cover that is capable of transferring load to the edge of the hole but does not contribute to the resistance of the slab.

7.13 UNIFORMLY LOADED CIRCULAR AND RING SLABS

The case of a uniformly loaded circular slab with fixed edges was treated in Example 7.6, leading to Eq. 7.30. It is of interest to consider further solutions for circular slabs.

7.13.1 Circular Slab Supported on n Columns and Subjected to Uniform Loading

The yield line pattern for a uniformly loaded circular slab of radius R symmetrically supported on n columns near the slab edge on a circle of radius r is shown in Fig. 7.59. Let the slab be reinforced isotropically with ultimate positive moments of resistance per unit width m_u, and the ultimate uniformly distributed load per unit area be w_u. The columns are assumed to provide simple point supports.

Consider segment A of the yield line pattern (see Fig. 7.59). The distance of the centroid of the segment from the slab center is given by

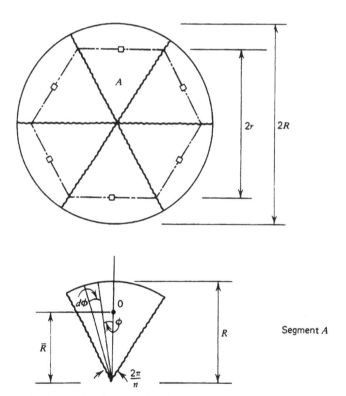

Figure 7.59 Uniformly loaded circular slab on columns with positive-moment yield lines.

$$\bar{R} = \frac{2 \int_0^{\pi/n} \frac{1}{2} R^2 \frac{2}{3} R \cos \phi \, d\phi}{2 \int_0^{\pi/n} \frac{1}{2} R^2 \, d\phi} = \frac{2}{3} R \frac{n}{\pi} \sin \frac{\pi}{n} \tag{7.85}$$

Also, the chord length across the top of the segment is $2R \sin (\pi/n)$. If the slab is given a small downward displacement δ at its center, the external work done by the loading is

$$n \frac{\pi R^2}{n} w_u \left(r - \frac{2}{3} R \frac{n}{\pi} \sin \frac{\pi}{n} \right) \frac{\delta}{r}$$

and the internal work done is

$$n m_u \frac{\delta}{r} 2R \sin \frac{\pi}{n}$$

Therefore, from the virtual work equation,

$$w_u = \frac{m_u}{(Rr/2) \left[(\pi/n)/\sin(\pi/n) \right] - (R^2/3)} \tag{7.86}$$

as obtained by Jones.[7.6] For example, if the slab is supported on six columns and $r = 0.8R$, then $\pi/n = \pi/6$ rad, and from Eq. 7.86, $w_u = 11.69 m_u/R^2$. Also, if the slab is supported around the whole of its edge, $r = R$ and because $(\pi/n)/\sin(\pi/n) \rightarrow 1$ when $n \rightarrow \infty$, Eq. 7.86 gives $w_u = 6m_u/r^2$, which is identical to Eq. 7.30 for a simply supported slab.

Finally, note that Eq. 7.86 is apparently applicable only if

$$\frac{Rr}{2} \frac{\pi/n}{\sin(\pi/n)} > \frac{R^2}{3} \quad \text{or} \quad \frac{r}{R} > \frac{2}{3} \frac{\sin(\pi/n)}{\pi/n}$$

This is because the equation is derived assuming that the centroid of each segment lies inside the column. If the centroid of segment lies outside the column, negative-moment yield lines form instead of positive-moment yield lines as in Fig. 7.60, and the ultimate load equation becomes

$$w_u = \frac{m_u'}{(R^2/3) - (Rr/2)[(\pi/n)/\sin(\pi/n)]} \tag{7.87}$$

where m_u' is the ultimate negative moment of resistance per unit width in all directions. For example, for a circular slab supported by a single central column, $r = 0$ and $w_u = 3m_u'/R^2$.

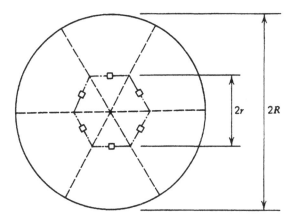

Figure 7.60 Uniformly loaded circular slab on columns with negative-moment yield lines.

A lower load capacity might be obtained from the mechanisms shown in Fig. 7.24 or 7.25, with Eq. 7.26 leading to w_u. If this occurs, the negative-moment yield lines occur along the axes of rotation shown in Fig. 7.59. The strength of the overhanging cantilever portions of the slab could also limit w_u.

7.13.2 Ring Slabs

Outside Edge Fixed and Inside Edge Free. Figure 7.61 shows a ring slab fixed around the outside edge and free (unsupported) around the inside edge, of outside radius R and inside radius r. The slab carries an ultimate uniformly distributed load per unit area w_u over its area and an ultimate line load per unit length p_u around the inside edge. The ultimate moments of resistance per unit width are m_u for positive circumferential moment and m'_u for radial negative moment. Note that these moments of resistance can be provided by

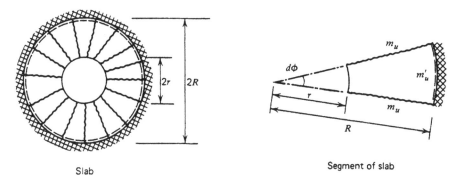

Slab   Segment of slab

Figure 7.61 Uniformly loaded ring slab fixed at outside edge and free at inside edge.

bottom reinforcement placed circumferentially and top reinforcement placed radially.

The yield line pattern and one segment of the slab are shown in Fig. 7.61. The nodal forces at the free edge are zero because the yield lines enter at 90° to the edge. Considering the equilibrium of a segment by taking moments about the support gives

$$m'_u R \, d\phi + m_u (R - r) \, d\phi = w_u \left[\frac{1}{2} R^2 \, d\phi \frac{R}{3} - \frac{1}{2} r^2 \, d\phi - \left(R - \frac{2}{3} r \right) \right]$$
$$+ p_u r \, d\phi (R - r)$$

$$m'_u R + m_u (R - r) = \frac{1}{6} w_u (R^3 - 3Rr^2 + 2r^3) + p_u r (R - r) \quad (7.88)$$

If the outside edge is simply supported, $m'_u = 0$.

Inside Edge Fixed and Outside Edge Free. Figure 7.62 shows a ring slab fixed around the inside edge and free (unsupported) around the outside edge, of inside radius r and outside radius R. The slab carries an ultimate uniformly distributed load per unit area w_u over its area and an ultimate line load per unit length p_u around the outside edge. The ultimate negative moments of resistance per unit width are m''_u for circumferential moment and m'_u for radial moment. Note that these moments of resistance can be provided for by top reinforcement placed circumferentially and radially.

The yield line pattern and a segment of the slab are shown in Fig. 7.62. The nodal forces at the free edge are zero because the yield lines enter at 90° to the edge. Considering the equilibrium of the segment by taking moments about the support gives

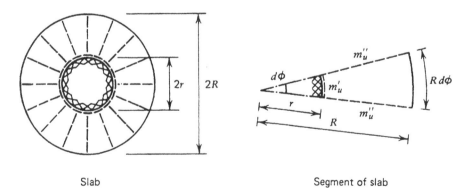

Slab   Segment of slab

Figure 7.62 Uniformly loaded ring slab fixed at inside edge and free at outside edge.

$$m'_u r \, d\phi + m''_u (R - r) \, d\phi = w_u \left[\frac{1}{2} R^2 \, d\phi \left(\frac{2}{3} R - r \right) + \frac{1}{2} r^2 \, d\phi \frac{r}{3} \right]$$

$$+ \, p_u R \, d\phi \, (R - r)$$

$$m'_u r + m''_u (R - r) = \frac{1}{6} w_u (2R^3 - 3R^2 r + r^3) + p_u R (R - r) \qquad (7.89)$$

If the inside edge is simply supported, $m'_u = 0$.

Outside and Inside Edges Fixed. Figure 7.63 shows a ring slab fixed around both the inside and outside edges, of inside radius r and outside radius R. The slab carries an ultimate uniformly distributed load per unit area w_u. The ultimate negative moments of resistance per unit width are m'_u for circumferential moment and radial moment at the inside edge and m''_u for radial moment at the outside edge. The ultimate positive moment of resistance per unit width is m_u for circumferential and radial moments.

The yield line pattern is shown in Fig. 7.63. There is no simple way of finding the nodal forces acting at the internal junction point of the yield lines (see Section 7.4.3). A virtual work solution can be obtained. Johansen[7.3] gives as the solution

$$m'_u + 2m_u + m''_u = \frac{1}{15} w_u (R - r)^2 \left(2.75 + \frac{r}{R} \right) \qquad (7.90)$$

7.14 SKEW SLABS

In skew slabs it is often preferable to place the reinforcement parallel to the edges to avoid excessive cutting of bars to different lengths. If the bars are

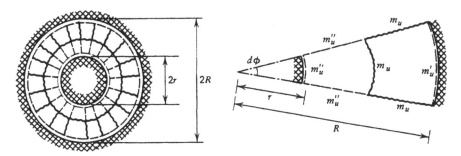

Figure 7.63 Uniformly loaded ring slab fixed at outside and inside edge.

placed parallel to the edges of the slab, they will not be at right angles in the two directions, and to obtain the ultimate moments of resistance per unit width along a yield line in a general direction in such a slab it is necessary to go back to first principles.

Figure 7.64a shows the reinforcement placed in directions x and s inclined at angle β. Figure 7.64b shows a yield line, with normal n inclined at angle α to the x-direction reinforcement, crossing each set of reinforcement bars. The actual yield line can be replaced by a stepped yield line consisting of small steps perpendicular to the reinforcement directions. As is assumed in

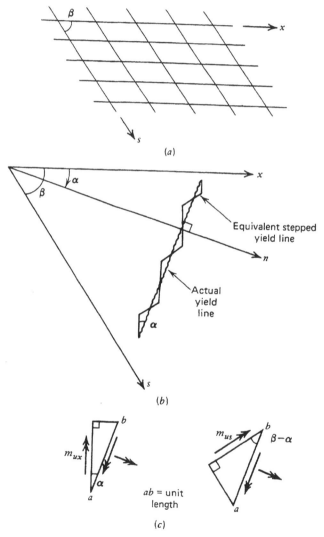

Figure 7.64   Yield line at general angle to skew reinforcement: (a) reinforcement; (b) yield line; (c) slab elements at yield line.

the Johansen yield criterion, the torsional moments on the steps of each yield line are assumed to be zero. The ultimate normal resisting moment per unit width m_{un} can be determined by considering the equilibrium of the small triangular elements shown in Fig. 7.64c by taking moments about the sides ab and considering the moment components from the x- and s-direction bars. The ultimate resisting moments per unit width in the x- and s-directions are m_{ux} and m_{us}, respectively. For the x-direction bars, the contribution to the ultimate normal resisting moment per unit width is $m_{ux} \cos^2\alpha$; for the s-direction bars, the contribution is $m_{us} \cos^2 (\beta - \alpha)$. Thus, the ultimate normal resisting moment per unit width from the x- and s-direction reinforcement is

$$m_{un} = m_{ux} \cos^2\alpha + m_{us} \cos^2(\beta - \alpha) \qquad (7.91)$$

Note that when $\alpha = 90°$ the case of bars at right angles, the s-axis becomes the y-axis and the second term on the right-hand side becomes $m_{us} \cos^2(90 - \alpha) = m_{uy} \sin^2\alpha$, which agrees with Eq. 7.2.

Similarly, taking moments about an axis perpendicular to ab shows that the torsional moment per unit width acting along the yield line is

$$m_{unt} = m_{ux} \sin \alpha \cos \alpha - m_{us} \sin(\beta - \alpha) \cos(\beta - \alpha) \qquad (7.92)$$

Again, when $\alpha = 90°$, the case of bars at right angles, the s-axis becomes the y-axis and the second term on the right-hand side becomes $-m_{ux} \sin \alpha \cos \alpha$, which agrees with Eq. 7.3.

To avoid consideration of nodal forces and torsional moments, the virtual work method can be used.

Example 7.15. A skew slab is simply supported at two opposite edges over a right span of l, as shown in Fig. 7.65. The reinforcing bars are placed parallel to the edges in the x- and s-directions and provide ultimate positive moments of resistance per unit width of m_{ux} and m_{us} in the x- and s-directions, respectively. Calculate the ultimate uniformly distributed load per unit area w_u for the slab.

SOLUTION. Compatibility of deformations of the yield line pattern requires that the yield line be parallel to the supports, and symmetry shows that the yield line lies at midspan. Hence, the angle δ between the normal to the yield line and the x-axis is $90°$. Therefore, from Eq. 7.91,

$$m_{un} = m_{ux} \cos^2 90° + m_{us} \cos^2(\beta - 90°)$$

$$= m_{us} \sin^2 \beta$$

Hence, the x-direction steel makes no contribution to the flexural strength of

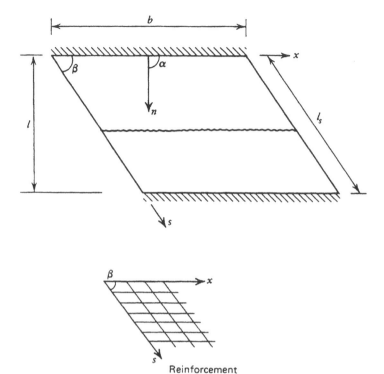

Figure 7.65   Uniformly loaded skew slab of Example 7.15.

the yield line in this case, which is to be expected since that steel is parallel to the yield line.

Using the virtual work method, let the yield line be given a small downward displacement δ. Then the internal work done is

$$m_{un}\theta_n l_0 = (m_{us} \sin^2\beta)\,\frac{2\delta}{0.5l}\,b = 4m_{us}\sin^2\beta\,\frac{\delta b}{l}$$

and the external work done is

$$\Sigma\, W_u\Delta = w_u lb\,\frac{\delta}{2}$$

Therefore, from the virtual work equation,

$$w_u = 8m_{us} \frac{\sin^2\beta}{l^2}$$

$$= \frac{8m_{us}}{l_s^2} \qquad\qquad (7.93)$$

where l_s is the span parallel to the edges of the slab.

7.15   APPROXIMATE YIELD LINE PATTERNS FOR UNIFORMLY LOADED RECTANGULAR SLABS

7.15.1   Use of Approximate Yield Line Patterns

It is evident that in cases where one or more unknown dimensions need to be determined to locate the yield line pattern, the amount of algebraic work required to achieve a yield line theory solution can be significant. In the virtual work method, this means solving simultaneously a number of equations obtained by differentiating the ultimate load expression with respect to each of the unknown dimensions in turn. In the equilibrium method, the equilibrium equations are solved simultaneously to find the unknown dimensions. In many cases the equations to be solved are nonlinear. An unfortunate aspect of yield line theory is that because it is an upper bound approach, failure to use the correct values of these dimensions in the ultimate load equation will mean that the ultimate load of the slab is overestimated and could therefore lead to an unsafe situation. This would appear to discourage the use of approximate yield line patterns. Fortunately, however, it can be shown that a yield line pattern that differs only slightly from the correct pattern will give a close estimate of the correct ultimate load if the *virtual work* method is used. Therefore, reasonable approximations regarding the yield line pattern may be used to simplify the ultimate load equations in many cases. The accuracy of approximate patterns for rectangular slabs is examined next. This work is based on papers published by Park.[7.28,7.29]

7.15.2   Uniformly Loaded Rectangular Slabs with All Edges Supported

In the general case of a uniformly loaded rectangular slab with all edges supported, the exact expression for the ultimate uniform load given by Eq. 7.59 is lengthy. A good simplifying approximation is to assume that the positive moment yield lines enter the corners of the slab at 45° to the edges (i.e., $l_1 = l_2 = l_3 = 0.5l_y$ in Fig. 7.39) and to use the virtual work method to determine the ultimate load. If the slab in Fig. 7.39 has the corner yield lines at 45° to the edges and if the center of the slab is given a small downward displacement δ, the virtual work equation for slabs with $l_x \geq l_y$ is

$$w_u \left[l_y^2 \frac{\delta}{3} + (l_x - l_y)l_y \frac{\delta}{2} \right] = (m'_{ux} + m_{ux}) \frac{\delta}{0.5l_y} l_y + (m''_{ux} + m_{ux}) \frac{\delta}{0.5l_y} l_y$$

$$+ (m'_{uy} + m_{uy}) \frac{\delta}{0.5l_x} l_x + (m''_{uy} + m_{uy}) \frac{\delta}{0.5l_y} l_x$$

$$w_u = \frac{12}{l_y^2[3(l_x/l_y) - 1]} \left[m'_{ux} + m''_{ux} + 2m_{ux} + \frac{l_x}{l_y} (m'_{uy} + m''_{uy} + 2m_{uy}) \right] \quad (7.94)$$

If the top steel is such that $m'_{ux} = m''_{ux}$ and $m'_{uy} = m''_{uy}$, Eq. 7.94 can be rewritten in the following convenient form, where $l_x \geq l_y$

$$w_u = \frac{12}{l_y^2[3(l_x/l_y) - 1]} \left[2\left(m_{ux} + \frac{l_x}{l_y} m_{uy} \right) + m_1 \right] \quad (7.95)$$

where m_1 depends on the support conditions and is given below for the various cases shown in Fig. 7.66.

Case	m_1	Case	m_1	Case	m_1
1	0	4	$m'_{ux} + \dfrac{l_x}{l_y} m'_{uy}$	7	$2m'_{ux} + \dfrac{l_x}{l_y} m'_{uy}$
2	m'_{ux}	5	$2m'_{ux}$	8	$m'_{ux} + 2\dfrac{l_x}{l_y} m'_{uy}$
3	$\dfrac{l_x}{l_y} m'_{uy}$	6	$2\dfrac{l_x}{l_y} m'_{uy}$	9	$2\left(m'_{ux} + \dfrac{l_x}{l_y} m'_{uy} \cdot 2 \right)$

Table 7.2 compares the accuracy of the approximate ultimate load obtained from Eq. 7.95 with the exact ultimate load obtained from Eq. 7.59 for practical slabs that have $m'_{ux} = m''_{ux}$, $m'_{uy} = m''_{uy}$, $m'_{ux}/m_{ux} = m'_{uy}/m_{uy} = 2.0$, and various l_x/l_y and m_{uy}/m_{ux} ratios and support conditions. The accuracy of the approximate equation is shown to be reasonable in most cases. The accuracy obtained when using Eq. 7.95 to analyze a given slab could be improved if the calculated ultimate load were decreased by the percentage error indicated in Table 7.2 for the appropriate case. The advantage of using Eq. 7.95 is that it is a much simpler expression than Eq. 7.59.

An even simpler yield line pattern is shown in Fig. 7.67 for the case of all edges fixed. The ultimate load given by this approximate yield line pattern is

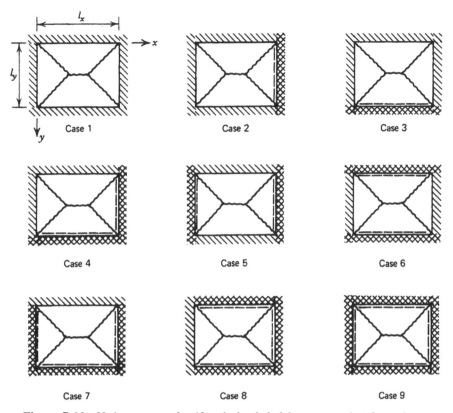

Figure 7.66   Various cases of uniformly loaded slabs supported at four edges.

$$w_u = \frac{6}{l_x^2}\,(2m_{ux} + m'_{ux} + m''_{ux}) + \frac{6}{l_y^2}\,(2m_{ux} + m'_{ux} + m''_{ux}) \qquad (7.96)$$

The various cases of simple supports can be obtained by putting the negative moment capacity at the appropriate edge or edges equal to zero in Eq. 7.96. The ultimate load given by Eq. 7.96 may be higher than that given by Eq. 7.95 in some cases, but the difference should not be large, especially if $l_x < 2l_y$.

7.15.3   Uniformly Loaded Rectangular Slabs with Three Edges Supported and One Edge Free

In analysis the determination of the ultimate uniform load of a particular slab from Eqs. 7.60 to 7.63 is not very convenient because the governing yield line pattern has to be established. Hence, some means of simplification is

TABLE 7.2   Accuracy of the Assumption of 45° Corner Yield Lines for Uniformly Loaded Rectangular Slabs Supported at All Edges[a]

$\dfrac{l_x}{l_y}$	$\dfrac{m_{uy}}{m_{ux}}$	Case								
		1	2	3	4	5	6	7	8	9
1.0	1	1.00	1.06	1.06	1.07	1.04	1.04	1.04	1.04	1.00
1.5	1	1.01	1.10	1.05	1.08	1.11	1.01	1.08	1.02	1.01
	2	1.00	1.04	1.08	1.07	1.03	1.05	1.05	1.03	1.00
	3	1.01	1.02	1.11	1.08	1.01	1.07	1.05	1.04	1.01
	5	1.03	1.02	1.14	1.11	1.00	1.10	1.07	1.07	1.03
2.0	1	1.02	1.11	1.06	1.09	1.15	1.01	1.10	1.02	1.02
	2	1.00	1.04	1.07	1.07	1.05	1.03	1.06	1.02	1.00
	3	1.01	1.02	1.09	1.08	1.02	1.05	1.06	1.03	1.01
	5	1.02	1.01	1.12	1.09	1.00	1.07	1.07	1.05	1.02
3.0	1	1.02	1.12	1.06	1.10	1.17	1.00	1.12	1.02	1.02
	2	1.00	1.04	1.07	1.07	1.06	1.02	1.07	1.01	1.00
	3	1.00	1.02	1.08	1.07	1.02	1.03	1.06	1.01	1.00
	5	1.01	1.01	1.10	1.08	1.00	1.04	1.07	1.03	1.01

[a] Approximate w_u/exact w_u when $m'_{ux} = m''_{ux}$, $m'_{uy} = m''_{uy}$, $m'_{ux}/m_{ux} = m'_{uy}/m_{uy} = 2$ for the various cases of support conditions shown in Fig. 7.66.

desirable. A good approximation is to assume that the corner yield lines are at 45° to the edges of the slab and to use the virtual work equation to find the ultimate uniform load. For mode 1 (Fig. 7.40a) this means assuming that $l_1 = l_y$, and for mode 2 (Fig. 7.40b) it means assuming that $l_1 = 0.5l_x$. It is

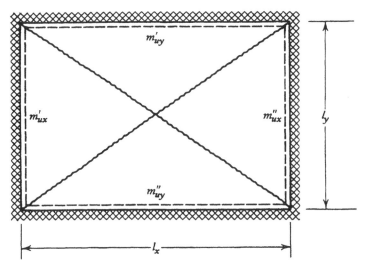

Figure 7.67   Alternative approximate yield line pattern for a uniformly loaded slab supported at four edges.

evident that with these assumed values for the yield line dimensions, mode 1 is applicable only when $l_x \geq 2l_y$ and that mode 2 is applicable only when $l_x \leq 2l_y$. Hence, as a result of the approximation, the governing mode now depends only on the slab dimensions.

If for mode 1 (Fig. 7.40a) $l_1 = l_y$ and the slab is given a small downward displacement δ at the center of the free edge, the virtual work equation when $l_x \geq 2l_y$ is

$$w_u \left[2l_y^2 \frac{\delta}{3} + l_y(l_x - 2l_y) \frac{\delta}{2} \right] = 2(m'_{ux} + m_{ux}) \frac{\delta}{l_y} l_y + (m'_{uy}l_y + 2m_{uy}l_y) \frac{\delta}{l_y}$$

$$w_u = \frac{6}{l_y^2[3(l_x/l_y) - 2]} \left[2(m'_{ux} + m_{ux}) + \frac{l_x}{l_y} m'_{uy} + 2m_{uy} \right] \qquad (7.97)$$

If for mode 2 (Fig. 7.40b), $l_1 = 0.5l_x$ and the slab is given a small downward displacement δ at the center of the free edge, the virtual work equation when $l_x \leq 2l_y$ is

$$w_u \left[0.5l_x^2 \frac{\delta}{3} + l_x(l_y - 0.5l_x) \frac{\delta}{2} \right] = 2(m'_{ux} + m_{ux}) \frac{\delta}{0.5l_y} l_y$$

$$+ (m'_{uy} + m_{uy}) \frac{\delta}{0.5l_x} l_x$$

$$w_u = \frac{12}{l_x^2[3(l_y/l_x) - 0.5]} \left[2 \frac{l_y}{l_x} (m'_{ux} + m_{ux}) + m'_{uy} + m_{uy} \right] \qquad (7.98)$$

Equations 7.97 and 7.98 can be rewritten in the following convenient form: If $l_x \geq 2l_y$, from Eq. 7.97,

$$w_u = \frac{12}{l_y^2[3(l_x/l_y) - 2]} [2(m_{ux} + m_{uy}) + m_2] \qquad (7.99)$$

where m_2 depends on the support conditions and is given below for the various cases shown in Fig. 7.68:

Case	1a	2a	3a	4a
m_2	0	$2m'_{ux}$	$\dfrac{l_x}{l_y} m'_{uy}$	$2m'_{ux} + \dfrac{l_x}{l_y} m'_{uy}$

If $l_x \leq 2l_y$, from Eq. 7.98,

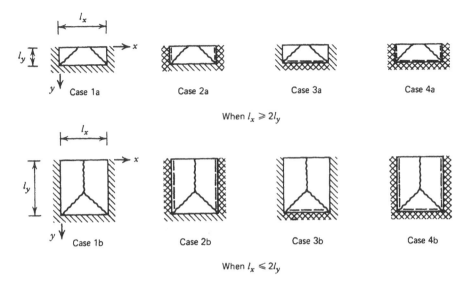

Figure 7.68 Various cases of uniformly loaded slabs supported at three edges.

$$w_u = \frac{12}{l_x^2[3(l_y/l_x) - 0.5]} \left(2\frac{l_y}{l_x} m_{ux} + m_{uy} + m_3 \right) \qquad (7.100)$$

where m_3 depends on the support conditions and is given below for the various cases shown in Fig. 7.68.

Case	1b	2b	3b	4b
m_3	0	$2\dfrac{l_y}{l_x} m'_{ux}$	m'_{uy}	$2\dfrac{l_y}{l_x} m'_{ux} + m'_{uy}$

The ultimate load is given by either Eq. 7.99 or 7.100, depending on the dimensions of the slab. It is to be noted that the dimension l_x is always taken parallel to the free edge. Table 7.3 compares the accuracy of the approximate ultimate load obtained from Eqs. 7.99 and 7.100 with the exact ultimate load obtained from Eqs. 7.60 to 7.63 for practical slabs that have $m'_{ux}/m_{ux} = m'_{uy}/m_{uy} = 2$, various l_y/l_x and m_{ux}/m_{uy} ratios, and the support conditions shown in Fig. 7.68. The assumed yield line patterns are shown to give reasonable accuracy in most cases. Where necessary, the accuracy of the approximate equations can be improved by decreasing the calculated ultimate load by the percentage error indicated in the table for the appropriate case.

TABLE 7.3 Accuracy of the Assumption of 45° Corner Yield Lines for Uniformly Loaded Rectangular Slabs Supported at Three Edges[a]

$\dfrac{l_y}{l_x}$	$\dfrac{m_{ux}}{m_{uy}}$	Case			
		1a or 1b	2a or 2b	3a or 3b	4a or 4b
3.0	1	1.02	1.00	1.14	1.02
	2	1.00	1.01	1.05	1.00
	3	1.00	1.01	1.02	1.00
	5	1.00	1.02	1.00	1.00
2.0	1	1.02	1.00	1.16	1.02
	2	1.00	1.01	1.06	1.00
	3	1.00	1.02	1.02	1.00
	5	1.01	1.03	1.00	1.01
1.0	1	1.02	1.01	1.15	1.02
	2	1.00	1.03	1.05	1.00
	3	1.01	1.05	1.02	1.01
	5	1.02	1.07	1.00	1.02
0.5	1	1.08	1.04	1.13	1.01
	2	1.02	1.10	1.04	1.02
	3	1.04	1.14	1.01	1.04
0.333	1	1.03	1.02	1.06	1.00
	2	1.00	1.13	1.01	1.03
	3	1.02	1.21	1.00	1.06

[a] Approximate w_u/exact w_u when $m'_{ux}/m_{ux} = m'_{uy}/m_{uy} = 2$ for the cases of support conditions shown in Fig. 7.68.

7.15.4 Uniformly Loaded Rectangular Slabs with Two Adjacent Edges Supported and the Remaining Edges Free

Equation 7.66 of mode 2 (Fig. 7.41b) is already in convenient form and in any case cannot be simplified. Equations 7.64 and 7.65 of mode 1 (Fig. 7.41a) can be simplified, however, by assuming that the positive moment yield line is at 45° to the edge of the slab and using the virtual work method to find the ultimate load. If for the slab of Fig. 7.41a, $l_1 = l_y$ and the slab is given a downward virtual displacement δ at the free corner, the virtual work equation when $l_x \geq l_y$ is

$$w_u \left[l_y^2 \frac{\delta}{3} + l_y(l_x - l_y) \frac{\delta}{2} \right] = (m'_{ux} + m_{ux}) \frac{\delta}{l_y} l_y + (m'_{uy} l_x + m_{uy} l_y) \frac{\delta}{l_y}$$

$$w_u = \frac{6}{l_y^2 [3(l_x/l_y) - 1]} (m_{ux} + m_{uy} + m_4) \qquad (7.101)$$

where m_4 depends on the support conditions and is given below for the various cases shown in Fig. 7.69.

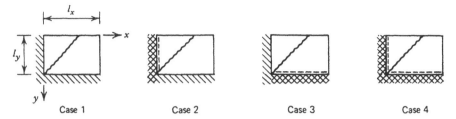

Figure 7.69 Various cases of uniformly loaded slabs supported at two adjacent edges.

Case	1	2	3	4
m_4	0	m'_{ux}	$\dfrac{l_x}{l_y} m'_{uy}$	$m'_{ux} + \dfrac{l_x}{l_y} m'_{uy}$

Table 7.4 compares the accuracy of the approximate ultimate load obtained from Eq. 7.101 with the exact ultimate load obtained from Eqs. 7.64 and 7.65 for practical slabs that have $m'_{ux}/m_{ux} = m'_{uy} = 2$, various l_x/l_y and m_{uy}/m_{ux} ratios, and the support conditions shown in Fig. 7.69, and shows that the approximate equation gives reasonable accuracy in most cases. Again, the ac- of the approximate equation can be improved by decreasing the cal- culated ultimate load by the percentage error indicated in the table.

It is to be noted that both Eqs. 7.101 and 7.66 have to be used to analyze a given slab. The ultimate load is the minimum of the two ultimate loads calculated from the equations.

TABLE 7.4 Accuracy of Assumption of 45° Corner Yield Line for Mode 1 of Uniformly Loaded Rectangular Slab Supported at Two Adjacent Edges[a]

$\dfrac{l_x}{l_y}$	$\dfrac{m_{uy}}{m_{ux}}$	Case			
		1	2	3	4
1.0	1	1.08	1.13	1.13	1.01
2.0	1	1.02	1.04	1.04	1.00
	2	1.13	1.00	1.10	1.01
	3	1.26	1.02	1.14	1.04
3.0	1	1.01	1.07	1.02	1.01
	2	1.10	1.00	1.06	1.01
	3	1.22	1.01	1.09	1.02

[a]Approximate w_u/exact w_u when $m'_{ux}/m_{ux} = m'_{uy}/m_{uy} = 2$ for the cases of support conditions shown in Fig. 7.69.

7.16   TRIAL-AND-ERROR METHOD FOR APPROXIMATE YIELD LINE PATTERN

For uniformly loaded rectangular slabs supported at four edges it was evident in Section 7.15 that the ultimate load could be determined with good accuracy using the virtual work equation with approximate yield line patterns with corner yield lines at 45° to the slab sides. For more general cases of slabs, it is more difficult to decide on approximate positions of the yield lines that result in sufficient accuracy. For these general cases, to avoid the lengthy algebra involved in an exact solution, a trial-and-error procedure may be adopted.

In a trial-and-error solution, a probable yield line pattern and positions of the yield lines are assumed and the equilibrium equations are used to calculate the ultimate load carried by each segment of the assumed yield line pattern. If the ultimate loads so found for each segment of the yield line pattern are not in reasonable agreement (they will be equal if the assumed yield line pattern is correct), the yield line pattern is altered until the agreement between ultimate loads is better. Inspection will indicate the manner in which the yield line pattern should be altered to obtain better agreement. When good agreement between the ultimate loads of the segments is obtained, the virtual work equation is then written for the final yield line pattern assumed and a very good approximation for the ultimate load will be obtained. Thus, the method uses the equilibrium equations to determine good approximations for the positions of the yield lines and then obtains a close estimate of the ultimate load from the virtual work equation.

The development of small computer programs using a spreadsheet, for instance, will greatly aid the solution using trial-and-success methods because new trials can so easily be completed. In some cases it is possible to use the load w_u as calculated from the virtual work solution to back-calculate the variable dimensions, using the equilibrium equations. If these dimensions are then used in a new virtual work solution, the load w_u so found will be very close to the final, fully converged solution.

To apply the equilibrium equations to segments of yield line patterns, the first moments of area of trapezoids about their bases are often required. The two common cases are shown in Fig. 7.70. It may be shown that

$$\text{first moment of area of segment 1 about X–X} = \frac{(2b_1 + b_2)h^2}{6}$$

$$(7.102)$$

$$\text{first moment of area of segment 2 about X–X} = \frac{(h_1^2 + h_1 h_2 + h_2^2)b}{6}$$

$$(7.103)$$

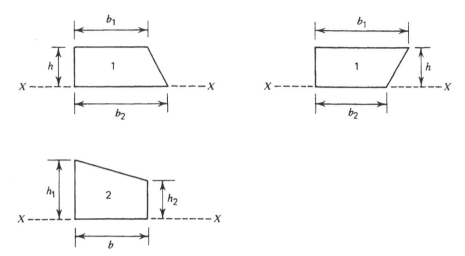

Figure 7.70   First moments of area of trapezoids about their bases.

Also, in the solutions, the work done by loading on a segment of general shape when rotating about a specified axis is often required. Consider a general segment carrying load W_u with the centroid of the load at distance $\bar{n}$ from the axis of rotation as shown in Fig. 7.71, where $\bar{n}$ is measured in the direction of the normal to the axis. If the segment is given a small rotation θ_n about the axis, the external work done by the total load on the segment is $W_u \bar{n} \theta_n$, because $\bar{n}\theta_n$ is the vertical displacement of the centroid of the loading W_u. Also, if $\bar{x}$ and $\bar{y}$ are the projected lengths of $\bar{n}$ in the $\bar{x}$- and $\bar{y}$-directions, Δ is the vertical displacement of the centroid, and Δ_0 is the vertical displacement of point O shown in Fig. 7.71, then

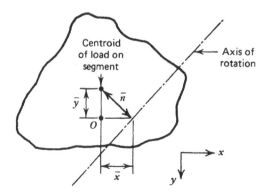

Figure 7.71   General segment of yield line pattern rotating about an axis.

$$\bar{n}\theta_n = \Delta = \Delta_0 + (\Delta - \Delta_0) = \frac{\Delta_0}{x}\bar{x} + \frac{\Delta - \Delta_0}{y}\bar{y} = \bar{x}\theta_x + \bar{y}\theta_y$$

Hence, the work done by the loading is, for all the segments,

$$\Sigma\,W_u\Delta = \Sigma\,W_u\bar{n}\theta_n = \Sigma\,W_u(\bar{x}\theta_x + \bar{y}\theta_y) \tag{7.104}$$

Note that because the equilibrium equations have been written before the virtual work solution, the values of $W_u\bar{x}$ and $W_u\bar{y}$ will have already been obtained. Therefore, the external work done by the loading can readily be found.

The trial-and-error procedure will be illustrated by two examples.

Example 7.16. The slab shown in Fig. 7.72*a* is simply supported on two columns and along one edge. The slab is 8 in. (203 mm) thick and is reinforced by No. 4 (12.7-mm-diameter) bars on 12-in. (305-mm) centers both ways in the top and bottom. The clear cover to the short direction bars is $\frac{3}{4}$ in. (19.1 mm). The concrete has a cylinder strength of 3000 lb/in² (20.7 N/mm²), and the steel has a yield strength of 60,000 lb/in² (414 N/mm²). Calculate the ultimate load applied uniformly to the entire area of the slab.

SOLUTION. The ultimate moment per unit width is given by[7.12]

$$\phi A_s f_y \left(d - 0.59 A_s \frac{f_y}{f_c'} \right)$$

where ϕ is the strength reduction factor, taken by the ACI Code[7.21] as 0.9, A_s the area of tension steel per unit width, f_y the steel yield strength, f_c' the concrete compressive cylinder strength, and d the effective depth of the tension steel. For No. 4 bars with $\frac{3}{4}$ in. of cover in the short direction of the slab, $d = 8 - \frac{3}{4} - \frac{1}{4} = 7$ in. In the long direction, $d = 7 - \frac{1}{2} = 6.5$ in. Assume that the mean d of 6.75 in. is in both directions; then the slab is isotropic with ultimate negative and positive moments of resistance per unit width, including the strength reduction factor, of

$$m_u' = m_u = 0.9 \times \frac{0.2}{12} \times 60,000 \left(6.75 - 0.59 \times \frac{0.2 \times 60,000}{12 \times 3000} \right)$$

$$= 5898 \text{ lb-ft/ft width}$$

The slab has four possible failure modes, each of which will be examined. *MODE 1* (Fig. 7.72*b*). In this failure mode the slab is divided into three segments. Segment A rotates about the simple support line as axis, and each

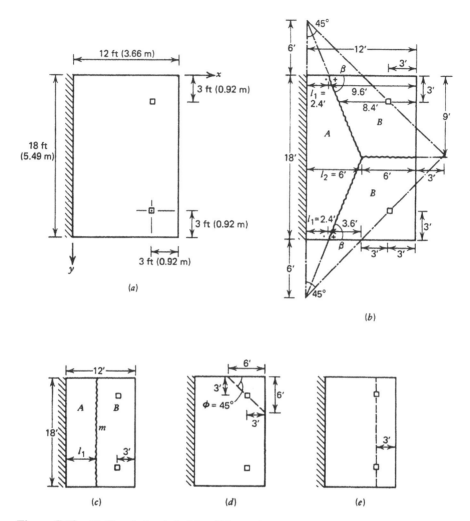

Figure 7.72 Uniformly loaded slab of Example 7.16: (a) slab; (b) mode 1 (first trial values for dimensions defining positions of yield lines shown); (c) mode 2; (d) mode 3; (e) mode 4.

of the identical segments B rotates about an axis through the column. The positions of the yield lines are defined by two unknown dimensions, shown as l_1 and l_2 in the figure.

As a first trial assume that the axes of rotation which pass through the columns are at 45° to the sides of the slab and that the junction of the yield lines is at the center of the slab (i.e., $l_1 = 2.4$ ft and $l_2 = 6.0$ ft). The nodal forces at the free edge between segments A and B are, according to Eq. 7.20, $\pm m_u \cot \beta = \pm 5898 \times 3.6/9 = \pm 2359$ lb, acting in the directions shown in the figure (a dot is an upward force and a cross is a downward force). The other nodal forces are zero.

The exact solution of the slab would involve writing three equilibrium equations to determine l_1, l_2, and the ultimate load. These three equilibrium equations will now be written using the assumed values of l_1 and l_2. For segment A, taking moments in the x-direction about the simple support and referring to Eq. 7.103,

$$(5898 \times 18) + (2 \times 2359 \times 2.4) = w_u \times \frac{2(2.4^2 + 2.4 \times 6 + 6^2)9}{6}$$

$$117,490 = 168.48w_u$$

$$w_u = 697 \text{ lb/ft}^2 \tag{i}$$

For segment B, taking moments in the x-direction about an axis passing through the column parallel to the long sides and referring to Eq. 7.103,

$$(5898 \times 9) - (2359 \times 6.6) = w_u \frac{(6.6^2 + 6.6 \times 3 + 3^2)9}{6}$$

$$- w_u(3 \times 9 \times 1.5)$$

$$37,510 = 68.04w_u$$

$$w_u = 551 \text{ lb/ft}^2 \tag{ii}$$

For segment B, taking moments in the y-direction about an axis passing through the column parallel to the short sides and referring to Eq. 7.102,

$$(5898 \times 9.6) + (2359 \times 3) = w_u \frac{(2 \times 6 + 8.4)6^2}{6}$$

$$- w_u \frac{(2 \times 9.6 + 8.4)3^2}{6}$$

$$63,700 = 81.00w_u$$

$$w_u = 786 \text{ lb/ft}^2 \tag{iii}$$

Had the values chosen for l_1 and l_2 been correct, the value for w_u obtained from each of Eqs. i, ii, and iii would have been identical. The true value for w_u lies in the range 551 to 786 lb/ft². The next step would be to adjust the assumed values for l_1 and l_2 so as to obtain closer agreement between the three values for w_u. Inspection of the assumed yield line pattern and the values for w_u obtained for each segment would normally indicate the direction in which the unknown dimensions should be changed to obtain better agreement.

However, the agreement between the w_u values is already such that a reasonably accurate value will be given by the virtual work equation on the basis of the yield line pattern with $l_1 = 2.4$ ft and $l_2 = 6.0$ ft.

Give the junction point of the yield lines a small downward displacement δ.

Segment	Components of Rotation θ_x	θ_y	Components of Internal Work Done $m_{ux}\theta_x y_0$	$m_{uy}\theta_y x_0$
A	$\delta/6$	0	$5898 \times \dfrac{\delta}{6} \times 18 =$ $17,649\delta$	0
B	$\delta/9$	$\delta/9$	$5898 \times \dfrac{\delta}{9} \times 9 =$ 5898δ	$5898 \times \dfrac{\delta}{9} \times 9.6 =$ 6291δ

Therefore,

$$\Sigma\, m_{ux}\theta_x y_0 + \Sigma\, m_{uy}\theta_y x_0 = (17{,}694 + 2 \times 5898)\delta + (2 \times 6291)\delta$$

$$= 42{,}072\delta \qquad\qquad \text{(iv)}$$

The work done by the loading may be found from Eq. 7.104 as

$$\Sigma\, W_u \Delta = \Sigma\, W_u(\bar{x}\theta_x + \bar{y}\theta_y)$$

Now values for $W_u\bar{x}$ and $W_u\bar{y}$ for each segment have already been obtained when writing the equilibrium equations (see Eqs. i, ii, and iii).

Segment	Components of Rotation θ_x	θ_y	Components of External Work Done $W_u\bar{x}\theta_x$	$W_u\bar{y}\theta_y$
A	$\delta/6$	0	$168.48w_u\dfrac{\delta}{6} =$ $28.08w_u\delta$	0
B	$\delta/9$	$\delta/9$	$68.04w_u\dfrac{\delta}{9} =$ $7.56w_u\delta$	$81.00w_u\dfrac{\delta}{9} =$ $9.00w_u\delta$

Therefore,

$$\Sigma\, W_u(\bar{x}\theta_x + \bar{y}\theta_y) = (28.08 + 2 \times 7.56)w_u\delta + (2 \times 9.00)w_u\delta$$

$$= 61.20w_u\delta \qquad\qquad\qquad (v)$$

Hence, from the virtual work equation, Eq. iv equals Eq. v, which gives

$$42{,}072 = 61.20w_u\delta$$

$$w_u = 687 \text{ lb/ft}^2$$

Note: The exact solution is $l_1 = 2.97$ ft and $l_2 = 5.64$ ft, giving

$$w_u = \frac{m_u}{8.73} = \frac{5898}{8.73} = 676 \text{ lb/ft}^2$$

It is obvious that the approximate solution has given a very good estimate of the ultimate load.

MODE 2 (Fig. 7.72c). This alternative yield line pattern has a single positive moment yield line parallel to the supported edge and the slab collapses as a wide beam. The exact ultimate load is easily calculated by the equilibrium method, considering strips of unit width. For segment A, taking moments about the simply supported edge;

$$\frac{w_u l_1^2}{2} = 5898 \qquad\qquad\qquad (vi)$$

For segment B, taking moments about the axis through the columns,

$$w_u(12 - l_1)\left(\frac{12 - l_1}{2} - 3\right) = 5898 \qquad\qquad (vii)$$

Equating the values of $5898/w_u$ given by Eqs. vi and vii results in

$$\frac{l_1^2}{2} = (12 - l_1)\left(\frac{12 - l_1}{2} - 3\right) \qquad l_1 = 4 \text{ ft}$$

Substituting $l_1 = 4$ ft into Eq. vi gives

$$w_u = 2 \times \frac{5898}{4^2} = 737 \text{ lb/ft}^2$$

Note that live load placed only between the simply supported edge and the columns would give a worse case of loading for this mechanism than the case of live load placed over the entire slab considered above.

MODE 3 (Fig. 7.72d). This alternative yield line pattern has a single negative moment yield line across each corner over the column and the corner portions of the slab fold down. It can be shown that $\phi = 45°$ in the figure gives the minimum ultimate load for this yield line pattern with isotropic reinforcement. Using the equilibrium method, taking moments about the yield line gives

$$w_u \frac{6 \times 6}{2} \frac{6}{3\sqrt{2}} = 5898 \times 6\sqrt{2}$$

$$w_u = 1966 \text{ lb/ft}^2$$

MODE 4 (Fig. 7.72e). The final alternative pattern considered has a single negative moment yield line over the columns parallel to the supported side, and the cantilever portion of the slab folds down. Using the equilibrium method, taking moments about the yield line gives

$$\frac{w_u \cdot 3^2}{2} = 5898$$

$$w_u = 1311 \text{ lb/ft}^2$$

Examining the ultimate loads of the four modes shows that mode 1 is the most critical. Therefore, the ultimate load is, with very good accuracy, 687 lb/ft², including the dead load of the slab, as long as the entire area is loaded.

Example 7.17. A uniformly loaded isotropically reinforced square slab is simply supported along two adjacent edges and on a column in the opposite corner as shown in Fig. 7.73. Show that at the ultimate load the unknown dimension l_1 of the yield line pattern is $0.75l$. Determine the relationship between the ultimate uniform load per unit area w_u and the ultimate positive moment of resistance per unit width m_u.

SOLUTION. Assuming that $l_1 = 0.75l$, the remaining dimensions of the yield line pattern can be calculated and are shown on segments A and B in Fig. 7.73. The nodal forces acting at the free edge between segments A and B according to Eq. 7.20 are $\pm m_u \cot \beta = \pm m_u (0.15l/0.25l) = \pm 0.6m_u$, acting in the directions shown in the figure (a dot is an upward force and a cross is a downward force). The nodal forces at the junction of the yield lines are zero. Consider the equilibrium of the segments. Because there was only one unknown dimension, l_1, only two equilibrium equations need to be written. For segment A, taking moments in the x-direction about the simple support and referring to Eq. 7.103:

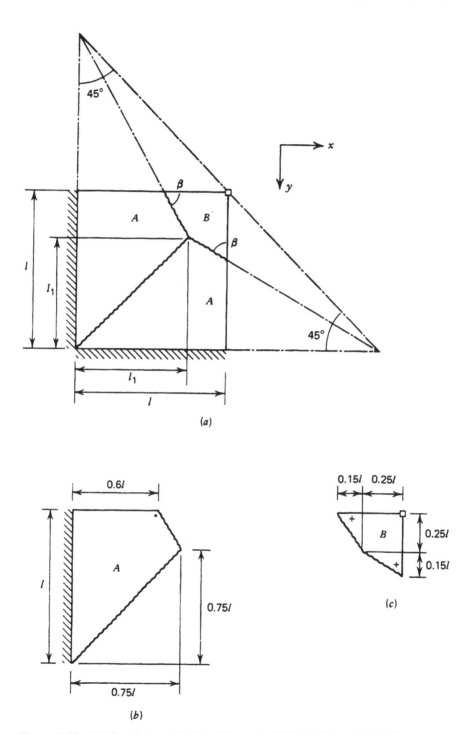

Figure 7.73   Uniformly loaded slab of Example 7.17: (*a*) slab and yield line pattern; (*b*) segment A; (*c*) segment *B*.

$$m_u l + 0.6 m_u (0.6l) = w_u \frac{0.75^2 l^2}{2} \frac{0.75l}{3}$$

$$+ w_u [0.6^2 l^2 + 0.6l(0.75l) + 0.75^2 l^2] \frac{0.25l}{6}$$

$$1.36 w_u l = 0.1275 w_u l^3$$

$$w_u = 10.67 \frac{m_u}{l^2}$$

For segment B, taking moments in the x-direction about the simple column support and referring to Eq. 7.103:

$$m_u \cdot 0.4l - 0.6 m_u \cdot 0.4l = w_u \frac{0.25l(0.15l)}{2} \frac{0.25l}{3}$$

$$+ w_u [0.25^2 l^2 + 0.25l(0.4l) + 0.4^2 l^2] \frac{0.25l}{6}$$

$$0.16 m_u l = 0.015 w_u l^3$$

$$w_u = 10.67 \frac{m_u}{l^2}$$

Hence, the ultimate load given by the two equilibrium equations is the same. Therefore, the chosen dimensions of the yield line pattern are correct and $w_u = 10.67 m_u l^2$.

7.17   COMPARISON WITH TEST RESULTS

It is obviously essential that the validity of the assumptions made in yield line theory be thoroughly checked by tests before the method is used with confidence in analysis and design. It is particularly important that the method give a safe estimate of the ultimate load of the slab and that slabs designed by the method behave satisfactorily at the service loads. A large number of tests have been conducted in the past and the results of some of these tests are summarized below.

7.17.1   Tests Conducted by the Deutscher Ausschuss für Eisenbeton

The Deutscher Ausschuss für Eisenbeton (German Reinforced Concrete Board) conducted an extensive series of tests at the Stuttgart Materials Testing Establishment under the direction of C. Bach and O. Graf during the period 1911–1925. Johansen has compared the results of these tests with the theoretical ultimate loads calculated by yield line theory; the calculations are reported in full in Johansen's book.[7.2] The results were as follows:

1. *Isotropic square slabs.* Forty-two slabs with simply supported edges were tested under uniform loading simulated by 16-point loads or a circular array of eight-point loads. The spans were 2.00 m (78.7 in.) and the span/thickness ratios were in the range 17 to 30. The steel ratio varied between 0.32 and 2.00%. The ratio of measured ultimate to theoretical ultimate loads for the slabs varied between 0.90 and 1.15. Figure 7.74 shows the yield line pattern of two of the slabs after testing.

2. *Isotropic rectangular slabs.* Six slabs with simply supported edges were tested under simulated uniformly distributed loading. The spans were either 2.00 m (78.7 in.) by 3.00 m (118.1 in.), or 2.00 m (78.7 in.) by 4.00 m (157.5 in.), and the short span/thickness ratio was 28.6. The steel ratio was 0.37%. The ratio of measured ultimate to theoretical ultimate loads for the slabs varied between 1.00 and 1.07.

3. *Isotropic square slabs with a continuous edge.* Three pairs of square slabs with a common fixed edge and the remaining edges simply supported were tested under simulated uniformly distributed loading. The spans were 2.00 m (78.7 in.) and the span/thickness ratio was 16.4. The steel ratio was

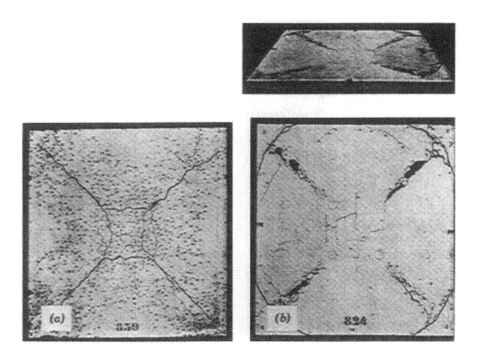

Figure 7.74   Yield line pattern at failure of simple supported uniformly loaded square slabs:[7.2] (*a*) bottom surface; (*b*) top surface.

0.61% for negative moment and 0.37% for positive moment. The ratio of measured ultimate to theoretical ultimate loads for the slabs varied between 0.97 and 1.03.

4. *Orthotropic rectangular slabs.* Five slabs with simply supported edges were tested under simulated uniformly distributed loading. The spans were 2.00 m (78.7 in.) by 4.00 m (157.5 in.) and the short span/thickness ratio was 16.7. The ratio of the ultimate moment of resistance per unit width in the direction of short span to that of the long span (μ) varied between 1.09 and 10.2. The ratio of the measured ultimate to theoretical ultimate loads for the slabs varied between 1.03 and 1.08. Figure 7.75 shows the yield line pattern of one of the slabs after testing. The yield strength of the long-span bars in two of the slabs with large μ values was not given and it was assumed to be the same as the short-span bars in the calculations.

5. *Orthotropic slabs supported at two opposite edges.* Fifteen rectangular slabs with two opposite edges simply supported were tested by applying point

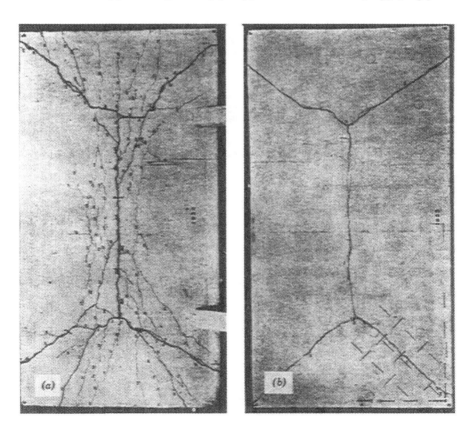

Figure 7.75 Yield line pattern at failure of simply supported uniformly loaded rectangular slab:[7.2] (*a*) bottom surface; (*b*) top surface.

loads near the slab center. The slabs had a span of 2.00 m (78.7 in.) and a width of 3.00 m (118.1 in.). The span/thickness ratio varied between 19.8 and 10.9. The ratio of the ultimate moment of resistance per unit width in the direction of the span to that in the direction of the width varied over a wide range and various amounts of steel were placed. The ratio of measured ultimate to theoretical ultimate loads for the slabs varied between 0.88 and 1.09.

Johansen[7.2] concluded that the series of tests discussed above gave ample evidence of the real nature of yield lines and the validity of the theory. The differences between the theory and tests were small and mainly on the conservative side. For these tests the load on the slab at first yielding of the steel was approximately two-thirds of the ultimate load when distributed loading was applied.

7.17.2 Tests Conducted by IRABA

The French delegation to the European Concrete Committee has had series of tests conducted by the Institute of Applied Research on Reinforced Concrete in Paris. The tests were mainly to verify the application of yield line theory in cases where cold-worked steel is used. The results obtained from nine slabs may be seen reported in a bulletin of the European Concrete Committee.[7.8]

1. *First series.* Three 2.50-m (98.4-in.) square slabs with a span/thickness ratio of 25 were tested. The edges were simply supported and four concentrated loads were applied at the $\frac{1}{4}$-span points. The slabs were reinforced isotropically: the first by 0.84% of mild steel, the second by 0.50% of indented bars, and the third by 0.5% of Tor 40 bars. The ratio of measured ultimate to cracking loads was 2.97, 3.10, and 4.11 for the three slabs and the ratio of measured ultimate to theoretical ultimate loads was 1.10, 0.90, and 1.06. Because of the lack of travel of the loading jacks, there was some doubt whether the second slab had reached its full ultimate load.

2. *Second series.* Three 2.50-m (98.4-in.) by 5.00 m (196.9-in.) rectangular slabs with a short span/thickness ratio of 25 were tested. The edges were simply supported and eight concentrated loads were applied arrayed symmetrically over the surface of the slab. The first slab was reinforced by mild steel, the second by indented bars, and the third by Tor 40 bars. The steel percentage in the short spans was identical to the slabs of the first series, and in the long span the steel percentages were approximately two-thirds that of the short span. The ratio of measured ultimate to cracking loads for the three slabs was 4.10, 3.48, and 4.53, and the ratio of measured ultimate to theoretical ultimate loads was 1.12, 0.94, and 1.04.

3. *Third series.* Three square slabs of identical dimensions and steel content to the first series were tested. The slabs were supported at the corners and loaded by four concentrated loads applied at the $\frac{1}{4}$-span points. The ratio of the measured ultimate to theoretical ultimate loads was 0.99, 1.02, and 0.87. In the case of the low result, a failure in punching shear occurred before the full flexural strength of the slab was reached.

The IRABA concluded from these tests that the failure pattern of a slab is predicted accurately using yield line theory. The types of steel used and the comparisons of the ultimate loads showed that the theory is applicable to steel with a short yield range or no definite yield stress as well as to mild steel. It was also evident that high-bond bars distributed the cracking and produced finer cracks than did plain mild steel bars.

7.17.3   Tests Conducted at the Technical University of Berlin

An extensive experimental program of research into the collapse of reinforced concrete slabs has been conducted at the Technical University of Berlin by Jaeger.[7.30] A total of 134 rectangular slabs were tested under uniform or concentrated loading with various combinations of the possible boundary conditions of fixed, simply supported or free edges, or column supports. Orthotropic or isotropic mild or high-tensile steel was used. Excellent agreement between the theoretical and experimental yield line patterns and the ultimate loads was obtained. Figures 7.76 to 7.82 give an indication of this agreement. The dimensions of the slabs in millimeters are shown on the figures. In his conclusions Jaeger points out that yield line theory gives a safe estimate of the ultimate load of a slab but that designers should never lose sight of serviceability requirements. Therefore, reinforcement arrangements that are very different from elastic theory are not desirable because they may lead to extensive cracking at the service load.

7.17.4   Tests Conducted at the TNO Institute for Building Materials and Structures

For the purpose of studying the development of yield line patterns the TNO Institute for Building Materials and Structures in Holland tested a series of model slabs of dimensions $500 \times 300 \times 10$ mm ($19.7 \times 11.8 \times 0.39$ in.).[7.31] The slabs were made from microconcrete and reinforced by thin spot-welded wire mesh. The slabs had various shapes, boundary conditions, and loading. Good agreement was found between predicted and experimental patterns. In the case of concentrated loads, there was a tendency for simple straight-line patterns to form rather than fan mechanisms. On the whole, the yield line patterns conformed with theory.

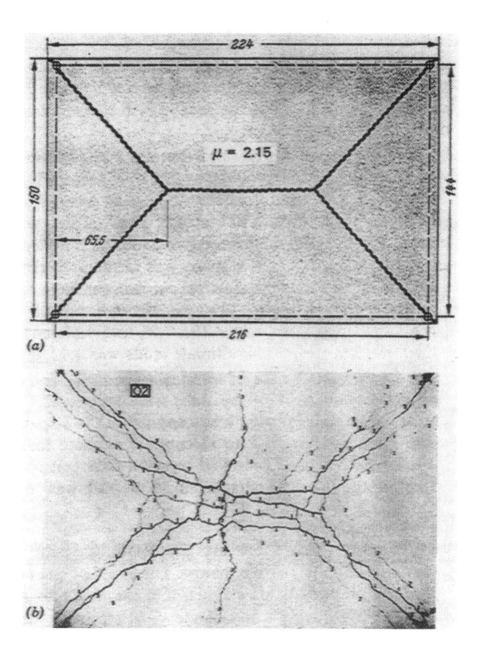

Figure 7.76 Uniformly loaded slab simply supported at all edges[7.30]: (*a*) theoretical yield line pattern; (*b*) actual crack pattern on bottom surface ($w_{u\,\text{test}}/w_{u\,\text{theory}} = 1.03$).

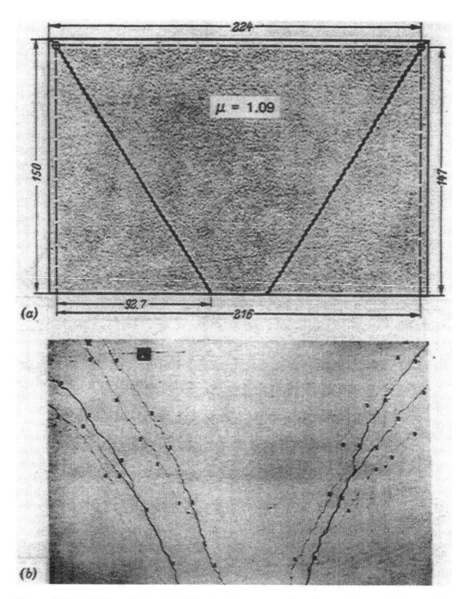

Figure 7.77 Uniformly loaded slab simply supported at three edges and free at the fourth[7.30]: (*a*) theoretical yield line pattern; (*b*) actual crack pattern on bottom surface ($w_{u\,\text{test}}/w_{u\,\text{theory}}$ = 1.05).

7.17.5 Tests Conducted at the University of Manchester

In Section 6.5 we referred to a series of tests conducted by Taylor et al.[7.32] at the University of Manchester on 10 uniformly loaded square slabs simply

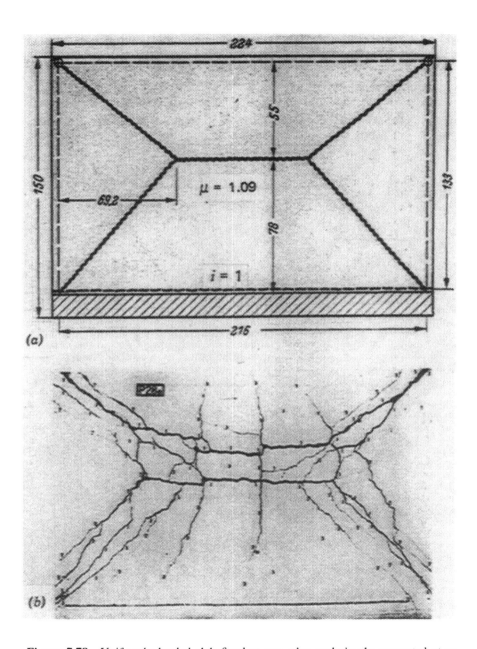

Figure 7.78 Uniformly loaded slab fixed at one edge and simply supported at remaining edges[7.30]: (*a*) theoretical yield line pattern; (*b*) actual crack pattern on bottom surface ($w_{u\,\text{test}}/w_{u\,\text{theory}} = 1.15$).

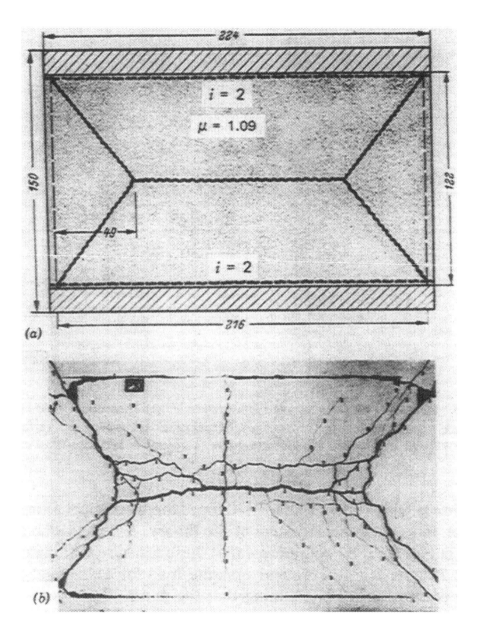

(a)

(b)

Figure 7.79 Uniformly loaded slab fixed at two opposite edges and simply supported at the remaining edges[7.30]: (*a*) theoretical yield line pattern; (*b*) actual crack pattern on bottom surface ($w_{u\,\text{test}}/w_{u\,\text{theory}} = 1.13$).

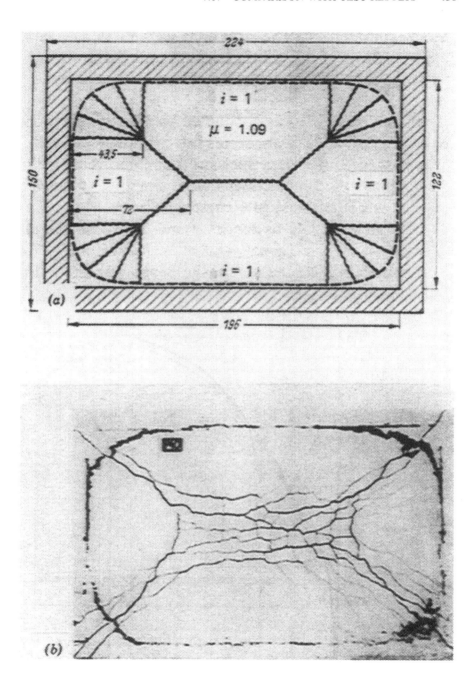

Figure 7.80   Uniformly loaded slab fixed at all edges[7.30]: (*a*) theoretical yield line pattern; (*b*) actual crack pattern on bottom surface; (*c*) actual crack pattern on top surface ($w_{u\,test}/w_{u\,theory}$ = 1.37); (*d*) load central deflection curve (1 kp/cm^2 = 2.05 kips/ft^2, 1 in. = 2.54 cm).

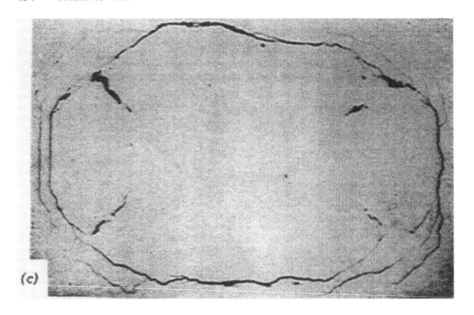

(c)

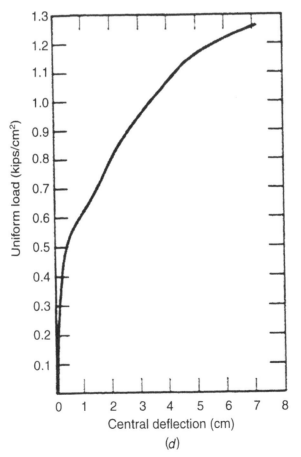

(d)

Figure 7.80 (*Continued*).

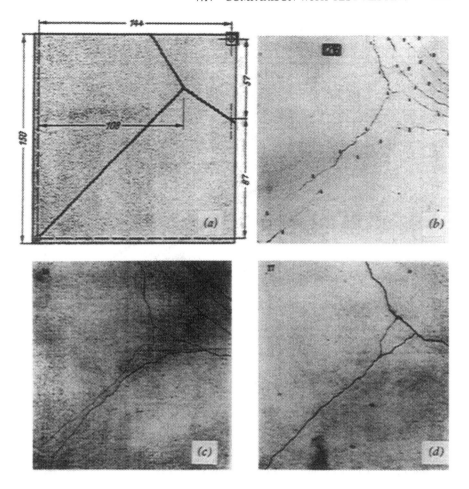

Figure 7.81 Uniformly loaded slabs simply supported at two adjacent edges and at a corner column[7.30]: (*a*) theoretical yield line pattern; (*b*), (*c*), (*d*) actual crack patterns on bottom surface of three slabs ($w_{u\,test}/w_{u\,theory}$ = 1.01, 1.01, and 1.00).

supported at all edges over spans of 6 ft (1.83 m). The measured ultimate load of these slabs was 1.26 to 1.80 times the theoretical yield line theory ultimate load. The large enhancement of strength was attributed to strain hardening of the reinforcement and tensile membrane action that developed at large deflections. The slabs behaved satisfactorily at the service loads.

A second series of tests[7.33] was conducted on 10 uniformly loaded square slabs of the same spans as the first series but cast monolithically with edge beams. The panels were 2 in. (51 mm) thick and the beams were 3 in. (76 mm) wide and 6 in. (152 mm) deep. The specimens were simply supported at the four corners. The tests were to investigate composite beam–slab collapse modes of the type shown in Fig. 7.42 as modes 1 and 2, and of the

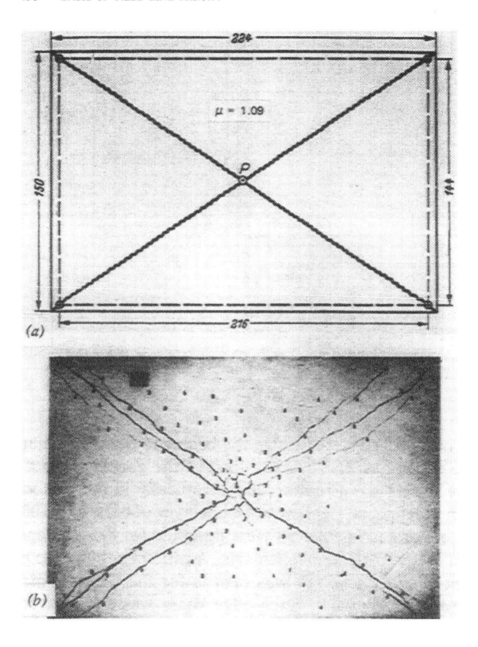

Figure 7.82   Central concentrated load on a slab simply supported at all edges[7.30]: (*a*) theoretical yield line pattern; (*b*) actual crack pattern on bottom surface ($w_{u\,\text{test}}/w_{u\,\text{theory}} = 1.05$).

effect of the beams on the strength of the diagonal panel mode shown in Fig. 7.42 as mode 3. Four of the slabs failed by the composite beam–slab mode at measured ultimate loads that were 1.00 to 1.10 times the theoretical ultimate load for that mode, indicating the accuracy of the theory. A typical slab after testing is shown in Fig. 7.83. It is of interest that these measured ultimate loads were 0.99 to 2.28 times the theoretical ultimate load for a diagonal panel mode (ignoring corner fans and assuming that the panel edges were simply supported on the beams), indicating that the panel strength had been enhanced by the restraint at the edges due to the presence of the beams. It is also of interest that one of these panels developed wide cracking in the top of the slab along the lines of the beams due to negative moment in the panel, but that this cracking apparently did not appreciably affect the flexural strength of the L beams (it had been assumed that some of the slab acted as

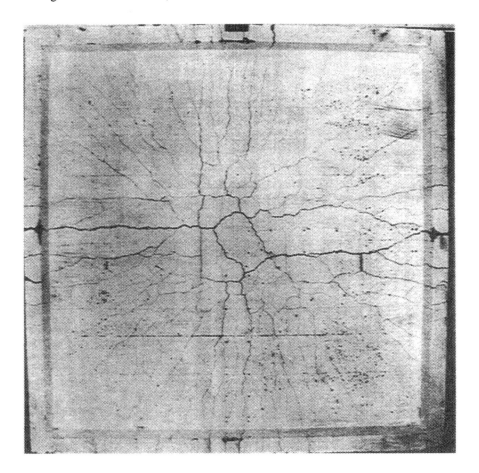

Figure 7.83 Uniformly loaded panel with integral beams showing composite beam–slab collapse mode.[7.33]

the compression flange of the beam). This supported the conclusion reached by Taylor et al.[7.34] in earlier tests on isolated T-beams (in which yield lines were induced in the slab along the beam at the slab–beam rib junctions) that the presence of a yield line has some effect on the flexural strength of the beam, because of the reduction in the effectiveness of the slab flange, but that the effect is only slight. The remaining six slabs failed by the diagonal panel mode (mode 3 of Fig. 7.42) at measured ultimate loads that were 1.37 to 2.31 times the theoretical yield line load for that mode, ignoring corner fans and assuming that the slabs were simply supported on the beams. A typical slab after testing is shown in Fig. 7.84. The enhancement of ultimate load was due partly to the panel edge restraint resulting from the torsional strength of the beams. This edge restraint varied during the tests because of differences in the strength of the connection between the beams at the corners. The beams

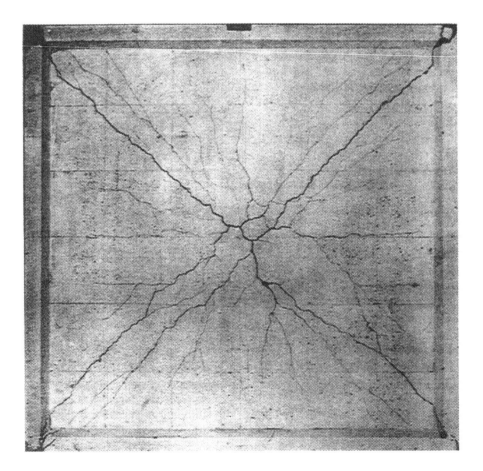

Figure 7.84 Uniformly loaded panel with integral beams showing diagonal panel collapse mode.[7.33]

in one specimen were not connected, and in others torsional failure occurred in the corners. The enhancement of load was also due to tensile membrane action in the slab. These tests again indicated the reserve of strength available in panels when acting compositely with beams in a floor due to negative-moment restraint at the edge beams and to membrane action.

7.17.6 Tests Conducted at the University of Canterbury

Two two-panel $\frac{1}{4}$-scale reinforced concrete model beam and slab floors have been constructed and tested under uniform loading at the University of Canterbury to investigate the effect of monolithic beams on the strength of panels and to check the safety and serviceability of beam and slab floors designed by limit design. The results of these two tests are described briefly below.

The first model floor[7.35] consisted of two continuous panels arrayed side by side, each approximately 5 ft (1.52 m) square and $1\frac{1}{2}$ in. (38 mm) thick. The supporting beams were approximately $4\frac{1}{2}$ in. (114 mm) deep and 3 in. (76 mm) wide. The panels and the beams were cast together. The floor was simply supported on reaction points at the six beam junctions. The panels were designed as if fixed-edge, assuming the beams to be strong enough in torsion and flexure to support the ultimate load of the panels without failing themselves. The panels were reinforced isotropically in the bottom and around the edges in the top; the steel area of each layer of bars was 0.002 of the gross area of the concrete. The beams were designed for flexure considering composite beam–slab collapse mechanisms. A ratio of negative to positive bending moment of close to unity was chosen for the ultimate moments of the beams as well as the slabs. Each edge beam was designed for the torsion introduced by the negative moments in a panel when the yield line develops along the panel edge. The theoretical ultimate loads of the floor for the panel mechanism, composite beam–slab mechanism 1, and composite beam–slab mechanism 2 shown in Fig. 7.85 were 360, 386, and 353 lb/ft² (17.2, 18.5, and 16.9 kN/m²), respectively. The measured ultimate load was 442 lb/ft² (21.1 kN/m²), which was 15 to 25% greater than the theoretical ultimate loads. The floor after testing is shown in Fig. 7.86.

Although one of the composite beam–slab mechanisms was predominant at the ultimate load, the steel had in fact yielded at all the critical sections in the beams. The redistribution of bending moments required to reach the bending moment configuration adopted in the limit design was achieved satisfactorily. For the continuous beams parallel to the long edges of the floor, 22% redistribution of the elastic negative moment was necessary to achieve the distribution of ultimate moments. This ductile behavior of the floor was to be expected, however, since the reinforcement ratios in the slab and the beams were relatively small, as indicated by the small depth of the rectangular concrete compressive stress blocks, which varied between 0.03 and 0.14 of the effective depth of the critical sections. The design for torsion and shear sat-

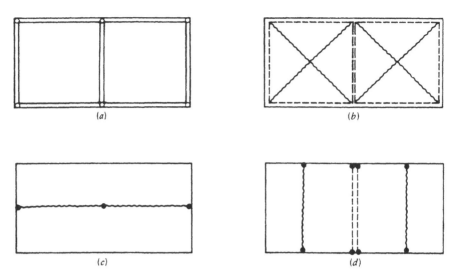

Figure 7.85 Uniformly loaded two-panel beam-and-slab floor and possible collapse modes[7.35]: (a) floor; (b) panel mechanisms; (c) composite beam-slab mechanism 1; (d) composite beam-slab mechanism 2.

isfactorily prevented failure of the edge beams of the floor, and diagonal tension cracks were kept small. Only some of the top steel in the panels over the edge beams reached the yield strength, however, and hence the torsional moments actually present were less than the design values. This reduced torsional moment is due to the very large reduction in the torsional stiffness of the edge beams after diagonal tension cracking and meant that the full ultimate negative-moment capacity of the panels could not be developed at the edge beams even for the small area of steel in the panels (0.002 of the gross area of concrete). The behavior of the floor at the service load was satisfactory in that crack widths and deflections were at an acceptable level. The maximum crack widths at service load were 0.002 in. (0.05 mm) in the beams and 0.003 in. (0.08 mm) at the bottom of the panels. Cracking would have been more severe in the panels over the top of the interior beam, but the top of the slab could not be observed because of the presence of the loading bags. However, the strains indicated that the crack widths there were probably approximately 1.3 times the maximum width at the bottom of the panels. These crack widths need to be scaled to obtain the likely widths in the prototype structure. If the crack width is assumed to be directly proportional to the scale factor (see Chapter 9), the measured widths should be multiplied by approximately 4 to obtain the likely prototype values. The crack widths so obtained come within allowable values for interior exposure according to the 1995 ACI Code limitations (see Chapter 9). The deflections at the service load varied between span/3000 and span/800 at the midspan of the beams and span/540 at the center of the panels. It is considered that the test showed that the limit design method had led to a safe and satisfactory structure.

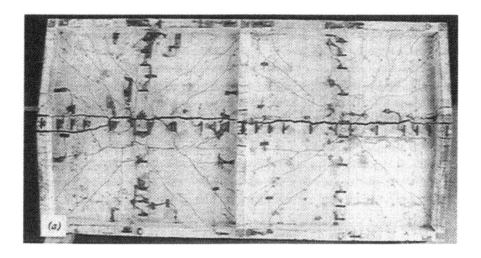

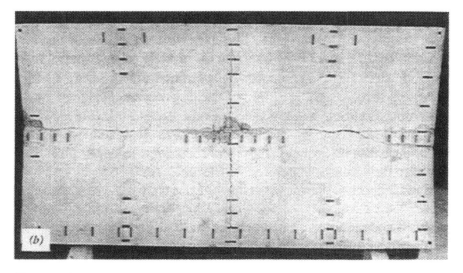

Figure 7.86 Uniformly loaded two-panel beam-and-slab floor showing collapse mode[7.35]: (*a*) bottom surface; (*b*) top surface.

The second model floor[7.36] was similar to the first model floor except that two stiff and strong cantilever regions were cast beside the long sides of the floor to simulate the effect of adjacent beams and panels of a multipanel floor. The stiff and strong regions were held horizontal by reactive forces from the test frame during the test. Thus, the test regions of the model floor could be regarded as two adjacent panels of a multipanel floor, each panel having three continuous edges and one exterior edge. The two panels and the weak direction beams were designed by the same limit design procedure as was used for the first model floor. For this floor the theoretical ultimate loads of the panel mechanism and composite bean–slab mechanism were 358 and 349 lb/ft^2 (17.1 and 16.7 kN/m^2), respectively. The measured ultimate load was 657 lb/ft^2 (31.4 kN/m^2), which was 84 to 88% greater than the theoretical ultimate loads. The very significant load enhancement was due to membrane action and indicated the substantial reserve of strength that is available when failure is restricted to only parts of a floor structure. (The stiff and strong cantilever portions simulating adjacent panels were not loaded during the test except by the reactive forces at the edges, which held them horizontal.) Full redistribution of bending moments occurred in the floor before maximum load. The behavior of the floor at the service load was satisfactory, in that cracking and deflections were at an acceptable level. The test showed that the limit design method led to a safe and satisfactory structure.

A $\frac{1}{4}$-scale nine-panel beam and slab floor was also tested to investigate membrane behavior in multipanel floors. The details of this floor and the test results are described in Chapter 12.

7.17.7   Tests Conducted at the University of Illinois

In 1956, the University of Illinois began an extensive research program into the design methods for floor slabs. The experimental phase of the investigation has included the testing of five $\frac{1}{4}$-scale models of multipanel reinforced concrete floor slabs. Each of the floors contained nine 60-in. (1.52-m) square panels arrayed three by three. Four of the floors were designed by the 1956 ACI Code methods for slabs and the working stress of the steel was taken to be 20,000 lb/in^2 (138 N/mm^2). The other floor was designed by a limit design method. The results of these tests had a considerable bearing on the procedures for slab design recommended in the ACI Codes from 1971 on and were discussed in Chapter 4. A brief summary of the tests has been given by Sozen and Siess.[7.37] A comparison of the yield line theory ultimate loads with the measured ultimate loads is given here, as it gives an interesting insight into the strength of multipanel floors.

1. *Flat plate floor.*[7.38] Specimen F1 consisted of nine bays arranged three by three with spans of 60 in. (1.52 m) supported internally on columns without enlarged heads and supported at the edges of the floor on beams. The slab was 1.75 in. (44 mm) thick. The floor had been designed by the 1956 ACI Code empirical method for a total service load (dead plus live) of 155 lb/ft^2 (7.41 kN/m^2). At the service load, only very small cracks at the columns were observed. The theoretical ultimate load of the floor was computed as 320 lb/ft^2 (15.3 kN/m^2) for failure mechanism 1 of Fig. 7.87a, which is a yield line pattern involving failure of the exterior panels of the floor alone. The alternative yield line pattern, failure mechanism 2 of Fig. 7.87b, involving plastic hinges in the edge beams, had a computed ultimate load of 400 lb/ft^2 (19.1 kN/m^2). The floor actually failed at a load of 360 lb/ft^2 (17.2 kN/m^2)

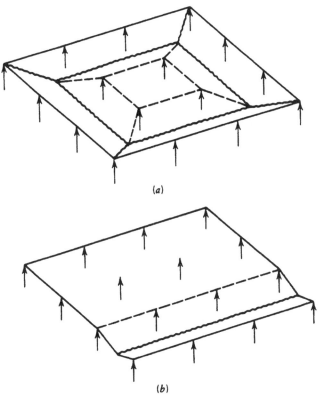

(a)

(b)

Figure 7.87 Failure mechanisms for uniformly loaded nine-panel flat plate and flat slab floors with edge beams: (a) collapse mechanism 1; (b) collapse mechanism 2.

by punching shear. At this load extensive yielding of steel had occurred and the pattern of cracking was in excellent agreement with the idealized yield line pattern. Thus, the failure in punching shear was a secondary failure, occurring after the theoretical ultimate flexural load was reached.

2. *Flat slab floor.*[7.39] Specimen F2 was of similar dimensions to the flat plate floor. It was supported internally on column capitals, and 20-in. (508-mm) square drop panels, and at the edges by beams. The floor had been designed using the 1956 ACI Code empirical method for a total service load of 285 lb/ft² (13.6 kN/m²). The theoretical ultimate load of the floor was 565 lb/ft² (27.0 kN/m²) for a yield line pattern that involved failure of the exterior panels alone (failure mechanism 1 of Fig. 7.87a). The measured load at the failure of the floor was 550 psf (26.3 kN/m²). At this load extensive yielding of the steel had occurred, and there was serious distress in the edge beams due to torsion and in the edge columns.

3. *Flat slab floor.*[7.40] Specimen F3 was of similar dimensions to the other flat slab floor and was designed by the 1956 ACI Code empirical method for

a total service load of 285 lb/ft² (13.6 kN/m²). Instead of using bar rein-forcement, as in all the other four floors, the slab was reinforced with welded wire fabric. Cracking was extensive on both the top and the bottom at service load. The theoretical ultimate load of the floor was 775 lb/ft² (37.1 kN/m²), based on the 0.2% offset stress of the reinforcement, for a yield line pattern that involved failure of the exterior panels alone (failure mechanism 1 of Fig. 7.87a). The measured failure load was 995 lb/ft² (47.6 kN/m²) and was accompanied by fracture of reinforcement. The considerable difference be-tween the failure load and the service load was due to the yield strength (0.2% offset stress) of the steel fabric being 70,000 lb/in² (483 N/mm²), whereas the yield strength of the bars of the other flat slab floor was 42,000 lb/in² (290 N/mm²) and both had been designed for a steel working stress of 20,000 lb/in² (138 N/mm²).

4. *Beam-and-slab floor with stiff beams.*[7.41] Specimen T1 consisted of nine panels arrayed three by three with spans 60 in. (1.52 m) by 60 in. (1.52 m) supported internally on 6-in. (152-mm) by 3-in. (76-mm) beams and exter-nally on $4\frac{1}{4}$-in. (108-mm) by 3-in. (76-mm) beams. The slab was $1\frac{1}{2}$ in. (38 mm) thick. The floor had been designed by the 1956 ACI Code Method 1 for two-way slabs for a total service load of 145 lb/ft² (6.94 kN/m²). At the service load, no cracks were observed. Slight cracking was observed on the top of the slab at dead load plus twice the live load. A yield line analysis of the floor indicated that the theoretical ultimate load was 393 lb/ft² (18.8 kN/m²) for failure of the central panel. For a folding type of composite beam–slab collapse mechanism involving the interior beams, the theoretical ultimate load was 467 lb/ft² (22.3 kN/m²). For a folding type of composite beam–slab collapse mechanism involving the exterior beams as well, the theoretical ultimate load was 529 lb/ft² (25.3 kN/m²). The measured ultimate load when the floor was loaded uniformly over the whole area was actually 537 lb/ft² (25.7 kN/m²). At this load yielding was general and failure occurred by a folding type of composite beam–slab mechanism involving the beams and exterior panels along one side of the floor, as in Fig. 7.87b, and also a tor-sional failure of the edge beams. A further test was conducted in which the interior panel was loaded alone, and a load of 829 lb/ft² (39.7 kN/m²) was reached when two of the supporting beams failed in combined shear and torsion. The enhancement of the ultimate strength of the panels was due to the tensile membrane action, which was enforced by the lateral restraint of the surrounding beams and panels. This was illustrated by the fact that when loading was applied over the entire floor, the panels did not fail before the composite beam–slab folding type of collapse mechanism occurred, and also that the interior panel when loaded alone was carrying almost twice the the-oretical yield line load when the beams failed.

5. *Beam-and-slab floor with shallow beams.*[7.42] Specimen T2 was of sim-ilar dimensions to the beam-and-slab floor with stiff beams apart from the fact that the beams were all 3 in. (76 mm) by 3 in. (76 mm) in cross section.

The floor was designed by apportioning half the total moment to the beams and half to the slab. The total static moment to be carried by the slab and the beams in each span was allocated to the sections so that the negative moment/positive moment ratio of the interior span was 2:1, and 1.4:1 for the interior negative moments on the end spans. The total service load was 145 lb/ft^2 (6.94 kN/m^2). At the service load of the floor, no cracks could be observed. At dead load plus twice the live load, a few cracks were visible. The theoretical failure load was 388 psf (18.6 kN/m^2) and involved a composite beam–slab folding mechanism of an edge row of beams and panels. The measured ultimate load of the floor was 466 lb/ft^2 (22.3 kN/m^2). The failure of the structure was due to general yielding at the critical sections and to torsional failure at the junction of the edge beams and the columns. This enhanced ultimate load was probably caused by the increase in the flexural strength of the beams due to the compressive forces induced by tensile membrane action in the panels.

7.17.8   Tests Conducted at the Portland Cement Association

The Portland Cement Association tested[7.43] a $\frac{3}{4}$ scale model of a flat plate floor that had an identical prototype to the $\frac{1}{4}$-scale model flat plate F1 tested at the University of Illinois. The deflections, crack patterns, distribution of service load moments, mode of failure, and ultimate load were similar to those of the $\frac{1}{4}$-scale model, clearly indicating the value of relatively small-scale models in structural research. The theoretical ultimate load of the floor was 350 lb/ft^2 (16.7 kN/m^2) for failure mechanism 1 of Fig. 7.87a. The actual failure load of the floor was 369 lb/ft^2 (17.7 kN/m^2) and failure occurred by punching shear.

REFERENCES

7.1   A. Ingerslev, "The Strength of Rectangular Slabs," *J. Inst. Struct. Eng.,* Vol. 1, No. 1, January 1923, pp. 3–14.

7.2   K. W. Johansen, *Brudlinieteorier,* Jul. Gjellerups Forlag, Copenhagen, 1943, 191 pp. (*Yield Line Theory,* translated by Cement and Concrete Association, London, 1962, 181 pp.)

7.3   K. W. Johansen, *Pladeformier,* Polyteknisk Forening, Copenhagen, 1946, 186 pp.; 2nd ed., 1949, 186 pp.; 3rd ed., 1968, 240 pp. (*Yield-Line Formulae for Slabs,* translated by Cement and Concrete Association, London, 1972, 106 pp.)

7.4   E. Hognestad, "Yield Line Theory for the Ultimate Flexural Strength of Reinforced Concrete Slabs," *Proc. ACI,* Vol. 24, March 1953, pp. 637–656.

7.5   R. H. Wood, *Plastic and Elastic Design of Slabs and Plates,* Thames and Hudson, London, 1961, 344 pp.

7.6   L. L. Jones, *Ultimate Load Analysis of Reinforced and Prestressed Concrete Structures,* Chatto & Windus, London, 1962, 248 pp.

7.7   R. H. Wood and L. L. Jones, *Yield-Line Analysis of Slabs,* Thames and Hudson, Chatto & Windus, London, 1967, 400 pp.

7.8   *Dalles et planchers dalles: applications de la théorie des lignes de rupture aux calculs de résistance en flexion,* Bulletin d'Information 35, Comité Européen du Béton, 1962. (*The Application of the Yield-Line Theory to Calculations of the Flexural Strength of Slabs and Flat-Slab Floors,* Information Bulletin 35, European Concrete Committee, translated by C. V. Amerongen, Cement and Concrete Association, London, 180 pp.)

7.9   "Dalles et Structures Planes," *Recommendations Comité Européen du Béton,* Tome 3, 1972. ("Slabs and Plane Structures," *Recommendations of the European Concrete Committee,* Vol. 3, 1972.)

7.10 *De berekening van platen volgens de vloeilijentheorie,* C.U.R. Rapport 26A, Commissie voor Uitvoering van Research ingesteld door de Betonvereniging, 1962, 79 pp. (*The Analysis of Slabs by the Yield-Line Theory,* Report 26A, Dutch Committee for Concrete Research, Translation 135, Cement and Concrete Association, London, 36 pp.)

7.11 S. N. Shukla, *Handbook for Design of Slabs by Yield-Line and Strip Methods,* Structural Engineering Research Centre, Roorkee, India, 1973, 180 pp.

7.12 R. Park and T. Paulay, *Reinforced Concrete Structures,* Wiley, New York, 1975, 769 pp.

7.13 K. O. Kemp, "The Evaluation of Nodal and Edge Forces in Yield-Line Theory," in *Recent Developments in Yield-Line Theory,* Magazine of Concrete Research Special Publication, Cement and Concrete Association, London, May 1965, pp. 3–12.

7.14 L. L. Jones, "The Use of Nodal Forces in Yield-Line Analysis," in *Recent Developments in Yield-Line Theory,* Magazine of Concrete Research Special Publication, Cement and Concrete Association, London, May 1965, pp. 63–74.

7.15 C. T. Morley, "Equilibrium Methods for Least Upper Bounds of Rigid-Plastic Plates," in *Recent Developments in Yield-Line Theory,* Magazine of Concrete Research Special Publication, Cement and Concrete Association, London, May 1965, pp. 13–24.

7.16 M. P. Nielsen, "A New Nodal-Force Theory," in *Recent Developments in Yield-Line Theory,* Magazine of Concrete Research Special Publication, Cement and Concrete Association, London, May 1965, pp. 25–30.

7.17 R. H. Wood, "New Techniques in Nodal-Force Theory for Slabs," in *Recent Developments in Yield-Line Theory,* Magazine of Concrete Research Special Publication, Cement and Concrete Association, London, May 1965, pp. 31–62.

7.18 E. C. Demsky and D. S. Hatcher, "Yield Line Analysis of Slabs Supported on Three Sides," *Proc. ACI,* Vol. 66, No. 9, September 1969, pp. 741–744.

7.19 R. H. Wood, *Studies in Composite Construction,* Part 2, *The Interaction of Floors and Beams in Multistorey Buildings,* Research Paper 22, DSIR National Building Studies, Her Majesty's Stationery Office, London, 1955, 125 pp.

7.20 R. Park, "Limit Design of Beams for Two-Way Reinforced Concrete Slabs," *J. Inst. Struct. Eng.,* Vol. 46, No. 9, September 1968, pp. 269–274.

7.21 R. Park, "Some Irregular Cases of Beam Loading in Limit Design of Slab and Beam Floors," *J. Inst. Struct. Eng.,* Vol. 48, No. 1, January 1970, pp. 17–19.

7.22 *Building Code Requirements for Structural Concrete,* ACI 318-95 and *Commentary,* ACI 318R-95, American Concrete Institute, Farmington Hills, Mich., 1995, 371 pp.

7.23 H. Gesund and O. P. Dikshit, "Yield Line Analysis of the Punching Problem at Slab/Column Intersections," in *Cracking, Deflection, and Ultimate Load of Concrete Slab Systems,* SP-30, American Concrete Institute, Detroit, 1971, pp. 177–201.

7.24 R. Park and S. Islam, "Strength of Slab–Column Connections with Shear and Unbalanced Flexure," *J. Struct. Div., ASCE,* Vol. 102, No. ST9, September 1976, pp. 1879–1901. "Discussion" by H. Gesund in Vol. 103, No. ST6, June 1977, pp. 1321–1323; "Closure," by R. Park and S. Islam in Vol. 103, No. ST12, December 1977, pp. 2423–2425.

7.25 S. D. Lash and A. Banerjee, "Strength of Simply Supported Square Plates with Central Square Openings," *Trans. Eng. Inst. Can.,* Vol. 10, No. A-5, June 1967, pp. 3–11.

7.26 A. Zaslavsky, "Yield-Line Analysis of Rectangular Slabs with Central Openings," *Proc. ACI,* Vol. 64, December 1967, pp. 838–844.

7.27 S. Islam and R. Park, "Yield-Line Analysis of Two Way Reinforced Concrete Slabs with Openings," *J. Inst. Struct. Eng.,* Vol. 49, No. 6, June 1971, pp. 269–276.

7.28 R. Park, "Design of Reinforced Concrete Slabs by Yield-Line Theory," *N. Z. Eng.,* Vol. 18, No. 2, February 1963, pp. 56–65.

7.29 R. Park, "Yield-Line Design of Concrete Slabs with Free Edges," *N. Z. Eng.,* Vol. 21, No. 2, February 1966, pp. 63–68.

7.30 A. Sawczuk and T. Jaeger, *Grenztragfähigkeits-Theorie der Platten,* Springer-Verlag, Berlin, 1963, 522 pp.

7.31 *Experimentele onderzoekingen betreffende het plastisch gedrag van platen,* Commissie voor Uitvoering van Research ingesteld door de Betonvereniging, C.U.R. Rapport 26B, 1963, 68 pp. (*Experimental Investigations of the Plastic Behaviour of Slabs,* Report 26B, Dutch Committee for Concrete Research, Translation 136, Cement and Concrete Association, London, 35 pp.)

7.32 R. Taylor, D. R. H. Maher, and B. Hayes, "Effect of the Arrangement of Reinforcement on the Behaviour of Reinforced Concrete Slabs," *Mag. Concr. Res.,* Vol. 18, No. 55, June 1966, pp. 85–94.

7.33 B. Hayes and R. Taylor, "Some Tests on Reinforced Concrete Beam–Slab Panels," *Mag. Concr. Res.,* Vol. 21, No. 67, June 1969, pp. 113–120.

7.34 R. Taylor, B. Hayes, and R. C. Wallin, "Some Tests on the Effect of Yield-Lines on the Behaviour of Reinforced Concrete T Beams," *Mag. Concr. Res.,* Vol. 17, No. 52, September 1965, pp. 131–134.

7.35 R. Park, "The Behaviour of a Model Slab and Beam Floor Designed by Limit Design," *Civ. Eng. Trans., Inst. Eng., Aust.,* Vol. CE 12, No. 1, April 1970, pp. 1–6.

7.36 R. Park, "Further Test on a Reinforced Concrete Floor Designed by Limit Procedures," in *Cracking, Deflection and Ultimate Load of Concrete Slab Systems,* ACI SP-30, American Concrete Institute, Detroit, 1971, pp. 251–269.

7.37   M. A. Sozen and C. P. Siess, "Investigation of Multi-panel Reinforced Concrete Floor Slabs: Design Methods—Their Evolution and Comparison," *Proc. ACI,* Vol. 60, August 1963, pp. 999–1028.

7.38   D. S. Hatcher, M. A. Sozen, and C. P. Siess, "Test of a Reinforced Concrete Flat Plate," *J. Struct. Div., ASCE,* Vol. 91, No. ST5, May 1965, pp. 205–231.

7.39   D. S. Hatcher, M. A. Sozen, and C. P. Siess, "Test of a Reinforced Concrete Flat Slab," *J. Struct. Div., ASCE,* Vol. 95, No. ST6, June 1969, pp. 1051–1072.

7.40   J. O. Jirsa, M. A. Sozen, and C. P. Siess, "Test of a Flat Slab with Welded Wire Fabric," *J. Struct. Div., ASCE,* Vol. 92, No. ST3, March 1966, pp. 199–224.

7.41   W. L. Gamble, M. A. Sozen, and C. P. Siess, "Tests of a Two-Way Reinforced Concrete Slab," *J. Struct. Div., ASCE,* Vol. 95, No. ST6, June 1969, pp. 1073–1096.

7.42   M. D. Vanderbilt, M. A. Sozen, and C. P. Siess, "Tests of a Modified Reinforced Concrete Two-Way Slab," *J. Struct. Div., ASCE,* Vol. 95, No. ST6, June 1969, pp. 1097–1116.

7.43   S. A. Guralnick and R. W. LaFraugh, "Laboratory Study of a 45-Foot Square Flat Plate Structure," *Proc. ACI,* Vol. 60, September 1963, pp. 1107–1186.

8 Design by Yield Line Theory

8.1 INTRODUCTION

The basis of yield line theory was outlined in Chapter 7 and used to derive ultimate load formulas for reinforced concrete slabs with a range of boundary conditions and types of loading. It is evident that such formulas are suitable for the ultimate load analysis of given slabs and can also be used for design. Denmark and Sweden appear to be the countries with the greatest experience with slabs designed using yield line theory. Yield line theory has been allowed by the codes of those two countries for many years. The British code has allowed yield line theory method since 1957. The 1997 British code[8.1] states that the method may be used "providing the ratio between support and span moments are similar to those obtained by the use of elastic theory." The European Concrete Committee–International Federation of Prestressing[8.2] recommends yield line theory as one possible approach to slab design provided that serviceability states are checked. The 1995 ACI Code[8.3] states that "a slab system may be designed by any procedure satisfying conditions of equilibrium and geometrical compatibility if shown that the design strength at every section is at least equal to the required strength . . . and that all serviceability conditions, including specified limits on deflections, are met." The 1995 New Zealand Standard[8.4] contains a statement similar to that in the 1995 ACI Code.

There is no doubt that design by yield line theory will lead to slabs with the required strength, but the method may be held suspect by some designers because of possible serviceability problems. However, serviceability problems can be avoided by checking deflections and crack widths during the design process. The many years of use of yield line theory in Europe and elsewhere should give designers confidence in the method. It is likely that with more experience of slab systems in service that have been designed by yield line theory, the approach will gain more acceptance in North America.

Yield line theory (and the strip method of Chapter 6) has its greatest application in the design of slab systems that are not covered by the detailed procedures of codes. Detailed code procedures normally apply only to uniformly loaded rectangular arrays of panels. In practice, the designer is often faced by slab systems with panels of various shapes and boundary conditions, including free edges, the presence of openings, and complex loading configurations. Limit design methods provide the designer with a particularly useful approach for these situations.

It was emphasized in Chapter 7 that yield line theory is an upper bound approach, and therefore the designer should be careful to examine all possible yield line patterns to ensure that the one giving the lowest ultimate load is used; otherwise, the strength of the slab may be overestimated. However, the comparison of test results from a wide range of slabs with predictions by yield line theory discussed in Section 7.17 indicates the real nature of yield lines and demonstrates that yield line theory gives a safe estimate of the ultimate load of slabs provided that the critical yield line pattern is used. In many cases there is a substantial reserve of strength not predicted by the theory, which gives added safety.

The critical yield line patterns and ultimate load formulas for slabs with various shapes, boundary conditions, and loading are available in the literature. A range of formulas were derived in Chapter 7. The English translation of Johansen's publications[8.5,8.6] covers a wide variety of slabs. The European Concrete Committee have published volumes[8.7,8.8] that give ultimate load formulas, recommendations, and design examples. Translations of two Dutch reports[8.9,8.10] contain formulas, recommendations, and describe 10 practical designs that are also useful. Many other references in English (e.g., Wood[8.11,8.13] and Jones,[8.12, 8.13] and others quoted in Chapter 7), give a useful range of design information. A publication in French by Bernaert et al.[8.14] gives a comprehensive range of formulas for slabs of most shapes, boundary conditions, and loading, including flat slab and flat plate floors. This extremely useful French publication was written to constitute the basis for an appendix to the unified practical recommendations of the European Concrete Committee.

8.2   STRENGTH AND SERVICEABILITY PROVISIONS

8.2.1   Design Load and Moment of Resistance

In design the problem is to determine the design (ultimate) moments of resistance per unit width required for a slab with known dimensions, boundary conditions, and factored (ultimate) load, the factored load being the required service loads multiplied by the load factors. For gravity loads, according to the ACI Code,[8.3] the factored (ultimate) load is

$$U = 1.4D + 1.7L \tag{8.1}$$

where D is the service dead load and L the service live load.

Reinforcement is provided for the design moments. If the ultimate resisting moment per unit width in a particular direction is to be m_u, the design equation for steel in that direction is[8.15]

$$m_u = \phi A_s f_y \left(d - 0.59 A_s \frac{f_y}{f_c'} \right) \qquad (8.2)$$

where ϕ is the strength reduction factor, taken by the ACI Code as 0.9; A_s the area of tension steel per unit width; f_y the steel yield strength; d the effective depth to the tension steel; and f_c' the compressive cylinder strength of the concrete. The effect of compression steel on the flexural strength is negligible and may be neglected.

The ultimate load equations derived for slabs in the general case of ortho-tropic reinforcement are in terms of the dimensions of the slab and the neg-ative and positive ultimate resisting moments per unit width m'_{ux} and m_{ux}, respectively, in the x-direction, and m'_{uy} and m_{uy}, respectively, in the y-direction. Hence, for a given ultimate load and spans, it is necessary to assume ratios for m'_{ux}/m_{ux}, m'_{uy}/m_{uy}, and m_{ux}/m_{uy}, or values for some moments, before the ultimate moments can be found. It is evident that an infinite number of reinforcement arrangements could be used. The factors influencing the choice of reinforcing arrangement are discussed below.

8.2.2   Reinforcement Ratios

According to the ACI Code,[8.3] the spacing of bars at critical sections should not exceed twice the slab thickness or 18 in. (500 mm), whichever is smaller, and the minimum amount of steel placed in the slab in the directions of the spans should not be less than that required for shrinkage and temperature reinforcement. This minimum amount is either 0.002 of the gross concrete area if grade 40 [$f_y = 300$ N/mm^2] or grade 50 [$f_y = 350$ N/mm^2] deformed bars are used, or 0.0018 where grade 60 [$f_y = 420$ N/mm^2] deformed bars or welded wire fabric are used, or $0.0018 \times 60{,}000/f_y$, but not less than 0.0014 where reinforcement with $f_y > 60{,}000$ lb/in^2 (420 N/mm^2) is used.

When determining the maximum amount of steel that can be placed in the slab, due regard to the requirements of section ductility should be given. For the general lower bound design method and the strip method, it was recom-mended in Chapters 5 and 6 that the reinforcement ratio A_s/d should not exceed $0.5\rho_b$, where ρ_b is the balanced failure steel ratio.[8.3,8.15] In the general lower bound method and the strip method the reinforcement is uniformly spaced in bands, and moment strengths quite close to the elastic distributions of moments may be obtained. Yield line theory makes use of uniformly spaced reinforcement across the entire slab in many cases (although with different spacing in the top and bottom and different spacing in each direction) and the amount of moment redistribution required to achieve the design con-figuration of moments may be more significant than for the general lower bound method and the strip method. The 1978 CEB-FIP model code[8.2] re-quires for both the strip and the yield line design methods that the tension reinforcement ratio should not exceed one-half of that which corresponds to the value for which the tension steel strain reaches yield and the extreme concrete compressive fiber strain reaches 0.0035 simultaneously. Using the CEP-FIP partial safety coefficients for material strengths and concrete compressive stress block parameters, one-half of the tension steel for this limiting strain profile represents a ρ value of $0.38\rho_b$ when $f'_c \leq 4000$ lb/in^2 (27.6 N/mm^2) and $f_y = 40,000$ lb/in^2 (276 N/mm^2) or 60,000 lb/in^2 (414 N/mm^2), or $0.40\rho_b$ when $f'_c = 5000$ lb/in^2 (34.5 N/mm^2) and $f_y = 40,000$ lb/in^2 (276 N/mm^2), or $0.41\rho_b$ when $f'_c = 5000$ lb/in^2 and $f_y = 60,000$ lb/in^2 (414 N/mm^2), where ρ_b is the balanced steel ratio as defined in the ACI Code.[8.3,8.15] It should also be noted that the slab tests discussed in Section 7.17 have invariably shown sufficient ductility to achieve the moment configuration required by the yield line pattern. Also, the biaxial compression condition that exists in the compression zone of much of the slab area will increase the ductility of the concrete as compared with the uniaxial compression case and allow greater compressive strains to be reached before crushing of concrete commences. In view of all these factors, it would seem reasonable in yield line theory design to require that the tension steel ratio not exceed $0.4\rho_b$. The steel ratio $0.4\rho_b$ for some commonly used steel and concrete strengths is given in Table 8.1.

8.2.3   Reinforcement Arrangements

Yield line theory allows the designer freedom to choose arrangements of reinforcement that lead to simple detailing. However, it cannot be overemphasized that the arrangements of reinforcement chosen should be such that the resulting distribution of ultimate moments of resistance at the various sections throughout the slab does not differ widely from the distribution of moments given by elastic theory. If large differences between the distribution of ultimate resisting moments and the elastic moments do exist, it may mean that cracking at the service load will be excessive because low steel ratios at highly stressed sections may lead to high steel stresses and hence large crack widths. Such regions of high steel stress at service load may also result in large deflections. Hence, it is important that the designer should maintain a feeling for the elastic distribution of bending moments and use it to help

TABLE 8.1   Steel Ratio $0.4\rho_b$ for Various Steel and Concrete Strengths

f'_c (lb/in^2 [N/mm^2])	f_y (lb/in^2 [N/mm^2])	$0.4\,\rho_b$
3000 [20]	40,000 [300]	0.0148 [0.0128]
3000 [20]	60,000 [420]	0.0086 [0.0081]
4000 [30]	40,000 [300]	0.0198 [0.0193]
4000 [30]	60,000 [420]	0.0114 [0.0121]

decide the ratios of negative to positive ultimate resisting moments and ratios of the ultimate resisting moments to be used in the two directions. Just how far the reinforcement arrangement can differ from the elastic bending moments and still result in a serviceable slab has not been determined conclusively, but the tests that have been carried out (e.g., see Section 7.17) do indicate that sensible arrangements of steel result in serviceable slabs. It is recommended that ratios of negative to positive ultimate moments of resistance per unit width between 1 and 2 should be used. (The CEB-FIP model code[8.2] allows this ratio to be between 0.5 and 2.0.) Some account should be taken of the degree of restraint at the edges. For example, if full restraint against rotation is anticipated, a value in the range 1.5 to 2.0 could be used, but if some rotation is expected, a value in the range 1.0 to 1.5 would be more appropriate. Ratios of the ultimate moments of resistance per unit width in the two directions should take account of the direction of maximum elastic bending moment. For example, in two-way slabs the greatest ultimate resisting moment per unit width should act in the direction of the short span.

In many cases, uniformly spread steel (i.e., constant bar spacing across the whole width or length of slab) will not be acceptable. This is particularly true in a flat plate with small columns, because of the very high negative moments at and near the columns. If the steel is uniformly spread across the plate, yielding could occur at or below service load in the vicinity of the columns, even if the overall ratio of negative to positive moments provided is reasonable. Thus, the reinforcement in some slabs will need to be more heavily concentrated in some portions of a yield line than in others to follow the elastic bending moment diagram more closely.

At edges that have been considered as simply supported, care should be taken to provide top reinforcement to control cracking due to fortuitous restraining moments. Such reinforcement should be for approximately 0.33 to 0.5 of the maximum positive ultimate moment of resistance per unit width.

Elastic theory shows that large torsional moments may exist in slabs in the corner regions. Yield line theory shows a strength reduction due to the formation of fans in slab corners which can be significant if top steel is absent. Both top and bottom steel should be present in the corner regions of all slabs. The reinforcement present in the top and the bottom should be provided for a distance of 0.2 of the longer span in each direction from the corner and should provide an ultimate positive and negative resisting moment per unit width equal to the maximum positive ultimate moment per unit width in the slab.

8.2.4 Serviceability Checks

For satisfactory behavior at the service load, the deflections and crack widths should not be excessive. Methods for checking deflections and crack widths are discussed in detail in Chapter 9. The thickness of the slab in most practical cases will be set by the need for adequate stiffness for deflection control. The thickness should not be less than the minimum values specified in the ACI Code[8.3] unless the deflections are computed and shown to be within the limiting values allowed by the code (see Chapter 9).

Excessive cracking should not be a problem provided that a reasonable arrangement of reinforcement is used, as discussed previously. The maximum steel stresses at service load are required for a check of crack widths. The steel stresses at service load may be calculated using elastic theory for sections[8.15] from the distribution of bending moments given by elastic theory for slabs. The elastic theory distribution of bending moments for slabs is discussed in Chapters 2 and 3. For a serviceability check, the maximum elastic theory bending moments may be estimated from tables published in texts, or from standard finite element or finite difference computer programs that are available for slab systems. A text by Barês[8.16] gives extensive sets of tables covering the elastic theory bending moments for a wide range of slab shapes, loadings, and boundary conditions. Other references to texts and computational procedures are given in Chapters 2 and 3. It is suggested that the maximum steel stress calculated from the elastic theory bending moments at the service load should not exceed $0.8f_y$, where f_y is the specified yield strength of the reinforcement, and that Eq. 9.20 (see Chapter 9) should also be satisfied if f_y exceeds 40,000 lb/in² (276 N/mm²). These suggested requirements should ensure that steel yielding does not occur at the service load due to unforeseen behavior and that the crack widths do not exceed the limiting values accepted by the ACI Code. This serviceability check may lead to a rearrangement of the reinforcing steel to obtain service load steel stresses and crack widths of an acceptable level.

A further serviceability consideration concerns the length of top steel. If a low ratio of ultimate negative to positive moments is assumed in design, the length of top steel found by considering alternative collapse mechanisms (as in Section 8.4.1) may be quite small. It should be checked that the top steel extends far enough to provide for the regions of negative moment that elastic theory indicates will exist at the service load. This will ensure that cracking due to negative moments at service load is controlled.

8.2.5   Other Design Aspects

Other aspects of design involve the effects of superposition of loading on the required ultimate resisting moments, the extent of negative moment steel and reinforcement arrangements giving minimum weight of reinforcement in two-way slabs, the methods of calculating design bending moments in beams supporting two-way slabs, and the allocation of reinforcement between column and middle strips in beamless slabs. These aspects and some design examples are considered in the next sections.

8.3 SUPERPOSITION OF LOADING

It is sometimes necessary to determine the ultimate moments of resistance per unit width required for a slab that may be subjected to various types of loading: for example, combined concentrated and uniform loading. Suppose that the ultimate moments required for each load applied separately are known. Although the principle of superposition is strictly applicable only to linear-elastic structures, it may be shown that summing the ultimate moments of the separate cases gives a result that is on the safe side, even though in yield line theory the structure is behaving nonlinearly. This is analogous to the case of analysis considered in Section 7.6. Note, however, that in yield line analysis the lowest ultimate load for given ultimate moments is sought, whereas in design the greatest ultimate moments for a given ultimate load is sought. That is, the correct yield line solution is obtained when the ratio of ultimate load to ultimate moments of resistance is a minimum.

To illustrate combined load cases, consider two loading cases, designated W_{u1} and W_{u2}. The exact yield line pattern for load W_{u1} applied alone to the slab requires ultimate moments m_{u1}, and all the other yield line patterns will require ultimate moments smaller than m_{u1}. The exact yield line pattern for load W_{u2} applied alone to the slab requires yield moments m_{u2}, and all other yield line patterns will require ultimate moments smaller than m_{u2}. Now if loads W_{u1} and W_{u2} are applied to the slab simultaneously, the exact yield line pattern for the combined loading case may either be one of the exact patterns for separate loading given above or some other pattern. Say that the ultimate moments required for W_{u1} and W_{u2} for this combined loading pattern are $\mathfrak{M}_{u1}$ and $\mathfrak{M}54_{u2}$, respectively. Then the correct ultimate moments for the combined loading case $W_{u1} + W_{u2}$ are $\mathfrak{M}_{u1} + \mathfrak{M}_{u2}$. But since $m_{u1} \geq \mathfrak{M}_{u1}$ and $m_{u2} \geq \mathfrak{M}_{u2}$, then $m_{u1} + m_{u2} \geq \mathfrak{M}_{u1} + \mathfrak{M}_{u2}$. Hence, the reinforcement provided for the sum of the loads on the basis of the sum of the ultimate moments found from the yield line patterns for each load applied separately will always be equal to or greater than the reinforcement actually required. The reinforcement will be exactly equal only when all the yield line patterns are identical. This reasoning is valid for the superposition of any number of loads.

Thus, adding the moments strengths required for the loads applied separately will always give a conservative design. This approximate approach can be very conservative if the yield line patterns for the separate loading cases are significantly different.

Example 8.1. Determine the ultimate negative moments of resistance per unit width m_u' required for the cantilever slab shown in Fig. 8.1 if the slab is isotropically reinforced in the top and there is no bottom steel. The slab is loaded by a uniformly distributed service dead load of 125 lb/ft² (5.98 kN/m²), a uniformly distributed live load of 60 lb/ft² (2.87 kN/m²), and a concentrated live load of 15,000 lb (66.8 kN) at one unsupported corner.

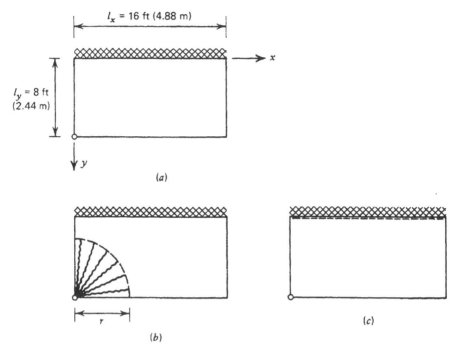

Figure 8.1   Cantilever slab with uniform and concentrated loading of Example 8.1: (a) slab; (b) mode 1; (c) mode 2.

APPROXIMATE SOLUTION.   From Eq. 8.1, the factored uniform loading is

$$w_u = (1.4 \times 125) + (1.7 \times 60) = 277 \text{ lb/ft}^2$$

and the factored concentrated load is

$$P_u = 1.7 \times 15,000 = 25,500 \text{ lb}$$

For the load P_u acting alone, mode 1 of Fig. 8.1b is the critical collapse mode. The required ultimate negative moment of resistance per unit width for P_u alone for this mechanism according to Eq. 7.52 is

$$m'_{u1} = \frac{2P_u}{\pi}$$

$$= 2 \times \frac{25,500}{\pi} = 16,230 \text{ lb-ft/ft width}$$

For the load w_u acting alone, mode 2 of Fig. 8.1c is the critical collapse mode. The required ultimate negative moment of resistance per unit width for w_u alone for this mechanism is

$$m'_{u2} = \frac{w_u l_y^2}{2}$$

$$= 277 \times \frac{8^2}{2} = 8860 \text{ lb-ft/ft width}$$

Therefore, using the principle of superposition, the required moment strength is

$$m'_u = m'_{u1} + m'_{u2} = 16{,}230 + 8860 = 25{,}090 \text{ lb-ft/ft width}$$

EXACT SOLUTION. If both P_u and w_u are on the slab, either of the modes shown in Fig. 8.1b and c could form. If both loads are on the slab and mode 1 develops, the virtual work equation for a downward displacement δ of the loaded corner is

$$P_u \delta + w_u \frac{\pi}{4} r^2 \frac{\delta}{3} = m'_u \delta \frac{\pi}{2}$$

$$m'_u = \frac{2P_u}{\pi} + \frac{w_u r^2}{6}$$

where r is the radius of the failure cone. For maximum m'_u, $r = l_y$. Therefore,

$$m'_u = \left(2 \times \frac{25{,}500}{\pi} \right) + \left(277 \times \frac{8^2}{6} \right)$$

$$= 19{,}180 \text{ lb-ft/ft width}$$

If both loads are on the slab and mode 2 develops, the virtual work equation for a downward displacement δ of the loaded corner is

$$P_u \delta + w_u l_x l_y \frac{\delta}{2} = m'_u l_x \frac{\delta}{l_y}$$

$$m'_u = P_u \frac{l_y}{l_x} + w_u \frac{l_y^2}{2}$$

$$= \left(25{,}500 \times \frac{8}{16} \right) + \left(277 \times \frac{8^2}{2} \right)$$

$$= 21{,}610 \text{ lb-ft/ft width}$$

Mode 2 requires the greatest ultimate moment and therefore is more critical than mode 1. Therefore, the correct design moment is $m'_u = 21{,}610 \text{ lb-ft/ft}$ width.

Note that the approximate approach of adding the required moment strengths for the loads applied separately gives a required m'_u that is 16% greater than the exact value in this example. Thus, the approximate solution is safe, but for better accuracy it is best to calculate the required ultimate moments from the yield line pattern that actually develops when the loads are applied simultaneously.

8.4 DESIGN OF UNIFORMLY LOADED TWO-WAY SLABS

Ultimate load formulas for slabs with various boundary conditions were derived in Chapter 7, and general strength and serviceability provisions for design were discussed in Section 8.2. Further aspects of interest in design are the extent of negative-moment steel and the arrangements of steel that lead to minimum weight design.

8.4.1 Extent of Top Steel in Uniformly Loaded Rectangular Slabs

So far in the theory it has been assumed that the reinforcement is placed uniformly over the entire area of the slab. In most cases the top steel may be curtailed. The cutoff points may be calculated[8.17,8.18] by examining the alternative collapse mechanisms that could form as a result and ensuring that the ultimate load is not decreased by the curtailment. The top steel must also extend far enough to provide for the elastic distribution of moments that exists at the service loads.

Uniformly Loaded Rectangular Slab with All Edges Supported. To illustrate how the required extent of top steel can be calculated, the minimum length of top steel for a uniformly loaded slab at the ultimate load with all edges fixed will be determined. The positive and negative ultimate moments of resistance per unit width are m_{ux} and $m'_{ux} = im_{ux}$, respectively, in the x-direction, and m_{uy} and $m'_{uy} = im_{uy}$, respectively, in the y-direction. Figure 8.2 shows the reinforcement placed in such a slab. The length of the top steel as a proportion of the spans is defined by the coefficient λ, and λ is taken to be the same for both spans for convenience. Figure 8.3a shows the basic type of yield line pattern for the slab that would be used to design the slab. Figure 8.3b, c, and d show other possible yield line patterns that could form as a result of the curtailment of the steel. In the regions of the slab where there is no top steel, the ultimate negative moments of resistance must be considered to be zero; that is, the tensile strength of the concrete is neglected. Hence, in the modes of Fig. 8.3b, c, and d, the interior parts of the slab which collapse may be considered to be simply supported along those lines that coincide with the lines of cutoff of the top steel. The length of top steel is satisfactory if the modes of Fig. 8.3b, c, and d have ultimate loads which are at least equal to that of the mode of Fig. 8.3a.

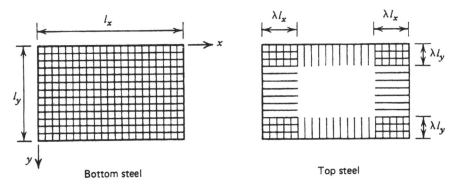

Bottom steel   Top steel

Figure 8.2 Reinforcement for a uniformly loaded fixed-edge slab.

From Eq. 7.59, the ultimate uniform load of the basic yield line pattern shown in Fig. 8.3a is

$$w_u = \frac{24\mu m_{uy}(1 + i)}{l_y^2(l_y/l_x)^2\{[1 + 3\mu(l_x/l_y)^2]^{1/2} - 1\}^2} \tag{8.3}$$

where $\mu = m_{uy}/m_{ux}$. The ultimate uniform load of the alternative yield line

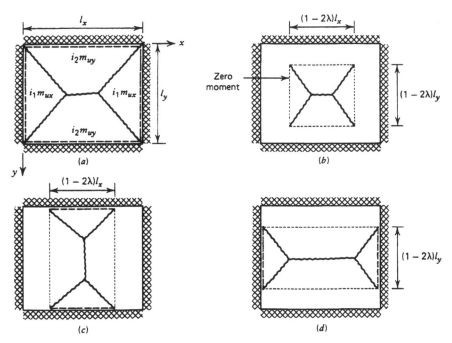

Figure 8.3 Yield line patterns for a uniformly loaded fixed-edge slab with curtailed top steel.

pattern shown in Fig. 8.3b is obtained from Eq. 8.3 with l_x put equal to $l_x(1 - 2\lambda)$, l_y put equal to $l_y(1 - 2\lambda)$, and $i = 0$, and is

$$w_u = \frac{24\mu m_{uy}}{l_y^2(1 - 2\lambda)^2(l_y/l_x)^2\{[1 + 3\mu(l_x/l_y)^2]^{1/2} - 1\}^2} \tag{8.4}$$

For the ultimate load given by Eq. 8.4 to at least equal that given by Eq. 8.3 requires that

$$\frac{1}{(1 - 2\lambda)^2} \geq 1 + i$$

In the limit the minimum value for λ is given by

$$(1 + i)(1 - 2\lambda)^2 = 1$$

which on solving the quadratic equation for λ gives

$$\lambda = 0.5 \left(1 - \frac{1}{\sqrt{1 + i}}\right) \tag{8.5}$$

(Note the similarity to Eq. 6.9 for a uniformly loaded fixed-end beam.)

In practice it is not necessary to examine the collapse modes of Fig. 8.3c and d, since it is unlikely that the ultimate loads of these two mechanisms will be more than 5% lower than that of the mode of Fig. 8.3b. Hence, the cutoff length for uniformly loaded slabs with all edges fixed may be calculated from Eq. 8.5 for ultimate load conditions. It should also be checked to ensure that these λ values provide for the distribution of elastic bending moments at the service load.

The λ values for top steel at the fixed edges of slabs with one, two, or three simply supported edges at the ultimate load may be calculated by a similar procedure. In these cases the interior portion that may collapse as a simply supported slab will extend to the simply supported edge or edges, and λ is found to vary with both the l_x/l_y and m_{uy}/m_{ux} ratios. Hence, the algebra to be manipulated in the general case is lengthy, even if the ultimate load of the slab is expressed using the simpler approximate equation 7.95. The value of λ in the general case is given by the solution of a cubic equation. Table 8.2 gives values for λ calculated for practical slabs in which $i = m'_{ux}/m_{ux} = m'_{uy}/m_{uy} = 2$, using Eq. 7.95 for the cases of support conditions shown in Fig. 8.4. The length of negative-moment reinforcement, measured from support to theoretical cutoff points, in the x- and y-directions at fixed edges is λl_x and λl_y, respectively. At a simply supported edge, λ and i are both zero. It is to be noted that the bars should extend a distance d or 12 bar diameters, which-

TABLE 8.2 Coefficients Defining Lengths of Negative-Moment Reinforcement at Fixed Edges of Uniformly Loaded Rectangular Slabs Supported at Four Edges[a]

$\dfrac{l_x}{l_y}$	$\dfrac{m_{uy}}{m_{ux}}$	Case								
		1	2	3	4	5	6	7	8	9
1.0	1	0	0.30	0.30	0.29	0.22	0.22	0.23	0.23	0.21
1.5	1	0	0.33	0.28	0.29	0.23	0.21	0.24	0.22	0.21
	3	0	0.26	0.31	0.29	0.19	0.22	0.24	0.22	0.21
	5	0	0.21	0.32	0.29	0.17	0.22	0.24	0.22	0.21
2.0	1	0	0.36	0.28	0.29	0.25	0.21	0.25	0.22	0.21
	3	0	0.28	0.30	0.29	0.21	0.22	0.25	0.22	0.21
	5	0	0.23	0.31	0.29	0.18	0.22	0.25	0.22	0.21
3.0	1	0	0.38	0.28	0.29	0.27	0.21	0.26	0.21	0.21
	3	0	0.30	0.30	0.29	0.22	0.22	0.27	0.22	0.21
	5	0	0.24	0.30	0.29	0.19	0.22	0.27	0.22	0.21

[a] λ values when $i = m'_{ux}/m_{ux} = m'_{uy}/m_{uy} = 2$ for the various cases of support conditions shown in Fig. 8.4.

ever is greater, beyond the theoretical point of cutoff to ensure that the bars are anchored satisfactorily, where d is the effective depth of the slab.[8.3]

It is apparent[8.11] that there is another mode of collapse that may govern the length of top reinforcement in panels of a continuous slab-and-beam floor. This mode may occur when the live load is placed only on alternate panels to give a checkerboard live load distribution. Then the panels that carry dead load only may be forced up and fail if the top steel is not carried far enough into the panel. This collapse mechanism for part of a continuous slab-and-beam floor is shown in Fig. 8.5. For this collapse mechanism to occur, the supporting beams must undergo the same rotation as the panel segments of the mechanism. Inspection of adjacent panels of the mechanism shows that since the center of one panel moves up and the center of the adjacent panel moves down, the supporting beams must twist in opposite directions along the same length, necessitating the formation of torsional hinges in the beams at the corners of the panel. Figure 8.6 shows the collapse mechanism of a panel loaded by dead load only of a floor with checkerboard live load distribution. The panel develops zero-moment yield lines along the lines of cutoff of the top steel within the panel and negative-moment yield lines connect the corners of the zero-moment lines with the corners of the panel.

Let the uniformly distributed dead load per unit area be w_{ud} and the uniformly distributed live load per unit area on the adjacent panels be w_{ul}. In the limit the restraining moments around the edge of the panel that force the panel up are m'_{ux} and m'_{uy}. Let T_{ux} and T_{uy} be the torsional strength of the x- and y-direction beams, respectively. Let the center of the panel be given a small displacement δ upward. The virtual work equation may be written as

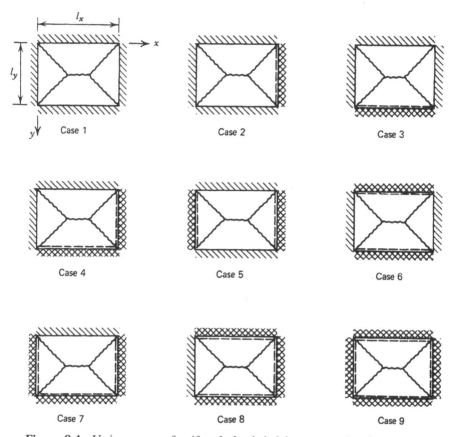

Figure 8.4   Various cases of uniformly loaded slabs supported at four edges.

$$2\left(m'_{ux}l_y\,\frac{\delta}{\lambda l_x} + m'_{uy}l_x\,\frac{\delta}{\lambda l_y}\right) = 4\left(m'_{ux}\lambda l_y\,\frac{\delta}{\lambda l_x} + m'_{uy}\lambda l_x\,\frac{\delta}{\lambda l_y}\right)$$

$$+ 4\left(T_{ux}\,\frac{\delta}{\lambda l_y} + T_{uy}\,\frac{\delta}{\lambda l_x}\right)$$

$$+ w_{ud}\left[4\lambda^2 l_x l_y\,\frac{\delta}{3} + 2\lambda l_x(1-2\lambda)l_y\,\frac{\delta}{2}\right.$$

$$\left.+ 2\lambda l_y(1-\lambda)l_x\,\frac{\delta}{2} + l_x l_y(1-2\lambda)^2\delta\right]$$

where the term on the left-hand side is due to the edge moments caused by the downward movement of the adjacent panel, the first term on the right-hand side is due to the negative-moment corner yield lines, the second term on the right-hand side is due to the torsional moments in the beams, and the

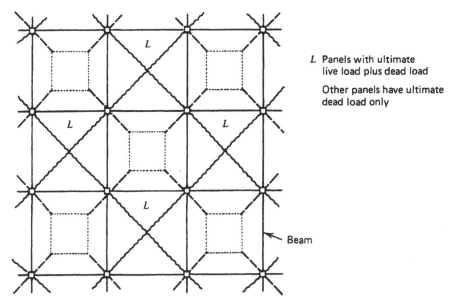

L Panels with ultimate
   live load plus dead load

Other panels have ultimate
dead load only

Beam

Figure 8.5 Possible mode of collapse of a uniformly loaded continuous slab-and-beam floor with checkerboard live load distribution.

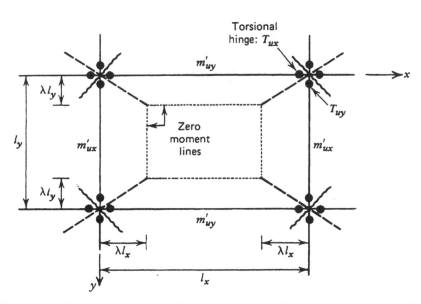

Figure 8.6 Collapse mechanism of a panel under dead load with checkerboard live loading on the floor.

third term on the right-hand side is due to the dead load. The equation simplifies to

$$\left(\frac{2}{\lambda} - 4\right)\left(m'_{ux}\frac{l_y}{l_x} + m'_{uy}\frac{l_x}{l_y}\right) - \frac{4}{\lambda}\left(\frac{T_{ux}}{l_y} + \frac{T_{uy}}{l_x}\right) - \frac{w_{ud}l_x l_y}{3}(4\lambda^2 - 6\lambda + 3) = 0$$

(8.6)

The values of $m'_{ux} = im_{ux}$ and $m'_{uy} = im_{uy}$ required in Eq. 8.6 can be obtained from Eq. 7.95:

$$m_{ux} = \frac{(w_{ud} + w_{ul})[3(l_x/l_y) - 1]l_y^2}{24(1 + i)[1 + \mu(l_x/l_y)]}$$

(8.7)

where $\mu = m_{uy}/m_{ux}$ and $i = m'_{ux}/m_{ux} = m'_{uy}/m_{uy}$. Table 8.3 shows values of λ obtained from Eqs. 8.6 and 8.7 for the panel shown in Fig. 8.6 when $i = 2$ for various values of l_x/l_y, m_{uy}/m_{ux}, and w_{ul}/w_{ud} and assuming that $T_{ux} = T_{uy} = 0$. The λ values of Table 8.3 are generally greater than those of Table 8.2 and therefore are more critical. However, Table 8.3 has been calculated assuming that the beams have zero torsional strength. Thus, Table 8.3 applies strictly only to the case of a continuous slab resting on but not connected to walls or on steel beams. When reinforced concrete beams of normal size are used, the torsional strength will be sufficient to reduce the values of λ given by Eqs. 8.6 and 8.7 to less than those of Table 8.2, making Table 8.2 more critical than Table 8.3. However, some doubt may be felt regarding the duc-

TABLE 8.3 Coefficients Defining Lengths of Negative-Moment Reinforcement for a Uniformly Loaded Rectangular Slab Continuous at All Edges with Checkerboard Live Load Distribution, Beams of Zero Torsional Strength, and $i = 2$

		λ		
$\dfrac{l_x}{l_y}$	$\dfrac{m_{uy}}{m_{ux}}$	$\dfrac{w_{ul}}{w_{ud}} = 1$	$\dfrac{w_{ul}}{w_{ud}} = 3$	$\dfrac{w_{ul}}{w_{ud}} = 5$
1.0	1	0.20	0.32	0.38
1.5	1	0.21	0.32	0.38
	3	0.22	0.34	0.39
	5	0.22	0.34	0.39
2.0	1	0.21	0.33	0.38
	3	0.23	0.34	0.39
	5	0.23	0.35	0.40
3.0	1	0.22	0.34	0.39
	3	0.24	0.35	0.40
	5	0.25	0.36	0.40

tility of plastic torsional hinges in beams. Table 8.3 then applies to slabs supported on reinforced concrete beams of uncertain torsional ductility or on uncased steel I sections or resting on walls.

It is to be noted that Tables 8.2 and 8.3 give values of λ for $i = 2$. Should smaller values of i be used in design, the λ values can theoretically be reduced, but the tabulated values of λ for $i = 2$ can be used conservatively. Table 8.3 is also based on the ultimate dead load, w_{ud}, on the unloaded panels. A more critical situation is the case of only service dead load w_d on the unloaded panels, and in this case the table should be entered using w_{ul}/w_d in place of w_{ul}/w_{ud}.

A similar problem can occur if the slab is subjected to a strip loading, as shown in Fig. 3.18b. It can be treated almost the same as a continuous beam case, provided that the effects of the column restraints are properly taken into account.

Uniformly Loaded Rectangular Slab with Three Edges Supported and One Free. A slab that has the supported edges fixed and is reinforced as shown in Fig. 8.7 could develop any of the collapse modes shown in Fig. 8.8. The lengths of the top steel measured from the supports have been taken as λl_x and $2\lambda l_y$ in the x- and y-directions, respectively, since elastic theory shows that the y-direction steel should be fairly extensive. In regions of the slab where there is no top steel, the ultimate negative moments of resistance are considered to be zero, and hence in the modes of Fig. 8.8b, c, and d the interior parts of the slab that collapse must be considered to be simply supported at those lines that coincide with the lines of the cutoff of the top steel. The length of top steel is satisfactory if the ultimate loads of the modes of Fig. 8.8b, c, and d are at least equal to the ultimate load of the basic mode

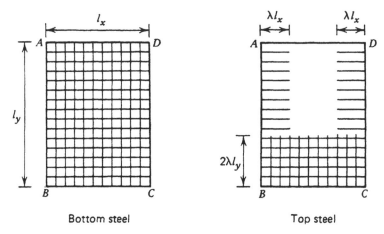

Figure 8.7 Reinforcement for a uniformly loaded slab with three edges fixed and one edge (AD) free.

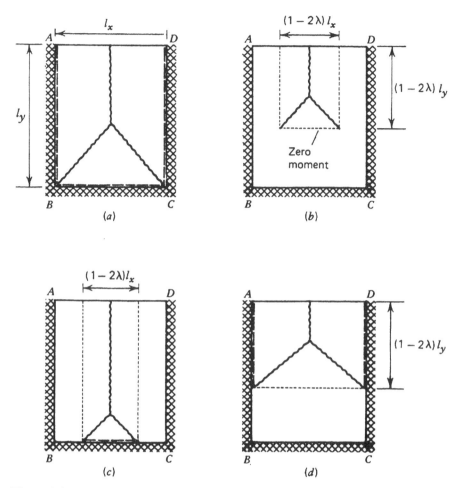

Figure 8.8   Yield line patterns for uniformly loaded slab with three edges fixed, one edge (AD) free, and curtailed top reinforcement.

of Fig. 8.8a. The minimum λ values for the top reinforcement at the fixed edges of slabs with the support conditions shown in Fig. 8.9 have been calculated and are tabulated in Table 8.4 for slabs with $i = m'_{ux}/m_{ux} = m'_{uy}/m_{uy} = 2$ and various l_y/l_x and m_{ux}/m_{uy} ratios. For slabs with three fixed edges, the λ values were found by equating the collapse loads of the modes of Fig. 8.8a and b, since it was found that the remaining modes were critical in very few cases, and when critical had collapse loads that never differed by more than 5% from those of the mode of Fig. 8.8b. For slabs with one or two simply supported edges there are only two possible modes, since the collapsing portions will extend to the simply supported edge or edges and will cause some of the modes of Fig. 8.8 to become identical. The collapse

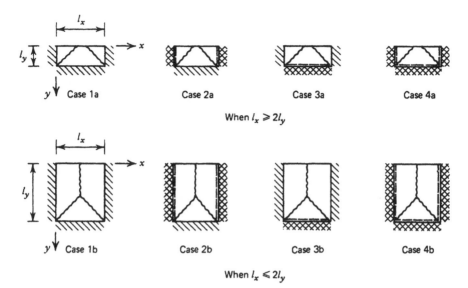

Figure 8.9 Various cases of uniformly loaded slabs supported at three edges.

TABLE 8.4 Coefficients Defining Lengths of Negative-Moment Reinforcement at Fixed Edges of Uniformly Loaded Rectangular Slabs Supported at Three Edges[a]

$\dfrac{l_y}{l_x}$	$\dfrac{m_{ux}}{m_{uy}}$	Case			
		1a or 1b	2a or 2b	3a or 3b	4a or 4b
3.0	1	0	0.21	0.28	0.21
	3	0	0.21	0.24	0.21
	5	0	0.21	0.20	0.21
2.0	1	0	0.21	0.28	0.21
	3	0	0.21	0.23	0.21
	5	0	0.22	0.20	0.21
1.0	1	0	0.21	0.25	0.21
	3	0	0.22	0.21	0.21
	5	0	0.22	0.18	0.21
0.5	1	0	0.22	0.31	0.21
	2	0	0.23	0.26	0.21
	3	0	0.23	0.23	0.21
0.333	1	0	0.21	0.33	0.23
	2	0	0.23	0.29	0.23
	3	0	0.24	0.25	0.22

[a] λ when $i = m'_{ux}/m_{ux} = m'_{uy}/m_{uy} = 2$ for the cases of support conditions shown in Fig. 8.9.

loads of the two modes were equated to find the λ values at the one or two fixed edges. The approximate ultimate load equations, 7.99 and 7.100, were used in these calculations. In Table 8.4 the lengths of top steel measured from a fixed support to the theoretical cutoff point are λl_x and $2\lambda l_y$ in the x- and y-directions, respectively, as in Fig. 8.7. At a simply supported edge, λ is zero. In addition, bars should extend a distance d or 12 bar diameters, whichever is greater, beyond the theoretical point of cutoff to ensure that the bars are anchored satisfactorily, where d is the effective depth of the slab.[8.3]

It is to be noted that Table 8.4 gives values of λ for $i = 2$. Should smaller values of i be used in design, the λ values can theoretically be reduced, but the tabulated values of λ can be used conservatively. It should also be checked to ensure that the λ values provide for the distribution of elastic bending moments at the service load.

Uniformly Loaded Rectangular Slab with Two Adjacent Edges Supported and Two Edges Free. The yield line patterns of Fig. 7.41 show that top and bottom steel should be placed over the entire area since resistance to negative and positive moment is required over most of the slab. The content of top steel could be reduced toward the free corner by cutting alternate bars.

8.4.2 Minimum-Weight Design

The ultimate load equations derived for slabs in the general case of orthotropic reinforcement are in terms of the dimensions of the slab and the ultimate moments of resistance per unit width m'_{ux}, m_{ux}, m'_{uy}, and m_{uy}. Hence, for a given ultimate load and spans, it is necessary to assume ratios or values for some moments before the required ultimate moments can be found. It has already been pointed out that theoretically, an infinite number of reinforcement arrangements could be used. In many cases isotropic reinforcement is adequate for strength, but it is also possible[8.17,8.18] to determine the allocation of the reinforcement for the various ultimate moments which results in the minimum amount of steel for a given thickness of slab. The distribution of reinforcement that results in a minimum weight of steel for various cases of rectangular slabs is considered next.

Uniformly Loaded Rectangular Slab with All Edges Supported

Uniformly Spread Reinforcement. Consider the case of a uniformly loaded slab with all edges fixed and with negative and positive ultimate moments of resistance per unit width m'_{ux} and m_{ux}, respectively, in the x-direction, and m'_{uy} and m_{uy}, respectively, in the y-direction. Let the reinforcement be placed as in Fig. 8.2. The area of steel per unit width for each moment is equal to β times the moment, where β is the reciprocal of the product of the yield strength of the steel and the internal lever arm. Strictly, β will not have the same value for each ultimate moment, but the assumption of constant β is sufficiently accurate for a study of economy. Examination of Fig. 8.2 shows that the volume of steel in the slab, neglecting the d or 12-bar-diameter extension to the top steel, is

$$V = \beta l_x l_y [m_{ux} + m_{uy} + 2\lambda(m'_{ux} + m'_{uy})] \tag{8.8}$$

To examine the effect of varying the ratio of the ultimate moments, let

$$\mu = \frac{m_{uy}}{m_{ux}} \quad \text{and} \quad i = \frac{m'_{ux}}{m_{ux}} = \frac{m'_y}{m_{uy}}$$

On substituting for m'_{uy}, m'_{ux}, and m_{ux} in terms of m'_{uy}, μ, and i, and putting λ equal the value given by Eq. 8.5, the volume of steel given by Eq. 8.8 becomes

$$V = \beta l_x l_y \left(1 + \frac{1}{\mu}\right)[1 + i - i(1 + i)^{-1/2}]m_{uy} \tag{8.9}$$

Also, from Eq. 8.3,

$$m_{uy} = \frac{w_u l_y^2}{24(1 + i)}\left\{3 + \frac{2}{\mu}\left(\frac{l_y}{l_x}\right)^2 - 2\frac{l_y}{l_x}\left[\left(\frac{l_y}{\mu l_x}\right)^2 + \frac{3}{\mu}\right]^{1/2}\right\}$$

which on substituting into Eq. 8.9 gives

$$V = \frac{w_u \beta l_y^4}{24}[1 - i(1 + i)^{-3/2}]\left(1 + \frac{1}{\mu}\right)\left\{3\frac{l_x}{l_y} + \frac{2l_y}{\mu l_x} - 2\left[\left(\frac{l_y}{\mu l_x}\right)^2 + \frac{3}{\mu}\right]^{1/2}\right\}$$

$$\tag{8.10}$$

For a slab with given dimensions and ultimate load, Eq. 8.10 gives the volume of steel in terms of i and μ. The values of i and μ for a minimum volume of steel will now be examined.

1. Considering i, $\partial V/\partial i = 0$ from Eq. 8.10 gives $i = 2.0$ for the minimum volume of steel. Using Eq. 8.10, it can be shown, however, that use of $i = 1.0$ in design will mean an increase in the steel content of only 4% above the minimum volume, and therefore that the volume of steel is relatively insensitive to changes in i.

2. Considering μ, $\partial V/\partial \mu = 0$ from Eq. 8.10 gives the following values of μ for a minimum volume of steel for slabs with various l_x/l_y ratios:

l_x/l_y	1.0	1.2	1.4	1.6	1.8	2.0	2.5	3.0
μ	1.0	2.3	3.9	5.7	7.7	10.0	16.7	25.0

Figure 8.10 shows the variation of the ratio $V/(V$ when $\mu = 1)$ with μ and l_x/l_y. That is, the figure shows the steel volume required for various μ values and l_x/l_y ratios as a proportion of the steel volume for the isotropic case. Figure 8.10 shows that as l_x/l_y increases from unity, the value of μ for minimum steel volume increases rapidly. It is evident that for many slabs with a large l_x/l_y ratio, the value of μ for minimum steel volume is so large that it cannot be used in practice because of the code requirement for minimum steel in one direction. For such slabs the best available design is obtained when the μ value used is as large as possible, that is, when the minimum steel

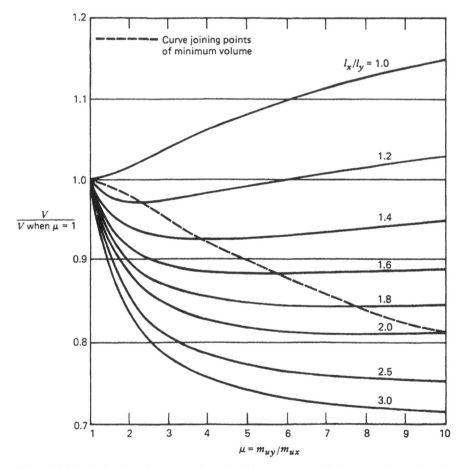

Figure 8.10 Variation of volume of steel with m_{uy}/m_{ux} and l_x/l_y ratios for uniformly loaded slabs with fixed edges.

allowable is placed in the direction of the long span. However, for slabs that are nearly square, use of the μ value for minimum volume is practical. It is of interest that Fig. 8.10 also shows that the volume of steel is relatively insensitive to changes in μ over a significant range of μ values, depending on the l_x/l_y ratio. For instance, if a 5% tolerance on the volume of steel were allowed, Wood[8.11] has shown that values of μ anywhere in the following ranges could be used:

l_x/l_y	1.0	1.25	1.5	2.0
μ	1–3.3	1–10	1.6–16	2.8–50

It becomes clear that for large l_x/l_y ratios, orthotropic steel is much more economical than isotropic steel, but that a wide range of m_{uy}/m_{ux} ratios will result in a nearly minimum volume.

It was emphasized previously that the distribution of the ultimate moments of resistance provided by the reinforcement should not differ too greatly from the distribution of moments given by elastic theory; otherwise, there may be excessive cracking at the service loads. It is fortunate that the values of i and μ for close to the minimum volume of steel do not show wide differences from elastic theory. In practical design the code requirements for minimum steel, and the suggested limitation on maximum steel ratio, are such that large values for μ will not be possible. In any case, from the point of view of serviceability, large values for m'_{uy}/m'_{ux} are undesirable for slabs with l_x/l_y ratios significantly greater than unity. Elastic theory (see Chapter 3) shows that for a constant short span l_y, as the long span l_x gets longer, the long-span positive moments reduce and the short-span positive and negative moments increase. However, the long-span negative moment does not reduce and remains about the same value as the negative moment in a square slab having the same span as the short span. Thus, in the design of rectangular slabs, to control cracking and for economy of steel, the long-span bottom steel content should be the minimum allowed by the code, the short-span bottom steel content should be greater (as required for strength), and i should be in the range 1 to 2 in the short span but greater in the long span.

Although in the discussion above only the case of a slab with all edges fixed has been considered, it is evident that for rectangular slabs with various combinations of fixed and simply supported edges, in order to control cracking and for economy of steel, the foregoing rules also apply approximately.

Banded Reinforcement. Wood[8.11] has shown that if a slab is reinforced non-uniformly to follow the elastic bending moment diagram exactly, the optimum minimum-weight solution is almost reached. A reinforcement arrangement that follows the elastic theory distribution of bending moments exactly is impracticable, however. One of the great advantages of limit design is that full use can be made of simple reinforcement arrangements because of moment redistribution. A compromise between elastic theory and uniformly distributed reinforcement is to place the bottom steel in bands.

For example, consider a simply supported uniformly loaded square slab. Let the length of each side be l. For uniform isotropic steel over the entire slab, Eq. 8.10 indicates that the volume of steel is

$$V = 0.0833 w_u \beta l^4 \tag{8.11}$$

Now consider the slab to have an edge band that has less steel in it than is present in the central region. Figure 8.11a shows the reinforcement; both meshes are isotropic. Let the positive ultimate moments of resistance per unit width be m_{u1} in the edge bands and m_{u2} in the central region. Figure 8.11b and c show the possible yield line patterns. Mode 1 is the basic mechanism. Mode 2, in which the square yield line forms at the edge of the central mesh, is the alternative mechanism that could form as a result of the banding.

Consider the center of the slab to be given a small downward displacement δ and apply the virtual work equation to both modes to find their ultimate loads:

Mode 1: $$w_u l^2 \frac{\delta}{3} = 4[m_{u1}(l - 2x) + m_{u2}2x] \frac{\delta}{0.5l}$$

$$w_u = \frac{24}{l^2} \left[m_{u1} \left(1 - \frac{2x}{l} \right) + m_{u2} \frac{2x}{l} \right] \tag{i}$$

Mode 2: $$w_u \left[(l - 2x)^2 \frac{\delta}{3} + 4(l - 2x)x \frac{\delta}{2} + 4x^2 \delta \right] = 4m_{u1} \frac{\delta}{0.5(l - 2x)} l$$

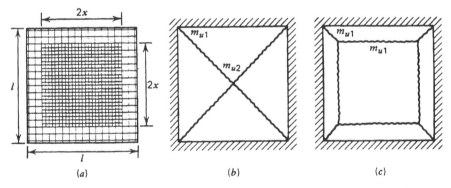

Figure 8.11 Simply supported square uniformly loaded slab with banded reinforcement: (a) Reinforcement; (b) mode 1; (c) mode 2.

$$w_u = \frac{24m_{u1}}{l^2[1 - 8(x/l)^3]} \tag{ii}$$

For a given x, the minimum volume of steel is when the ultimate loads of the two modes are the same. From Eq. ii,

$$m_{u1} = \frac{w_u l^2}{24} \left[1 - 8 \left(\frac{x}{l} \right)^3 \right] \tag{8.12}$$

Substituting m_{u1} into Eq. i and noting that the two ultimate loads are the same gives

$$m_{u2} = \frac{w_u l^2}{24} \left[1 + \left(\frac{x}{0.5l} \right)^2 - \left(\frac{x}{0.5l} \right)^3 \right] \tag{8.13}$$

Now the volume of steel in the banded slab is

$$V = 2\beta [m_{u1}(l^2 - 4x^2) + m_{u2}(2x)^2]$$

Substituting m_{u1} and m_{u2} from Eqs. 8.12 and 8.13 into V gives

$$V = \frac{\beta w_u l^4}{12} \left[1 - \left(\frac{x}{0.5l} \right)^3 + \left(\frac{x}{0.5l} \right)^4 \right]$$

For a minimum volume of steel, $dV/dx = 0$, giving $x = 0.375l$. Then

$$V = \frac{\beta w_u l^4}{12} \left[1 - \left(\frac{3}{4} \right)^3 + \left(\frac{3}{4} \right)^4 \right]$$
$$= 0.0745 w_u \beta l^4 \tag{8.14}$$

Comparison of Eqs. 8.11 and 8.14 shows that the banded slab requires only 89% of the steel of the nonbanded slab to carry the same ultimate load. Also, for a banded slab, when $x = 0.375l$ the ultimate moments of resistance per unit width are

from Eq. 8.12: $\qquad m_{u1} = 0.0241 w_u l^2 \tag{8.15}$

from Eq. 8.13: $\qquad m_{u2} = 0.0475 w_u l^2 \tag{8.16}$

Banded reinforcement must be used with caution, however, to ensure that an alternative yield line pattern does not govern the design.

Uniformly Loaded Rectangular Slab with Three Edges Supported and One Free. The ratios between the ultimate moments of resistance should be selected to give the least weight of steel, but they should not differ too much from the elastic theory distribution of bending moments, to ensure that cracking at the service load is not excessive. Such considerations indicate that $m'_{ux}/m_{ux} = m'_{uy}/m_{uy} = 1$ to 2 satisfy both requirements fairly well. For a minimum weight of steel, for l_x/l_y between 3 and 2 (see Fig. 7.40a and b), the m_{uy}/m_{ux} ratio should be about 0.5 or 1, but for $l_x/l_y < 2$, the m_{uy}/m_{ux} ratio becomes very small, and m_{uy} should usually be made as small as possible by putting the minimum steel allowable in the y-direction. (Note that m'_{uy}/m_{uy} should then be increased to ensure serviceability; that is, m'_{uy} and m'_{ux} should not be significantly different.)

Uniformly Loaded Rectangular Slab with Two Adjacent Edges Supported and the Remaining Edges Free. It is suggested that at fixed edges, values of $m'_{ux}/m_{ux} = m'_{uy}/m_{uy} = 1$ to 2 could be used. For square slabs, use $m_{ux} = m_{uy}$ for maximum economy. For slabs with $l_x > l_y$ in Eq. 7.98, use $m_{ux} = m_{uy}$ for case 1 of Fig. 7.69, and $m_{uy}/m_{ux} > 1$ (but not greater than 3) for cases 2, 3, and 4 of Fig. 7.69. Equation 7.66 should be used to check that the magnitude of the negative ultimate moments is sufficient. With one or two simply supported edges, Eq. 7.66 will usually indicate that additional top steel is required. As previously, to ensure serviceability, the values of m'_{ux} and m'_{uy} should not be too different.

8.4.3 Design Examples

Some example designs of two-way slabs are given below to illustrate the design procedure. These three slabs have also been designed using the strip method in Chapter 6 (see Examples 6.1, 6.2, and 6.3). In view of the conservative nature of test results compared with yield line theory predictions (see Section 7.17), the theoretical reduction in ultimate load due to corner fans, discussed in Section 7.7, will be ignored.

Example 8.2. A rectangular interior panel of a continuous slab and beam floor system has clear spans of 16 ft (4.88 m) and 24 ft (7.32m), as in Fig. 8.12. The panel carries a uniformly distributed service live load of 150 lb/in² (7.18 kN/m²). The concrete is of normal weight with a cylinder strength of 4000 lb/in² (27.6 N/mm²), and the steel has a yield strength of 60,000 lb/in² (414 N/mm²). Design a suitable panel.

SOLUTION

STIFFNESS REQUIREMENTS. Assuming stiff beams, the minimum slab thickness according to ACI 318-95[8.3] is given by Eq. 9.7, which for a slab aspect ratio of $\beta = 24/16 = 1.5$ gives $h = l_n/46.4 = 24 \times 12/46.4 = 6.21$ in. Use, say, a $6\frac{1}{2}$-in.-thick slab.

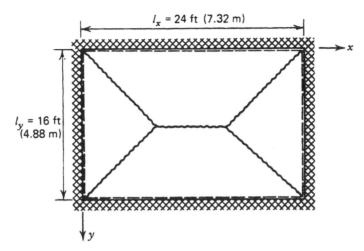

Figure 8.12 Design Examples 8.2 and 8.3.

STRENGTH REQUIREMENTS. Assuming that the unit weight of the concrete is 150 lb/ft³, the service dead load is $D = (6.5/12)150 = 81$ lb/ft². The service live load is $L = 150$ lb/ft². Therefore, the factored (ultimate) load according to the ACI Code, Eq. 8.1, is

$$w_u = 1.4 \times 81 + 1.7 \times 150 = 368 \text{ lb/ft}^2$$

Figure 8.12 shows the yield line pattern for the slab. The ultimate load of the slab is given with good accuracy by Eq. 7.95 as

$$w_u = \frac{24}{l_y^2[3(l_x/l_y) - 1]}\left[m'_{ux} + m_{ux} + \frac{l_x}{l_y}(m'_{uy} + m_{uy}) \right] \qquad (8.17)$$

The minimum amount of steel allowable is 0.0018 of the gross concrete section. This leads to $A_s = 0.0018 \times 6.5 = 0.0117$ in²/in width, requiring No. 3 bars on $0.11/0.0117 = 9.4$ in., say 9-in. centers, giving $A_s = 0.0122$ in²/in width. Using No. 3 bars, with $\frac{3}{4}$ in. of cover in the *y*-direction, $d = 6.5 - 0.94 = 5.56$ in., and in the *x*-direction $d = 6.5 - 1.31 = 5.19$ in. Placing minimum steel in the *x*-direction in the bottom of the slab, from Eq. 8.2,

$$m_{ux} = 0.9 \times 0.0122 \times 60{,}000 \left(5.19 - 0.59 \times 0.0122 \times \frac{60{,}000}{4000} \right)$$

$$= 3348 \text{ lb-ft/ft width}$$

If all slab ultimate moments have this value, then from Eq. 8.17,

$$w_u = \frac{24}{16^2(3 \times 1.5 - 1)} (2 \times 3348 + 1.5 \times 2 \times 3348) = 448 \text{ lb/ft}^2$$

which is higher than the required design load of 368 lb/ft². Therefore, minimum steel meets ultimate load requirements. The use of No. 3 bars on 9-in. centers both ways in the top and bottom would satisfy the required factored (ultimate) load.

The bottom steel is over the whole panel, but the top steel may be curtailed. From Eq. 8.5, $i = 1$ and

$$\lambda = 0.5 \left(1 - \frac{1}{\sqrt{2}}\right) = 0.146$$

Therefore, $\lambda l_x = 0.146 \times 24$ ft $= 3$ ft 6 in. and $\lambda l_y = 0.146 \times 16$ ft $= 2$ ft 4 in. Additional length beyond theoretical point of cutoff is d or 12 bar diameters, whichever is greater, giving 6 in. Hence, for the factored load required, the top steel should extend into the slab from each side at least a distance of 4 ft 0 in. in the x-direction and 2 ft 10 in. in the y-direction.

For the l_x/l_y ratio of 1.5 of this slab, Table 7.2 shows that the error from using the approximate equation 7.95 (i.e., Eq. 8.17) is only 1% in this case. Also, reference to Fig. 8.10 indicates that the value of $\mu = 1$ used does not result in the minimum volume of steel. The design of the slab has been governed very much by minimum steel requirements.

SERVICEABILITY CHECK FOR CRACKING. A serviceability check for cracking is particularly important because the ratios of steel areas in the top and bottom of the slab, and in the two directions used in this example (both unity), are so different from the elastic theory ratio of moments. The elastic theory distribution of moments for a fixed-edge slab of these dimensions (see Chapters 2 and 3) carrying a service load of 231 psf gives maximum x-direction moments of -3380 and $+880$ lb-ft/ft width and maximum y-direction moments of -4480 and $+2090$ lb-ft/ft width. These moments were found from the tabulation in Ref. 8.16 and assume Poisson's ratio to be 0.15. The critical moment is the y-direction negative moment. For the steel ratio of 0.0022 and a modular ratio of 8, elastic theory[8.15] for cracked sections indicates that the lever arm is $kd = 0.94d$. Hence, the maximum steel stress at service load for the foregoing reinforcement distribution is

$$f_s = \frac{M}{jdA_s} = \frac{4480}{(0.94 \times 5.56)(0.11/9)} = 70,100 \text{ psi}$$

which is greater than the specified yield strength. For the steel stress at service load not to exceed $0.8f_y$ (see Section 8.2.4) requires the service load moment not exceed $(4480 \times 0.8 \times 60,000/70,100) = 3070$ lb-ft/ft width. Hence, reduce the spacing of top steel in the short-span direction from 9 in. to $(9 \times 0.8 \times 60,000)/70,100 = 6.2$ in., say 6 in. For the top steel in the long-span direction the spacing needs to be reduced from 9 in to $9 \times 3070/3380 = 8.2$ in., say 8 in. For all other reinforcing positions, 9-in. spacing is adequate.

Since $f_y > 40,000 \text{ lb/in}^2$, a check to ensure that crack widths are not excessive is necessary (see Section 8.2.4) using Eq. 9.21. The maximum $t_b A$ for the slab is for the x-direction steel, where $t_b A = (1.31 \times 2)9 = 23.6 \text{ in}^3$. Now,

$$\frac{145,000}{\sqrt[3]{t_b A}} = \frac{145,000}{\sqrt[3]{23.6}} = 50,600 \text{ lb/in}^2$$

Therefore, according to Eq. 9.20*b*, the slab is satisfactory for exterior (or interior) exposure, since the maximum steel stress at service load is $0.8f_y$, which is smaller than 50,600 lb/in².

A further serviceability check concerns the length of top steel. The ratio of maximum negative to positive elastic (service load) bending moments is $4480/2090 = 2.14$ in the y-direction. The length of top steel in the y-direction may be found from the distribution of elastic bending moments. Approximately, assuming a parabolic distribution of bending moments in the y-direction, the value of $i = 2.14$ would, according to Eq. 6.9 (which is identical to Eq. 8.5), require a λ value of 0.218, leading to $\lambda l_y = 0.218 \times 16 = 3$ ft 6 in. Adding 6 in. for the additional length required beyond theoretical cutoff (d or 12 bar diameters), it is evident that top steel should extend into the slab in the y-direction from each side a distance of 4 ft. This requirement is greater than the length of top steel required for the factored load found previously and hence will govern. It is evident that the length of top steel in the x-direction required for the factored load (4 ft 0 in.) will satisfy serviceability, since the maximum x-direction negative moment is less than the maximum y-direction negative moment.

SUMMARY. The x-direction (long-span) reinforcing should be No. 3 bars on 8-in. centers in the top and 9-in. centers in the bottom. The y-direction reinforcing should be No. 3 bars on 6-in. centers in the top and No. 3 bars on 9-in. centers in the bottom. The top steel should extend 4 ft into the slab from each edge. Note that minimum steel in the bottom both ways, and $m'_{ux}/m_{ux} = 1.10$ and $m'_{uy}/m_{uy} = 1.45$, satisfy both strength and serviceability requirements in this example. Comparison with Example 6.1 shows that in the strip design, minimum steel governed everywhere except in the middle region of the short-span direction.

Example 8.3. Recalculate the reinforcement required for the panel of Example 8.2 if the live load to be carried is 400 lb/ft² (19.1 kN/m²).

SOLUTION

STRENGTH REQUIREMENTS. For a $6\frac{1}{2}$-in.-thick slab, the factored load becomes

$$w_u = 1.4 \times 81 + 1.7 \times 400 = 793 \text{ lb/ft}^2$$

Figure 8.12 shows the yield line pattern for the slab. As previously, minimum steel requirements are $A_s = 0.0117 \text{ in}^2/\text{in}$ width, equivalent to No. 3 bars on 9-in. centers.

Using No. 4 bars with $\frac{3}{4}$ in. of cover in the top in the y-direction, $d = 6.5 - 1.0 = 5.5$ in. Using No. 4 bars in the top in the x-direction, $d = 5.5 - 0.5 = 5.0$ in. Using No. 3 bars with $\frac{3}{4}$ in. of cover in the bottom in the y-direction, $d = 6.5 - 0.94 = 5.56$ in. Using No. 3 bars in the bottom in the x-direction, $d = 5.56 - 0.38 = 5.18$ in.

Following the considerations of Section 8.4.2, place the minimum steel in the direction of the long span in order to make μ as large as possible, and use an i value of 2.0. Placing minimum steel in the x-direction at the bottom of the slab (No. 3 bars on 9-in. centers), from Eq. 8.2,

$$m_{ux} = 0.9 \times 0.0122 \times 60,000 \left(5.18 - 0.59 \times 0.0122 \times \frac{60,000}{4000} \right)$$

$$= 3341 \text{ lb-ft/ft width}$$

Then $m'_{ux} = 2m_{ux} = 6682$ lb-ft/ft width and, from Eq. 8.2,

$$6682 = 0.9 \times A_s \times 60,000 \left(5.00 - 0.59 \times \frac{60,000}{4000} \right)$$

giving $A_s = 0.0259 \text{ in}^2/\text{in}$, requiring No. 4 bars on $0.2/0.0259 = 7.7$ in., say 7-in. centers, for x-direction top steel. Also, $m'_{uy} = 2m_{uy}$ and therefore, from Eq. 8.17,

$$793 = \frac{24}{16^2(3 \times 1.5 - 1)} (6682 + 3341 + 1.5 \times 3 \, m_{uy})$$

$$m_{uy} = 4352 \text{ lb-ft/ft width}$$

$$m'_{uy} = 8704 \text{ lb-ft/ft width}$$

Therefore, from Eq. 8.2 for y-direction bottom steel,

$$4352 = 0.9 \times A_s \times 60,000 \left(5.56 - 0.59 \times A_s \times \frac{60,000}{4000} \right)$$

giving $A_s = 0.0148 \text{ in}^2/\text{in}$, requiring No. 3 bars on $0.11/0.0148 = 7.4$ in., say 7-in. centers. And from Eq. 8.2 for y-direction top steel,

$$8704 = 0.9 \times A_s \times 60,0000 \left(5.5 - 0.59 \times A_s \times \frac{60,000}{4000} \right)$$

giving $A_s = 0.0308$ in^2/ in, requiring No. 4 bars on $0.20/0.0308 = 6.5$ in., say 6-in. centers.

As in Example 8.2, the bottom steel is placed over the whole panel, but the top steel may be curtailed. From Eq. 8.5, $i = 2$ and

$$\lambda = 0.5 \left(1 - \frac{1}{\sqrt{3}} \right) = 0.211$$

Therefore, $\lambda l_x = 0.211 \times 24 = 5$ ft 1 in. and $\lambda l_y = 0.211 \times 16 = 3$ ft 5 in. Additional length beyond the theoretical point of cutoff is d or 12 bar diameters, giving 6 in. Hence, the top steel should extend into the slab from each side at least a distance of 5 ft 7 in. in the x-direction and 3 ft 11 in. in the y-direction, for the required factored load.

Equation 8.10 could be used to show that when $l_x/l_y = 1.5$, for maximum economy of steel $\mu = 4.7$. Obviously, this value could not be used in Example 8.2 or 8.3. However, the actual value used in Example 8.3, $\mu = 4352/3341 = 1.30$, will only lead to about a 7% increase in steel volume, as shown by Fig. 8.10. Hence, the steel is still used fairly effectively. With heavily loaded slabs there is more opportunity to optimize the steel arrangement.

The maximum steel ratio used in Example 8.3 is $\rho = A_s/d = 0.0308/5.5 = 0.0056$, which is less than $0.4\rho_b = 0.0114$ according to Table 8.1.

SERVICEABILITY CHECK FOR CRACKING. The elastic theory distribution of moments for a fixed-edge slab of these dimensions (see Chapters 2 and 3) carrying a service load of 481 lb/ft^2 gives maximum x-direction moments of -7040 and $+1830$ lb-ft/ft width and maximum y-direction moments of -9320 and $+4350$ lb-ft/ft width. These moments were found from the tabulation in Ref. 8.16 and assume Poisson's ratio to be 0.15. The lever arm jd may be found using elastic theory[8.15] for cracked sections. A value for j of 0.9 is a reasonable approximation. The more exact lever arms found for the actual steel ratios and a modular ratio of 8 are used below. The maximum steel stresses found from $f_s = M/jdA_s$ are:

Top steel in x-direction: $f_s = \dfrac{7040}{(0.92 \times 5)(0.2/7)} = 53,600$ lb/in^2

Bottom steel in x-direction: $f_s = \dfrac{1830}{(0.94 \times 5.18)(0.11/9)} = 30,700$

Top steel in y-direction: $f_s = \dfrac{9320}{(0.87 \times 5.5)(0.2/6)} = 58,400$

Bottom steel in y-direction: $f_s = \dfrac{4350}{(0.94 \times 5.56)(0.11/7)} = 53,000$

Most of these maximum steel stresses exceed $0.8f_y = 0.8 \times 60,000 = 48,000$

lb/in^2 (see Section 8.2.4). Hence, reduce the spacing of steel to reduce the steel stresses. That is, make the reinforcing as follows:

Top steel in x-direction:	No. 4 bars on $7 \times \dfrac{48,000}{53,600}$
	= 6.3 in., say 6-in. centers
Bottom steel in x-direction:	No. 3 bars on 9-in centers
Top steel in y-direction:	No. 4 bars on $6 \times \dfrac{48,000}{58,400}$
	= 4.9 in., say 5-in. centers
Bottom steel in y-direction:	No. 3 bars on $7 \times \dfrac{48,000}{53,000}$
	= 6.3 in., say 6-in. centers

Since $f_y > 40,000$ lb/in^2, a check to ensure that crack widths are not excessive is necessary (see Section 8.2.4) using Eq. 9.20. The worst case for crack widths will be when $f_s \sqrt[3]{t_b A}$ is maximum, that is, for x-direction top steel where $t_b A = 1.5 \times 2 \times 6 = 18$ in^3. or for x-direction bottom steel where $t_b A = 1.3 \times 2 \times 9 = 23.6$ in^3. Therefore, maximum $f_s \sqrt[3]{t_b A} = 48,000 \times \sqrt[3]{23.6} = 137,600$ lb/in., which is adequate for both interior and exterior exposure according to Eq. 9.20.

Since the ratio of maximum elastic negative to positive moments in the y-direction is $9320/4350 = 2.14$, the length of top steel found from the factored load with $i = 2$ will be not quite sufficient at service load. Therefore, increase the length of top steel in the y-direction to 4 ft, as in Example 8.2. Inspection of the x-direction elastic bending moment diagram indicates that the required length of top steel in the x-direction for the elastic bending moments is satisfied by the length of top steel found for the factored load. Hence, the top steel should extend into the slab from each side a distance of 5 ft 7 in. in the x-direction and 4 ft in the y-direction.

SUMMARY. The x-direction (long-span) reinforcing should be No. 4 bars on 6-in. centers in the top extending 5 ft 7 in. into the slab from the edges and No. 3 bars on 9-in. centers in the bottom. The y-direction reinforcing should be No. 4 bars on 5-in. centers in the top, extending 4 ft into the slab from the edges and No. 3 bars on 6-in. centers in the bottom. Note that minimum steel in the bottom in the long span (x-direction), and $m_{uy}/m_{ux} = 1.44$, $m'_{ux}/m_{ux} = 2.55$, and $m'_{uy}/m_{uy} = 2.20$, satisfy both strength and service-ability requirements in this example.

Example 8.4. A 4-m (13.12-ft) by 6-m (19.68-ft) rectangular panel at the edge of a floor system is free (unsupported) along one long side and is continuous with adjacent panels at supporting beams along the other three sides.

The slab is shown in Fig. 8.13. The panel carries a uniformly distributed service live load of 5 kN/m² (104 lb/ft²). The concrete is of normal weight with a cylinder strength of 30 N/mm² (4350 lb/in²), and the steel has a yield strength of 300 N/mm² (43,500 lb/in²). Design a suitable panel. Grade 300 reinforcement is used instead of the more common Grade 420 to reduce the influence of the minimum steel area requirements. The metric bar series contained in the ASTM A615-96 specification will be used.

SOLUTION

STIFFNESS REQUIREMENTS. The minimum slab thickness can be computed conservatively by referring to Table 9.3. For both edges continuous and a span of 6.0 m, the minimum thickness allowed is $l/28 = 6000/28 = 214$ mm. Use, say, an 220-mm-thick slab.

STRENGTH REQUIREMENTS. Assuming that the unit weight of the concrete is 2400 kg/m³, the service dead load is $D = (0.22)2400(9.806$ N/kgf$) = 5178$ N/m² $= 5.2$ kN/m². The service live load is $L = 5.0$ kN/m². Therefore, the factored (ultimate) load according to the ACI Code, Eq. 8.1, is

$$w_u = 1.4 \times 5.2 + 1.7 \times 5.0 = 15.78 \text{ kN/m}^2$$

Figure 8.13 shows the yield line pattern for the panel. For the panel $l_x/l_y = 6/4 = 1.5$, and the ultimate load is given with good accuracy by Eq. 7.97 as

$$w_u = \frac{12}{l_x^2[(3(l_y/l_x) - 0.5]} \left[2\frac{l_y}{l_x}(m'_{ux} + m_{ux}) + m'_{uy} + m_{uy} \right] \quad (8.18)$$

Following the considerations of Section 8.4.2, place the minimum steel in

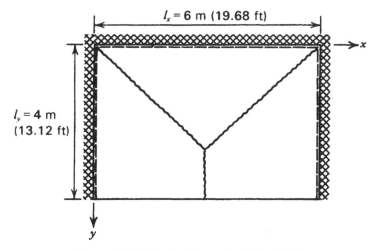

$l_x = 6$ m (19.68 ft)

$l_y = 4$ m (13.12 ft)

x

y

Figure 8.13   Design Examples 8.4 and 8.5.

the y-direction (perpendicular to the free edge) since the load is carried more effectively in the x-direction between the supported edges. Using 13-mm bars (area = 129 mm², diameter = 12.7 mm) with 20 mm clear cover in the x-direction, d = 220 − 20 − 6 = 194 mm. In the y-direction, d = 194 − 13 = 181 mm. The minimum amount of steel permitted is 0.002 of the gross section, giving A_s = 0.002 × 220 = 0.440 mm²/mm width, requiring 13-mm bars on 129/0.44 = 293 mm, say 300-mm centers, giving 0.430 mm²/mm.

Placing minimum steel in the y-direction at the bottom of the slab, from Eq. 8.2,

$$m_{uy} = 0.9 \times 0.43 \times 300 \left(181 - 0.59 \times 0.43 \times \frac{300}{30} \right)$$

$$= 20{,}720 \text{ N-mm/mm} = 20.72 \text{ kN-m/m width}$$

If all the panel ultimate moments have this value, then from Eq. 8.18,

$$w_u = \frac{12}{6^2[(3/1.5) - 0.5]} \left(\frac{2}{1.5} \times 2 \times 20.72 + 2 \times 20.72 \right) = 21.49 \text{ kN/m}^2$$

which is higher than the required design load of 15.78 kN/m². Therefore, minimum steel meets ultimate load requirements. The use of 13-mm bars on 300-mm centers both ways in the top and bottom would satisfy the required factored (ultimate) load.

The bottom steel is placed over the entire panel, but the top steel may be curtailed. From Table 8.4, case 4b, it is apparent that λ = 0.21 if i = 2. This value of λ is conservative for smaller i values. Therefore, λl_x = 0.21 × 6 m = 1.26 m, and $2\lambda l_y$ = 2 × 0.21 × 4 m = 1.68 m may be used. Additional length beyond the theoretical point of cutoff is d or 12 bar diameters, whichever is greater, giving 194 mm. Hence, the x-direction top steel extends into panel from each side a distance of 1.45 m, and the y-direction top steel extends into the panel from the fixed edge a distance of 1.87 m.

For the l_x/l_y ratio of 1.5 used for this slab, Table 7.3, case 4b, shows that the error from using the approximate Eq. 7.97 is only 2% in this case.

SERVICEABILITY CHECK FOR CRACKING. The elastic theory distribution of moments for a slab of these dimensions with one long side free (unsupported) and fixed along the other three sides (see Chapters 2 and 3) carrying a service load of 10.2 kN/m² gives maximum x-direction moments of −30.19 and +13.55 kN-m/m width and maximum y-direction moments of −20.07 and +3.10 kN-m/m width. These moments were found from the tabulation in Ref. 8.16 and assume Poisson's ratio to be 0.15. The critical moment is the x-direction negative moment. For the steel ratio of 0.0024 and a modular ratio of 8, elastic theory[8.15] for a cracked section indicates that the lever arm is $0.943d$. Hence, the steel stress at the service load for the foregoing reinforcement distribution is

$$f_s = \frac{M}{jdA_s} = \frac{30190}{(0.943 \times 194)(0.43)} = 383.8 \text{ N/mm}^2$$

which is greater than the specified yield strength. For the steel stress at service load not to exceed $0.8f_y$ (see Section 8.2.4) requires the service load moment not to exceed $(30.19 \times 0.8 \times 300/383.8) = 18.88$ kN-m/m width. Hence, reduce the spacing of the top steel in the x-direction from 300 mm to $(300 \times 0.8 \times 300/383.8) = 188$ mm, say 185 mm. The distribution of x-direction negative moment shows that this reduced spacing (i.e., where the moment exceeds 18.88 kN-m/m width) is necessary over the 2-m length of each fixed side adjacent to the unsupported edge. It can also be shown that the y-direction negative moment causes $f_s = 274$ N/mm^2, which is more than $0.8f_y$. This leads to a requirement for 13-mm bars at 260 mm for the y-direction top steel within the central 2-m width of the slab. It is evident that the steel stress at service load is equal to or less than $0.8f_y$ elsewhere in the x- and y-directions in the slab. Since $f_y = 300$ N/mm^2, a check to ensure that crack widths are not excessive is unnecessary (see Section 8.2.4).

A further serviceability check concerns the length of top steel. The ratio of the maximum negative to positive (elastic) service load bending moments is $30.19/13.55 = 2.23$ in the x-direction. Approximately, assuming a parabolic distribution of bending moments in the x-direction, the value of $i = 2.23$ would, according to Eq. 6.9 (which is identical to Eq. 8.5), require that $\lambda l_x = 0.222 \times 6 = 1.33$ m. Adding 194 mm for the additional length required beyond theoretical cutoff (d or 12 bar diameters), it is evident that the x-direction top steel should extend into the slab from each support a distance 1.53 m, which exceeds the length found for factored load requirements. The elastic bending moment diagram shows that the length of top steel in the y-direction found for factored load requirements is satisfactory.

SUMMARY. The reinforcing should be 13-mm bars on 300-mm centers both ways in the top and the bottom, except that in the x-direction in the top over the 2-m length of each fixed edge adjacent to the unsupported edge, the spacing should be 185 mm. The y-direction top steel should be spaced at 260 mm in the central 2-m width of the slab. The top steel should extend 1.53 m into the slab in the x-direction and 1.87 into the slab in the y-direction from the fixed edge. Note that minimum steel in the bottom both ways, and maximum $m'_{ux} = /m_{ux} = 1.5$ and $m'_{uy}/m_{uy} = 1.0$ at the fixed edges, satisfy both strength and serviceability requirements in this example. Comparison with Example 6.2 shows that the strip design has resulted in steel requirements that are close to those given above, and in a concentration of reinforcement in a strong band along the free edge (in the x-direction).

Example 8.5. Recalculate the reinforcement required for the panel of Example 8.4 if the live load to be carried is 20 kN/m².

SOLUTION

STRENGTH REQUIREMENTS. For an 220-mm-thick slab, the factored (ultimate) load becomes

$$w_u = 1.4 \times 5.2 + 1.7 \times 20 = 41.28 \text{ kN/m}^2$$

Figure 8.13 shows the yield line pattern for the slab. In the top of the slab, using 19-mm bars (area = 284 mm², diameter = 19.1 mm) with 20 mm cover, in the *x*-direction $d = 220 - 20 - 10 = 190$ mm and in the *y*-direction $d = 190 - 19 = 171$. In the bottom of the slab, using 13-mm bars with 20 mm of cover in the *x*-direction $d = 220 - 20 - 6 = 194$ in and in the *y*-direction $d = 194 - 13 = 181$ mm.

Following the considerations of Section 8.4.2, place the minimum steel in the *y*-direction to make the *x*-direction moment as large as possible, and use $m'_{uy} = 2m_{uy}$ and $m'_{ux} = 2\ m_{ux}$. Therefore, the bottom steel in the *y*-direction is 13-mm bars on 300-mm centers, giving from Eq. 8.2, $m_{uy} = 20.72$ kN-m/m width, as previously. Therefore, $m'_{uy} = 2 \times 20.72 = 41.44$ kN-m/m = 41,440 N-mm/mm width and, from Eq. 8.2,

$$41{,}440 = 0.9 \times A_s \times 300 \left(171 - 0.59 \times A_s \times \frac{300}{30} \right)$$

giving $A_s = 0.927$ mm²/mm. Hence, the *y*-direction top steel is 19-mm bars on 284/0.927 = 306 mm, say 300-mm centers.

Substituting for m'_{uy} and m_{uy} and putting $m'_{ux} = 2\ m_{ux}$ in Eq. 8.18 gives

$$41.28 = \frac{12}{6^2(3/1.5 - .05)} \left(\frac{2}{1.5} 3m_{ux} + 41.44 + 20.72 \right)$$

requiring $m_{ux} = 30.90$ kN-m/m width, and therefore $m'_{ux} = 2 \times 30.90 = 61.80$ kN-m/m width.

From Eq. 8.2, for the *x*-direction bottom steel,

$$30{,}900 = 0.9 \times A_s \times 300 \left(194 - 0.59 \times A_s \times \frac{300}{30} \right)$$

giving $A_s = 0.601$ mm²/mm, requiring 13-mm bars on 129/0.301 = 215-mm centers.

From Eq. 8.2, for the x-direction top steel,

$$61,800 = 0.9 \times A_s \times 300 \left(190 - 0.59 \times A_s \times \frac{300}{30}\right)$$

giving $A_s = 1.253$ mm^2/mm, requiring 19-mm bars on $284/1.253 = 227$ mm, say 225-mm centers.

As in Example 8.4, the bottom steel is over the whole panel, but the x-direction top steel extends into the panel from each side a distance of 1.53 m and the y-direction top steel extends into the panel from the fixed edge a distance of 1.87 m.

Since $m'_{ux} = 2m_{ux}$, $m'_{uy} = 2m_{uy}$, and $m_{ux}/m_{uy} = 30.90/20.72 = 1.49$, the steel arrangement given above is reasonable from the point of view of the elastic distribution of moments and of economy. The maximum steel ratio used is $\rho = A_s/d = 1.253/190 = 0.00660$, which is less than $0.4\rho_b = 0.0198$ according to Table 8.1.

SERVICEABILITY CHECK FOR CRACKING. The elastic theory distribution of moments for a slab of these dimensions with one long side free (unsupported) and fixed along the other three sides (see Chapters 2 and 3), carrying a service load of 25.2 kN/m^2, gives maximum x-direction moments of -74.59 and $+33.47$ kN-m/m width and maximum y-direction moments of -49.59 and $+7.66$ kN-m/m width. These moments were found from the tabulation in Ref. 8.16 and assume Poisson's ratio to be 0.15. The lever arm jd may be found from elastic theory[8.15] for cracked sections. A value of $j = 0.9$ is a reasonable approximation. The maximum steel stresses found from $f_s = M/0.9dA_s$ are:

Top steel in x-direction: $\quad f_s = \dfrac{74,590}{(0.9 \times 190)(1.262)} = 346$ N/mm^2

Bottom steel in x-direction: $\quad f_s = \dfrac{33,470}{(0.9 \times 194)(0.601)} = 319$ N/mm^2

Top steel in y-direction: $\quad f_s = \dfrac{49,590}{(0.9 \times 171)(0.947)} = 340$ N/mm^2

Bottom steel in y-direction: $\quad f_s = \dfrac{7,660}{(0.9 \times 181)(0.430)} = 109$ N/mm^2

Most of these steel stresses exceed $0.8f_y = 0.8 \times 300 = 240$ N/mm^2 (see Section 8.2.4). Hence, reduce the spacing of the steel where necessary to reduce the steel stresses. That is, make the reinforcing as follows:

Top steel in x-direction: 19-mm bars on $225 \times \dfrac{240}{346}$

$$= 156 \text{ mm, say 150-mm centers}$$

Bottom steel in x-direction: 13-mm bars on $215 \times \dfrac{240}{319}$

$$= 169 \text{ mm, say 170-mm centers}$$

Top steel in y-direction: 19-mm bars on $300 \times \dfrac{240}{340}$

$$= 212 \text{ mm, say 210-mm centers}$$

Bottom steel in y-direction: 13-mm bars on 300-mm centers

Examination of the elastic bending moment distribution shows that this reduced spacing in the x-direction need take place only over the end 2 m nearest the free edge. Elsewhere, the spacing found from the factored load requirements results in steel stresses less than $0.8f_y$ at service load. Since $f_y = 300 \text{ N/mm}^2$, a check to ensure that crack widths are not excessive is unnecessary (see Section 8.2.4).

A further serviceability check concerns the length of top steel. As in Example 8.4, the elastic theory moments in the x-direction give $i = 2.23$, which will govern the length of top steel and require top steel extending 1.53 m from each edge. In the y-direction the length of top steel found for the factored load requirements is satisfactory.

SUMMARY. The reinforcing in the top should be 19-mm bars in the x-direction on 150-mm centers and in the y-direction on 210-mm centers; this spacing could be increased to 300-mm centers for the x-direction bars outside the end 2 m of edge nearest the free edge. The reinforcing in the bottom should be 13-mm bars in the x-direction on 170-mm centers and in the y-direction on 300-mm centers; this spacing could be increased to 215-mm centers for the x-direction bars outside the end 2 m of slab nearest the free edge. The top steel should extend 1.53 m into the slab in the x-direction and 1.87 m into the slab in the y-direction from the fixed edge. Note that minimum steel in the bottom in the y-direction and maximum $m_{ux}/m_{uy} = 2.0$, and maximum $m'_{ux}/m_{ux} = 2.35$ and $m'_{uy}/m_{uy} = 2.87$ at the fixed edges, satisfy both strength and serviceability requirements in this example.

Example 8.6. A corner panel of a floor system is continuous with adjacent panels at supporting beams along two edges and simply supported at the other two edges except for a rectangular opening that is unsupported at its edges. Figure 8.14a shows the panel. The service loads are a uniformly distributed live load of 100 lb/ft² (4.78 kN/m²) and a line load of 300 lb/ft (4.38 kN/m) positioned as shown in the figure. The concrete is of normal weight with

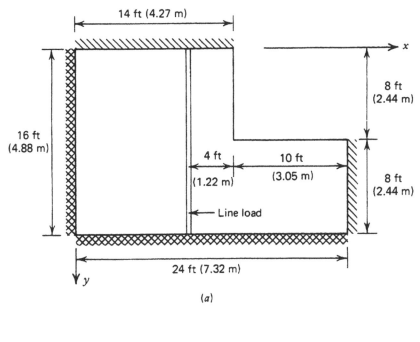

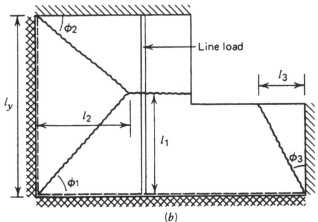

Figure 8.14 Design Example 8.6: (*a*) slab; (*b*) general form of yield line pattern; (*c*) assumed yield line pattern 1. (*d*) assumed yield line pattern 2.

a cylinder strength of 4000 lb/in² (27.6 N/mm²), and the steel has a yield strength of 40,000 lb/in² (276 N/mm²). Design a suitable panel.

SOLUTION
STIFFNESS REQUIREMENTS. The minimum thickness can be computed conservatively by referring to Table 9.3. For one edge continuous and a span

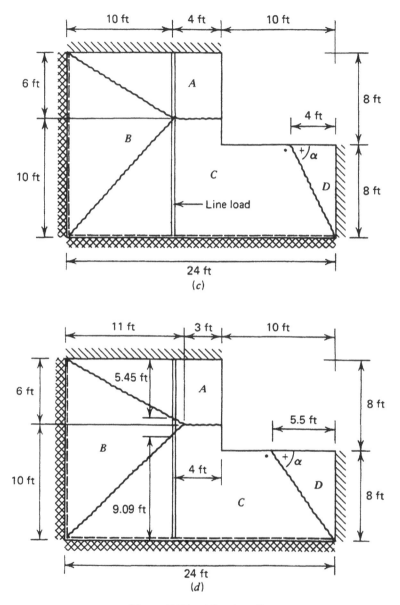

Figure 8.14 (*Continued*).

of 16 ft, the minimum thickness allowed is $l/24 = 16 \times 12/24 = 8$ in. Use, say, an 8-in.-thick slab.

STRENGTH REQUIREMENTS. Assuming that the unit weight of the concrete is 150 lb/ft³, the service dead load is $D = (8/12)150 = 100$ lb/ft². The service live loads L are 100 psf uniformly distributed and a 300-lb/ft line load. Therefore, the factored (ultimate) loads according to ACI 318-95, Eq. 8.1, are

$$w_u = 1.4 \times 100 + 1.7 \times 100 = 310 \text{ lb/ft}^2$$

and a line load

$$P_u = 1.7 \times 300 = 510 \text{ lb/ft}$$

No standard solution exists for a slab with these boundary conditions, opening size, and loading arrangement. Hence, the ultimate load and moments need to be related by an analysis from first principles. The general form of the yield line pattern is shown in Fig. 8.14b. There are three unknown dimensions, which locate the positions of the yield lines, l_1, l_2, and l_3. As a general rule, openings (being regions of zero strength in the slab) tend to attract yield lines (see Section 7.12), and hence l_1 is such that the horizontal yield line runs into the opening rather than passing below the opening. The combination of fixed and simply supported edges also forces l_1 to be greater than $l_y/2$. A direct design solution would involve either writing four equilibrium equations and solving them simultaneously to eliminate the unknown dimensions and to determine the ultimate moments required to support the design load, or writing the virtual work equation and determining the values of the unknown dimensions that lead to the maximum ultimate moments for the design load, again requiring the solution of four simultaneous equations. To avoid the lengthy algebra involved in such solutions, the trial-and-error method outlined in Section 7.16 forms a more convenient design approach. In this approach the positions of the yield lines are estimated and the equilibrium method is used to investigate the correctness of the assumed yield line pattern. When the yield lines have been found to lie reasonably close to the exact location, the virtual work equation is used to obtain a close estimate of the required design moments. The trial-and-error method is demonstrated below. The slab will be assumed to be isotropically reinforced with equal ultimate positive and negative moments of resistance per unit width $m_u' = m_u$, respectively.

As a first trial, assume that the yield line pattern is as shown in Fig. 8.14c. The dimension l_1 is chosen to be $\frac{5}{8}l_y$ since the simply supported edge tends to attract the positive-moment yield line and a fixed edge to repel it. Dimensions l_1 and l_2 are made the same, because symmetry suggests that because the two adjacent edges are fixed, the corner angle ϕ_1 is approximately equal to 45° Note that ϕ_2 is less than 45° because a corner yield line between a fixed and simply supported edge will tend to be attracted toward the simply supported edge. Similarly, the corner angle ϕ_3 is also less than 45°. The assumed values for l_1, l_2, and l_3 are shown on Fig. 8.14c. The nodal forces at the free edge between segments C and D are, according to Eq. 7.20, $\pm m_u$

$\cot \alpha = \pm m_u \,(4/8) = \pm 0.5 m_u$, acting downward in the acute angle and upward in the obtuse angle. The nodal forces elsewhere within the slab are zero. The equilibrium of each segment will be considered by taking moments about the supports as axes. Use will be made of Eq. 7.99, and $m_u' = m_u$ will be substituted.

Segment A: $m_u \cdot 14 = 310(2 \times 4 + 14)\dfrac{6^2}{6} + 510 \times \dfrac{6^2}{2}$

$$m_u = 3579 \text{ lb-ft/ft}$$

Segment B: $(m_u' + m_u)16 = 310 \times 16 \times \dfrac{10^2}{6}$

$$m_u = 2583 \text{ lb-ft/ft}$$

Segment C: $m_u' \cdot 24 + m_u(14 + 4) + 0.5\, m_u \cdot 8 = 310(2 \times 4 + 14)\dfrac{10^2}{6}$

$$+ 310\,(2 \times 6 + 10)\dfrac{8^2}{6} + 510 \times \dfrac{10^2}{2}$$

$$m_u = 4607 \text{ lb-ft/ft}$$

Segment D: $m_u \cdot 8 - 0.5 m_u \cdot 4 = 310 \times 8 \times \dfrac{4^2}{6}$

$$m_u = 1102 \text{ lb-ft/ft}$$

If the correct positions of the yield lines had been assumed, the m_u value calculated for each of the four segments would be identical.

As a second trial the yield lines will be shifted so as to obtain better agreement between the m_u values from each segment. The m_u value from segment C needs to be decreased, and those from segments B and D need to be increased. Assume the dimensions shown in Fig. 8.14d. The nodal force at the free edge between segments C and D is $\pm m_u\,(5.5/8) = \pm 0.688 m_u$. The equilibrium equations are rewritten below.

Segment A: $m_u \cdot 14 = 310(2 \times 3 + 14)\dfrac{6^2}{6} + 510 \times \dfrac{5.45^2}{2}$

$$14 m_u = 44{,}770$$

$$m_u = 3198 \text{ lb-ft/ft}$$

Segment B: $(m_u' + m_u)16 = 310 \times 16 \times \dfrac{11^2}{6} + 510 \times 1.46 \times 10$

$$32 m_u = 107{,}470$$

$$m_u = 3358 \text{ lb-ft/ft}$$

Segment C: $m'_u \cdot 24 + m_u(14 + 5.5) + 0.688m_u \times 8$

$$= 310(2 \times 3 + 14)\,\frac{10^2}{6} + 310(2 \times 4.5 + 10)\,\frac{8^2}{6}$$

$$+ 510 \times \frac{9.09^2}{2}$$

$$49.00m_u = 187{,}230$$

$$m_u = 3821 \text{ lb-ft/ft}$$

Segment D: $m_u \cdot 8 - 0.688m_u(5.5) = 310 \times 8 \times \dfrac{5.5^2}{6} = 12{,}500$

$$4.216m_u = 12{,}500$$

$$m_u = 2965 \text{ lb-ft/ft}$$

The agreement between the m_u values is close enough for a very good estimate of the exact m_u to be obtained from the virtual work method using the second assumed yield line pattern. From Eqs. 7.7 and 7.101, the virtual work equation may be written as

$$\Sigma\, m_{ux}\theta_x y_0 + \Sigma\, m_{uy}\theta_y x_0 = \Sigma\, W_u(\bar{x}\theta_x + \bar{y}\theta_y)$$

Note that the numerical values for the terms $W_u\bar{x}$ and $W_u\bar{y}$ have already been obtained in the equilibrium equations. Let the junction point of segments A, B, and C be given a small downward displacement δ. The tip of segment D deflects downward by 0.8δ. Some of the virtual work terms are tabulated below.

Segment	θ_x	θ_y	$W_u\bar{x}$	$W_u\bar{y}$
A	—	$\dfrac{\delta}{6}$	—	44,770
B	$\dfrac{\delta}{11}$	—	107,470	—
C	—	$\dfrac{\delta}{10}$	—	187,230
D	$\dfrac{0.8\delta}{5.5}$	—	12,500	—

Therefore, the virtual work equation is

$$(m'_u + m_u) \left(\frac{\delta}{11}\right) 16 + m_u \left(\frac{0.8\delta}{5.5}\right) 8 + m_u \left(\frac{\delta}{6}\right) 14 + (m'_u \cdot 24 + m_u \cdot 19.5) \frac{\delta}{10}$$

$$= 107{,}470 \frac{\delta}{11} + 12{,}500 \frac{0.8\delta}{5.5} + 44{,}770 \frac{\delta}{6} + 187{,}320 \frac{\delta}{10}$$

$$m'_u = m_u = 3510 \text{ lb-ft/ft width}$$

Using No. 4 bars with $\frac{3}{4}$ in. of cover in the y-direction, $d = 8 - 1 = 7$ in. In the x-direction, $d = 8 - 1.5 = 6.5$ in. The minimum amount of steel permitted is 0.002 of the gross section, requiring $A_s = 0.002 \times 8 = 0.016$ in.²/in. width and giving, from Eq. 8.2,

$$m_u = 0.9 \times 0.016 \times 40{,}000 \left(6.5 - 0.59 \times 0.016 \times \frac{40{,}000}{4000}\right)$$

$$= 3690 \text{ lb-ft/ft width}$$

Therefore, minimum steel suffices for reinforcement, that is, No. 4 bars on $0.2/0.016 = 12.5$ in., say 12-in. centers. Therefore, place No. 4 bars on 12-in. centers both ways over the entire bottom of panel and in the top extending into the span at the two fixed edges.

As an approximation, case 4 of Table 8.2 indicates that the top steel should extend into the span by 0.29 times the span. To this length should be added d or 12 bar diameters, whichever is greater. Therefore, the length of x-direction top steel from the edge is 0.29×24 ft $+ 7$ in. $= 7$ ft 7 in. and the length of y-direction top steel from the edge is 0.29×16 ft $+ 7$ in. $= 5$ ft 3 in.

SERVICEABILITY CHECK FOR CRACKING. The serviceability check for cracking is particularly important since the ratios of the steel areas in the top and bottom of the slab, and in the two directions, used in this example (both unity) are so different from the elastic theory ratios of moments. The elastic theory distribution of moments for this particular slab carrying the service load of 200 lb/ft² uniformly distributed and 300-lb/ft line load were given in Chapter 2 (see Section 2.5 and Fig. 2.11). The maximum service load x-direction moments are -3860 and $+ 1590$ lb-ft/ft width, and the maximum service load y-direction moments are -5740 and $+3750$ lb-ft/ft width, at the service load. Hence, the critical moment is the y-direction negative moment. For a steel ratio of 0.0024 and a modular ratio of 8, elastic theory[8.14] for a cracked section indicates that the lever arm is $0.94d$. Hence, the maximum steel stresses at the service load for the foregoing reinforcement distribution found from $f_s = M/0.94dA_s$ are

Top steel in x-direction: $f_s = \dfrac{3860}{(0.94 \times 6.5)(0.20/12)}$

$= 37{,}900 \ \text{lb/in}^2$

Bottom steel in x-direction: $f_s = \dfrac{1590}{(0.94 \times 6.5)(0.20/12)}$

$= 15{,}600 \ \text{lb/in}^2$

Top steel in y-direction: $f_s = \dfrac{5740}{(0.94 \times 7)(0.20/12)}$

$= 52{,}300 \ \text{lb/in}^2$

Bottom steel in y-direction: $f_s = \dfrac{3750}{(0.94 \times 7)(0.20/12)}$

$= 34{,}200 \ \text{lb/in}^2$

Most of these steel stresses exceed $0.8f_y = 0.8 \times 40{,}000 = 32{,}000 \ \text{lb/in}^2$ (see Section 8.2.4). Hence, reduce the spacing of the steel where necessary to reduce the steel stresses. That is, make the reinforcing No. 4 bars at the following spacings:

Top steel in x-direction: $12 \times \dfrac{32{,}000}{37{,}900} = 10.1$ in., say 10-in. centers

Bottom steel in x-direction: 12-in. centers

Top steel in y-direction: $12 \times \dfrac{32{,}000}{52{,}300} = 7.3$ in., say 7-in. centers

Bottom steel in y-direction: $12 \times \dfrac{32{,}000}{34{,}200} = 11.2$ in., say 11-in. centers

Examination of the elastic bending moment distribution (see Fig. 2.11b) shows that this reduced top steel spacing for the x-direction reinforcement could be placed over the central 6 ft of the fixed short side and for the y-direction reinforcement could be placed over the central 18 ft of the fixed long side. Elsewhere, the spacing found from the factored load requirements results in steel stresses less than $0.8f_y$ at service load. Since $f_y = 40{,}000 \ \text{lb/in}^2$, a check to ensure that the crack widths are not excessive is unnecessary (see Section 8.2.4).

A further serviceability check concerns the length of top steel. Inspection of the elastic bending moment distribution shows that the length of top steel at the fixed edges given by $\lambda = 0.29$, chosen in the factored load solution, is satisfactory.

SUMMARY. The reinforcing in the bottom should be No. 4 bars on 12-in. centers for the *x*-direction bars and 11-in. centers for the *y*-direction bars. The reinforcing in the top in the *x*-direction at the fixed side is No. 4 bars on 10-in. centers; this spacing could be increased to 12-in. centers outside the central 6 ft of fixed side. The reinforcing in the top in the *y*-direction at the fixed side is No. 4 bars on 7-in. centers; this spacing could be increased to 12-in. centers outside the central 18 ft of the fixed side. The top steel should extend 7 ft 7 in. into the slab in the *x*-direction, and 5 ft 3 in. into the slab in the *y*-direction, from the fixed edges. Note that minimum steel in the bottom in the *x*-direction and $m_{ux}/m_{uy} = 0.92$, and maximum $m'_{ux}/m_{ux} = 1.2$ and $m'_{uy}/m_{uy} = 1.57$ at the fixed edges, satisfy both strength and serviceability requirements in this example.

Comparison with Example 6.3 shows that the strip design resulted in concentrations of reinforcement in strong bands across the slab and around the edges of the opening, and a concentration of *y*-direction negative-moment steel along part of the fixed edge, but elsewhere the steel is No. 4 bars on 12-in. centers.

Concluding Remarks Concerning Ratios of Ultimate Moments in Design Examples 8.2 to 8.6. These design examples illustrate that if the ratios of the ultimate moments per unit width follow the values recommended in Sections 8.2.3 and 8.4.2 (i.e., ratio of negative to positive moment of about 2 for fixed edges in the direction of greatest moment, the negative moments in the *x*- and *y*-directions at fixed edges about equal, and positive moments in the *x*- and *y*-directions to follow approximately the elastic moments with one of them provided by minimum steel), the slabs should satisfy serviceability requirements for steel stresses and cracking without difficulty. Hence, if those recommendations are followed, it should not be necessary to check steel stresses and crack widths at service loads except in unusual cases.

8.5 DESIGN OF BEAMLESS SLABS

Yield line patterns for uniformly loaded beamless floors were considered in Section 7.11. It is evident that these patterns could be used to design flat slab and flat plate floors. For the folding type of collapse mechanisms shown in Fig. 8.15, the approximate ultimate load equations are given by analogy with Eqs. 7.70 and 7.71 as

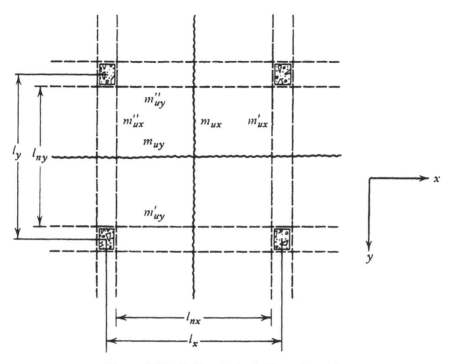

Figure 8.15 Uniformly loaded beamless slab.

$$\frac{1}{2}(m''_{ux} + m'_{ux})l_y + m_{ux}l_y = w_u l^2_{nx} \frac{l_y}{8} \qquad (8.19)$$

$$\frac{1}{2}(m''_{uy} + m'_{uy})l_x + m_{uy}l_x = w_u l^2_{ny} \frac{l_x}{8} \qquad (8.20)$$

where the reinforcement is uniformly spread and the ultimate moments per unit width are as shown on the yield lines in the figure.

Equations 8.19 and 8.20 could be used to design the reinforcement for the slabs. However, if the reinforcement is uniformly spread, the structure may not be serviceable, because elastic theory shows that the negative moments in the slab near the columns are very high, and yielding of steel may occur there under the service load. To ensure that yielding of steel does not occur in the slab at the service load, the reinforcement should be spread to follow the elastic bending moment diagram approximately. The European Concrete Committee[8.7] has recommended that to approximate the elastic moment diagram, the ultimate moments of resistance per unit width of slab in each direction should be distributed to follow either arrangement 1 or arrangement 2 of Fig. 8.16. Thus, the positive moment steel is spread uniformly or nearly

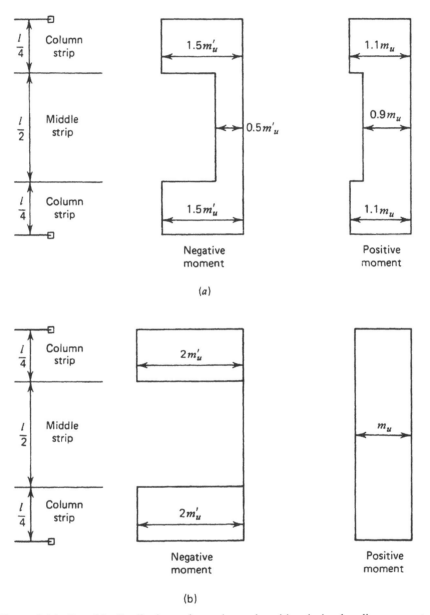

Figure 8.16 Possible distributions of negative and positive design bending moments: (*a*) arrangement 1; (*b*) arrangement 2.

uniformly across the panel, whereas the negative moment steel is concentrated entirely or primarily in the column strips. The European Concrete Committee[8.7] has also recommended that the ratio of average ultimate negative to average ultimate positive moments should be in the range 1.0 to 1.5. Choice of the ratio of average ultimate negative to positive moment allows the absolute values of these average ultimate moments to be found from Eqs. 8.19 and 8.20 for a particular design load and spans. Then these average ultimate moments are modified to follow one of the arrangements shown in Fig. 8.16.

Although there have been tests that generally support the use of arrangement 2 of Fig. 8.16, this distribution of steel in the negative-moment regions will be regarded by many designers to be extreme. If cracking occurs due to negative moment in the middle strip, it will be uncontrolled. Hence, for negative-moment steel, arrangement 1 should be preferred, but the uniform distribution of positive moment steel is reasonable. Also, the ratio of average ultimate negative to average ultimate positive moment should preferably be 1.5 or more rather than 1.0.

It is evident that with the concentration of reinforcement in column strips recommended above, the possibility of failure by one of the local yield line patterns discussed in Section 7.11.2 is remote. Nevertheless, the designer should check this possibility and also, if edge beams are present, check the possibility of failure by one of the yield line patterns discussed in Section 7.11.4. The shear strength of the slab–column connection is often critical and may govern the slab thickness and column size. The shear strength of slabs is discussed in Chapter 10.

No further discussion of this design approach for beamless slabs will be given here because it in fact is similar in principle to the direct design method of the ACI Code.[8.3] Equations 8.19 and 8.20 give the static moment M_0 for each panel based on the clear span as in the ACI Code, and the rest of the procedure is concerned with how this static moment should be distributed between the negative- and positive-moment sections and across the panels. For floors on a square column grid, the ACI distribution of moments is similar to arrangement 1 of Fig. 8.16 except that for positive moments, $1.2m_u$ is distributed to the column strips and $0.8m_u$ to the middle strips. The ACI Code also recommends that the ratio of average ultimate negative to average ultimate positive moment should be 1.86. However, 10 percentage redistribution from these recommended moments is allowed by the ACI Code. Yield line theory for beamless slabs is valuable in giving the designer an insight into the possible failure mechanisms that could occur.

Example 8.7. A square beamless slab with sides of length $2l$, forming the roof of a tank, is simply supported around all four sides on walls and at the center by a column. The slab carries a uniformly distributed design load per unit area w_u. Determine for the slab the design negative and positive moments per unit width m_u' and m_u, respectively.

SOLUTION. The slab and the yield line pattern are shown in Fig. 8.17. The central column is assumed to have zero width. Note that the nodal forces acting at the junction of the three positive-moment yield lines and the negative-moment yield line cannot be found by the rules derived in Section 7.4. The virtual work method will be used to relate the design load and moments.

The yield line pattern has one unknown dimension, shown in Fig. 8.17b as l_1. Let $ABCD$ be given a small displacement δ downward. Then

$$\text{internal work done} = 4m_u \frac{\delta}{l_1} 2l + 4(m'_u + m_u) \frac{\delta}{l - l_1} 2(l - l_1)$$

$$= 8m_u \delta l / l_1 + 8(m'_u + m_u)\delta$$

$$\text{external work done} = 4w_u l_1^2 \frac{\delta}{3} + 4w_u 2l_1(l - l_1) \frac{\delta}{2} + 4w_u(l - l_1)^2 \frac{2\delta}{3}$$

$$= \frac{4}{3} w_u \delta l(2l - l_1)$$

Therefore, from the virtual work equation,

$$w_u = 6 \frac{m'_u l_1 + m_u(l + l_1)}{l l_1(2l - l_1)} \tag{8.21}$$

For minimum w_u, $dw_u/dl_1 = 0$, which gives

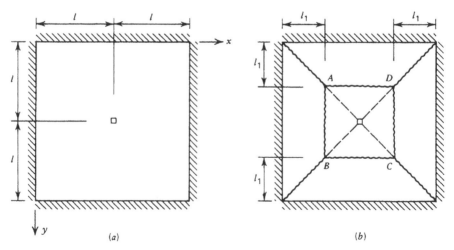

Figure 8.17   Design Example 8.7: (a) slab; (b) yield line pattern.

$$\frac{l_1}{l} = \frac{\sqrt{m_u^2 + 2m_u(m_u' + m_u)} - m_u}{m_u' + m_u} \tag{8.22}$$

If in the design $m_u' = 2m_u$ is used, then from Eq. 8.22, $l_1/l = 0.549$ and from Eq. 8.21,

$$w_u = 19.94 \frac{m_u}{l^2}$$

or

$$m_u' = w_u \frac{l^2}{9.97} \quad \text{and} \quad m_u = w_u \frac{l^2}{19.94}$$

Negative- and positive-moment steel can be designed on the basis of these moments. It is evident that the negative-moment steel should be placed with a greater concentration in the vicinity of the column than away from the column. In addition, top steel is required in the corners.

8.6   DESIGN OF SUPPORTING BEAMS FOR UNIFORMLY LOADED TWO-WAY SLABS

The design of beams supporting slabs that have been designed by yield line theory can be carried out by either an approach based on composite beam–slab collapse mechanisms or an approach based on the loading transferred to the beams. Both of these methods are considered.

8.6.1   Approach Based on Composite Beam–Slab Collapse Mechanisms

In the derivation of the ultimate load equations for slabs it was assumed that the supporting system is strong enough to support the ultimate load of the slab. In Section 7.10 consideration was given to possible composite beam–slab collapse mechanisms that could occur if the beams are relatively weak. These composite mechanisms form the basis of a possible design method for beams.

   The method is to design the panels as if the beams are strong enough to support the ultimate load of the slab and then to design the supporting beams by a limit design approach based on composite beam–slab collapse mechanisms so that the required ultimate load is reached. This approach had been suggested previously by Wood,[8.11] Wood and Jones,[8.13] and Park.[8.19,8.20]

   The method assumes that full redistribution of bending moments can occur in the beams, as well as in the slabs, before the ultimate load is reached. This assumption may not be justified if the steel ratio of the reinforced concrete beams is high. However, in the few tests that have been conducted, in which composite beam–slab collapse modes have occurred, full redistribution of bending moments has occurred in the beams at the ultimate load. Test results from floors were discussed in Section 7.17.

To illustrate the design approach, consider the uniformly loaded continuous slab and beam floor shown in Fig. 8.18. The floor is supported by columns at the junctions of the beams. The panels may be designed using either Eq. 7.59 or 7.95. The collapse mechanisms in which both the beams and the slabs participate are shown in Fig. 8.19. In these mechanisms the yield lines run across the floor and plastic hinges form in the beams. The floor fails by folding along the lines of the columns and within the spans. In Fig. 8.19 the positive and negative ultimate moments of resistance of the beams are shown as M_u and M'_u or M''_u, respectively, with the appropriate subscript, and the positive and negative ultimate moments of resistance of the slab per unit width are shown as m_u and m'_u or m''_u, respectively, with the appropriate subscript. Members of zero width will be assumed, to simplify the equations, so all dimensions are to centerlines of columns.

In Fig. 8.19a the folds run across the x-direction beams. Consideration of the equilibrium of the two parts of the collapse mechanism by taking moments about the support lines gives

$$0.5w_u l_y l_1^2 = (m''_{ux} + m_{ux})l_y + M''_{ux} + M_{ux} \qquad (8.23)$$

$$0.5w_u l_y (l_x - l_1)^2) = (m'_{ux} + m_{ux})l_y + M'_{ux} + M_{ux} \qquad (8.24)$$

For interior spans, usually $M''_{ux} = M'_{ux}$ and $m''_{ux} = m'_{ux}$, and hence $l_1 = 0.5l_x$. Then, from Eqs. 8.23 and 8.24,

$$M'_{ux} + M_{ux} = \tfrac{1}{8}w_u l_x^2 l_y - (m'_{ux} + m_{ux})l_y \qquad (8.25)$$

In the design of the panels, m'_{ux} and m_{ux} have been determined (from either Eq. 7.59 or 7.95 and the chosen ratio of m'_{ux}/m_{ux}), and hence the required ultimate moments of resistance for the x-direction beams may be found from Eq. 8.25. It is to be noted that the designer must choose the ratio of M'_{ux}/M_{ux} because Eq. 8.25 only gives the required free (static) moments of the beams, $M'_{ux} + M_{ux}$.

In Fig. 8.19b the folds run through the y-direction beams. Consideration of the equilibrium of the two parts of the mechanism gives

$$0.5w_u l_x l_2^2 = (m''_{uy} + m_{uy})l_x + M''_{uy} + M_{uy} \qquad (8.26)$$

$$0.5w_u l_x (l_y - l_2)^2 = (m'_{uy} + m_{uy})l_x + M'_{uy} + M_{uy} \qquad (8.27)$$

For interior spans, usually $M''_{uy} = M'_{uy}$ and $m''_{uy} = m'_{uy}$, and hence $l_2 = 0.5\, l_y$. Then, from Eqs. 8.26 and 8.27,

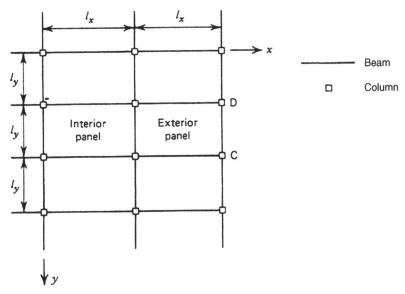

Figure 8.18 Slab-and-beam floor.

$$M'_{uy} + M_{uy} = \tfrac{1}{8} w_u l_y^2 l_x - (m'_{uy} + m_{uy})l_x \qquad (8.28)$$

Again, the ratio of M'_{uy}/M_{uy} must be chosen before Eq. 8.28 may be solved.

It is to be noted that the equations above apply to the interior beams of the floor. Also, it has been assumed that each interior beam has the same ultimate moments of resistance. At an edge beam running along the perimeter of the floor (such as DC of Fig. 8.18), the panels lie to one side only of the beam. Therefore, the ultimate moment of resistance of the edge beam is required to be one-half of the ultimate moment of resistance of the interior beam calculated by the equations above. Also, for the exterior panels of the floor, the strength of the beam–column junction and the torsional strength and stiffness of the edge beams will determine whether the negative-moment plastic hinge will form at the junctions of the beams intersecting the edge beams of the floor and whether the negative-moment yield lines form in the slab along the edge beams. For example, for the x-direction beams of the exterior panels of Fig. 8.18, it will be necessary to solve Eqs. 8.23 and 8.24 simultaneously with $M'_{ux} + m'_{ux}l_y$ put equal to the actual flexural strength of the beam plus slab as implied, or the flexural strength of the beam M'_{ux} plus twice the torsional strength of the y-direction edge beam at the column, or the sum of the upper and lower column flexural strengths, whichever is least.

At the collapse of a floor designed on the basis of either Eq. 7.59 or 7.95 for the slabs and Eqs. 8.23 to 8.28 for the beams, all the yield line patterns will occur simultaneously. That is, the yield lines of Fig. 7.39 and the yield lines and plastic hinges of Fig. 8.19a and b will all occur at the ultimate load. Thus, a large proportion of the reinforcement will be at the yield strength at

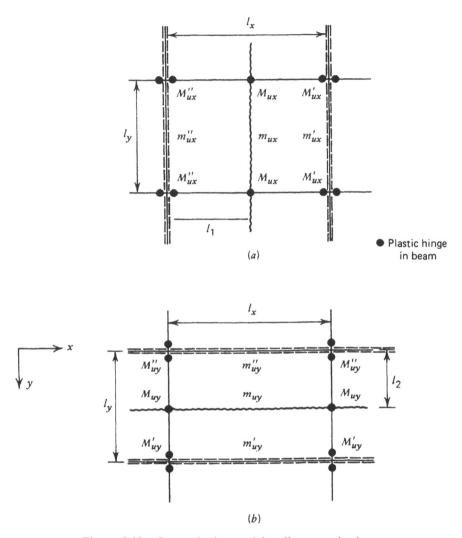

Figure 8.19 Composite beam–slab collapse mechanisms.

the ultimate load and the steel content will be close to that for minimum-weight design.

8.6.2 Approach Based on Loading Transferred to the Beams

Exact Loading Shapes. An alternative approach to the design of the beams is to consider the loading transferred to the beams from the slabs. A difficulty with yield line theory is that the manner in which the reactions are distributed along the edges of panels is not indicated by the theory. For instance, segment *ADFE* of the yield line pattern of Fig. 7.39 carries the uniformly distributed load per unit area w_u over the segment and the nodal forces acting downward at A and D (the nodal forces at E and F are zero). Thus, the total reaction applied to the beam AD by the panel could be determined from the loading and the nodal forces on the segment, but the manner in which this reaction is distributed along the beam is unknown.

It can be shown, however, that if the loading distribution on the beams at collapse is taken to follow the shape of the segments of the yield line pattern (triangular loading distribution for the short-span beams and trapezoidal loading distribution for the long-span beams), the maximum ultimate moments calculated for the beams are identical to those calculated considering composite collapse mechanisms. This was first shown by Wood,[8.11] Wood and Jones,[8.13] and Park.[8.19,8.20]

Consider the case of continuous (fixed-edge) panels. Figure 8.20 shows the yield line pattern for interior panels with the ultimate negative moments of resistance per unit width at opposite edges equal but not necessarily the same in each direction. The loading, which is assumed to be transferred to the short- and long-span beams from the adjacent panels, is shown in the figure.

For a short-span beam, the total load carried is

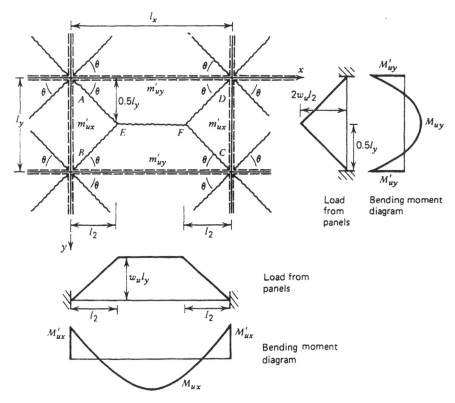

Figure 8.20 Assumed loading distribution on beams.

$$W_u = w_u l_y l_2$$

and this triangular loading causes a free (static) bending moment of magnitude

$$M'_{uy} + M_{uy} = \tfrac{1}{6} w_u l_y^2 l_2 \tag{8.29}$$

For a long-span beam, the total load carried is

$$W_{ux} = w_u l_y (l_x - l_2)$$

and this trapezoidal loading causes a free (static) bending moment of magnitude

$$M'_{ux} + M_{ux} = \tfrac{1}{8} w_u l_y l_x^2 - \tfrac{1}{6} w_u l_y l_2^2 \tag{8.30}$$

But taking moments about the support line BC for segment $BCFE$ shows that

$$\tfrac{1}{6} w_u l_y^2 l_2 = \tfrac{1}{8} w_u l_y l_x^2 - (m'_{uy} + m_{uy}) l_x \tag{8.31}$$

And taking moments about the support line AB for segment ABE shows that

$$\tfrac{1}{6} w_u l_y l_2^2 = (m'_{ux} + m_{ux}) l_y \tag{8.32}$$

On substituting Eq. 8.32 into Eq. 8.30, and Eq. 8.31 into Eq. 8.29, the following equations result:

$$M'_{ux} + M_{ux} = \tfrac{1}{8} w_u l_x^2 l_y - (m'_{ux} + m_{ux}) l_y \tag{8.33}$$

$$M'_{uy} + M_{uy} = \tfrac{1}{8} w_u l_y^2 l_x - (m'_{uy} + m_{uy}) l_x \tag{8.34}$$

It can be seen that Eqs. 8.33 and 8.25 are identical and that Eqs. 8.34 and 8.28 are identical. Hence, the two approaches result in beams of the same strength, and it may be concluded that if the beams are designed to support a loading distribution that follows the shape of the adjacent segments of the yield line pattern for collapse of the panels alone, the collapse load of the floor will equal that for the composite mechanisms involving the panels and the beams.

In the general case, the support conditions may not be symmetrical and the corner angles θ of the yield lines of Fig. 8.20 may not be equal. The load distribution may still be taken as that given by the shape of the adjacent segments of the yield line pattern in many cases. For example, consider the design of the edge beam AD for the exterior panel shown in Fig. 8.21. The collapse mechanisms involving failure of the panel and of the x-direction beams are shown. The shear force at G in the beam is zero. Taking moments about A for the portion AG of the beam gives

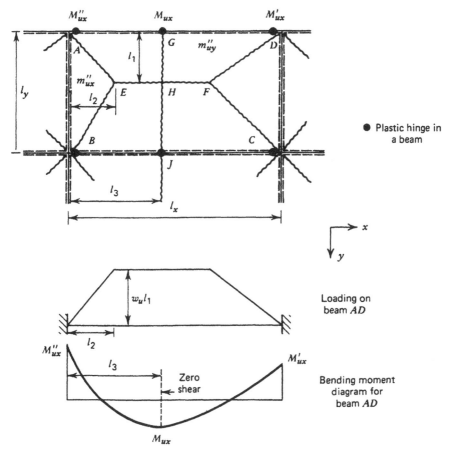

Figure 8.21 Combined collapse mechanism.

$$M''_{ux} + M_{ux} = \tfrac{1}{2}w_u l_1 l_3^2 - \tfrac{1}{6}w_u l_1 l_2^2 \tag{8.35}$$

Now taking moments about AB of segment ABE of the slab gives

$$\tfrac{1}{6}w_u l_y l_2^2 = (m''_{ux} + m_{ux})l_y \tag{8.36}$$

On substituting Eq. 8.36 into Eq. 8.35, the following equation results:

$$M''_{ux} + M_{ux} = \tfrac{1}{2}w_u l_1 l_3^2 - (m''_{ux} + m_{ux})l_1 \tag{8.37}$$

Similarly, an equation for $M'_{ux} + M_{ux}$ may be found by considering the equilibrium of the portion GD of the beam. It is evident that Eq. 8.35 gives the free bending moment due to the loading of the shape of the adjacent segment and that this equation can be arranged in the form of Eq. 8.37, which is the equation for the composite collapse mechanism of the type shown in Fig. 8.19a. The appearance of l_1 in Eq. 8.37 rather than $0.5l_y$ means that the required free (static) bending moments may differ from beam to beam, but it is evident that the total strength of all the beams will be identical to that required by the composite collapse mechanism equations.

It is to be noted, however, that this comparison of design approaches for beam AD of Fig. 8.21 assumes that the yield line GJ passes between E and F. There will be a small difference between the free (static) moments given for beam AD by the assumed loading and the composite collapse mechanism approaches if the yield line GJ does not pass between E and F (e.g., if $l_3 < l_2$). Similarly, in the case of beam AB, small differences will arise if in the composite collapse mechanism involving the y-direction beam the positive moment yield line does not pass through E and F. The reason for the difference between the free (static) moments in these situations is that additional loads (somewhat similar to nodal forces) need to be considered on the beam if exact agreement between the two approaches is to be obtained. However, it has been found[8.19] that if these additional loads are ignored, the errors induced may not be serious and may even be insignificant. In situations where the designer is in doubt, it would seem advisable to design the beams using the composite beam–slab collapse mechanism method.

Also, in the design of the exterior panels it can be assumed that the exterior edges of the panels are fixed at the edge beams of the floor if the edge beams are sufficiently strong and stiff in torsion to transfer the negative moment at the slab edges into the beam junctions. However, it should be noted that this will mean that a positive bending moment of magnitude $m''_{uy}l_x$ will be introduced into the connection at A, which must be considered in addition to the loading from the adjacent segments of the yield line pattern. Neglect of this end moment in the beam induced by torsion can result in large errors on the unsafe side.[8.20] For example, consider the uniformly loaded single square panel of span l shown in Fig. 8.22, which is supported on beams around all

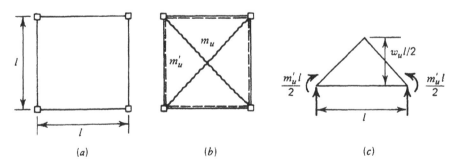

Figure 8.22   Uniformly loaded single panel floor: (a) floor; (b) panel mechanism; (c) beam loading.

edges and is simply supported on columns at the corners. The panel can be designed as fixed-edge (Fig. 8.22*b*), provided that the beams are sufficiently strong and stiff in torsion to sustain a maximum torsional moment at each end of $m'_u \, l/2$ induced by the negative ultimate moment of resistance per unit width m'_u along the panel edges. Then when designing the beams, the load distribution to be taken is the triangular shape of the adjacent segment of the yield line pattern plus a positive moment of $m'_u \, l/2$ at each end induced by the beam torsion (Fig. 8.22*c*). If these end moments are ignored, the design moment of the beams will be too low, for example, 25% too low if for the panel $m'_u/m_u = 2$.

Although it appears that in many cases the beams may be designed to support loading of the shape of the adjacent segments of the yield line patterns for collapse of the panels alone, it should be noted that it has only been shown that this loading gives the correct maximum ordinate of the free (static) bending moment diagram. The shape of the bending moment diagram given by this loading is not necessarily correct. However, an estimate of the ratios of the bending moment M'_{ux}/M_{ux} and M'_{uy}/M_{uy} to be used in design could be obtained by an elastic theory analysis of the bending moments in the frame of the structure with the beams supporting the load distribution found. Alternatively, the ratios of the beam moments could differ from those found by elastic theory if it could be shown that the plastic hinges have sufficient rotation capacity to allow full redistribution of bending moments before collapse, and that serviceability requirements for cracking and deflections have been met at service load levels.

Approximation to Loading on the Beams. Codes of practice sometimes recommend that for approximate methods of design of slabs and beams by elastic theory, the distribution of loading on the beams can be taken as that given by triangles and trapezoids with base angles of 45° ($\theta = 45°$ in Fig. 8.20). Using elastic theory, it can be shown that for rectangular panels on very stiff beams, this assumed reaction distribution is quite accurate, but if the stiffness of the beams is small, this assumed distribution may lead to large errors (see Chapter 3). It is of interest to check the accuracy of this assumed loading distribution for beams designed by limit design. Consider the case of an interior panel of the floor with $m'_{ux}/m_{ux} = m'_{uy}/m_{uy}$. The approximate loading distribution on the beams amounts to assuming that the yield lines enter the corners of the panel at 45° to the edges. The resulting approximate free (static) bending moments for the beams may be obtained by substituting $l_2 = 0.5l_y$ into Eqs. 8.30 and 8.29 and the exact free bending moments are given by Eqs. 8.25 and 8.28 with the value of m_{ux} and m_{uy} substituted from Eq. 7.59. Table 8.5 shows the approximate and exact free (static) bending moments compared for the *x*- and *y*-direction beams supporting panels with various l_x/l_y and m_{uy}/m_{ux} ratios. Similarly, the effect that this approximate design loading has on the ultimate load of the floor may be examined by comparing

TABLE 8.5   Comparison of Exact and Approximate Design Using 45° Beam Loading for Uniformly Loaded Floors with $i_1 = i_2 = i_3 = i_4$

$\dfrac{l_x}{l_y}$	$\dfrac{m_{uy}}{m_{ux}}$	Approximate Free Moments		w_u from Approximate Free Moments	
		Exact Free Moments		w_u Required	
		y-Direction Beams	x-Direction Beams	Failure of y-Direction Beams	Failure of x-Direction Beams
1.0	1	1.00	1.00	1.00	1.00
1.5	1	0.84	1.08	0.92	1.06
	3	1.25	0.94	1.09	0.95
	5	1.54	0.91	1.15	0.91
2.0	1	0.77	1.07	0.90	1.06
	3	1.18	0.98	1.05	0.98
	5	1.47	0.95	1.11	0.96
3.0	1	0.70	1.04	0.90	1.04
	3	1.12	0.99	1.02	0.99
	5	1.41	0.98	1.06	0.98

the required ultimate load with that actually obtained when the approximate free moments are substituted into Eqs. 8.25 and 8.28. Table 8.5 shows these ultimate loads compared.

Comparison of the free (static) bending moments in Table 8.5 shows that the use of the approximate 45° loading distribution results in some beams being weaker and others stronger than required. The comparison of the ultimate loads in the table shows that this has the effect of lowering the collapse load of the floor because failure will occur by the composite beam–slab collapse mechanism involving the weaker beams at a load less than the required ultimate load and before the failure loads of the other collapse mechanisms are reached. For example, for an isotropic ($m_{uy}/m_{ux} = 1.0$) slab with $l_x/l_y = 2.0$, Table 8.5 shows that use of the approximate loading distribution results in the short-span beams having only 77% of the required flexural strength, and this would mean that if loaded to collapse, the floor would fail by the composite collapse mechanism of Fig. 8.19b at 90% of the required ultimate load. It should be noted that Table 8.5 is for the symmetrical case of an interior panel. For exterior panels with unsymmetrical boundary conditions, the error involved in using the approximate 45° distribution of loading to the beams will be greater than that shown in Table 8.5, and hence the approximate loading assumption should be used with caution. This is a strong argument for always considering the slab–beam system as a whole rather than separating the two components and trying to design slabs and beams as isolated elements when they are, in fact, strongly interactive.

8.6.3 Other Arrangements of Beams and Columns

Only slabs supported at all edges by beams with columns at all the beam junctions have so far been considered. Other cases are discussed below.

Secondary Beams. If secondary beams are present in the floor, all beams may still be designed on the basis of loading of the shape of the adjacent segments of the yield line pattern of the panels. This may be shown by considering the portion of the uniformly loaded slab and beam floor shown in Fig. 8.23, in which the column grid is l_x by $2l_y$ and the panels span l_x by l_y. The yield line patterns for failure of the panels are shown in the figure.

The loading on the y-direction beams from the adjacent segments and the secondary beams is taken to be as illustrated in Fig. 8.23. It is to be noted that the load carried by the secondary beams is assumed to be transferred to the beam under consideration as a concentrated load. The free (static) bending moment caused by the loading on the y-direction beam may be shown to be of magnitude

$$M'_{uy} + M_{uy} = \tfrac{1}{2}w_u l_y(l_x - l_2)l_y + w_u l_2 l_y(l_y/2)$$

$$= \tfrac{1}{2}w_u l_x l_y^2 \tag{8.38}$$

It is to be noted that the assumed loading is a reasonable representation of the elastic conditions if the beams are quite stiff, and for any conditions if $l_x \gg l_y$. If the secondary beam is not stiff, the negative-moment yield lines parallel to that beam may not form.

Consider now a composite collapse mechanism that involves plastic hinges developing in the y-direction beams and yield lines running across the slab in the x-direction at the columns and at midspan. The positive moment yield line will run alongside the secondary beam of Fig. 8.23 and hence should be assumed to have zero strength because the slab may only be reinforced for negative moment there. The ultimate load of this composite mechanism is given by Eq. 8.28 with $2l_y$ substituted for l_y and m_{ux} put equal to zero:

$$M'_{uy} + M_{uy} = \tfrac{1}{2}w_u l_x l_y^2 - m'_{uy}l_x \tag{8.39}$$

Comparison of Eqs. 8.38 and 8.39 shows that the design of beams by Eq. 8.38 using the assumed loading distribution is slightly conservative. However, the difference is small. For instance, if $l_x = l_y$ in Fig. 8.23, $M'_{uy} + M_{uy}$ given by Eq. 8.38 is only approximately 4% larger than that given by Eq. 8.39 for slabs with equal top and bottom reinforcement.

For the x-direction beams of Fig. 8.23, the loading from the adjacent segments gives the same strength beams as consideration of the composite beam–slab collapse mechanism involving the x-direction beams, because the beams are of the same span as the panels in that direction. Therefore, it is evident

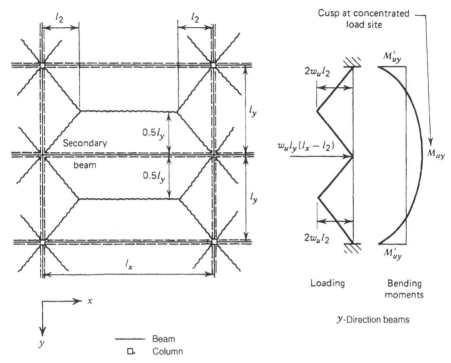

Figure 8.23   Uniformly loaded floor with secondary beams.

that floors with secondary beams can also be designed using the concept of loading of the shape of the adjacent segments of the yield line patterns for the panels.

Free Exterior Edges.   If the exterior edges of the floors are not supported by beams and columns, the beams of the exterior panels cannot be designed accurately by considering the uniform loading on the adjacent segments. This may be shown by considering the part of the floor with the free (unsupported) edge shown in Fig. 8.24. Consider the design of x-direction beam AB. Let the yield line patterns involving failure of the panels alone, and the composite beam–slab failure of the panels and the x-direction beams, occur simultaneously. It is evident that for the yield line pattern shown for the panels in Fig. 8.24, nodal forces will exist at the intersection of the positive-moment yield lines and the free edge in the directions shown. The magnitude of these nodal forces, according to Eq. 7.20, is $\pm m_{ux} \cot \alpha = \pm m_{ux} l_1/l_x$. Now considering the equilibrium of the segment CBD by taking moments about a y-direction axis passing through B gives

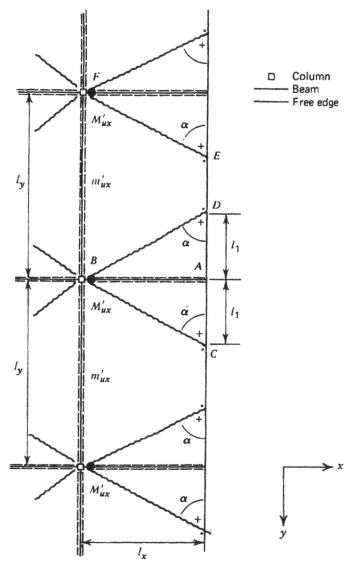

Figure 8.24 Uniformly loaded floor with free (unsupported) exterior edge.

$$M'_{ux} = w_u l_1 l_x \frac{2}{3} l_x + 2m_{ux} \frac{l_1}{l_x} l_x + m_{ux} \cdot 2l_1$$

$$= \frac{2}{3} w_u l_1 l_x^2 + 4m_{ux} l_1 \qquad (8.40)$$

Equation 8.40 gives the ultimate moment required for the x-direction beams and it is evident that this ultimate moment is greater than that obtained by considering only the external uniform loading acting on the segment, since the slab moment and the nodal forces both act to increase the beam moment. Therefore, use of the equivalent loading concept will lead to unsafe design in this case unless these additional forces are also included. That the ultimate moment given by Eq. 8.40 is equal to that found by considering the composite collapse mechanism may be shown by considering the equilibrium of segment *BDEF* by taking moments about *BF*.

$$m'_{ux}l_y + 2m_{ux}l_1 + 2m_{ux}\frac{l_1}{l_x}l_x = w_ul_xl_y(l_x/2) - w_ul_xl_1\frac{2}{3}l_x$$

$$\frac{2}{3}w_ul_x^2l_1 + 4m_{ux}l_1 = \frac{1}{2}w_ul_x^2l_y - m'_{ux}l_y \qquad (8.41)$$

Substituting Eq. 8.41 into Eq. 8.40 gives

$$M'_{ux} = \frac{1}{2}w_ul_x^2l_y - m'_{ux}l_y \qquad (8.42)$$

Equation 8.42 is also the ultimate moment found by considering the composite collapse mechanism.

In the case where the positive-moment yield lines in the panels intersect before reaching the free edge and run out to the edge as a single yield line, it can also be shown that the external uniform load on the adjacent segments gives an incorrect and unsafe design ultimate moment for the beam *AB*. In this case, Eq. 8.42 may be used or M'_{ux} calculated considering only the equilibrium of the adjacent segment when the panel and composite mechanisms form simultaneously in the manner used to derive Eq. 8.40.

For the y-direction beams of Fig. 8.24, the free moments $M'_{uy} + M_{uy}$ should also be found from the slab–beam composite collapse mechanism or from the equilibrium of the adjacent segments when both mechanisms form simultaneously. It is to be noted, however, that for the y-direction beams, design on the basis of the external uniform loading on the adjacent segments will lead to beam strengths that are either greater than or equal to the required strength.

8.6.4 Summary of Design Method for Beams

The design of continuous slab-and-beam floors involves the examination of the collapse mechanism involving failure of the panels alone (as if on strong beams) and of the composite beam–slab collapse mechanisms involving failure of both the panels and the beams. It is shown that in practice, except for floors with unsupported exterior edges and some cases of unsymmetrical yield line positions, only the collapse mechanism involving failure of the panels alone needs to be considered, since if the beams are designed to carry the loading that is on the adjacent segments of the yield line pattern of this mechanism, the ultimate load of the floor will be equal to that given by the composite beam–slab collapse mechanism. The load distribution on the beams is taken to follow the shape of the segments of the yield line pattern and gives the required maximum free (static) moments in the beams. Load distributions on the beams found on the basis of an approximate yield line pattern that has the yield lines intersecting the corners at 45° may result in the ultimate load of the floor being up to 10% lower than required in the case of interior panels with symmetrical support conditions and will lead to greater errors in the case of exterior panels with unsymmetrical support conditions.

For floors with free (unsupported) exterior edges, composite collapse mechanisms should be used to design the beams, since in this case use of loading of the shape of the adjacent segments of the yield line pattern leads to unsafe values for ultimate moments required in the beams perpendicular to the free edge. Care should be taken to include as loading any bending moment induced at the ends of beams as a result of torsional moments in the edge beams of floors.

It is to be noted that the approach for the design of floors with beams detailed in the 1995 ACI Code[8.3] has similarities to the composite beam–slab collapse mechanism approach. The ACI code method distributes the total panel moment found between the beam and the slab column and middle strips in a manner that depends on the relative flexural stiffness of the beams and slab and the ratio of the spans in the two directions. The resulting reinforcement arrangement will apparently ensure that a composite beam–slab collapse mechanism will always occur before a panel mechanism in a slab-and-beam floor.

REFERENCES

8.1   *Structural Use of Concrete*, BS 8110, Part 1:1997, *Code of Practice for Design and Construction*, British Standards Institution, London, 1997, 121 pp.

8.2   *International System of Unified Standard Codes of Practice for Structures*, Vol. II, *CEB-FIP Model Code for Concrete Structures*, Comité Euro-International du Béton/Fédération Internationale de la Précontrainte, Paris (English translation), April 1978, 348 pp.

8.3   *Building Code Requirements for Structural Concrete*, ACI 318-95, and *Commentary*, ACI 318R-95, American Concrete Institute, Farmington Hills, Mich., 1995, 371 pp.

8.4   *The Design of Concrete Structures*, NZS 3101:1995, Standards New Zealand, Wellington, New Zealand, 1995.

8.5   K. W. Johansen, *Brudlinieteorier,* Jul. Gjellerups Forlag, Copenhagen, 1943, 191 pp. (*Yield-Line Theory*, translated by Cement and Concrete Association, London, 1962, 181 pp.)

8.6 K. W. Johansen, *Pladeformler,* Copenhagen, 1946, 186 pp.; 2nd ed., Copenhagen, 1949, 186 pp.; 3rd ed., 1968, 240 pp. (*Yield-Line Formulae for Slabs,* translated by Cement and Concrete Association, London, 1972, 106 pp.)

8.7 *Dalles et planchers dalles; applications de la théorie des lignes de rupture aux calculs de résistance en flexion,* Bulletin d'Information 35, Comité Européen du Béton, 1962. (*The Application of the Yield-Line Theory to Calculations of the Flexural Strength of Slabs and Flat Slab Floors,* Information Bulletin 35, European Concrete Committee, translated by C. V. Amerongen, Cement and Concrete Association, London, 180 pp.)

8.8 "Dalles et structures planes," *Recommendations Comité Européen du Béton,* Tome 3, 1972. ("Slabs and Plane Structures," *Recommendations of the European Concrete Committee,* Vol. 3, 1972.)

8.9 *De berekening van platen volgens de vloeilijentheorie,* C.U.R. Raport 26A, Commissie voor Uitvoering van Research ingesteld door de Betonvereniging, 1962, 79 pp. (*The Analysis of Slabs by the Yield-Line Theory,* Report 26A, Translation 135, Dutch Committee for Concrete Research, Cement and Concrete Association, London, 36 pp.)

8.10 *Voorbeelden van plaatberekeningen met de vloeilijnentheorie,* C.U.R. Rapport 26C, Commissie voor Uitvoering van Research ingesteld door de betonvereniging, 1963. [*Examples of Slab Design by the Yield-Line Theory,* Report 26C, Translation No. 121 (by C. V. Amerongen), Dutch Committee for Concrete Research, Report 26C, Cement and Concrete Association, London, 117 pp.]

8.11 R. H. Wood, *Plastic and Elastic Design of Slabs and Plates,* Thames and Hudson, London, 1961, 344 pp.

8.12 L. L. Jones, *Ultimate Load Analysis of Reinforced and Prestressed Concrete Structures,* Chatto & Windus, London, 1962, 248 pp.

8.13 R. H. Wood and L. L. Jones, *Yield-Line Analysis of Slabs,* Thames and Hudson, Chatto & Windus, London, 1967, 400 pp.

8.14 S. Bernaert, A. M. Haas, and G. A. Steinmann, *Le Calcul aux états-limités des dalles et structures planes,* Supplément aux Annales de l'Institut Technique du Batiment et des Travaux Publics, Vingt-Deuxième Année, Paris, Mai 1969, No. 257, pp. 759–830. Série: Béton. Béton Armé 104. (*The Calculation of the Limit States of Slabs and Plane Structures,* Supplement of the Annals of the Technical Institute of Building and Public Works, 22nd year, Paris, May 1969, No. 257, pp. 759–830. Series: Concrete. Reinforced Concrete 104.)

8.15 R. Park and T. Paulay, *Reinforced Concrete Structures,* Wiley, New York, 1975, 769 pp.

8.16 R. Barês, *Tables for the Analysis of Plates, Slabs and Diaphragms Based on the Elastic Theory* (English translation by C. van Amerongen), Bauverlag GmbH, Wiesbaden, Germany, 1969, 579 pp. (A 1978 edition was also available.)

8.17 R. Park, "Design of Reinforced Concrete Slabs by Yield-Line Theory," *N. Z. Eng.,* Vol. 18, No. 2, February 1963, pp. 56–65.

8.18 R. Park, "Yield-Line Design of Concrete Slabs with Free Edges," *N. Z. Eng.,* Vol. 21, No. 2, February 1966, pp. 63–8.

8.19 R. Park, "Limit Design of Beams for Two-Way Reinforced Concrete Slabs," *J. Inst. Struct. Eng.,* Vol. 46, No. 9, September 1968, pp. 269–274.

8.20 R. Park, "Some Irregular Cases of Beam Loading in Limit Design of Slab and Beam Floors," *J. Inst. Struct. Eng.,* Vol. 48, No. 1, January 1970, pp.17–19.

9 Serviceability of Slabs

9.1 INTRODUCTION

The performance of structures at the service load is an important design consideration. If a slab system is designed to strength requirements alone, there is a danger that although the degree of safety against collapse may be adequate the performance of the structures at the service load may be unsatisfactory. For example, the slab system under service loads may show excessive cracking or the deflections may be unacceptably large. Therefore, the structure should be designed with reference to several limit states, the most important being strength at overloads, deflections at service loads, and crack widths at service loads. Other possible limit states are vibrations, and fatigue of reinforcing steel, at service loads. The aim in design must be to ensure an adequate margin of safety against collapse and against the possibility that the structure will become unfit for use at service loads.

The ACI Code[9.1] emphasizes design based on strength with serviceability checks. Deflections are controlled by specifying minimum allowable slab thicknesses, to ensure adequate stiffness, or alternatively, by requiring that deflections be computed to ensure that they do not exceed specified maximum allowable values. Cracking in one-way slabs is judged to be of concern only if the design yield strength of the reinforcement exceeds 40,000 lb/in² [300 N/mm²]. For steel in this higher-strength range, limiting arrangements of reinforcement are specified to ensure that the crack widths are not excessive. For two-way slabs, no explicit crack control criteria are specified in either the code[9.1] or the commentary. The ACI Code does not give criteria for the control of vibrations or the fatigue of steel, but designers should give these aspects attention when potentially they may be troublesome. Recent research on fatigue of reinforced concrete is reported in Ref. 9.2. In this chapter we discuss in detail methods for the control of deflections and flexural cracking in slab systems.

9.2 DEFLECTIONS

9.2.1 General Comments on Deflections

The prediction of deflections in reinforced concrete structures is always a problem for the designer, but it becomes even more problematic when the structure is a slab. There are two separate problems that lead to this difficulty. The first is the analysis leading to a deflection function, such as the familiar $\delta = wl^4/384EI$ for a prismatic beam of span l with fixed ends carrying a uniform load per unit length w. The second is the determination of the appropriate flexural rigidity, EI, to use once the deflection function has been found.

The elastic theory analysis of a slab, described in Chapter 2, is often difficult, but the growing availability of large computer programs with built-in capability for finite element analysis is easing the problem to some extent. In addition, deflection functions for many cases have been computed and tabulated. The more serious problem is that of the flexural rigidity EI. The determination of a reasonable flexural rigidity is more difficult for a slab than a beam, largely because the reinforcement ratios in slabs are usually quite low, because of relatively low bending moments. The reinforcement ratio is often governed by minimum steel area or maximum bar spacing requirements.

As a direct result of the low steel ratios, the ratio of uncracked to fully cracked flexural rigidity, EI_g/EI_{cr}, is very large, and approaches 10 for a slab with reinforcement ratio $\rho = 0.002$. Consequently, the amount of cracking has a very important influence on the magnitude of the deflection. As a further consequence of the low moments leading to the low steel ratios, much of the area of a slab will be uncracked at service load levels. Nevertheless, the concrete tensile stresses will be an appreciable fraction of the modulus of rupture, and cracking may occur at some time after first loading as a result of sustained tensile stresses, minor overloads, restrained shrinkage stresses, temperature effects, small support settlements, and other ordinarily negligible factors.

Because of this spread of cracking with time, the long-term deflections may be large relative to the initial deflection accompanying stripping of the formwork and initial loading. The ACI Code[9.1] specifies in Secs. 9.5.2.5 and 9.5.3.4 (ACI Code Eq. 9–10) that the *time-dependent* component of deflection be computed as the initial deflection times,

$$\lambda = \frac{\xi}{1 + 50\rho'} \tag{9.1}$$

where ρ' is the compression steel, A_s'/bd, and ξ is a time variable taken as 2.0 for ages of five years or more. Values of ξ are tabulated as follows:

3 months: $\xi = 1.0$
6 months: $\xi = 1.2$
12 months: $\xi = 1.4$
5 years: $\xi = 2.0$

If there is no compression steel (i.e., $A_s' = 0$), the total long-term deflection is then $(1 + \lambda) = 3$ times the initial deflection. The data quoted below clearly demonstrate that this provision may be wrong and misleading in some common cases, since this is applied equally to beams and one- and two-way slabs without making any distinctions about other characteristics. The ACI Code Sec. 9.5.2.5 gives an escape from its provisions, as the section starts as: "Unless values are obtained by a more comprehensive analysis . . . ."

In an early test of a very heavily loaded flat slab,[9.3] the initial deflection of 0.475 in. (12.1 mm) increased to 0.825 in. (20.9 mm) in one year, which agrees well with the trends of beam data, as is reasonable since the reinforcement ratios were comparable to those in beams. However, in one of the few reports of in-service measurements of deflections of flat plate structures, Taylor and Heiman[9.4,9.5] found that the midspan deflections at 2.3 years were as much as 6.5 to more than 7.5 times the initial deflections. The initial measured value might have been slightly too low on one slab, because of deflections of the supporting formwork, but the long-term multiplier would still be much greater than 3.0. The aggregate used in the concrete was apparently one that leads to high shrinkage and creep strains, which would be a contributing factor. However, the slab described in Ref. 9.5 was known never to have been overloaded, and it supported only insulation, roofing, and plaster weighing about 5% of the weight of the slab, so the loading conditions were extremely favorable. Blakey[9.6] also found long-term deflections of up to seven times the initial deflections in a lightweight concrete flat plate that was subjected to loading tests. More information on long-term deflection is badly needed before a general picture can be formed.

ACI Committee 435[9.7] discussed these and other cases of large deflections and made a number of observations and recommendations about controlling deflections. They noted that camber is an effective means of limiting some of the effects of deflection, although of course camber does not change the deflection.

Flat plate structures will be more likely to undergo very large time-dependent deflections than will flat slabs, since flat slabs will ordinarily be designed for considerably heavier loads and hence will be more heavily reinforced. As a result of the higher steel ratios, the loss of stiffness accompanying cracking will not be so great. Also, flat slabs are more likely to be relatively extensively cracked at the service load level, so time-dependent cracking should not be so important.

Two-way slabs supported on beams apparently have not experienced many problems with excessive deflections. The beams are very effective in reducing deflections, and the beams do not lose stiffness so greatly with cracking, since they will have at least moderate reinforcement ratios.

It is as hard to define an acceptable deflection as it is to compute the deflections. The largest acceptable deflection is obviously a function of the use of the structure and the nature of the other components, structural and nonstructural, of the building. Perhaps the largest source of difficulty in buildings that can be traced to excessive deflection is interference with the function of some nonstructural components. If the floor deflects too much, rigid masonry walls crack, glass may crack or shift in its frames, and doors and windows may no longer fit properly, all of which may cause clients to doubt the proficiency of the designer. In more extreme cases, furniture acquires a distinctive tilt, bookcases require leveling, and balls unerringly roll to the low points of the floor.

The 1995 ACI Code[9.1] puts limits on computed deflections ranging from span/180 to span/480, depending on the use and nature of the elements supported by the slab. The limits are reproduced as Table 9.1. The code is not clear about which span is to be considered for a slab, but presumably the longer of the two clear spans should be used to be consistent with the span/thickness limit expressions also given in the code.

TABLE 9.1   Maximum Permissible Computed Deflections

Type of Member	Deflection to Be Considered	Deflection Limitation
Flat roofs not supporting or attached to nonstructural elements likely to be damaged by large deflection.	Immediate deflection due to live load, L	$\dfrac{l}{180}$ [a]
Floors not supporting or attached to structural elements likely to be damaged by large deflections	Immediate deflection due to live load, L	$\dfrac{l}{360}$
Roof or floor construction supporting or attached to nonstructural elements likely to be damaged by large deflections	That part of the total deflection occurring after attachment of nonstructural elements (the sum of the long-time deflection due to all sustained loads and the immediate deflections due any additional live load)[b]	$\dfrac{l}{480}$ [c]
Roof or floor construction supporting or attached to nonstructural elements not likely to be damaged by large deflections		$\dfrac{l}{240}$ [d]

Source: After Ref. 9.1.

[a]Limit not intended to safeguard against ponding. Ponding should be checked by suitable calculations of deflection, including added deflections due to ponded water, and considering long-time effects of all sustained loads, camber, construction tolerances, and reliability of provisions for drainage.

[b]Long-time deflection shall be determined in accordance with Eq. 9.1, or more comprehensive analysis, but may be reduced by amount of deflection calculated to occur before attachment of nonstructural elements. This amount shall be determined on basis of accepted engineering data relating to time-deflection characteristics of members similar to those being considered.

[c]Limit may be exceeded if adequate measures are taken to prevent damage to supported or attached elements.

[d]But not greater than tolerance provided for nonstructural clements. Limit may be exceeded if camber is provided so that total deflection minus camber does not exceed limit.

Blakey[9.8] studied the deflections of flat plate structures and suggested that if cracking of masonry partitions is to be avoided, initial deflections may have to be limited to about span/1500. He suggested that the minimum plate thickness should be span/32, using the long-span center to center of columns, and furthermore that the positive-moment sections should not crack under full service load. This was to be accomplished by keeping the maximum stresses due to positive moments lower than the modulus of rupture.

Just how well these recommendations work out in practice has not been determined. Taylor's slab,[9.5] which was slightly thicker [8 in. (203 mm) opposed to 7.8 in. (198 mm)] than Blakey's recommendation, had an initial maximum deflection of long span/2300 but still developed a deflection of long span/350 without being subjected to live loads. After 2.3 years, the slab was continuing to deflect with time, and the trends of the curves suggest that the sag will eventually exceed long span/300. The computed maximum midspan tensile stresses were in the range 0.6 to 0.7 of the modulus of rupture, and the report makes no mention of positive-moment cracking, although the lower surface was covered with sprayed-on plaster, so any cracking might not be readily visible.

If, as seems apparent from the discussion above, cracking is an important factor in excessive long-term deflections, it is also clear that measures should be taken to minimize cracking. Shrinkage strains may often lead to cracking some time after completion of a slab. The shrinkage is restrained by the reinforcement in the concrete and also by other elements in the structure, such as columns, walls, and beams. The restraint of the shrinkage strains causes tensile stresses that may combine with the stresses caused by loads to produce cracking at much lower loads than otherwise expected.

Any measure that reduces the shrinkage should aid the floor in maintaining small deflections. Careful selection of the materials for the concrete and of the mix proportions, and careful attention to the moist curing of the concrete before it is allowed to dry will be helpful. In addition, nearly all measures that reduce shrinkage also reduce the creep potential of the same concrete, leading to further control of long-term deformations. The judicious use of construction and expansion joints will help to allow the shrinkage to occur while reducing the induced stresses by reducing the restraints.

Excessive deflections may also be traced to design and construction faults. In addition to outright errors, overestimation of the stiffness of exterior supporting elements would lead to both underestimation of the deflection and to more positive-moment cracking than expected. The current ACI Code approach using the equivalent column should make this a less serious problem than in the past.

A common construction problem is the misplacement of the negative-moment steel. If it is not very securely supported, it may be knocked out of position during the concreting operation, since the workers normally walk on it, resulting in the steel being too low in the section and having inadequate effective depth. A slab is very sensitive to this because of the small effective depths being used, and a 20% reduction is both very harmful and not hard to accomplish. There are at least three consequences of such placing the top reinforcement too low: (1) negative moment cracks are wider; (2) negative moment stiffness is substantially reduced, leading to higher positive moments and cracking, and greater deflection; and (3) the shear strength, of beamless slabs especially, is reduced. This reinforcement must also have adequate stiffness to resist being bent by the construction operations and larger bars than normally used for slab reinforcement may be needed.

Rain on a freshly concreted slab can also lead to deflection problems because of the increased water/cement ratio leading to a lower modulus of elasticity and to higher creep and shrinkage strains. Every successful contractor works with one eye on the weather. The sequence of events during construction may be very important to the serviceability of a floor. Overloads are perhaps more likely during construction than at any other time during the life of a structure, especially if the service live loads are small. A slab may be overloaded by materials stored on it, although this can be avoided by careful job planning and control.

A potentially more serious and less easily remedied situation arises when the floor being cast is supported by shoring resting on lower floors. No matter how carefully planned and controlled the job is, each of the lower floors is successively heavily loaded as the building goes up. This problem has been studied by many authors[9.9–9.16] (and others who could cited), and it appears that when three sets of props are used (i.e., each newly cast slab is supported by the three floors immediately below it), each floor except the top eventually has to resist from 1.8 to 2.3 (or more) times its own weight. The loads are slightly worse if two sets of props are used. The implications of such loads are reasonably serious, especially in flat plates, where the dead load often exceeds the service live load by a reasonably large margin, and where punching shear stresses are often critical. The maximum construction load is resisted by young concrete, only 14 or 21 days old if floors are cast at the rate of one each week. Since the concrete is young and probably not up to design strength, creep will be high and the slabs will crack more extensively than it would if loaded later. The slab resists some load at less than 7 days age in a floor-per-week job, since the removal of the props two floors below causes some deformation and stresses.

Hurd and Courtois[9.13] presented a simple procedure for estimating the floor and prop loads. There are various schemes for shoring and reshoring, in which a limited number of props are removed in a predetermined pattern in some parts of a story to allow formwork stripping and then replaced and tightened to some nominal load imposed on the props. Studies have included the effects of age on the slab stiffness, and therefore on the distribution of forces in the shoring system. Nothing changes the conclusion that many floors will be subjected to about twice their own dead loads before construction is complete.

Webster[9.16] notes that design factors, construction procedure factors, and environmental factors, such as temperature and wind, that affect the curing of concrete and the loading of the structure by wind all enter an assessment of the reliability of the structure during construction. Needless to say, the construction procedures must be carefully planned to avoid damage to already cast components, and the plan must be adhered to. As an extreme example of what can go wrong, the apparent premature removal of formwork and props left part of the 24th-floor formwork of a flat plate building near Washington, D.C.[9.17–9.19] supported only by the 23rd floor, which was itself less than a week old. When most of one section of the 24th-floor concrete had been cast, the 23rd floor failed, bringing down the floor above and all the floors below, with a tragic loss of life. Cool temperatures during the period when the supporting slabs were cast was a factor in the failure.

In view of the comments above, it appears that any prediction of the deflections of slabs, no matter how carefully made, must be subject to considerable uncertainty.

9.2.2 Computation of Deflections

Deflection equations are generally expressed in the form

$$\delta = C \frac{w l_1^4}{D} = C_1 \frac{w l_1^4}{E h^3} \tag{9.2}$$

where w is the uniformly distributed load per unit area; and C and C_1 are constants depending on panel shape, support conditions, and Poisson's ratio; l_1 the long span, either as clear span, center-to-center span, or some average span; $D = E h^3 / [12(1 - \mu^2)]$, the slab stiffness per unit width; E the Young's modulus; and h the slab thickness. As noted earlier, there are significant difficulties in determining the value of C or C_1 for realistic support conditions and in determining the effective value of D when cracking and time-dependent effects must be taken into account.

The most comprehensive studies of deflections of reinforced concrete slab structures are those by Vanderbilt et al.,[9.20,9.21] Chang and Hwang,[9.22,9.23] and Jofriet.[9.24] The work by Vanderbilt and Chang and Hwang is directed to the general problem of slabs with and without beams, while Jofriet's is concerned with flat plate structures that may have spandrel beams. All considered both the elastic theory analysis for deflections and the effects of cracking. The elastic theory deflection values are considered first.

Vanderbilt provided an extensive table of elastic deflection coefficients, the C of Eq. 9.2, for typical interior panels supported on square columns considering the beam relative stiffnesses, support size, and panel shape. Part of his values are reproduced as Table 9.2, for the condition that the moments of inertia of the beams in the two directions are proportional to the spans. The spans are the distances between centers of supports. Two other combinations of beam stiffnesses are considered in the paper, but the slab midpanel deflec-

TABLE 9.2 Deflection Coefficients for Interior Panels[a]

			Center of Panel			Midspan of Long Beam			Midspan of Short Beam		
		c/L Ratio:	0.0	0.1	0.2	0.0	0.1	0.2	0.0	0.1	0.2
S/L	α_L	α_S									
1.0	0.0	0.0	0.00581	0.00441	0.00289	0.00435	0.00304	0.00173			
	0.2	0.2	0.00438	0.00340	0.00240	0.00299	0.00207	0.00122			
	0.25	0.25	0.00415	0.00324	0.00233	0.00277	0.00192	0.00114		Same as long beam	
	0.5	0.5	0.00331	0.00271	0.00205	0.00198	0.00141	0.00085			
	1.0	1.0	0.00260	0.00222	0.00179	0.00130	0.00092	0.00056			
	2.0	2.0	0.00206	0.00184	0.00158	0.00077	0.00054	0.00033			
	2.5	2.5	0.00196	0.00174	0.00153	0.00065	0.00045	0.00028			
	4.0	4.0	0.00174	0.00159	0.00144	0.00043	0.00030	0.00018			
	5.0	5.0	0.00162	0.00154	0.00141	0.00035	0.00024	0.00015			
0.8	0.0	0.0	0.00420	0.00301	0.00189	0.00378	0.00262	0.00155	0.00230	0.00131	0.00057
	0.2	0.128	0.00321	0.00240	0.00160	0.00274	0.00193	0.00116	0.00157	0.00093	0.00043
	0.5	0.32	0.00251	0.00193	0.00137	0.00198	0.00139	0.00085	0.00108	0.00065	0.00031
	1.0	0.64	0.00195	0.00156	0.00117	0.00136	0.00095	0.00059	0.00072	0.00043	0.00021
	2.0	1.28	0.00149	0.00125	0.00101	0.00084	0.00058	0.00036	0.00042	0.00026	0.00013
	4.0	2.56	0.00118	0.00104	0.00090	0.00048	0.00033	0.00020	0.00023	0.00014	0.00007
0.6	0.0	0.0	0.00327	0.00234	0.00143	0.00321	0.00228	0.00137	0.00099	0.00040	0.00008
	0.2	0.072	0.00260	0.00190	0.00119	0.00250	0.00178	0.00108	0.00070	0.00030	0.00007
	0.5	0.18	0.00204	0.00151	0.00098	0.00190	0.00135	0.00082	0.00049	0.00022	0.00006
	1.0	.036	0.00156	0.00116	0.00079	0.00137	0.00096	0.00059	0.00032	0.00015	0.00004
	2.0	0.72	0.00111	0.00085	0.00061	0.00088	0.00061	0.00037	0.00019	0.00009	0.00003
	4.0	1.44	0.00078	0.00063	0.00049	0.00051	0.00035	0.00022	0.00010	0.00005	0.00002
0.4	0.0	0.0	0.00284	0.00205	—	0.00284	0.00204	—	0.00031	0.00004	—
	0.5	0.08	0.00185	0.00131	—	0.00182	0.00128	—	0.00016	0.00003	—
	1.0	0.16	0.00139	0.00098	—	0.00135	0.00094	—	0.00011	0.00002	—
	2.0	0.32	0.00094	0.00066	—	0.00089	0.00061	—	0.00006	0.000013	—
	4.0	0.64	0.00059	0.00042	—	0.00053	0.00036	—	0.00003	0.000009	—

Source: After Ref. 9.21.

[a] All deflections are given as coefficients of wL^4/D. It is assumed that $I_S = (S/L)I_L$. *Notation:* c, width and depth of square column section; L, long span center to center of columns; S, short span; w, load per unit area; D, $Eh^3/12(1 - \mu^2)$; E, modulus of elasticity; h, slab thickness; μ, Poisson's ratio; I_L, moment of inertia of long beam; I_S, moment of inertia of short beam; $\alpha_L = EI_L/DS$; $\alpha_S = EI_S/DL$.

tions are nearly insensitive to changes in the stiffness of the short-span beams, for a given stiffness of long-span beams. The deflection values were obtained from finite difference solutions.

Chang and Hwang[9.22] analyzed a group of slabs approximately the same as Vanderbilt had considered, but using the finite element method. The deflection values were nearly identical for similar slabs. They attributed the small differences to the fact that the finite element analysis used beams of finite widths while the earlier analysis had used zero-width beams. The finite element analysis also included T-beam behavior. They then extended the results to include a semiempirical expression for the equivalent of the C of Eq. 9.2 plus two additional expressions to take into account cracking and the differing restraint conditions existing in end spans. Finally, they altered the equation to account for creep effects. Their equation for short-term deflection is*

$$\delta_s = K_a K_b K_c \frac{w l_{n1}^4}{D} \qquad (9.3)$$

where K_a is the elastic deflection coefficient, K_b the coefficient accounting for state of cracking, K_c the coefficient accounting for reduced restraint in an end span, and l_{n1} the clear span in the long-span direction.

Each of the K values is determined from a complicated semiempirical equation. However, each term is manageable in a spreadsheet or equation solver program. The coefficient K_a was demonstrated to be quite accurate. K_b is based on the same effective moment of inertia concept that is contained in the ACI Code (see below). The K_c term is based on modifying the end-span deflections using a beam deflection equation based on the end moments and the static moment. Compared to test data, their example calculations were quite accurate for interior spans but tended to underestimate end-span deflections. It seems likely that they overestimated the exterior negative moments in their calculations, which is consistent with the underestimation of the deflections.

If long-term deflections are to be determined, the applied load, w, is replaced by $(w_v + i w_s)$, where w_v is the variable part of the load that is assumed to cause no creep, w_s the sustained load consisting of the self-weight plus the sustained part of the live load, and $i = 1 + \lambda$, where λ is a creep multiplier in the sense of the λ from Eq. 9.1, but does not have to have that value. Hwang and Chang[9.23] then built on Ref. 9.22 to provide some simpler equations were intended to be used in selecting the thickness of a slab to limit the deflection to a predetermined value.

Jofriet[9.24] gives an equation for the midpanel deflection of a typical interior panel of a beamless slab as

---

*Equation 9.3 is reproduced by permission from the American Society of Civil Engineers.

$$\delta^i_{max} = C_i \frac{wl_l^4}{Eh^3} \tag{9.4}$$

where C_i is a coefficient depending on the panel shape and support size, E the modulus of elasticity of the concrete, h the slab thickness, w the uniformly distributed load per unit area, and l_l a weighted average of the long clear span l_n and the long center-to-center span l, where

$$l_l = \frac{l + 3l_n}{4} \tag{9.5}$$

$$C_i = 0.0285 + 0.0375 \left(\frac{l_s}{l_l}\right)^3 \tag{9.6}$$

The short-span l_s is also a weighted average. Deflections computed using Eqs. 9.4 to 9.6 do not agree exactly with those given in Table 9.2, but they are reasonably close to the same. They were derived from a study of deflections obtained by finite element and other solutions of slabs.

Elastic deflection coefficients for additional cases are given by Timoshenko and Woinowsky-Krieger,[9.27] by Brotchie and Wynn,[9.28] and in a number of other references listed in Chapter 3.

A recent state-of-the-art report has been written by ACI Committee 435.[9.29] This report discusses beam analogies for deflection which have been suggested by several authors. In these analogies, the long-span column strip deflection is computed as if it were a uniformly loaded beam with the end moments as determined for design purposes. The deflection of the short-span middle strip is then found in the same way, and the two deflections summed to find the approximate deflection at the center of the panel. There are detail differences among the various methods in such matters as whether the clear span or the center-to-center span is considered and in how cracking is taken into account. A weakness of the beam analogy methods for finding slab deflections lies in the assumption of a parabolic bending moment diagram for the middle strip. Although a parabola is reasonable for a slab with stiff beams on all column lines, the moment diagram has a distinctly different shape for beamless slabs, and the relationship between end moments, midspan moment, and deflection must be significantly different than in a uniformly loaded prismatic beam.

Deflections of edge and corner panels are usually more critical than deflections of interior panels, but because of the large number of variables involved when column flexural and edge beam torsional stiffnesses are added to those considered for interior panels, tabulations of solutions are not feasible.

Jofriet[9.24] and Chang and Hwang[9.22] considered end spans, starting with the interior span deflections. Jofriet increased the end-span deflections by a ratio depending on the stiffness of the exterior equivalent column and the panel aspect ratio. For an end span, the C_i of Eq. 9.4 is replaced with*

$$C_e = C_i(1.6 - 0.8\alpha_{ec}^* l_1 / l_2)$$ (9.7)

where $\alpha_{ec}^* = \alpha_{ec}/1 + \alpha_{ec})$ and $\alpha_{ec} = K_{ec}/\Sigma(K_s + K_b)$ (see Section 4.5). The torsional stiffness of the edge beam is included in the K_{ec} calculation, but the flexural stiffness is not included directly. For a corner panel, the modifier of Eq. 9.7 is computed separately for the two directions, and the two modifiers are then multiplied together. Chang and Hwang derived the semiempirical term K_c of Eq. 9.3 considering the deflections of uniformly load beams with various end-moment conditions and other factors, including the panel aspect ratio and the beam stiffnesses.

The choice of flexural rigidity EI to be used in the equations for deflections will remain one of the difficult problems in predicting deflections. Vanderbilt et al.[9.20,9.21] suggested two empirical approaches. Deflections for low loads, up to those causing the initial midspan cracks, would be computed using the gross section EI. The deflection at the load causing the first section to yield could be computed using the fully cracked, transformed section value of EI, which obviously must be an averaged value. Two different transitions were suggested to compute deflections at intermediate loads, based on the shapes of observed load–deflection curves. Jofriet[9.24] provided a simple equation providing a reduction in stiffness as the moments exceeded the cracking moments.

The finite element method provides a straightforward method of finding the deflection of an uncracked slab. A fully cracked slab can also be modeled fairly readily. However, the development of a rational transition from uncracked to partially cracked to fully cracked has been elusive. There is a strong tendency for the transition from uncracked to fully cracked to be much more abrupt in the analysis than in test structures, as can be seen in the paper by Bhatti et al.[9.25] Jofriet and McNeice[9.26] developed a transition that was much smoother and more gradual. In finite element jargon, this adaptation is often called *tension stiffening*.

The ACI Code[9.1] gives an equation for an effective moment of inertia to be used in deflection computations. This I_e is intermediate between the gross section and fully cracked section moments of inertia, depending on the applied moment relative to the cracking moment. This is stated as

$$I_e = I_g \left(\frac{M_{cr}}{M_a}\right)^3 + I_{cr} \left[1 - \left(\frac{M_{cr}}{M_a}\right)^3\right]$$ (9.8)

but not greater than I_g, where I_g is the moment of inertia of the unit width of

---

*Equation 9.7 is reproduced by permission from the American Society of Civil Engineers.

the gross (uncracked) section; I_{cr} the moment of inertia of the unit width of the cracked, transformed section; M_a the maximum applied moment in the span; $M_{cr} = f_r I_g/y_t$ the cracking moment; f_r the modulus of rupture of concrete $= 7.5\sqrt{f_c'}$ lb/in^2 $(0.62\sqrt{f_c'}$ N/mm$^2)$; and y_t the distance from the extreme tension fiber to the neutral axis of the gross (uncracked) section.

The applicability of Eq. 9.8 to the case of reinforced concrete slabs is not well established. The equation was derived from the results of tests and analyses of beams, which will generally have higher reinforcement ratios and consequently, potentially very different ratios of M_{cr}/M_a and I_{cr}/I_g.

In the case of continuous beams, the value of I_e used in a deflection computation is the average of the values of I_e obtained for the midspan and support sections. The code uses the same I_e expression for slabs. However, there is no comment on what value of M_{cr}/M_a to use in the calculation. There are many possibilities, since the unit moment varies continuously across sections as well as along spans. The maximum value of a local moment would not be suitable for use as M_a, but it is not clear that the average moment across a critical section would be reasonable, either.

ACI Committee 435[9.29] endorses the I_e concept but with a reduced modulus of rupture to reflect the influence of restrained shrinkage on the apparent tensile strength of the concrete. A value of $4\sqrt{f_c'}$ lb/in^2 $(0.33 \sqrt{f_c'}$ N/mm$^2)$ was suggested by several investigations. ACI Committee 435[9.29] also discussed the variability of both initial and long-term deflections. Deflections over the range of the average $\pm$ 60% were reported in one set of measurements of the deflections in 175 nominally identical bays of a building in Chicago. The investigation that led to this conclusion also reported the measured thicknesses of some slabs (Sbarounis[9.30]) in the same structure. Thicknesses measured at openings ranged from 6.75 to 8.0 in. (171 to 203 mm), with a standard deviation of 0.31 in. (7.9 mm). Since I_g varies as the thickness cubed, the thickest slab would have a moment of inertia about 66% larger than the thinnest slab. The thickness variation is an obvious source of at least some of the deflection variation.

9.2.3   ACI Code Provisions for Deflection Control

The thickness of a floor slab must be determined early in the design process so that the dead load of the floor system can be estimated. The most common limitation on thickness arises from the need to control deflections. The 1995 ACI Code, Sec. 9.5,[9.1] gives two methods for the control of deflections. These are the use of limiting thicknesses or the use of limiting computed deflections.

Limiting Thicknesses. For one-way slab construction, the minimum thickness stipulated in Table 9.3 [Table 9.5(a) of the ACI Code] applies unless the computation of deflections indicates that smaller thicknesses can be used without adverse effects. The table takes into account the yield strength of the

TABLE 9.3   Minimum Thickness of Nonprestressed Beams or One-Way Slabs Unless Deflections Are Computed[a]

Members Not Supporting or Attached to Partitions or Other Construction Likely to be Damaged by Large Deflections

Member	Simply Supported	One End Continuous	Both Ends Continuous	Cantilever
Solid one-way-slab	$\dfrac{l}{20}$	$\dfrac{l}{24}$	$\dfrac{l}{28}$	$\dfrac{l}{10}$
Beams or ribbed one-way slabs	$\dfrac{l}{16}$	$\dfrac{l}{18.5}$	$\dfrac{l}{21}$	$\dfrac{l}{8}$

Source: After Ref. 9.1.

[a] The values given shall be used directly for members with normal weight concrete (w_c = 145 lb/ft^3) [2300 kg/m^3] and grade 60 [grade 420] reinforcement. For other conditions, the values must be modified as follows: (1) For structural lightweight concrete having unit weights in the range 90 to 120 lb/ft^3 [1500 to 2000 kg/m^3], the values shall be multiplied by $(1.65 - 0.005w_c)$ [$1.65 - 0.0003w_c$] but not less than 1.09, where w_c is the unit weight in lb/ft^3 [kg/m^3]; (2) for f_y other than 60,000 lb/in^2 [420 N/mm^2], the values shall be multiplied by $(0.4 + f_y/100,000)$ [$0.4 + f_y/700$].

steel and the unit weight of the concrete. For slabs not built integrally with their supports, the span *l* may be considered as the clear span plus the depth of the slab but need not exceed the distance between centers of supports. For slabs continuous with their supports, the span *l* may be considered as the clear span.

For two-way slab construction the code gives a table and two equations for minimum slab thickness. The provisions take into account the yield strength of the steel, the shape of the panel, the relative stiffnesses of the supporting beams, and whether the panel is an interior or exterior panel. The provisions do not take into account the concrete strength or Young's modulus, or the applied load intensity. The thicknesses given may be inadequate for heavily loaded slabs, especially for beamless slabs.

Table 9.4 [Table 9.5(c) of the 1995 Code] gives the minimum thicknesses for slabs without interior beams.

The drop panel specified in the footnote to Table 9.4 is that specified in the ACI Code, Sec. 13.3.7, which has the qualifier "where a drop panel is used to reduce the amount of negative moment reinforcement. . . . " Drop panels of other dimensions are sometimes used to control shear stresses, but a smaller drop panel is not judged to add significantly to the overall flexural stiffness.

Two code equations for slab thickness are given for beam-supported panels with aspect ratios up to 2.0 (ACI Code Eqs. 9-11 and 9-12). For interior panels of slabs with beam stiffnesses α_m between 0.2 and 2.0, the minimum thickness is

TABLE 9.4   Minimum Thickness of Slabs Without Interior Beams

| Yield strength $f_y{}^a$ (lb/in² [N/mm²]) | Without Drop Panels[b] | | | With Drop Panels[b] | | |
| | Exterior Panels | | | Exterior Panels | | |
	Without Edge Beams	With Edge Beams[c]	Interior Panels	Without Edge Beams	With Edge Beams[c]	Interior Panels
40,000 [300]	$\dfrac{l_n}{33}$	$\dfrac{l_n}{36}$	$\dfrac{l_n}{36}$	$\dfrac{l_n}{36}$	$\dfrac{l_n}{40}$	$\dfrac{l_n}{40}$
60,000 [420]	$\dfrac{l_n}{30}$	$\dfrac{l_n}{33}$	$\dfrac{l_n}{33}$	$\dfrac{l_n}{33}$	$\dfrac{l_n}{36}$	$\dfrac{l_n}{36}$
75,000 [520]	$\dfrac{l_n}{28}$	$\dfrac{l_n}{31}$	$\dfrac{l_n}{31}$	$\dfrac{l_n}{31}$	$\dfrac{l_n}{31}$	$\dfrac{l_n}{34}$

Source: Adapted from Ref. 9.1. By permission from ACI International.

[a] For values of reinforcement yield strength between the values given in the table, minimum thickness shall be determined by linear interpolation.

[b] Drop panel shall extend not less than one-sixth of the center-to-center span on each side of the column. Projection of the drop panel below the slab shall be at least one-fourth of the thickness of the slab beyond the drop panel.

[c] Slabs with beams between columns along exterior edges. The value of α for the edge beam shall not be less than 0.8.

$$h = \frac{l_n(0.8 + f_y/200{,}000)}{36 + 5\beta(\alpha_m - 0.2)} \geq 5 \text{ in. [120 mm]} \tag{9.9}$$

where h is the slab thickness; l_n the clear span in the long direction of the panel, measured face to face of the columns in beamless slabs or face to face of beams (note that this is a different definition of l_n than was used in connection with the static moment computations of the direct design method in Chapter 4); f_y the yield strength of the reinforcement, psi units [if f_y is in N/mm² units, replace the term $f_y/200{,}000$ with $f_y/1500$]; α_m the average of the α values of the four beams at the edges of the panel, where α is the ratio of the EI of the beam section to the EI of the width of slab bounded laterally by centerlines of the adjacent panels on each side of the beam (where present); and β the aspect ratio of the panel = long clear span/short clear span. For slabs with higher beam stiffnesses, $\alpha_m \geq 2.0$, the minimum thickness is

$$h = \frac{l_n(0.8 + f_y/200{,}000)}{36 + 9\beta} \geq 3.5 \text{ in. [90 mm]} \tag{9.10}$$

This limiting value for h is the same as in Eq. 9.9 with $\alpha_m = 2.0$, so applies to the conventional two-way slab with relatively stiff beams.

Edge and corner panels are required either to have edge beams with $\alpha \geq$ 0.8 or to be made 10% thicker than the requirements of Eq. 9.9 or 9.10. The flexural stiffnesses of the supporting columns or walls is not taken into account in the ACI Code provisions.

A number of other limits are also placed on two-way slab thickness, including some absolute minimum values, apparently derived from simple construction considerations, as follows:

Slabs without beams or drop panels: 5 in. [120 mm]

Slabs with adequate drop panels: 4 in. [100 mm]

Slabs with beams having $\alpha_m \geq 2.0$:   3.5 in. [90 mm]

It is important to note that the heat-transmission portion of the fire resistance requirements will frequently lead to larger slab thicknesses than those just cited. Heat transmission is dependent on the aggregate in the concrete, with lightweight aggregates leading to the thinnest slabs, siliceous aggregates to the thickest slabs, and calcareous materials to intermediate thicknesses. Information on fire resistance is given in Chapter 13.

Limiting span/thickness ratios, l_n/h, for interior panels of two-way slabs are plotted versus β, the panel aspect ratio, in Fig. 9.1. This figure is specifically for the case of $f_y = 60,000$ lb/in^2. The thicknesses for $f_y = 420$ N/mm^2 are very slightly smaller for slabs with beams, due to the approximations made in the SI conversions. The relationships are straight lines and indicate h values ranging from about $l_n/33$ for a beamless slab to $l_n/49.1$ for a rectangular panel with $\beta = 2$ for a slab on stiff beams. For intermediate values of α_m, linear interpolations may be made using Fig. 9.1 or Eq. 9.9 can be solved directly. In practical terms and with grade 60 steel, a 7-in. (178-mm)-thick beamless two-way slab with suitable spandrel beams is adequate for a clear span of $l_n = 19$ ft (5.8 m). When stiff beams ($\alpha_m \geq 2.0$) are used, a square 6-in. (152-mm)-thick slab is adequate for a clear span of nearly 21 ft (6.3 m).

Limiting Computed Deflections. Thicknesses less than those specified by the ACI Code[9.1] may be used only when the deflections are computed and shown to be within acceptable bounds. They are to be calculated using a rational analysis taking into account all aspects of the geometry of the slab plus a recognition of the influence of cracking in reducing the effective moments of inertia of the slab and beam sections. The deflection calculations could be made using an elastic analysis such as described in Chapter 2 or an approximate analysis such as that described in Section 9.2.2. If this approach is adopted, the deflections must not exceed the limits set out in Table 9.1 [Table 9.5(b) of the ACI Code]. Other limits may be appropriate for some structural

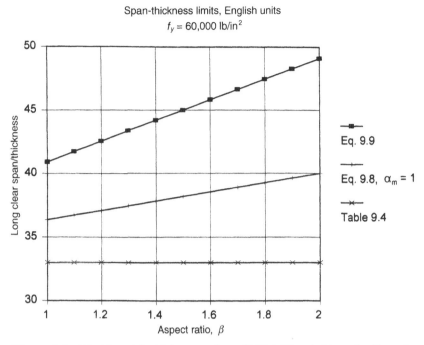

Figure 9.1   Limiting slab thicknesses from 1995 ACI code,[9.1] grade 60 steel.

uses, such as situations where machines must be kept to within very close alignment tolerances.

It should be clear from the material presented above that if limiting deflections to small values is important, the type of slab used should be considered carefully. Drop panels, column capitals (due to larger c distances and smaller l_n distances), and beams all lead to lower deflections for a given slab thickness and loading. The analytical and experimental results discussed previously provide numerical data that would be helpful in the design of deflection-limited structures.

9.3   CRACKING

9.3.1   Need for Crack Control

The occurrence of cracks in some regions of reinforced concrete slab systems is inevitable because of the low tensile strength of concrete. The tensile resistance of concrete is normally neglected in design. Slab systems designed with low steel stresses at the service load may serve their intended function with very limited cracking. In many cases no cracking is visible at all because many slabs are not subjected to their full service load and the concrete has some tensile strength. However, with high-service-load steel stresses, particularly as a result of the use of high-strength steel, and high service loads, some cracking must be expected at the service load. The cracking of a reinforced concrete slab at the service load should not be such as to spoil the appearance of the structure or to lead to corrosion of the reinforcement. These two requirements are considered next.

Aesthetic Considerations. The maximum size of a crack that may be considered nondetrimental to the appearance of a reinforced concrete slab, or nonconducive to feelings of alarm, depends on the position, length, width, illumination, and surface texture of the crack. The social background of the users and the type of slab system also exert an influence. The limits on aesthetic acceptability are difficult to set because of the variability of personal opinion. The maximum crack width that will neither impair a slab's appearance nor create public alarm is probably in the range 0.010 to 0.015 in. (0.25 to 0.38 mm), but larger crack widths may be tolerated.

Protection Against Corrosion. Portland cement concrete usually provides good protection for embedded reinforcing steel against corrosion. The protective value of the concrete is due primarily to its high alkalinity. If chemical agents such as carbon dioxide (producing carbonic acid) penetrate to the concrete surrounding the steel, the alkalinity is neutralized and the corrosion-inhibiting properties are reduced. Chlorides from deicing salts, sea spray, and so on, are also extremely active corrosion agents. Concrete of low permeability resists the penetration of corrosion agents. The main factors affecting the rate of diffusion of corrosion agents to the steel are the permeability of the concrete; the thickness of the concrete cover; the width, shape, and length of cracks; and the period of time the cracks are open. Once corrosion of steel begins, the rate of corrosion may increase because the product of corrosion is of greater volume than the steel corroded, thus causing increased concrete crack widths and possibly spalling of concrete cover.

The permeability of the concrete is a major factor affecting the corrosion of reinforcing steel. It is extremely important to avoid the presence of inferior concrete around the steel. The thickness of concrete cover also affects the rate of penetration of the corrosion agents. In many publications, cracking is assessed only in terms of crack widths on the concrete surface. However, it is evident that the shape of the crack (i.e., the variation in the crack width between the concrete surface and the bar surface) and the length of the crack are as important as the surface width of the crack in assessments of the reduction in the effectiveness of the cover concrete due to cracking. Thus, the importance of surface crack width has been overemphasized by many publications. Ideally, the durability of a reinforced concrete member should be assessed by estimating the rate of corrosion in terms of the thickness and permeability of the cover concrete; the width, shape, and length of cracks; the period of time the cracks are open; and in terms of the corrosive nature of the environment. The bar diameter is also a consideration in that for a given depth of corrosion in the bar, the percentage loss in bar area will be greater for small-diameter bars. However, full assessment appears to be impracticable at present, particularly because of the difficulty of determining the important parameters. The influence of cracking on corrosion of the reinforcement is still the subject of research, and conflicting data have been reported. It is possible that the effect of crack shape has not been appreciated in many cases because results have invariably been reported in terms of the crack width at the surface of the concrete. Some studies have indicated that surface crack widths of up to 0.016 in. (0.41 mm) have produced little or no corrosion, even in aggressive environments, whereas other reports have not been so optimistic. At present, cracking is controlled by specifying maximum allowance crack widths at the surface of the concrete for given types of environment.

9.3.2 Causes of Cracking

The causes of cracking in concrete are numerous, but most cracks occur as a result of one or more of the following actions.

Cracking Due to Settlement of Plastic Concrete. As concrete sets, it tends to settle slightly in the mold when in the plastic state. This causes the concrete to drop away slightly on each of the bars near the top surface of the concrete, because the bars are normally fixed in position. Lines of cracking following the reinforcement may result. Such cracks may sometimes be observed on the top surface of slabs and in beams over stirrups and other top steel. This type of cracking can be minimized by good mix design and by revibration and screeding of the plastic concrete.

Cracking Due to Volumetric Change. Drying shrinkage and thermal stresses cause volumetric changes that will introduce tensile stresses in the concrete if restrained, and therefore can lead to cracking. The restraint can arise in a number of ways. For example, concrete near the surface of members shrinks more than the concrete farther inside the member; therefore, the inner concrete will restrain the outer concrete, causing tensile stresses to develop near the surface, which may cause surface cracking. Also, shrinkage of slabs may be restrained by continuity with beams, columns, and foundations, and by the presence of reinforcing, thus introducing tension. For example, shrinkage cracks are often found along the junctions of slabs and beams in continuous slab and beam floors, because the slabs have shrunk more than the beams as a result of their greater surface area per unit volume of concrete and smaller reinforcement content. Similarly, temperature change will cause tension if the movements cannot occur unrestrained. Cracking due to shrinkage may be controlled by reducing the shrinkage of concrete by good mix design (e.g., by keeping the water content as low as possible) and by properly placed reinforcement. The reinforcement will not prevent cracking. Indeed, the restraint of the reinforcement will tend to encourage cracking, but the shrinkage strains are distributed along the bars by bond, and a number of fine cracks should occur (instead of a few wide cracks). The minimum amount and maximum spacing of reinforcement that may be used in slabs and walls is given in ACI 318-95, Secs. 7.12 and 13.3.2.[9.1] This reinforcement is intended to be adequate to control crack widths due to shrinkage and temperature stresses. Control joints in walls and slabs are an effective method of preventing unsightly shrinkage cracking in large expanses of concrete. Such joints consist normally of grooves in the concrete along which the concrete is encouraged to crack. This controlled cracking relieves the stresses elsewhere in the concrete. Sawed joints are commonly used in pavements for this purpose.

Cracking Due to Flexural Stresses Resulting from Applied Load or Reactions. Cracking will occur in the tension zone of slabs subjected to flexure arising from external loads or reactions once the modulus of rupture of the concrete is exceeded. The cracks may form perpendicular to the plane of the slab, as in the case of flexure without significant shear force; or when the shear force is significant, they may form inclined to the plane of the slab. Little information is available on the control of inclined (diagonal tension) cracks, but there is evidence that the control mechanism for diagonal tension cracking is similar to that for flexural cracking. Diagonal tension cracking is not normally a serious problem in slabs. Sections 9.3.3 to 9.3.5 deal with the mechanism of flexural crack formation and with flexural crack control.

9.3.3 Computation of Width of Flexural Cracks in One-Way Slabs

Many variables influence the width and spacing of flexural cracks in reinforced concrete beams and one-way slabs. Because of the complexity of the problem, a number of approximate, semitheoretical, and empirical approaches have been developed for the determination of the width of flexural cracks, each approach containing a selection of the variables. In early theories crack widths were believed to depend largely on the quality of bond between the concrete and steel; the spacing between cracks was determined from the tensile strength of the concrete and the rate of transfer of steel tension to the concrete by bond, and the crack width was generally postulated to be the elongation of the steel between two cracks. The crack width equations given by these basic considerations have been modified empirically by many research workers on the basis of experimental results. An alternative early approach was the *no-slip theory,* in which it was assumed that there was no slip of the steel relative to the concrete. The crack in this approach was therefore assumed to have zero width at the surface of the reinforcing bar and to increase in width as the surface of the concrete was approached (i.e., the crack was wedge-shaped). A review of methods for determining the width of flexural cracks in reinforced concrete beams and one-way slabs has been presented.[9.31] In the more recent methods, crack widths have been observed to be primarily a function of the stress in the steel, the thickness of concrete cover, the distribution of the steel reinforcement, and the relative distances from the neutral axis of the reinforcing steel and the extreme concrete tension fiber. However, crack width measurements are inherently subject to large scatter, even in careful laboratory work. Crack width is also influenced by shrinkage and other time-dependent effects, and by repeated loading. Thus, great accuracy cannot be expected from existing equations for crack width. Two of the recent methods of maximum crack width determination are discussed in some detail below.

Statistical Approach by Gergely and Lutz. The method most accepted in the United States for beams and one-way slabs is that due to Gergely and Lutz,[9.32] who subjected data from previous investigations to statistical analysis to determine the importance of the variables involved. Many combinations of variables were tried, and it was very difficult to obtain an equation that fitted all sets of data well. The important variables were found to be the effective area of concrete in tension, the number of bars, the side or bottom cover, the strain gradient from the level of the steel to the tension face, and the steel stress. Of these, steel stress was the most important. The following equations were developed for predicting the maximum crack widths on the surface of members reinforced by deformed bars. At the extreme tension fiber,

$$w_{max} = 0.076 \sqrt[3]{t_b A} \, \frac{h_2}{h_1} \, f_s \times 10^{-6} \text{ in.} \tag{9.11}$$

At the level of the reinforcement,

$$w_{max} = \frac{0.076 \sqrt[3]{t_b A}}{1 + \frac{2}{3} (t_s/h_1)} \, f_s \times 10^{-6} \text{ in.} \tag{9.12}$$

where t_b is the distance from extreme tension fiber to the center of the adjacent bar (in.), t_s the distance from the side of the beam to the center of the adjacent bar (in.), A the average effective area of concrete in tension around each reinforcing bar [$= A_e/n$, where A_e is the rectangular concrete area of width b and with same centroid as steel, b the beam width, and n the number of bars] (in.^{2}), f_s the steel stress (psi), h_1 the distance from centroid of steel to the neutral axis (in.), and h_2 the distance from the extreme tension fiber to the neutral axis (in.). Some of the notation is shown in Fig. 9.2. If all the dimensions are in millimeters and the stress is in N/mm^2, the numerical constant 0.076×10^{-6} in Eqs. 9.9 and 9.10 should be replaced by 1.1×10^{-5}, and the crack width will be given in millimeters. The maximum crack width, w_{max}, is about twice the average width and has a probability of occurrence of about 1%.

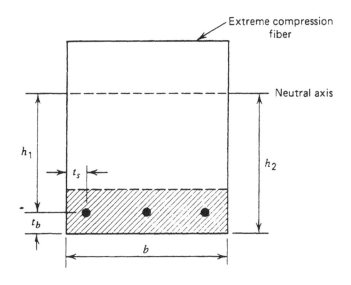

A_c = shaded rectangular
area of width b with
same centroid as
the steel

Figure 9.2 Notation for Gergely–Lutz crack width equation.[9.32]

Lloyd et al.[9.33] have measured maximum crack widths on one-way slabs reinforced by deformed bars, deformed wires, deformed wire fabric, and smooth wire fabric. Comparison of their experimental data with Eq. 9.12 is shown in Fig. 9.3. It is evident that considerable deviation from the maximum crack width predicted by the equation has occurred in some cases. It was concluded that deformed bars, deformed wires, deformed wire fabric, and smooth wire fabric control the crack width of one-way slabs equally well when the reinforcement is similarly placed, and that Eqs. 9.11 and 9.12 predicted the maximum crack widths satisfactorily in one-way reinforced concrete slabs.

General Approach by Beeby. No satisfactory theory exists to enable the accurate prediction of the cracking behavior of reinforced concrete one-way slabs, but the work of Beeby[9.34] at the Cement and Concrete Association has resulted in a clearer understanding of the mechanisms of cracking. Beeby measured crack widths and spacing at various points across the bottom of one-way reinforced concrete slabs, that is, for various values of c as in Fig. 9.4a. It was found that the crack spacing and width increased with distance from the bar and at some distance from the bar approached constant values, which were dependent on crack height rather than the distance from the bar.

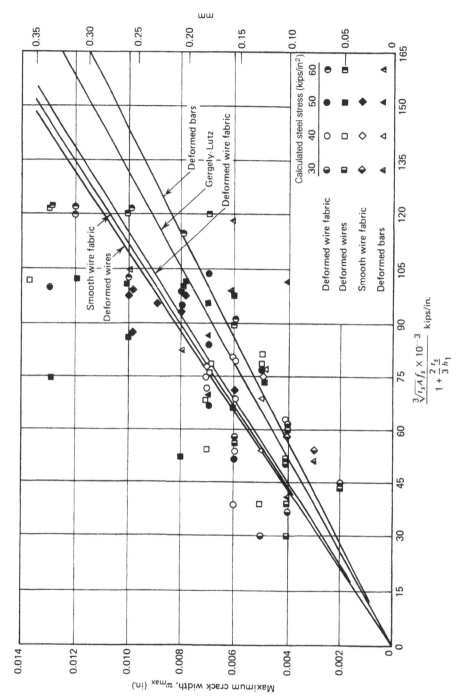

Figure 9.3 Maximum crack widths measured in one-way reinforced concrete slabs at level of reinforcement compared with Gergely–Lutz equation (1 kip/in² = 6.89 N/mm², 1 kip/in. = 175 kN/m).[9.32]

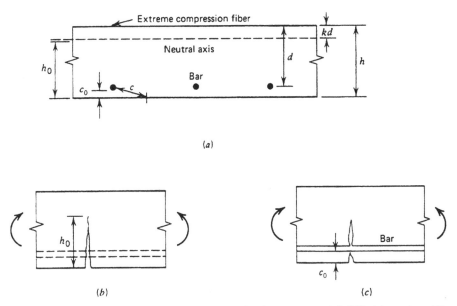

Figure 9.4 Effect of bar proximity on cracking in one-way slabs[9.34]: (*a*) section; (*b*) crack at distance from a bar, h_0 controlled; (*c*) crack at a bar, c_0 controlled.

Beeby therefore concluded that the crack pattern at any point was the result of interaction between two basic crack patterns.

Cracking at a point distant from a reinforcing bar is illustrated in Fig. 9.4*b*. The crack pattern in this case is controlled by crack height h_0. The crack will penetrate nearly to the neutral axis, and its height may be calculated by standard elastic theory using the steel content and the modular ratio. From St.-Venant's principle, it is evident that the concrete tensile stresses between the cracks are substantially unaffected by the crack at distances greater than h_0 from the crack. Hence, the next crack will form at a distance from the crack approximately equal to or greater than h_0. Hence, if the spacing between two existing cracks is $2h_0$ or greater, a new crack can form between them. If the spacing between two existing cracks is less than $2h_0$, a new crack cannot form between them. Therefore, the minimum spacing of the cracks is h_0 and the maximum is $2h_0$, giving a mean crack spacing of $1.5h_0$. A mean value of $1.33h_0$ was actually measured by Beeby in the tests. The crack width and the spacing were found to be directly proportional to the initial crack height h_0. Therefore. this type of cracking is controlled by the initial crack height h_0.

Cracking at a point directly below a reinforcing bar is illustrated in Fig. 9.4*c*. The no-slip theory mentioned previously originated from the Cement and Concrete Association and predicts wedge-shaped cracks with zero width at the bar. That is, a linear relationship is predicted between crack width and distance from the bar. Thus, directly below the bar the effective crack height is c_0; using the same reasoning as before, the crack spacing will vary between c_0 and $2c_0$, with a mean spacing of $1.5c_0$. Slip or deformations at the bar surface that occur before the crack pattern has fully developed will result in the crack having some width at the surface of the bar and will increase the effective crack height, causing larger crack spacings and widths. If there was no bond between concrete and steel, the crack pattern would be controlled by the initial crack height h_0. Thus, the effect of slip and internal deformations is to modify the c_0 controlled crack pattern toward the h_0 controlled crack pattern, and the widths of cracks in this general case will be a function of c_0, to take the wedge shape into account, and c_0/h_0, to take slip and internal fracturing at the bar surface into account.

Beeby found that the following equations gave the best fit to his experimental data. Maximum crack width directly below a bar:

$$\frac{w_{\text{max}0}}{\varepsilon_m} = K_1 c_0 + K_2 \frac{A}{d_b} e^{-K_3(c_0/h_0)} \tag{9.13}$$

Maximum crack width at a distance from a bar:

$$\frac{w_{\text{max}1}}{\varepsilon_m} = K_1 h_0 \tag{9.14}$$

Maximum crack width for intermediate positions:

$$w_{\text{max}} = \frac{c w_{\text{max}1} w_{\text{max}0}}{c_0 w_{\text{max}1} + (c - c_0) w_{\text{max}0}} \tag{9.15}$$

where ε_m is the average longitudinal strain at the level where cracking is being considered; K_1, K_2, and K_3 are constants that depend on the probability of the crack width being exceeded; c_0 the minimum cover to steel; A the effective area of concrete in tension surrounding one bar; d_b the bar diameter; e the base of the natural logarithm; h_0 the initial height of the crack; and c the distance from the point of measurement of the crack to the surface of the nearest bar.

Equations 9.13 to 9.15 are too complex for practical use. The equations can be simplified, as Beeby[9.35] has done, giving the crack width that will be exceeded by approximately 20% of the results as

$$w_{\text{max}} = \frac{3c\varepsilon_m}{1 + 2[(c - c_0)/(h - kd)]} \tag{9.16}$$

where h is the overall depth of the section, kd the neutral-axis depth, and

$$\varepsilon_m = \left[\varepsilon_s - \left(\frac{2.5bh}{A_s} \times 10^{-6} \right) \right] \frac{h - kd}{d - kd} \tag{9.17}$$

where ε_s is the strain in the steel at a crack, b the section width, h the overall depth of the section, A_s the area of tension steel, d the effective depth, and kd the neutral-axis depth. Equation 9.17 for ε_m is the steel strain at a crack less an empirical term due to the stiffening effect of concrete tension between cracks, modified by the strain gradient term to obtain the average strain at the extreme tension fiber of the member.

Example 9.1. A continuous one-way reinforced concrete slab is 10 in. (254 mm) thick. At the critical negative-moment section, the reinforcement consists of No. 5 [No. 16 mm] high-strength bars on 10-in. (254-mm) centers placed with ¾ in. (19 mm) of cover. At the service load the steel stress at the cracked sections is 52,000 lb/in² (359 N/mm²). The ratio of the modulus of elasticity of the steel to that of the concrete is $n = 9$, and the modulus of elasticity of the steel is 29×10^6 lb/in² (200,000 N/mm²). Calculate the maximum crack width at the extreme tension fiber of the concrete at the section.

SOLUTION. Consider a 10-in.-wide strip of slab containing one bar.

$$\frac{A_s}{b} = \frac{0.31}{10} = 0.031 \text{ in.}^2/\text{in.} \qquad d = 10 - 0.75 - 0.31 = 8.94 \text{ in.}$$

$$\rho = \frac{A_s}{bd} = \frac{0.031}{8.94} = 0.00347 \qquad \rho n = 0.00347 \times 9 = 0.0312$$

From elastic theory,[9.31]

$$k = \sqrt{(\rho n)^2 + 2\rho n} - \rho n$$
$$= \sqrt{0.0312^2 + 2 \times 0.312} - 0.0312 = 0.221$$

$$kd = 0.221 \times 8.94 = 1.98 \text{ in.}$$

$$h_1 = d - kd = 8.94 - 1.98 = 6.96 \text{ in.}$$

$$h_2 = h - kd = 10 - 1.98 = 8.02 \text{ in.}$$

From the Gergely–Lutz equation (Eq. 9.9),

$$t_b = 0.75 + 0.31 = 1.06 \text{ in.}$$

$$A = 2 t_b b = 2 \times 1.06 \times 10 = 21.2 \text{ in.}^2/\text{bar}$$

$$w_{max} = 0.076 \sqrt[3]{1.06 \times 21.2} \; \frac{8.02}{6.96} (52{,}000 \times 10^{-6})$$

$$= 0.0129 \text{ in. } (0.33 \text{ mm})$$

From the Beeby equations (Eqs. 9.14 and 9.15),

$$\varepsilon_s = \frac{52,000}{29,000,000} = 0.00179$$

$$\varepsilon_m = \left[0.00179 - \left(\frac{2.5 \times 10}{0.031} \times 10^{-6} \right) \right] \frac{8.02}{6.96}$$

$$= 0.000984 \times \frac{8.02}{6.96} = 0.00113$$

$$c_0 = 0.75 \text{ in.}$$

For maximum crack width, c is the distance from the bar surface to a point on the concrete surface midway between bars; that is,

$$c = \sqrt{t_b^2 + (0.5 \times \text{bar spacing})^2} - 0.5 \times \text{bar diameter}$$
$$= \sqrt{1.06^2 + 5^2} - 0.31 = 4.80 \text{ in.}$$
$$w_{max} = \frac{3 \times 4.80 \times 0.00113}{1 + [2(4.80 - 0.75)/8.02]}$$
$$= 0.0081 \text{ in. } (0.21 \text{ mm})$$

Note: The maximum crack width calculated from the Beeby equation is 37% less than that calculated from the Gergely–Lutz equation. For lightly reinforced members, the term in Eq. 9.17 which accounts for the stiffening effect of concrete between cracks reduces the average longitudinal steel strain to significantly less than the steel strain at the crack (average longitudinal steel strain = $0.55\varepsilon_s$ in this example), and hence Beeby's equations can be expected to give lower w_{max} values than the Gergely–Lutz equation in lightly reinforced slabs.

9.3.4   Computation of Width of Flexural Cracks in Two-Way Slabs

Much less attention has been given in past investigations to determination of the width of flexural cracks in reinforced concrete floors under two-way bending. The major work conducted in the United States is that by Nawy et al.,[9.36,9.37] whereas in the United Kingdom it is that by Clark.[9.38]

Approach by Nawy et al. Nawy et al.[9.36,9.37] have reported the results of tests on two-way concrete slabs reinforced by high-strength welded mesh reinforcement. Two distinct types of cracking were observed: an orthogonal pattern of cracks that followed the lines of the reinforcement, and a diagonal pattern which at higher loads eventually developed into the yield line pattern. Nawy et al. explained this behavior by assuming that the steel force was transferred mainly to the concrete at the node points of the crossing steel. The use of closely spaced small-diameter wires was found to result in the orthogonal crack pattern, whereas the use of widely spaced large-diameter wires led to the diagonal crack pattern. The width of cracks in the orthogonal pattern was found to be smaller than the width of cracks in the diagonal pattern, and thus the orthogonal pattern was to be preferred. A grid index factor was introduced to determine the effect of reinforcing steel spacing, diameter and content, and concrete cover thickness, on the crack width. Regression analysis was performed on the results of a range of two-way slab tests, and empirical equations for maximum crack width were proposed. The equations were also checked against crack widths measured by other investigators, including slabs reinforced by deformed bars rather than mesh reinforcement, and it was concluded that the equations gave good agreement with the experimental crack widths.[9.37] For two-way slabs, flat slabs, and plates, the Nawy–Orenstein equations for maximum crack width may be written as

$$w_{max} = K\beta f_s \sqrt{M_I} \quad \text{in.} \tag{9.18}$$

where K depends on the loading and boundary conditions and can be taken as 2.8×10^{-5} for uniformly loaded slabs and plates, $\beta = h_2/h_1 =$ ratio of distances to the neutral axis from the extreme tension fiber and from the centroid of reinforcing steel, f_s is the steel stress (ksi), and M_I is the grid index, given by

$$M_I = \frac{d_{b1}s_2}{\rho_{t1}} \tag{9.19}$$

where direction 1 is the direction in which the crack width is measured (normal to the crack direction), direction 2 the direction perpendicular to direction 1, d_{b1} the diameter of bar running in direction 1 (in.), s_2 the spacing of bars running in direction 2 (in.), and ρ_{t1} the area A_{b1} of bars running in direction 1 divided by the effective area of concrete in tension perpendicular to those bars (for all dimensions in mm and f_s in N/mm², replace the constant 2.8×10^{-5} with 4.06×10^{-6} to obtain crack widths in mm). The effective area of concrete in tension is the rectangular area extending across the slab that has the same centroid as the reinforcement and is equal to $2t_{b1}s_1$ for the direction 1 bars, where t_{b1} is the distance from direction 1 bar centroid to extreme tension fiber and s_1 is the spacing of direction 1 bars. Therefore, the grid index given by Eq. 9.19 can also be written as

$$M_I = \frac{d_{b1}s_2}{A_{b1}/2t_{b1}s_1} = \frac{2t_{b1}s_1s_2d_{b1}}{A_{b1}} \tag{9.20}$$

Most of the notation is illustrated in Fig. 9.5. If there is no transverse reinforcement, Nawy[9.37] suggests using $s_2 = 12$ in.; otherwise, the grid index is undefined.

Comparison of the Gergely–Lutz equation (Eq. 9.10) and the Nawy equation (Eq. 9.18) shows them to be quite different in form apart from the linear dependence on the steel stress f_s and position of the steel with respect to the

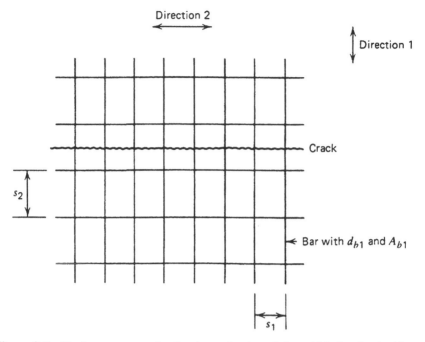

Figure 9.5   Steel arrangement for the determination of the grid index for the Nawy–Orenstein equation.[9.36]

neutral axis and the extreme tension fiber. Nawy and Blair[9.37] claim that Eq. 9.10 considerably underestimates the crack widths in one- and two-way slabs. However, an alternative view could be that Eq. 9.18 is unduly conservative in some cases. For example, in tests on eight reinforced concrete slabs supported on rectangular columns conducted by Hawkins et al.,[9.40] the average strain in the slab steel over the column face was measured when the crack width there was 0.016 in. (0.41 mm).

From these measured steel strains and the slab properties, the maximum crack widths were calculated for each specimen using the Gergely–Lutz equation (Eq. 9.10) and the Nawy–Orenstein equation (with $K\beta = 3.0 \times 10^{-5}$). The maximum crack widths from the Gergely–Lutz equation ranged between 0.011 and 0.016 in. with a mean of 0.014 in. and from the Nawy–Orenstein equation ranged between 0.010 and 0.023 in. with a mean of 0.018 in. It could be considered that Eq. 9.10 was as adequate as Eq. 9.18 as a design equation in this case. Also, it needs to be borne in mind that tests on multi-panel floor systems have shown that the critical cracking takes place principally in negative-moment regions above the faces of the beams in slab and beam floors, and near the columns in flat slab and flat plate floors, whereas the studies of two-way slabs reported by Nawy et al. have dealt principally with positive-moment regions. Certainly, it would appear reasonable to use the Gergely–Lutz equation in regions of slabs where one-way bending was predominant, for example in negative-moment regions when slabs are supported on stiff beams and in slabs at column faces and in the critical positive-moment regions of rectangular panels. The Nawy–Orenstein grid index M_I defined by Eq. 9.19 breaks down if s_2 becomes large or if the direction 2 bars are not present (i.e., $s_2 \rightarrow \infty$), as can be the case in the negative-moment regions in the middle strips of most slabs. As noted previously (Ref. 9.37), the use of $s_2 = 12$ in. is suggested for such cases.

Example 9.2. If the critical negative-moment section of the slab in Example 9.1 is part of a two-way slab and the reinforcement in each direction is identical, calculate the maximum crack width at the extreme tension fiber using the Nawy–Orenstein equation.

SOLUTION.

$$\beta = \frac{h_2}{h_1} = \frac{8.02}{6.96} = 1.15 \qquad K = 2.8 \times 10^{-5}$$

$$f_s = 52 \text{ kips/in}^2 \qquad d_{b1} = 0.625 \text{ in.} \qquad s_2 = 10 \text{ in.}$$

$$\rho_{t1} = \frac{0.31}{10 \times 2 \times 1.06} = 0.0146$$

Therefore, from Eq. 9.16,

$$w_{\max} = 2.8 \times 10^{-5} \times 1.15 \times 52 \sqrt{0.625 \times \frac{10}{0.0146}}$$

$$= 0.0346 \text{ in. (0.88 mm)}$$

Note that this calculated crack width is 2.7 times the width calculated by the Gergely–Lutzequation and 4.3 times the width calculated by the Beeby equation in Example 9.1. The crack width calculated by the Nawy–Orenstein equation in this example is so high that it is difficult to understand. The maximum plausible crack width is approximately $w = \varepsilon_s \times$ crack spacing $\times \beta$. For this example, with the implied crack spacing of $s_2 = 10$ in., $w = (52/29,000) \times 10 \times 1.15 = 0.021$ in. (0.52 mm). This is still much larger than the values from the Gergely–Lutzor Beeby equations. Or, turning the problem around, the implied crack spacing is at least $s_2 = w/(\varepsilon_s^* \beta) = 0.0346/ (52/29,000)(1/1.15) = 16.8$ in. (426 mm). Both of these calculations assume zero elongation in the concrete and that the bar is free to slip relative to the concrete except at the points midway between cracks. The maximum crack spacing for unbonded reinforcement tends toward $2h_0 = 2 \times 8.02 = 16.0$ in.

(406 mm) for this slab, according to Beeby's work outlined in Section 9.3.3. However, the bars are not unbonded, and therefore crack spacing should be smaller than calculated above. In fact, the transverse steel spacing s_2 is implicitly assumed by Nawy et al. to be the crack spacing. Therefore, the maximum crack width should be much smaller than calculated above. Hence, it would seem that the Nawy–Orenstein equation can very greatly overestimate the maximum crack widths under some common conditions.

One source of this disagreement may be size effects. The thickest slab considered in the derivation of these equations was 4.1 in. (104 mm), less than half the thickness of the example slab. Clark[9.39] tested scale models of some of Beeby's slabs[9.34] and found that the crack widths did not scale, probably for a combination of geometric (strain gradient) and material behavior reasons. The crack widths in the prototype slabs were somewhat larger than the scale factor times the widths in the models.

Approach by Clark. Clark[9.40] has reported the results of theoretical and experimental studies of one-way spanning slabs with bars at various angles to the direction of moment, thus simulating regions of two-way slabs with bending predominantly in one direction. In the proposed method of crack width prediction, the equivalent area of steel normal to the crack direction was found using the following relationship determined by Lenschow[9.41] for slabs cracked in one direction:

$$A_n = \sum_{i=1}^{m} A_i \cos^4\alpha_i \tag{9.21}$$

where m is the number of steel layers, A_i the area of the ith layer of steel, and α_i the angle between the normal to the crack and the direction of the ith steel layer. Once the equivalent area of steel normal to the crack is found using Eq. 9.21, the neutral-axis depth and steel stress can be found from elastic theory and then Beeby's equation used to predict the maximum crack width. Reasonable agreement with experimentally obtained crack widths was obtained by Clark. An interesting result from the tests was that for slabs with bars in various directions there was apparently little interaction between the bars apart from contributing to the equivalent area of steel normal to the crack. For example, the crack control obtained from a slab reinforced only perpendicular to the cracks was little different from that obtained when bars were also placed parallel to the cracks.

Calculation of Steel Stress. To apply the crack width equations to two-way slabs, the bending moment at the cracked sections is required so that the steel stresses may be calculated. Elastic theory for slabs may be used to determine the bending moment in the service load range, but it should be kept in mind that cracking will cause some redistribution of moments from that calculated for an uncracked slab even though the slab is in the elastic range.

9.3.5 Code Provisions for Crack Control

Permissible Crack Widths. The permissible values for the width of flexural cracks in practice depend mainly on the environment in which the structure has to serve, particularly from the point of view of the possibility of corrosion of the reinforcement. The permissible values recommended by ACI Committee 224[9.42] are listed in Table 9.5.

In comparison with these values the crack control equations of ACI 318-95[9.1] are based on only two maximum allowable crack widths, 0.016 in. (0.41 mm) for interior exposure and 0.013 in. (0.33 mm) for exterior exposure. According to the commentary, these limiting crack widths were chosen primarily to give reasonable reinforcing details in terms of practical experience with existing structures. However, the code does warn that these limiting values may not be sufficient for structures subjected to aggressive exposure or designed to be watertight, and that for such structures special investigations and precautions are required. Although clear experimental evidence is not available regarding the maximum crack widths beyond which danger of corrosion exists, it is evident that Table 9.5 could be taken as a guide. However, as is evident from the discussion in Section 9.3.1, protection against corrosion is not just a matter of limiting the surface crack width in the concrete. A reasonable thickness of good-quality, well-compacted concrete is also essential for durable structures.

It is also worth noting that the British code of practice BS 8110:1997[9.43] requires in general that the surface crack widths at service load not exceed 0.3 mm (0.012 in.). Smaller widths may be required where members are subjected to particularly aggressive environment. The recommendations of the European Concrete Committee–International Federation of Prestressing[9.44] require that the surface crack widths at the service load not exceed 0.1 mm (0.004 in.) in a very exposed (particularly aggressive) environment, 0.2 mm

TABLE 9.5 Permissible Crack Widths in Reinforced Concrete

Exposure Condition	Maximum Allowable Crack Width [in. (mm)]
Dry air or protective membrane	0.016 (0.41)
Humidity, moist air, soil	0.012 (0.30)
Deicing chemicals	0.007 (0.18)
Seawater and seawater spray, wetting and drying	0.006 (0.15)
Water-retaining structures	0.004 (0.10)

Source: After Ref. 9.42.

(0.008 in.) in an unprotected environment (external member in bad weather conditions or internal member in a damp or aggressive environment), or 0.3 mm (0.012 in.) in a protected environment (internal member in normal surroundings).

ACI 318-95 Method for One-Way Slabs. The ACI 318-95[9.1] method for beams and one-way slabs is based on the Gergely–Lutzequation (Eq. 9.10) with h_2/h_1 put equal to 1.2. The requirement may be written as permissible maximum crack width $\geq 0.076\ ^3\sqrt{t_b A} \times 1.2 f_s \times 10^{-6}$ in., with t_b in inches, A in square inches, and f_s in lb/in^2 (for t_b in mm, A in mm^2, and f_s in N/mm^2, replace the numerical constant with 1.1×10^{-5}). Substituting the permissible values for crack widths (0.013 in. [0.33 mm] for interior exposure and 0.016 in. [0.40 mm] for exterior exposure) into the equation gives, approximately

For interior exposure:
$$f_s \sqrt[3]{t_b A} \leq 175,000 \text{ lb/in.}$$
$$\leq 30,000 \text{ N/mm} = 30 \text{ MN/m} \qquad (9.21a)$$

For exterior exposure:
$$f_s \sqrt[3]{t_b A} \leq 145,000 \text{ lb/in.}$$
$$\leq 25,000 \text{ N/mm} = 25 \text{ MN/m} \qquad (9.21b)$$

where again the units are inches and pounds or millimeters and newtons. ACI 318-95 requires the section to be proportioned so that either Eq. 9.21a or 9.21b is satisfied. This check need only be carried out when the design yield strength for the reinforcement exceeds 40,000 lb/in^2 [300 N/mm^2]. In structures subjected to very aggressive environment or designed to be watertight, Eq. 9.21b does not apply, since a smaller maximum allowable crack width needs to be adopted.

To use Eq. 9.21a or 9.21b, the steel stress f_s at the service load is required. This steel stress may be found from $f_s = M/jdA_s$, where m is the service load bending moment, jd the lever arm of the internal moment, and A_s the steel area. Alternatively, f_s may be taken as 60% of the specified yield strength of the steel, provided that the moment redistribution from the elastic bending moment diagram assumed in design is not greater than that allowed in ACI 318-95.

The commentary to the ACI Code indicates that the average value for h_2/h_1 is about 1.35 for floor slabs rather than the value of 1.2, which really applies to beams, used to determine Eqs. 9.21a and 9.21b. Accordingly, it would be consistent for thin one-way slabs to reduce the limiting values on the right-hand side of Eqs. 9.21a and 9.21b to 156,000 lb/in. [27 MN/m] for interior exposure and 129,000 lb/in. [22 MN/m]for exterior exposure.

Example 9.3. Check the adequacy of the reinforcement arrangement at the critical negative moment section of the slab in Example 9.1 using the ACI 318-95 approach.

SOLUTION. With reference to Example 9.1, f_s = 52,000 lb/in², t_b = 1.06 in., and A = 21.2 in.²/bar. Therefore,

$$f_s \sqrt[3]{t_b A} = 52,000 \sqrt[3]{1.06 \times 21.2} = 147,700 \text{ lb/in.}$$

which according to Eqs. 9.22*a* and 9.22*b* is adequate for interior exposure and is almost adequate for exterior exposure.

ACI 318-77 Method for Two-Way Slabs. The commentary to ACI 318-71[9.45] recommended use of the Nawy–Orenstein equation (Eq. 9.18) for crack control in two-way slabs. However, there is no reference to crack control formulas for two-way slabs in either the 1995 ACI Code or its commentary. [9.1] In the absence of such code recommendations it would appear necessary to use an approximate equation, such as the Gergely–Lutzexpression, if checks on two-way slabs are necessary.

It is evident that crack widths will not normally be a problem in design unless the steel stresses at service load are very high or the crack widths are to be kept very small. In view of the wide scatter of measured crack widths on structural elements, great accuracy in calculations for crack control cannot be justified. The best crack control is obtained when the reinforcing bars are well distributed over the zone of concrete tension. The aim is to ensure that fine, closely spaced cracks form rather than a few wide cracks. The ACI Code[9.1] requires the spacing of the reinforcement at critical sections of solid two-way slabs not to exceed the smaller of twice the slab thickness or 18 in. [500 mm], to ensure that bars are not spaced widely apart.

REFERENCES

9.1   *Building Code Requirements for Structural Concrete*, (ACI 318-95) and *Commentary*, ACI 318R-95, American Concrete Institute, Farmington Hills, Mich., 1995, 371 p.

9.2   W. G. Corley, J. M. Hanson, and T. Helgason, "Design of Reinforced Concrete for Fatigue," *J. Struct. Div., ASCE,* Vol. 104, No. ST6, June 1978, pp. 921–932.

9.3   A. R. Lord, "Extensometer Measurements in a Reinforced Concrete Building over a Period of One Year," *Proc. ACI,* Vol. 13, 1917, pp. 45–60.

9.4   P. J. Taylor and J. L. Heiman, "Long-Term Deflection of Reinforced Concrete Flat Slabs and Plates," *Proc. ACI,* Vol. 74, November 1977, pp. 556–561.

9.5   P. J. Taylor, "Initial and Long-Term Deflections of a Reinforced Concrete Flat Plate Structure," *Civ. Eng. Trans., Inst. Eng. Aust.,* Vol. CE12, No. 1, April 1970, pp. 14–20.

9.6   F. A. Blakey, "Deformations of an Experimental Lightweight Flat Plate Structure," *Civ. Eng. Trans., Inst. Eng. Aust.,* Vol. CE3, No. 1, March 1961, pp. 18–22.

9.7   ACI Committee 435, "Observed Deflections of Reinforced Concrete Slab Systems, and Causes of Large Deflections," ACI 435.8R-85, in *Deflections of Concrete Structures*, ACI SP-86, American Concrete Institute, Detroit, 1985, pp. 15–61.

9.8   F. A. Blakey, "Deflection as a Design Criterion in Concrete Buildings," *Civ. Eng. Trans., Inst. Eng. Aust.,* Vol. CE5, No. 2, September 1963, pp. 55–58.

9.9   R. K. Agarwal and N. J. Gardner, "Form and Shore Requirements for Multistory Flat Slab Type Buildings," *Proc. ACI,* Vol. 71, November 1974, pp. 559–569.

9.10   P. Grundy and A. Kabaila, "Construction Loads on Slabs with Propped Formwork in Multistory Buildings," *Proc. ACI,* Vol. 60, December 1963, pp. 1729–1738.

9.11   P. J. Taylor, "Effects of Formwork Stripping Time on Deflections of Flat Slabs and Plates," *Aust. Civ. Eng. Constr.,* Vol. 8, No. 2, February 1967, pp. 31–35.

9.12   F. D. Beresford, discussion of Ref. 9.11, *Aust. Civ. Eng. Constr.,* Vol. 8, No. 5, May 1967, p. 61.

9.13   M. K. Hurd and P. C. Courtois, "Method of Analysis for Shoring and Reshoring in Multistory Buildings," in *Forming Economical Concrete Buildings*, ACI SP-90, American Concrete Institute, Detroit, 1985, pp. 283–294.

9.14   P. C. Stivaros and G. T. Halvorsen, "Shoring/Reshoring Operations in Multistory Buildings," *ACI Struct. J.,* Vol. 87, No. 5, September–October 1990, pp. 589–596.

9.15   X. L. Liu, W. F. Chen, and M. D. Bowman, "Construction Loads on Supporting Floors," *Concrete International,* Vol. 7, No. 12, December 1985, pp. 21–26.

9.16   F. A. Webster, "Reliability of Multistory Slab Structures Against Progressive Collapse During Construction," *J. ACI,* Vol. 77, No. 6, November–December 1980, pp. 449–457.

9.17   "Collapse Blamed on Premature Shore Removal," *Eng. News Rec.,* Vol. 190, No. 16, April 19, 1973, p. 11. Also March 8, 1973, p. 12; March 15, 1973, p. 12.

9.18   E. V. Leyendecker and S. G. Fattal, *Investigation of the Skyline Plaza Collapse in Fairfax County, Virginia*, Building Science Series 94, National Bureau of Standards, U.S. Department of Commerce, Washington, D.C., February 1977, 95 pp.

9.19   N. J. Carino, K. A. Woodward, E. V. Leyendecker, and S. G. Fattal, "A Review of the Skyline Plaza Collapse," *Concrete International,* Vol. 5, No. 7, July 1983, pp. 35–42.

9.20   M. D. Vanderbilt, "Deflections of Reinforced Concrete Floor Slabs," Ph.D. thesis, University of Illinois at Urbana–Champaign, 1963, 287 pp. Also issued as *Civil Engineering Studies*, Structural Research Series 263, Department of Civil Engineering, University of Illinois, Urbana, Ill.

9.21   M. D. Vanderbilt, M. A. Sozen, and C. P. Siess, "Deflections of Multiple-Panel Reinforced Concrete Floor Slabs," *J. Struct. Div., ASCE,* Vol. 91, No. ST4, August 1965, pp. 77–101.

9.22   K.-Y. Chang and S.-J. Hwang, "Practical Estimation of Two-Way Slab Deflection," *J. Struct. Eng., ASCE,* Vol. 122, No. 2, February 1996, pp. 150–159.

9.23   S.-J. Hwang and K.-Y. Chang, "Deflection Control of Two-Way Reinforced Concrete Slabs," *J. Struct. Eng., ASCE,* Vol. 122, No. 2, February 1996, pp. 160–168.

9.24   J. C. Jofriet, "Short Term Deflections of Concrete Flat Plates," *J. Struct. Div., ASCE,* Vol. 99, No. ST1, January 1973, pp. 167–182.

9.25   M. A. Bhatti, B. Lin, and J. P. Idelin Molinas Vega, "Effect of Openings on Deflections and Strength of Reinforced Concrete Floor Slabs," in *Recent Developments in Deflection Evaluation of Concrete,* ACI SP-161, edited by E. G. Nawy, American Concrete Institute, Farmington Hills, Mich., 1996, pp. 149–164.

9.26   J. C. Jofriet and G. M. McNeice, "Finite Element Analysis of Reinforced Concrete Slabs," *J. Struct. Div., ASCE,* Vol. 97, No. ST3, March 1971, pp. 785–806.

9.27   S. Timoshenko and S. Woinowsky-Krieger, *Theory of Plates and Shells,* 2nd ed., McGraw-Hill, New York, 1959, 580 pp.

9.28   J. F. Brotchie and A. J. Wynn, *Elastic Deflections and Moments in an Internal Panel of a Flat Plate Structure: Design Information,* Division of Building Research Technical Paper (Second Series) 4, Commonwealth Scientific and Industrial Research Organization, Melbourne, Victoria, Australia, 1975, 168 pp.

9.29   ACI Committee 435, "State-of-the-Art Report on Control of Two-Way Slab Deflections," ACI 345.9R-91, *ACI Struct. J.,* Vol. 88, No. 4, July–August 1991, pp. 501–514.

9.30   J. A. Sbarounis, "Multistory Flat Plate Buildings: Measured and Computed One-Year Deflections," *Concrete Int.,* Vol. 6, No. 8, August 1984, pp. 31–35.

9.31   R. Park and T. Paulay, *Reinforced Concrete Structures,* Wiley, New York, 1975, 769 pp.

9.32   P. Gergely and L. A. Lutz, "Maximum Crack Width in Reinforced Flexural Members," *Causes, Mechanism and Control of Cracking in Concrete,* ACI SP-20, American Concrete Institute, Detroit, 1968, pp. 87–117.

9.33   J. P. Lloyd, H. M. Rejali, and C. E. Kesler, "Crack Control in One-Way Slabs Reinforced with Deformed Welded Wire Fabric," *Proc. ACI,* Vol. 66, May 1969, pp. 366–376.

9.34   A. W. Beeby, *An Investigation of Cracking in Slabs Spanning One Way,* Technical Report TRA 433, Cement and Concrete Association, London, April 1970, 33 pp.

9.35   A. W. Beeby, "Prediction and Control of Flexural Cracking in Reinforced Concrete Members," *Cracking, Deflection and Ultimate Load of Concrete Slab Systems,* SP-30, American Concrete Institute, Detroit, 1971, pp. 55–75.

9.36   E. G. Nawy and G. S. Orenstein, "Crack Width Control in Reinforced Concrete Two-Way Slabs," *J. Struct. Div., ASCE,* Vol. 96, No. ST3, March 1970, pp. 701–721.

9.37   E. G. Nawy and K. W. Blair, "Further Studies on Flexural Crack Control in Structural Slab Systems," *Cracking, Deflection and Ultimate Load of Concrete Slab Systems,* ACI SP-30, American Concrete Institute, Detroit, 1971, pp. 1–41.

9.38   L. A. Clark, *Crack Similitude in 1:3.7 Scale Models of Slabs Spanning One Way,* Technical Report 42.455, Cement and Concrete Association, London, March 1972, 24 pp.

9.39   L. A. Clark, *Flexural Cracking in Slab Bridges,* Technical Report 42.479, Cement and Concrete Association, London, May 1973, 12 pp.

9.40   N. M. Hawkins, H. B. Fallsen, and R. C. Hinojosa, "Influence of Column Rectangularity on the Behavior of Flat Plate Structures," in *Cracking, Deflection and Ultimate Load of Concrete Slab Systems,* ACI SP-30, American Concrete Institute, Detroit, 1971, pp. 127–146.

9.41   R. J. Lenschow, "A Yield Criterion for Reinforced Concrete Under Biaxial Moments and Forces," Ph.D. Thesis, University of Illinois at Urbana–Champaign, 1966, 527 pp. Also issued as *Civil Engineering Studies,* Structural Research Series 311, Department of Civil Engineering, University of Illinois, Urbana, Ill.

9.42   ACI Committee 224, "Control of Cracking in Concrete Structures," *Proc. ACI,* Vol. 69, December 1972, pp. 717–753.

9.43   *Structural Use of Concrete,* BS 8110: Part 1, *Code of Practice for Design and Construction,* British Standards Institution, London, 1997, 121 pp.

9.44   *International System of Unified Standard Codes of Practice for Structures,* Vol. II, CEB-FIP Model Code for Concrete Structures, Comité Euro-International du Béton/Fédération International de la Précontrainte, Paris (English translation), April 1978, 348 pp.

9.45   *Commentary on Building Code Requirements for Reinforced Concrete,* ACI 318-71, American Concrete Institute, Detroit, 1971, 96 pp.

10 SHEAR STRENGTH OF SLABS

10.1 INTRODUCTION

In previous chapters only the flexural behavior of reinforced concrete slabs has been considered. Shear is generally not critical when slabs carry distributed loads or line loads and are supported by beams or walls, because in such cases the maximum shear force per unit length of slab is relatively small. However, shear can be critical in slabs in the vicinity of concentrated loads, because the maximum shear force per unit length of slab is relatively high around such loads. Concentrated loads can be applied to slabs by the transfer of forces: (1) from slab to columns in flat plate and flat slab floors, (2) from columns to footings, (3) from piles to pile caps, and (4) from applied loads such as wheel loads. The moments induced by concentrated loads on slabs in the elastic range were considered in Chapters 2 and 3. The flexural strength of slabs carrying concentrated loads was considered in Chapters 6, 7, and 8.

In many cases the shear stresses in slabs around concentrated loads can be more critical than the flexural stresses, and shear then governs the design. This is particularly true of slab-column connections in flat plate and flat slab floors where the size of the column, or column capital, and the slab thickness may be governed by the magnitude of the shear force to be transferred. The shear strength of slabs (or footings) in the vicinity of concentrated loads is governed by the more severe of two conditions, either beam action or two-way action.

Beam Action. In beam action the slab fails as a wide beam with the critical section for shear extending along a section in a plane across the entire width of the slab (or footing). In this case the slab should be treated as a wide beam and the shear strength equations for beams of the ACI Code[10.1] apply. It is to be noted that the code assumes that the critical section for shear in beam action is located at d from the face of the column or applied load (or from the face of a line load or supporting beam or wall), where d is the distance from the extreme compression fiber to the centroid of the longitudinal tension reinforcement. In fact, the critical section passes through the critical diagonal tension crack across which failure is considered to occur. Therefore, for this type of shear failure, conventional theory for beam shear applies (see, e.g., Ref. 10.2) and the theory will not be discussed further here.

Two-Way Action. In two-way action the slab fails in a local area around the concentrated load. The critical section extends around the concentrated load or column. A punching shear failure occurs along a truncated cone or pyramid caused by the critical diagonal tension crack around the concentrated load or column. In this case conventional theory for beam shear does not apply. The ACI Code[10.1] assumes that the critical section is located at $d/2$ from the perimeter of the column, or column capital, or applied load. It is to be noted that the once-held concept of the column (or applied load) being pushed through the slab as in Fig. 10.1a is incorrect. The critical section assumed in recent ACI codes, and the cracks that occur in the concrete in the actual failure mode, are shown in Fig. 10.1b. Figure 10.2 shows a reinforced concrete slab–column specimen after shear failure. The pyramid of concrete around the column left after failure along the diagonal tension crack is clearly visible.

Shear failure at slab–column connections can have disastrous consequences, as has been demonstrated clearly by some flat plate structures that have failed during construction. Shear failure at a slab–column connection can result in progressive failures of adjacent connections of the same floor, as the load is transferred elsewhere, causing the adjacent connections to become more heavily loaded. Also, the lower floors may fail progressively as they become unable to support the impact of material dropping from above. Hence, caution is clearly needed in shear strength calculations, and attention should be given to the low ductility associated with shear strength in order to avoid brittle failure conditions if possible.

Existing design procedures for shear strength, as recommended in the 1995 ACI Code,[10.1] are based primarily on the results of slab–column tests. The actual behavior of the failure region of the cracked slab is extremely complex, primarily because of the combined flexural and diagonal tension cracking and the three-dimensional nature of the problem. The design provisions used are of necessity derived from empirical simplifications of the real behavior.

A comprehensive state-of-the-art report on the shear strength of reinforced concrete slabs, reviewing knowledge up to the early 1970s, has been presented by ASCE-ACI Committee 426.[10.4] The commentary to the ACI code gives some of the background to the present code recommendations. ASCE-ACI Committee 426 has also published[10.5] suggested revisions to the shear provisions of the 1971 ACI Code. The suggested revisions introduce comprehensive design provisions for moment transfer at slab–column connections in flat plates and slabs and were meant for consideration by the Code Committee. The ACI Code[10.1] does not contain all of the suggested revisions of Ref. 10.5, but it is evident that Ref. 10.5 is a valuable supplementary design guide to the code. Much of the effort since these references were written has been concentrated on the seismic resistance of slab–column connections. This is generally beyond the scope of this book, but there is great emphasis on deformation capacity, or ductility, in addition to strength.

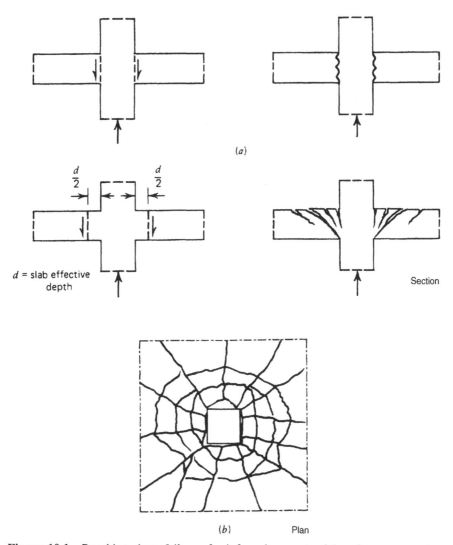

d = slab effective
depth

Section

(b)   Plan

Figure 10.1 Punching shear failure of reinforced concrete slab–column connection
with axial column load: (*a*) shear failure at column face (misconception—does not
occur); (*b*) assumed critical section and actual failure mode.

There are four general approaches to the analysis of the shear capacity of
slab–column connections. These are:

1. Calculating nominal shear stresses on a critical section located at some
   distance from the face of the column, as shown in Fig. 10.1*b*

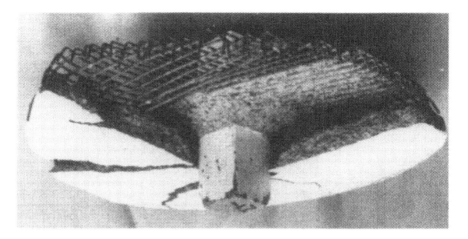

Figure 10.2 Reinforced concrete slab–column specimen after punching shear failure due to axial column load.[10.3]

2. Treating the slab strips framing into the faces of the columns as beam strips that are subjected to various combinations of moment, shear, and torsion
3. Using plate theory solutions, implemented with nonlinear finite element analyses, to examine and evaluate the state of stress near the column
4. Using truss analogies or strut-and-tie models to evaluate the forces near the column.

Each method has both rational and empirical or semiempirical components. Each of these except case 3 will be examined in some detail. The plate theory solutions present great difficulties in both the modeling of the connection between the slab and column, and in finding suitable failure criteria for the concrete, which is in a complex state of stress. The ACI and many other codes are based on the nominal stress analysis, case 1 in the list above.

10.2 SHEAR STRENGTH OF SLABS TRANSFERRING UNIFORM SHEAR

In this section we examine the punching shear strength of slabs with uniform shear around the critical section. That is, the load is considered to be applied without eccentricity with respect to the critical section of the slab.

10.2.1 Mechanism of Shear Failure of Slabs Without Shear Reinforcement

After diagonal tension cracking has occurred in the vicinity of the critical section of the slab around the perimeter of the loaded area, the slab carries the shear force by shear across the compression zone, aggregate interlock, and dowel action. However, where two-way bending occurs, the nominal ultimate shear stress that can be developed in a slab at the assumed critical section is much higher than in a beam. This increase in punching shear strength of slabs is due to the three-dimensional nature of the slab shear failure mechanism. The discussion below summarizes the mechanism of shear failure and draws largely on material reported in Ref. 10.4.

When the load is applied to the slab, the first crack to form is a roughly circular tangential crack around the perimeter of the loaded area due to negative bending moments in the radial direction. Radial cracks, due to negative bending moments in the tangential direction, then extend from that perimeter (see Fig. 10.1b). Because the radial moment decreases rapidly away from the loaded area, a significant increase in load is necessary before tangential cracks form around the loaded area some distance out in the slab. The diagonal tension cracks that develop in the slab tend to originate near middepth and are therefore more similar to web-shear cracks than to flexure-shear cracks. The stiffness of the slab surrounding the cracked region tends to control the opening of the diagonal tension cracks, thus preserving the shear transfer by aggregate interlock at higher loads. Such control is not present in beams. Punching shear failure may occur eventually, accompanied by general yielding of the slab reinforcement. However, yielding of the slab reinforcement is not necessary for the shear failure; nor does yielding of that reinforcement necessarily result in shear failure. Diagonal tension web-shear cracks first form at about one-half of the load at shear failure at approximately one-half of the slab depth from the periphery of the loaded area.

Note that the equilibrium requirements of the concrete compression forces and the slab steel tensile forces do not throw much light on the depth of the concrete compression zone of the slab at the critical diagonal tension crack. Figure 10.3 shows the horizontal forces acting on sections near the critical diagonal tension crack of a slab–column connection. For the slab, statics does not give a unique value for the concrete compressive force C_1 at one face of the column, because equilibrium requires that $C_1 + C_2 = T_1 + T_2$ but not necessarily that $C_1 = T_1$. Hence, the force C_1 can be redistributed to the surrounding concrete, depending on the distribution of slab stiffness. It may be thought that the shear strength could be increased by concentrating more slab tension reinforcement through the failure pyramid, thus leading to a greater depth of concrete in compression and an increased dowel action. However, such an increase in steel content has a limited effect on the shear strength because part of the force in the reinforcement can be balanced outside the failure pyramid and dowel action of the slab tension steel is not particularly effective, since the top cover concrete tends to split away from the slab. Alexander and Simmonds[10.6] investigated the effects of several variables in a series of tests and concluded that bond stress played a significant role in the shear problem. Decreasing the bar spacing to concentrate steel over the column increases the likelihood of local bond failures, which offsets the potential benefits of concentrating the reinforcement.

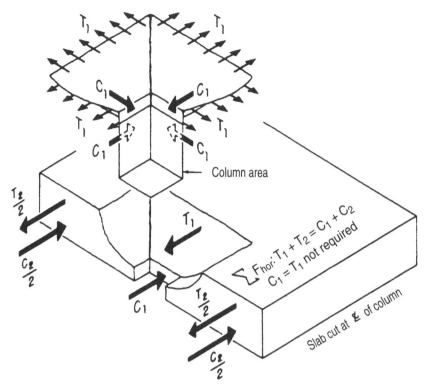

Figure 10.3   Horizontal forces acting on sections near critical diagonal tension cracks of a reinforced concrete slab–column connection.[10.4]

The stiffness of the slab surrounding the failure region means that in-plane outward displacements of the slab which tend to occur in the cracked sections are restrained, and in-plane compression forces are developed in the slab as a result. These in-plane forces increase the shear and flexural capacity of the critical sections, but they also reduce the ductility of the failure mode.

The critical sections of the slab for moment and shear are both at or close to the perimeter of the loaded area, and hence it would be expected that moment–shear interaction would occur. There is a change in the characteristics of the failure mode and load–deflection curves measured for slabs with different slab reinforcement ratios. For small slab steel ratios, general yielding of slab reinforcement can occur at failure and a yield line pattern forms either locally or extending out to the slab boundaries (see Sections 7.11.1 to 7.11.4). In such cases a ductile flexural failure occurs, although it is possible for a secondary punching shear failure to occur after the yield line pattern has developed. When the slab steel ratio is large, a brittle punching shear failure can occur accompanied either by yielding of the slab reinforcement over the column or by no yielding of the slab reinforcement. Figure 10.4 illustrates

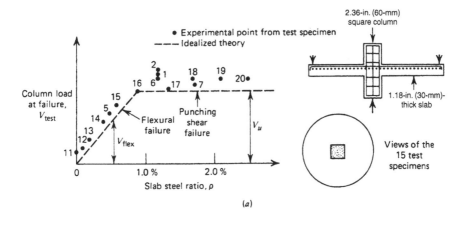

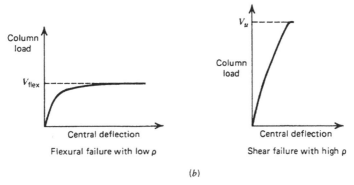

Figure 10.4   Effect of slab reinforcement ratio on shear strength and ductility of slab–column connection with axial column load: (*a*) effect of slab steel ratio on the slab failure load[10.3]; (*b*) effect of slab steel ratio on load–deflection curves.

these points. Figure 10.4*a* shows the results from some small-scale slabs tested in which the slab steel has a yield strength of 61,600 lb/in² (425 N/mm²) and the concrete compressive cube strength varied between 4000 and 5600 lb/in² (27.5 and 38.4 N/mm²). Figure 10.4*b* shows typical load–deflection curves for slabs failing in flexure and punching shear.

The variables that primarily affect the punching shear strength, as found by the extensive test results summarized in Ref. 10.4 and later observations by Alexander and Simmonds,[10.6–10.9] are:

1. *The quantity $\sqrt{f'_c}$, where f'_c is the compressive cylinder strength of the concrete.* This is because the tensile strength of the concrete is proportional to $\sqrt{f'_c}$ and shear failures are controlled primarily by concrete tensile strength.

However, the ratio of the nominal ultimate shear stress to $\sqrt{f'_c}$ can show significant scatter in practice. If the ultimate load is less than that for flexural collapse and the slab is not yielding over a wide area at collapse, the non-yielding portion of the slab can provide in-plane constraint, which considerably enhances the nominal ultimate shear stress. On the other hand, if the slab has a relatively low flexural capacity and develops large deflections before punching failure, the nominal ultimate shear stress can be reduced due to the smaller in-plane forces, the rapidly decreasing neutral axis depth, and the inelasticity of the compressed concrete.

2. *The ratio of the side length of the loaded area to the effective depth of the slab, c/d.* This is because for a given effective depth of slab and shape of loaded area, the length of the critical section becomes greater as the loaded area increases, resulting in an increase in shear strength. ACI-ASCE 326 Committee (later 426)[10.10] proposed, mainly on the basis of an extensive investigation by Moe,[10.11] that the shear strength of normal-weight concrete slabs with square columns be computed from

$$V_c = 4 \left(\frac{d}{c} + 1 \right) \sqrt{f'_c}\, bd \qquad (10.1)$$

in pound and inch units [in newton and millimeter units, replace the 4 with $\frac{1}{3}$], where b is the perimeter of column $= 4c$. If the critical section is taken at distance $d/2$ from the periphery of the column, Eq. 10.1 can be written as

$$V_c = 4\sqrt{f'_c}\, b_0 d \quad \text{lb} = (\sqrt{f'_c}/3)b_0 d \quad \text{N} \qquad (10.2)$$

where b_0 is the perimeter of critical section at $d/2$ from column periphery $= 4(c + d)$. Hence, the choice of a critical section at $d/2$ from the column periphery means that the shear strength equation becomes independent of the c/d ratio. However, for large values of $b_0 d$, the shear capacity diminishes slowly below that from Eq. 10.2 as $b_0 d > 20$, for interior columns.

3. *The ratio of the side lengths of the loaded area.* This is because it has been found that if the length of the column perimeter is held constant and the ratio of long side to short side is increased, the shear strength decreases because predominantly one-way bending, and therefore mainly beam action shear, tends to develop at the long faces of the loaded area. This also reflects the tendency for the shear force to concentrate at the "ends" of elongated column sections; that is, the shear stresses are not uniformly distributed around the column.

4. *The concrete aggregate.* This is because for the same concrete compressive strength f'_c, lightweight concrete has a lower splitting tensile strength than that of normal-weight concrete.

5. *The bar spacing and cover.* Small cover and small bar spacing tend to lead to lowered bond strength, which is an important factor in at least some

cases.[10.6-10.9] These papers are concerned with the development of truss or strut-and-tie models, which require extensive knowledge about bond and development of the reinforcement.

The shear force when punching shear failure occurs has been found to be relatively independent of the steel index ρf_y within the column area, where ρ is the ratio of tension reinforcement in slab and f_y the steel yield strength. However, a concentration of slab tension reinforcement in the column region is to be encouraged since it improves the flexural behavior of the slab in the service load range.

Also, compression (bottom) reinforcement in the slab continuous through the column is required by the ACI Code because it can act as a suspension net holding the slab to the column and thus support some load after punching failure has occurred. The possibility of the slab system surviving a punching failure by redistributing the remaining vertical forces increases accordingly. Therefore, properly detailed bottom reinforcement in the slab may prevent a catastrophic failure. Note that top steel is not effective in providing post-punching resistance because it tends to tear out of the slab when punching occurs, due to concrete cover over this steel splitting off. Various aspects of this problem have been studied by Regan,[10.12] Hawkins and Mitchell,[10.13] and Mitchell and Cook.[10.14]

The ACI Code requires that all bottom column strip bars or wires be continuous or adequately lap spliced and that at least two of these bars or wires pass through the column core. The Canadian Code[10.15] has a requirement for a calculation of the area of steel passing through the column core as*

$$\Sigma A_{sb} = \frac{2V_{se}}{f_y} \tag{10.3}$$

where ΣA_{sb} is the total area of steel passing through the column core, summed on the four sides of the column (for an interior column), V_{se} the service load (unfactored) shear force but not less than 2 × dead load shear, and f_y the yield stress of reinforcement.

Although relatively small quantities of reinforcement will be found to be adequate to satisfy this requirement, it will generally exceed the ACI requirement. The form of the equation appears to be closely related to an equation given by Mitchell and Cook.[10.14] The equation can be derived by assuming that the bottom bars passing through the column core are all bent down 30°

---

*With permission of CSA International, Eq. 10.3 is reproduced from CSA Standard A23.3-94, *Design of Concrete Structures*, which is copyrighted by CSA International, 178 Rexdale Boulevard, Etobicoke, Ontario, M9W 1R3. "While use of this material has been authorized, CSA International shall not be responsible for themanner in which the material is presented, nor for any interpretation thereof."

from the horizontal, and equating the vertical component of the bar yield forces to V_{se}.

Although the bottom steel requirements are given in Chapter 13 of ACI 318-95, they may be regarded as extensions of ACI Code, Sec. 7.13, "Requirements for Structural Integrity," which requires continuous adequately spliced bars in both the top and bottom of all perimeter beams, plus closed stirrups with 135° hooks to be placed along the full lengths of these beams. The intent is to provide some toughness to help a structure survive various unanticipated accidents against which fully rational design is impossible or impractical.

10.2.2 ACI Code Approach to Shear Strength Without Shear Reinforcement

The 1995 ACI Code[10.1] requires that the design of sections subject to shear without shear reinforcement should be based on

$$V_u \leq \phi V_c \tag{10.4}$$

where V_u is the factored shear force at section considered, ϕ the strength reduction factor for shear (= 0.85), and V_c the nominal shear strength provided by the concrete.

The nominal shear strength of normal-weight concrete slabs (or footings) for two-way action is given as the smaller of the values from Eq. 10.5 or 10.6:

$$V_c = \left(2 + \frac{4}{\beta_c}\right) \sqrt{f'_c}\, b_0 d \quad \text{lb} \qquad \text{but not greater than } 4\sqrt{f'_c}\, b_0 d \quad \text{lb}$$

$$\tag{10.5a}$$

where β_c is the ratio of long side to short side of concentrated load or reaction area, f'_c the compressive cylinder strength of concrete (lb/in²), b_0 the perimeter of critical section (in.), and d the distance from the extreme compression fiber to the centroid of tension reinforcement in the slab (in.). If the dimensions are in millimeters and f'_c is in N/mm², Eq. 10.5 is expressed as

$$V_c = \left(1 + \frac{2}{\beta_c}\right) \frac{\sqrt{f'_c}\, b_0 d}{6} \quad \text{N} \qquad \text{but not greater than } \frac{1}{3} \sqrt{f'_c}\, b_0 d \quad \text{N}$$

$$\tag{10.5b}$$

When the columns or capitals become very large, the following equation may govern

$$V_c = \left(\frac{\alpha_s d}{b_0} + 2\right) \sqrt{f'_c}\, b_0 d \qquad \text{lb} \qquad\qquad (10.6a)$$

where $\alpha_s = 40$ for interior columns, 30 for interior columns, and 20 for corner columns. In SI units, the equation is stated as

$$V_c = \left(\frac{\alpha_s d}{b_0} + 2\right) \frac{\sqrt{f'_c}\, b_0 d}{12} \qquad \text{N} \qquad\qquad (10.6b)$$

Equation 10.6 controls, for an interior column, when $b_0/d > 20$, as is normally the case for the critical section near the face of a drop panel and as may be the case near the face of a column capital. When lightweight concrete is used, the $\sqrt{f'_c}$ in Eqs. 10.5 and 10.6 is multiplied by 0.75 for all-lightweight concrete or by 0.85 for sand-lightweight concrete.

For a loaded area of general shape, the critical section is located so that its length is a minimum and that it is not closer than $d/2$ from the periphery of the loaded area. Critical sections for columns of different shapes, as illustrated by the ASCE-ACI Committee 426 report,[10.5] are shown in Fig. 10.5. Shear stresses must be checked at all applicable critical sections. In a flat slab with drop panels, stresses must be checked both at the section near the face of the column or capital and at near the face of the drop panel. Note that ASCE-ACI Committee 426 suggests that a loaded area of circular shape can be replaced by a loaded area of rectangular shape and equal perimeter.

Current codes ignore the difference, but it has been demonstrated conclusively that a circular loaded area is able to transmit appreciably more shear force than a square area having the same perimeter.[10.16] The improved shear strength is apparently a result of not having the stress concentrations that occur at the corners of rectangular columns. The same series of tests clearly demonstrated that increasing the flexural strength of the slab can increase the shear capacity, as demonstrated earlier by Moe[10.11] and others.

In Eq. 10.5 the nominal ultimate shear stress on the critical section is $4\sqrt{f'_c}$ lb/in^2 [$\sqrt{f'_c}/3$ N/mm^2] when the ratio β_c of long side to short side of a rectangular column is in the range 1 to 2. For $\beta_c > 2$, the nominal ultimate shear stress reduces linearly to $2\sqrt{f'_c}$ lb/in^2 [$\sqrt{f'_c}/6$ N/mm^2] when $\beta_c \to \infty$, which is the nominal ultimate shear stress for one-way action. The nominal ultimate shear stress when $\beta_c > 2$ is an averaging of the more complex actual situation in which the failure stress varies from about $4\sqrt{f'_c}$ lb/in^2 around the ends of the loaded area down to $2\sqrt{f'_c}$ lb/in^2 or less along the long sides between the two ends. For shapes of loaded area other than rectangular, β_c is taken to be the ratio of the longest overall dimension of the effective loaded area to the shortest overall dimension of the effective loaded area, the directions of the two dimensions being at right angles.[10.5] The effective loaded area is that area totally enclosing the actual loaded area for which the perimeter

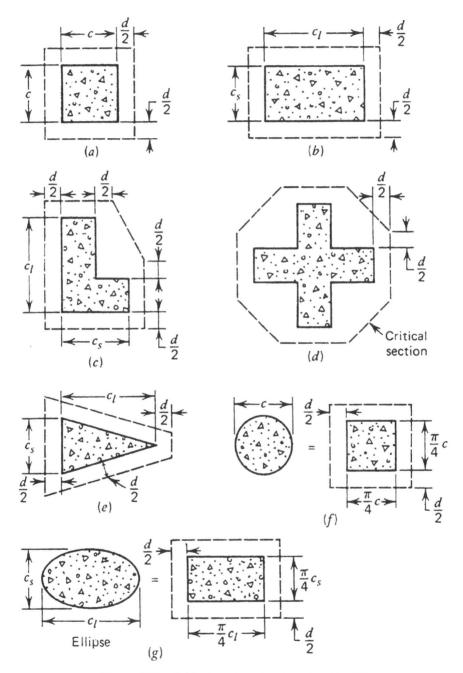

Figure 10.5   Critical sections for shear in slabs.[10.5]

is a minimum. The quantity β_c for an L-shaped loaded area is illustrated in Fig. 10.6.

The effective depth d in Eq. 10.4 is not explicitly defined for two-way reinforcement in the ACI Code,[10.1] but ASCE-ACI Committee 426[10.5] suggests that d can be taken as

$$d = \frac{A_{s1}d_1 + A_{s2}d_2}{A_{s1} + A_{s2}} \tag{10.7}$$

where A_{s1} and A_{s2} are the areas of tension reinforcement passing through the critical section in perpendicular directions 1 and 2, respectively; and d_1 and d_2 are the effective depths to reinforcement A_{s1} and A_{s2}, respectively. The quantity d equals the average effective depth of the steel if the loaded area is approximately square and the reinforcement ratios approximately equal for directions 1 and 2. The need to calculate d exactly increases as the β_c ratio increases and the slab thickness decreases.

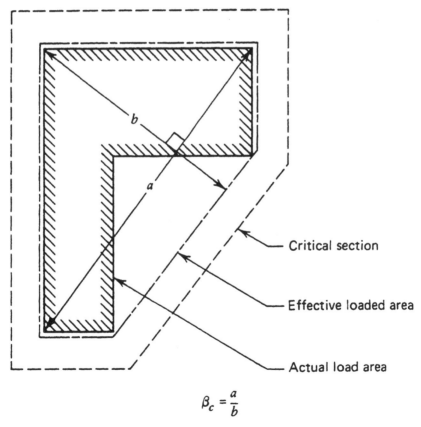

$$\beta_c = \frac{a}{b}$$

Figure 10.6 Value of β_c for a nonrectangular loaded area.[10.5]

10.2.3 Truss Models for Shear Strength

Although truss analogies have been used to represent the shear response of beams for most of the time that reinforced concrete structures have been built, their application to slabs is recent. Alexander and Simmonds[10.6–10.9] have presented two versions of a truss model applied to the flat plate–column connection problem. Figure 10.7 shows the basics of the earlier, simpler model.[10.6] Each bar reinforcing passing through (and near) the column holds an inclined strut in equilibrium; the vertical component of the force is

$$V_b = A_{\mathrm{bar}} f_y \tan \alpha \qquad (10.8)$$

The forces V_b are summed around the column perimeter to obtain the

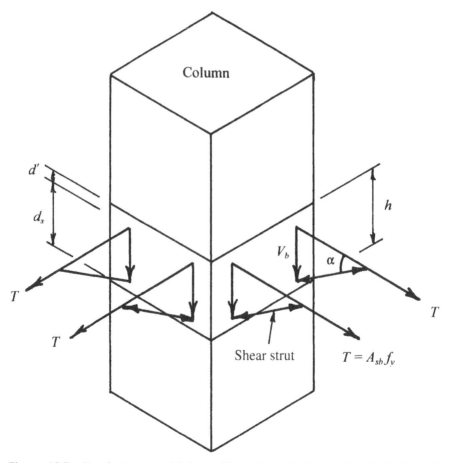

Figure 10.7 Simple truss model for uniform shear. (Redrawn from Ref. 10.6. By permission from ACI International.)

failure force V_u. The value of $\tan \alpha$ is determined from a semiempirical equation[10.6]:

English units: $\tan \alpha = 1.0 - e^{-0.85K}$

SI units: $\tan \alpha = 1.0 - e^{-2.25K}$

$$(10.9)$$

where

$$K = \frac{s_{\textit{eff}} \times d' \times \sqrt{f'_c}}{A_{\text{bar}} \times f_f \times (c/d_s)^{0.25}} \qquad (10.10)$$

where s_{eff} is the effective tributary width of the reinforcing bar = the sum of half the distances to the bar on each side of the bar being considered, $\leq 3d'$ on each side; d' the cover from the center of the mat to the tension face of the concrete; f'_c compressive strength of the concrete, in kips/in² or N/mm² units; A_{bar} the area of the bar considered; f_y the yield stress of the bar considered; c the width of the column face considered; and d_s the average effective depth of the two layers in the mat.

The inclusion of the bars passing through the column is obvious. Bars outside the column may make a contribution, depending on their proximity. It was recommended that bars within d_s of the column be included but at a linearly decreasing rate with distance. Thus, a bar at $0.9d_s$ from the column should make a contribution of $0.1 A_{\text{bar}}f_y \tan \alpha$. This truss analogy was extended to include unbalanced shear, that is, shear plus bending, and edge columns, with considerable success.

Although Fig. 10.7 shows straight struts, Alexander and Simmonds[10.8,10.9] later concluded that the struts must be curved, and they then derived their *bond model* as an alternative. They considered the equilibrium of the slab strips that frame into each face of the column. For a concentric loading on a square column supporting a slab with equal steel in the two orthogonal directions, their equation for punching shear is (Ref. 10.8, Eq. 7)

$$P_u = 8 \times \sqrt{M_s} \times w_{\text{ACI}} \qquad (10.11)$$

where M_s is the sum of the negative moment capacity at the column and the midspan positive moment capacity for a slab strip of width c, the column dimension, and w_{ACI} is the shear capacity of a strip of concrete slab of unit width, in one-way shear, computed as $d \times 2\sqrt{f'_c}$ lb/in = $d \times \sqrt{f'_c}/6$ N/mm (f'_c in lb/in² or N/mm² units).

The strip moment capacities are computed using special definitions of the steel ratio ρ from (Ref. 10.8, Eq. 8)

$$M = \rho f_y j d^2 c \qquad (10.12)$$

where $\rho = A_s/bd$, A_s is the area of steel within the strip width c plus half the areas of the first bars on the sides of the strip, and b is the distance between the first bars on the two sides of the strip.

The positive moment capacity is zero in many test specimens, where the end of the slab strip considered is at a simply supported edge. This second analysis has similarities to the beam analogy approaches which are discussed later, and its derivation contains elements of plate theory. It has not yet been extended to include shear and unbalanced moment.

10.2.4   ACI Code Approach to Shear Strength with Shear Reinforcement

The shear strength and ductility of slabs (and footings) can be increased by the use of shear reinforcement in the form of stirrups, bent bars, stud-shear reinforcement, or structural steel shearheads. Bent bars and stirrups must be very carefully detailed to ensure proper anchorage. Some forms of shear reinforcement used in the past have not been fully effective, because of anchorage deficiencies. An ACI Committee Report[10.18] discusses stud-shear and stirrup reinforcement. Figure 10.8 shows some effective types of shear reinforcement.

Stirrup and Bent Bar Shear Reinforcement. The 1995 ACI Code[10.1] recommends that the design of sections containing stirrups or bent bar shear reinforcement should be based on

$$V_u \leq \phi V_n \qquad (10.13)$$

where V_u is the factored shear force at the section considered, ϕ the strength reduction factor for shear ($= 0.85$), and V_n the nominal shear strength of the section.

The nominal shear strength of normal-weight concrete slabs (or footings) is given as

$$V_n = 2\sqrt{f'_c}\, b_0 d + V_s \quad \text{lb} \qquad \text{but not greater than } 6\sqrt{f'_c}\, b_0 d \quad \text{lb}$$

$$(10.14a)$$

In SI units this is expressed as

$$V_n = (\sqrt{f'_c}/6)b_0 d + V_s \quad \text{N} \qquad \text{but not greater than } (\sqrt{f'_c}/2)b_0 d \quad \text{N}$$

$$(10.14b)$$

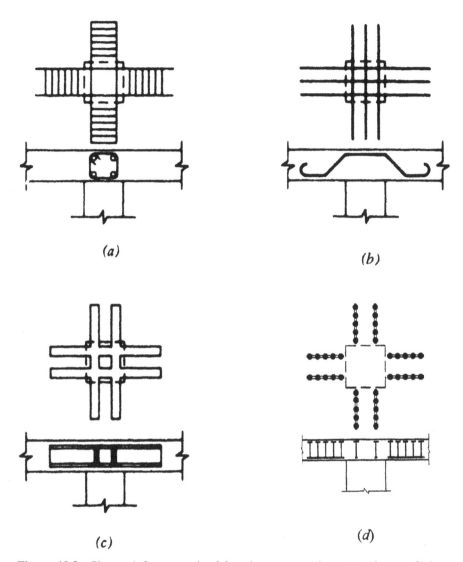

Figure 10.8 Shear reinforcement in slab–column connections: (*a*) stirrups; (*b*) bent bars; (*c*) structural steel shearhead; (*d*) stud-shear reinforcement.

where f_c' is the compressive cylinder strength of the concrete (lb/in²), b_0 the perimeter of the critical section located not closer than $d/2$ from the perimeter of the loaded area (in.), d the distance from the extreme compression fiber to the centroid of tension reinforcement in the slab (in.), and V_s the shear force carried by shear reinforcement, given by Eqs. 10.15 and 10.16.

The shear strength should be investigated at the critical section of length b_0 defined in Section 10.2.2 and at sections more distant from the loaded area. The shear that needs to be carried by V_s becomes smaller at sections more distant from the loaded area (i.e., for values of b_0 for sections farther than $d/2$ from the periphery of the loaded area), and shear reinforcement must be continued away from the loaded area until it is no longer required.

The nominal ultimate shear stress at the critical section acting with the shear reinforcement is taken as $2\sqrt{f_c'}$ lb/in² [$\sqrt{f_c'}/6$ N/mm²] because at approximately that stress, diagonal tension cracks (web-shear cracks) begin to form and the shear reinforcement is considered to carry all the shear force above that load. This code requirement originated from a code discussion by Carpenter et al.[10.17] The maximum nominal ultimate shear stress that can be carried by the concrete and shear reinforcement is not permitted to exceed $6\sqrt{f_c'}$ lb/in² [$0.5\sqrt{f_c'}$ N/mm²]. When lightweight concrete is used, the $\sqrt{f_c'}$ value in Eq. 10.14 should be multiplied by 0.75 for all-lightweight concrete or by 0.85 for sand-lightweight concrete.

When shear reinforcement perpendicular to the axis is used,

$$V_s = A_v f_y \frac{d}{s} \tag{10.15}$$

where A_v is the area of shear reinforcement within distance s, f_y the yield strength of shear reinforcement, d the effective depth of slab longitudinal steel, and s the spacing of shear reinforcement along the slab steel. The stirrup spacing s must not exceed $d/2$. A longitudinal slab bar must exist in each corner of the stirrups, and the stirrups must be effectively anchored at each bar to develop the yield strength of the stirrup. Figure 10.8a shows a typical arrangement of stirrups. For the particular arrangement shown in Fig. 10.8a, A_v in Eq. 10.15 would be the area of eight stirrup legs. The shear force carried by the stirrups is V_s at the critical section at $d/2$ from the periphery of the loaded area and also is V_s at all sections farther from the periphery of the loaded area until the stirrups are discontinued, assuming that s remains unchanged. Tests have shown that although multiple or single U stirrups may be effective in increasing the shear strength (provided that they are anchored adequately around longitudinal slab bars at all ends and pass around longitudinal slab bars at all bends), closed stirrups are to be preferred because of a greater increase in ductility of the slab at ultimate load. Inclined stirrups can also be used as shear reinforcement, in which case Eq. 10.15 is modified by multiplying the right-hand side by $(\sin \alpha + \cos \alpha)$, where α is the angle between inclined stirrups and longitudinal slab bars.

Where the shear reinforcement consists of a single group of bent bars,

$$V_s = A_v f_y \sin \alpha \tag{10.16}$$

where A_v is the area of shear reinforcement, f_y the yield strength of shear reinforcement, and α the angle between the inclined bar and longitudinal slab bars. The bent bar should cross the critical section and be anchored at each end to develop its yield strength. Figure 10.8b shows a typical arrangement. For the particular arrangement of three bars in each direction shown in Fig. 10.8b, A_v in Eq. 10.16 would be the area of 12 bars. The bent bars at 45° in the arrangement shown in Fig. 10.8b carry shear force across the critical section at $d/2$ from the column face. If shear reinforcement is required at critical sections farther from the column face, additional groups of bent bars would be required farther from the column, and this may become difficult because of congestion of reinforcement over the column. Closed stirrups are preferred as shear reinforcement rather than bent bars, since bent bars need to be anchored carefully to ensure that they intercept diagonal tension cracks effectively. Rather minor shifts in crack location and/or slope may allow the crack to evade the shear reinforcement when there is only a single line of bars, as in this case.

Structural Steel Shearheads. Shear reinforcement formed from structural steel I or channel sections may be used in slabs. The shearhead is fabricated by welding the sections into four identical arms at right angles. The arms must be continuous through the column sections. Figure 10.8c shows a typical arrangement. The ends of each shearhead arm may be cut at an angle of not less than 30° to the horizontal, providing that the moment strength of the tapered section can resist the shear force attributed to that arm. The ratio α_v of the flexural rigidity of each shearhead arm to the flexural rigidity of the surrounding fully cracked composite slab section (including the shearhead) of width $(c_2 + d)$ should not be less than 0.15, where c_2 is the size of rectangular or equivalent rectangular loaded area measured transverse to the arm and d the effective depth of the slab.

The design of the shearhead is based on two basic criteria. First, the shearhead must have adequate flexural strength to carry the shear acting along the arms. Second, the shear stresses in the concrete near the end of the arms must be limited. The presence of the shearhead means that the slab longitudinal reinforcement for negative moment in the column strip can be reduced.

The plastic moment of resistance required for each arm of the shearhead can be calculated from the following equation, which is based on reported test data:[10.19]

$$M_p = \frac{V_u}{2\eta\phi}\left[h_v + \alpha_v\left(l_v - \frac{c_1}{2}\right)\right] \tag{10.17}$$

where V_u is the total factored shear force to be carried by the connection, η the number of arms, ϕ the strength reduction factor for flexure = 0.9, h_v the depth of the shearhead arm cross section, l_v the length of the shearhead arm from the center of the shearhead, and c_1 the size of the rectangular or equivalent rectangular loaded area measured in the direction of the arm. Equation 10.17 was derived from the assumed distribution of vertical shear force along one arm of the shearhead shown in Fig. 10.9b, which is based implicitly on the applied forces shown in Fig. 10.9c (which are explicitly for the case of one shearhead arm on each side of the column). The shear force along each arm is taken to be $\alpha_v V_c /4$, where V_c is the shear force at diagonal tension cracking in the slab. The peak shear force at the face of the column is taken as $(V_u/4) - (V_c/4)(1 - \alpha_v)$, where the first term is the total shear force to be carried at the face of the column and the second term is the shear force considered to be carried to the column face by the concrete compression zone of the slab. The second term approaches zero for a heavy shearhead arm and approaches $V_c/4$ for a light shearhead arm. Equation 10.17 is given by this shear force distribution assuming that V_c is one-half of V_u. It is evident that M_p is the plastic moment of resistance of each shearhead arm necessary to ensure that the ultimate shear is attained. It should be noted that M_p and α_v are not independent variables since both are related to the same steel section.

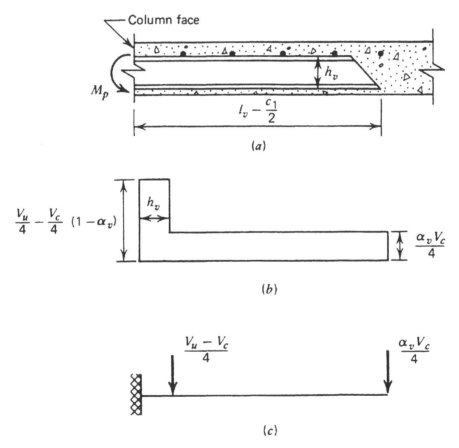

Figure 10.9 Idealized distribution of vertical shear force acting on shearhead arm[10.1]: (a) shearhead arm in slab; (b) assumed shear force diagram for shearhead arm; (c) assumed forces acting on shearhead arm.

The length l_v of each shear arm is calculated by extending the arms far enough so that V_u does not exceed $\phi 4\sqrt{f'_c}b_0 d$ lb $[\phi(\sqrt{f'_c}/3)b_0 d$ N] on the following critical section. The critical section is considered to cross each shearhead arm at $0.75[l - (c_1/2)]$ from the column face and is located so that b_0 is a minimum but need not approach closer than $d/2$ to the periphery of the column section. The critical section is illustrated in Fig. 10.10. The critical section does not extend to the end of the shearhead arms because some slabs tested failed when the nominal ultimate shear stress at the ends of the arms was less than $4\sqrt{f'_c}$ lb/in² lb/in² $[\sqrt{f'_c}/3$ N/mm²], and hence a more conservative critical section has been assumed. Also, when shearhead reinforcement is provided, V_u should not be taken as greater than $\phi 7\sqrt{f'_c}b_0 d$ lb $[\phi 0.6\sqrt{f'_c}b_0 d$ N] on the critical section defined in Section 10.2.2. Again, when lightweight concrete is used, the $\sqrt{f'_c}$ in the equations above should be multiplied by 0.75 for all-lightweight concrete or by 0.85 for sand-lightweight concrete.

A shearhead may be assumed to contribute a negative-moment resistance to the slab column strip of

$$M_v = \frac{\phi \alpha_v V_u}{8}\left(l_v - \frac{c_1}{2}\right) \tag{10.18}$$

but not to exceed any of: 30% of the total moment required for each slab column strip, the change in column strip moment over length l_v, or the value of M_p calculated from Eq. 10.17. Equation 10.18 is derived from the shear force distribution on the shearhead shown in Fig. 10.9, ignoring the peak shear and assuming that V_c is one-half of V_u.

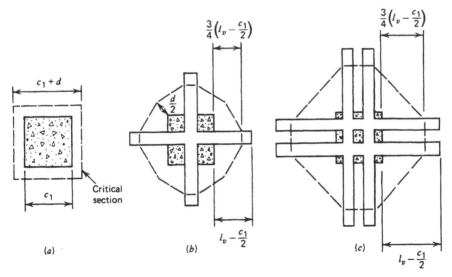

Figure 10.10 Location of critical section for concrete shear when structural steel shearhead is used[10.1]: (a) no shearhead; (b) small shearhead; (c) large shearhead.

Stud-Shear Reinforcement. Steel studs, similar to those used to enforce composite action between steel girders and concrete slabs, are being used as shear reinforcement.[10.20]–[10.24] The patented stud-shear reinforcement system uses studs welded to steel strips, as shown in Figs. 10.8d and 10.11. The studs are spaced along the strips at a designer-selected spacing of s, and several such strips can be placed in parallel rows. The studs have heads which are significantly larger than those used on the typical composite steel girder, and the total height of the stud plus strip is equal to the slab thickness less the sum of the required cover thicknesses top and bottom. Many series of tests have been completed, some of which are reported in Refs. 10.20 to 10.24. The spacing arrangement of the studs could be selected using the 1995 ACI Code. Reference 10.21 contains various suggested design criteria which in general liberalize several clauses of the ACI Code, but they have not been adopted by the ACI Code.

As compared with closed stirrups, installation of stud-shear reinforcement is simple. All of the studs can be of the same height, but the stirrups in one direction must be significantly shorter than those in the other because of the crossing flexural reinforcement, reducing their effectiveness. The head on the stud and the steel strip it is welded to provide good anchorage, and this plus that fact that all studs can be a major fraction of the slab thickness in total height are claimed to provide significant advantages in controlling shear cracking.

10.2.5 Influence of Openings, Free Edges, and Service Ducts in Slabs

When openings in slabs are located at a distance less than 10 times the slab thickness from the loaded area, or when openings in flat plates or flat slabs are located within the column strips, the ACI Code[10.1] requires that the critical slab section for shear be modified. The part of the critical section that is enclosed by radial projections from the extremities of the opening to the centroid of the loaded area should be considered ineffective in the calculation of $b_0 d$. The effective part of the critical section for loaded areas near typical openings and free edges are shown as dashed lines in Fig. 10.12. The perimeter around the circular support near the free corner will often control the thickness of a pile cap supported on a few high-capacity piles. The ACI Code[10.1] allows the ineffective portion of the perimeter to be one-half that defined above in the case of slabs with shearheads. It is of interest to note that ASCE-ACI Committee 426 suggests[10.5] that the effect of openings and free edges on the shear strength need be taken into account only if they are located closer to the periphery of the loaded area than four times the slab thickness or twice the development length l_d of the slab reinforcement that passes through the column or over the shearhead, whichever is larger. These suggested values by Committee 426 were based on the consideration that local distortions associated with either loading or openings dissipate within about

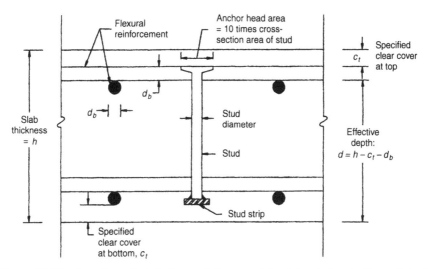

Figure 10.11 Location of stud-shear reinforcement in slab relative to flexural steel. (From Ref. 10.20. By permission from ACI International.)

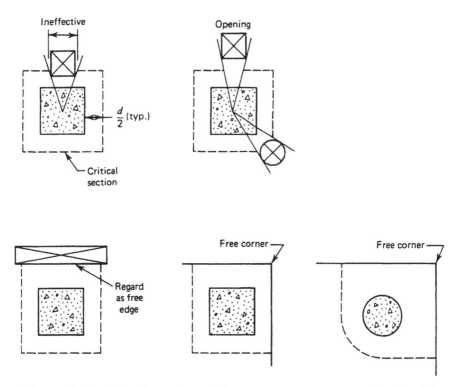

Figure 10.12 Effect of openings and free edges on critical section for shear.[10.1]

two slab thicknesses, and therefore a separation of four slab thicknesses gives a reasonable criterion. Also, to satisfy possible anchorage requirements associated with stress changes close to loadings and openings, a separation of two development lengths appears necessary.

The influence of service ducts embedded in slabs near columns of flat plate floors has been investigated by Hanson.[10.25] The slabs were 8 in. (203 mm) thick and located in some slabs was a two-level duct system consisting of ducts with cross sections of either $7\frac{1}{4}$ in. (184 mm) $\times$ $1\frac{3}{8}$ in. (35 mm) or $3\frac{1}{8}$ in. (79 mm) $\times$ $1\frac{3}{8}$ in. (35 mm). The ducts were placed in the horizontal plane of the slab close to middepth but entirely on the tension side of the neutral axis of the slab. The tests indicated that service ducts of the sizes listed above did not significantly alter the performance of the connection provided that the ducts were located at least two slab thicknesses from the column face. It was found that a duct placed near the end of the arm of a shearhead reduced the effectiveness of the shearhead, and hence it was also recommended that ducts not be placed near the ends of shearheads.

10.2.6 Special Problems with Pile Caps

Column footings supported on soil or rock do not normally present special problems, and Chapter 15 of the ACI Code[10.1] adequately defines the locations of the critical sections for flexure and shear and then refers back to Sec. 11.12 of the code for the details of the shear design. However, there are some potential problems with pile caps that apparently become most serious when there are only a few very high-capacity piles. As an example, given the right subsurface conditions, a pile cap may be supported on four 10-in. (250-mm)-diameter piles having service load capacities of 100 tons (900 kN) or higher, and the pile spacing may be as small as 3 ft (0.9 m).

Such a pile cap is illustrated in Fig. 10.13. With the proportions shown, the normal critical section for beam shear, located at d from the column face, will usually be outside the piles, and because of the treatment of the pile forces described in Sec. 15.5.3 of the ACI Code,[10.1] the design shear forces are zero. That section of the code provides that a pile whose center is located half its diameter or more inside the designated critical section contributes no shear force, while a pile located a similar distance outside the critical section makes a full contribution of the shear force on the critical section.

For the same footing, the critical section for punching shear, which is located $d/2$ from the column face may be at or outside the piles, and the nominal punching shear stress around the column may be very low. The pile cap shown in Fig. 10.13 is one in which the computed punching shear stress around the column is zero.

The low or nonexistent nominal shear stresses computed for the beam-shear case do not actually represent the conditions in the footing, however, and the entire shear force must be transmitted through the concrete in a stress

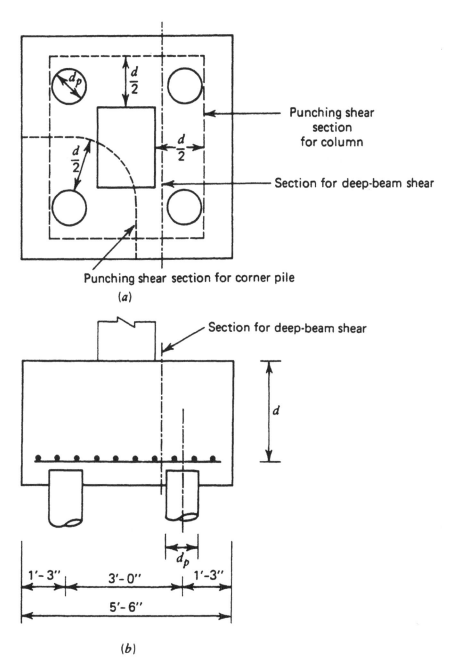

Figure 10.13   Pile cap, showing critical sections for shear: (*a*) plan of pile cap; (*b*) section through pile cap. (1 in = 25.4 mm)

state similar to deep-beam shear. The appropriate critical section is shown in Fig. 10.13, and the shear stress transmitted across this plane should not exceed that permitted by Sec. 11.8 of the ACI Code.[10.1] The ACI Code does not contain this instruction, but the authors strongly suggest that this should be done. The commentary to Code Sec. 15.5.3 suggests the use of the *CRSI Handbook*[10.26] for such cases. The *CRSI Handbook* contains tables and analyses for cap depths that were developed considering both deep-beam shear and punching shear in situations where the failure angle is restrained to be very steep.

The entire shear force must also be transmitted across a perimeter located at the face of the column. If the failure surface is forced to be nearly vertical rather than spreading at 45° or less from horizontal, the average shear stress that can be transmitted becomes very high, but it is not unlimited.

The shear stresses from Code Sec. 11.8 will generally be higher than those from the *CRSI Handbook*. The user of both sets of provisions is warned that although the equations in the two cases are very similar, the results may be quite different because of differences in the locations of the critical sections and definitions of terms in the equations.

The depths required for shear for the four-pile cap will be so great that the flexural requirements are often satisfied by minimum reinforcement. This steel, at 0.0018 times the gross area of the cap cross section for grade 60 or grade 420 material, is required in both directions.

The depth required for two-way shear at a corner pile will often control the depth of the cap, considering the punching shear perimeter for a corner pile shown in Fig. 10.12 or 10.13. Ignoring the interference between the potential failure surfaces for the corner pile punching up through the footing and for the column punching down through the footing, the minimum effective depths for various pile loads might be as follows (1 ton = 8.9 kN, 1 in. = 25.4 mm):

Pile Capacity (tons) $(D + L)$	Pile Diameter (in.)	Minimum d (in.)
100	10	28.5
80	10	24.2
60	8	19.9
40	8	14.4

These effective depths were computed with the following assumptions:

$$f'_c = 3000 \text{ lb/in}^2 \text{ (20.7 N/mm}^2)$$

distance from edge of cap to center of pile = 15 in. (380 mm)

average load factor = $1.6(D + L)$

$$\phi = 0.85$$

Greater depths will generally be required for caps supported on more than four piles. The normal shear critical sections have meaning for the larger caps, and stresses on those sections will generally govern. Experimental work on pile caps has been reported, and it is clear that the shear capacity problem is complex. The actual failure surfaces are not regular geometrical shapes, but are strongly influenced by the locations of the column and piles and may be as shown in Fig. 10.14.

A completely different approach to the design of pile caps is found in the use of strut-and-tie or truss models, as has been demonstrated by Adebar et al.[10.28] Figure 10.15 gives some of the basic concepts. Failure may occur due

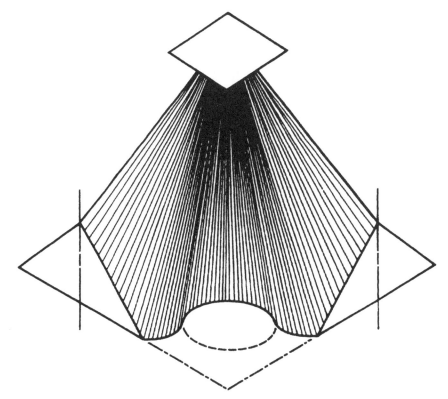

Figure 10.14 Suggested failure surface for four-pile cap.[10.27]

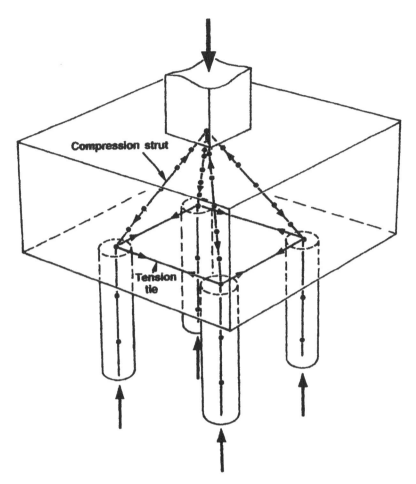

Figure 10.15   Simple three-dimensional truss model for a four-pile cap. (From Ref. 10.28. By permission from ACI International.)

to yielding of the tension ties, failure of the anchorage of the tension ties, failure of the compression struts, or failure due to excessive bearing stresses over the tops of the piles or under the column. An important feature of this model is the emphasis on the fact that the reinforcement must be anchored so that its full capacity can be developed over the piles, very near the ends of the bars. Comparisons with test results show the method to be somewhat conservative but considerably more consistent than the ACI Code. This model is incorporated in the Canadian Code,[10.15] where many details can be found. Other details are given in Ref. 10.28 and the references contained therein. Siao[10.29] discusses a similar model, with an emphasis on the strut capacity.

It must be noted that the strut-and-tie models are quite different from the truss models suggested by Alexander and Simmonds, since in these cases the locations and slopes of the struts are completely defined by the geometry of the pile cap.

It is also very clear that the designer must take into account the high probability that some of the piles will not be installed in their correct locations. The basic design should account for errors at least equal to the stated tolerance on pile locations. Piles too close to the column location normally do not cause problems unless the footing is to resist a substantial moment in addition to the thrust or unless a significant eccentricity is created. Piles too far from the column may significantly increase both the imposed bending moment and the shear force the cap must resist.

10.3 SHEAR STRENGTH OF SLAB-COLUMN CONNECTIONS TRANSFERRING SHEAR AND UNBALANCED MOMENT

In this section we examine the shear strength of slabs with nonuniform shear around the critical section. That is, both axial load and unbalanced bending moment are transferred at the slab–column connection. The term *unbalanced bending moment* is used to emphasize that this is the moment being transferred between slab and column at the connection. This can be compared with the situation at an interior column of a symmetrically loaded floor where the slab has a negative moment at the column, but that moment on one side of the column is balanced by the negative moment in the slab on the other side of the column and hence no unbalanced slab moment remains to be transferred to the column.

10.3.1 Behavior of Slab–Column Connections Transferring Shear and Unbalanced Bending Moment

In a flat plate or flat slab floor carrying gravity loading there will generally be transfer of both shear and unbalanced bending moment at edge columns and at some interior columns. This design aspect becomes particularly important when horizontal loading on the building due to wind or earthquake causes a substantial unbalanced bending moment to be transferred between the slab and the columns. The transfer of unbalanced bending moment causes the distribution of shear stress in the slab around the column to become nonuniform and reduces the shear strength of the connection. The shear force and unbalanced bending moment are transferred by combined bending, torsion, and shear at the faces of the critical section in the slab around the column.

If the shear strength of the slab is reached, the slab will fail in diagonal tension on the side of the column where the vertical shear stress is highest, resulting in the column punching through the slab and the top reinforcing bars in the slab splitting off the cover concrete. Figure 10.16*a* shows a slab–column connection in a test rig after shear failure due to transfer of shear and

Figure 10.16   Reinforced concrete slab–column specimen transferring shear and un-balanced moment after loading[10.30]: (*a*) general view; (*b*) failure region.

unbalanced bending moment. Figure 10.16b shows a closer view of the failure region of the slab with the broken concrete removed, demonstrating particularly the splitting away of a substantial area of the top concrete cover.

An alternative to a shear failure would be a flexural failure involving a local yield line pattern, as discussed in Section 7.11.3. The possibility of a yield line type of failure should be checked in analysis and design, particularly if the slab steel ratio is low. However, it should be borne in mind that yield line theory will very likely underestimate the upper limit to the unbalanced moment transfer strength of the connection because, as has already been discussed in Section 10.2, local failure of the slab around the connection can be accompanied by significant in-plane compressive forces in the slab induced by the lateral restraint of the surrounding unyielding slab portions. These in-plane membrane forces will enhance the moment capacities of the slab sections of the yield line pattern. The moment enhancement has not been taken into account in the yield line analysis of Section 7.11.3, because of the difficulty of estimating the membrane forces, taking into account the effects of the stiffness of the surround, including cracking, and the geometry of displacements of the slab segments.

Shear reinforcement can be used in the slab around the column to increase both the shear strength and the ductility of the connection when transferring unbalanced moment and shear. Such shear reinforcement can take the form of properly anchored stirrups, bent bars, structural steel shearheads, or stud-shear reinforcement.

During an earthquake the slab–column connections of flat plate and flat slab structures may be subjected to repeated reversals of unbalanced bending moment, which may lead to failure in the slab around the column due to degradation of the shear strength. Shear reinforcement should be incorporated to make such connections adequately ductile. Multistory flat plate or flat slab buildings have limited use as seismic-resistant structures without the presence of frames or walls to stiffen the building against excessive horizontal deflections. However, even with the presence of such stiffening elements to reduce interstory deflections, some unbalanced moment transfer will be necessary, and slab–column connections need to be made adequately ductile.

10.3.2 Methods of Analysis and Design

Existing methods of analysis and design of reinforced concrete slab–column connections transferring shear and unbalanced bending moment have been described previously.[10.4,10.5,10.6,10.7,10.31,10.32] The methods can be placed in four categories.

1. *Analyses based on a linear variation in shear stress.* Typical of this type of analysis is the method specified by the 1995 ACI Code.[10.1] The analysis assumes that shear stresses on a critical perimeter vary linearly with distance from the centroidal axis of the perimeter and are induced by the

shear force and part of the unbalanced bending moment. The remainder of the unbalanced bending moment is carried by flexure in the slab. The method is semiempirical, but the approach usually results in a conservative estimate of the measured strength. The method has only been developed for slabs without shear reinforcement.

2. *Analyses based on thin plate theory.* Elastic thin plate theory, incorporating finite element or finite difference analyses, has been used to determine shear and moment distributions in the slab around columns for use in design. However, with yielding, a substantial redistribution of actions can occur. Finite element analyses that take into account yielding, or a procedure proposed by Long,[10.33] can be used to include inelastic effects. Design for the determined shear and moment distribution can be carried out by providing sufficient shear, torsional, and flexural strength at all sections for the actions. The actions from analysis can be averaged over strips of small width for convenience in design.

3. *Beam analogies.* The slab adjacent to the column is considered to act as beams running in two directions at right angles framing into the column faces. The slab strips making up the beams are subjected to bending moment, torsional moment, and shear force, and redistribution of these actions is assumed to be able to occur between the beams. Each beam is assumed to be able to develop its ultimate bending moment, torsional moment, and shear force, making due allowance for interaction effects, at the critical sections near the column faces. The strength of the connection is calculated by summing the contributions of the strengths of the beams. The beam analogy of Hawkins et al.[10.5] assumes slab failure at an interior column when ultimate conditions are reached for at least three beams framing into a column. This beam analogy predicts up to eight possible limiting strength combinations of bending, torsion, and shear, which permits the development of an ultimate shear–moment interaction diagram for the capacity of an interior slab–column connection. The large number of possible limiting strength combinations for interior connections makes its application relatively difficult. For edge and corner slab–column connections of floors, the number of limiting strength combinations reduces to two or three, and the application becomes more simple for these cases. An alternative beam analogy has been developed by Park and Islam[10.32] in which the strength of the connection is obtained by summing the flexural, torsional, and shear strength of all the beams. This implies sufficient ductility in bending, torsion, and shear at the critical sections to allow the simultaneous development there of the ultimate capacities, and enables easier application for interior connections. These beam analogies are also capable of including the additional strength due to shear reinforcement in the slab. Other beam analogies, for example that of Stamenkovic[10.34] of semiempirical nature, exist. Beam analogies are not mentioned in the 1995 ACI Code[10.1] or its commentary but are recommended by ASCE-ACI Committee 426[10.5] as an alternative to the ACI Code procedure, which assumes a linear variation of shear stress.

4. *Truss analogies.* Alexander and Simmonds[10.6,10.7] developed a truss analogy that has already been mentioned in connection with punching shear in the absence of unbalanced moment. They applied it to the edge column case and developed complete $M–V$ interaction diagrams that conservatively reproduced test results. The starting point in the analysis is the strut arrangement shown in Fig. 10.7, where all the struts directly resist the gravity force. These

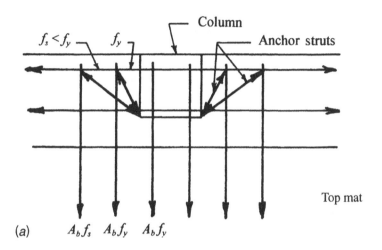

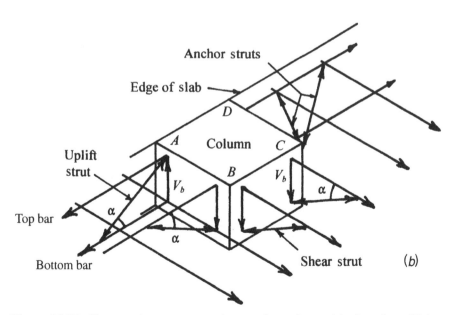

Figure 10.17   Truss analogy representation at edge column: (*a*) plan view; (*b*) isometric view. (Redrawn from Ref. 10.7. By permission from ACI International.)

struts are referred to as *gravity* or *shear struts*. As moment is introduced, forces in some of the bars are reversed, and different struts are anchored by bottom bars. These struts are referred to as *uplift* or *inverted shear struts*. A third, fundamentally different, strut known as an *anchor strut* is also introduced. A reinforcing bar may equilibrate a shear strut or an anchor strut, or the bar capacity may be divided between the two functions. The anchor struts essentially give a way to include more than merely the bars passing through the column in the direct moment resistance of the slab on one face of the column. Figure 10.17 shows isometric and plan views of an edge column with one particular set of struts acting. The geometry matches that of the example in Ref. 10.7. The combination of the one shear strut and one uplift strut on face *AB* can be taken as equivalent to a torque. The uplift strut can exist only if the equilibrating bottom bar is anchored adequately.

10.3.3 ACI Code Approach

The procedure recommended in the 1995 ACI Code[10.1] is based on investigations by Hanson and Hanson[10.35] and others reviewed in the ASCE-ACI Committee 426 report.[10.4] As mentioned earlier, the method has only been developed for slabs without shear reinforcement.

Let V_u and M_u be the factored (ultimate) shear force and unbalanced bending moment, acting about the centroidal axis of the column section, to be transferred. The critical section is located so that its perimeter is a minimum but need not approach closer than $d/2$ to the perimeter of the column, where d is the effective depth of slab reinforcement. Of the unbalanced bending moment M_u, $\gamma_v M_u = (1 - \gamma_f)M_u$ is assumed to be transferred by eccentricity of shear about the centroid of the critical slab section, and $\gamma_f M_u$ is assumed to be transferred by flexure of the slab, where

$$\gamma_f = \frac{1}{1 + \frac{2}{3}\sqrt{b_1/b_2}} \tag{10.19}$$

where $b_1 = c_1 + d$ or $c_1 + d/2$ (see Fig. 10.18), c_1 is the size of the rectangular or equivalent rectangular column or capital measured in direction of the moment, $b_2 = c_2 + d$ or $c_2 + d/2$, and c_2 is the size of the rectangular or equivalent rectangular column or capital measured transverse to the direction of the moment. Note that $\gamma_v = 0.4$ for square interior columns ($c_1 = c_2$ with d = average effective depth of the two layers of reinforcement). For rectangular columns, γ_v is greater than 0.4 when $b_1 > b_2$ (i.e., when the width of the face of the critical section resisting moment decreases); conversely, γ_v is less than 0.4 when $b_1 < b_2$ (i.e., when the width of the face of the critical section resisting moment increases).

The fraction of the unbalanced moment transferred by slab flexure, $\gamma_f M_u$, is considered to be transferred by the slab ultimate resisting moments over

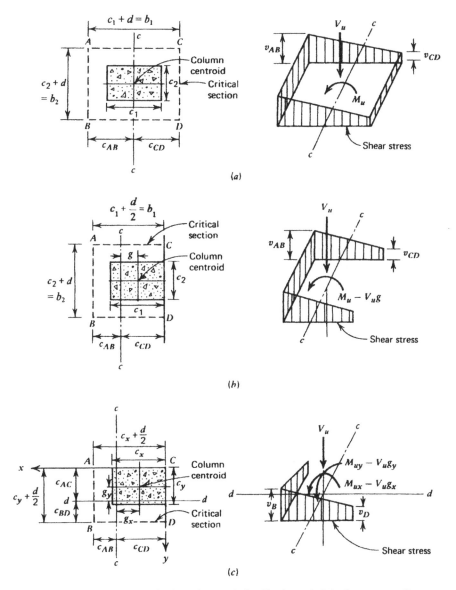

Figure 10.18 Assumed critical section and distribution of slab shear stress for connection transferring shear and unbalanced moment: (*a*) interior column connection; (*b*) edge column connection; (*c*) corner column connection.

an effective slab width between lines that are 1.5 slab or drop panel thicknesses (1.5*h*) outside opposite faces of the column or capital. Thus, sufficient slab reinforcement should be present between these lines to transfer the moment $\gamma_f M_u$. This may require reinforcement to be added to the slab through and adjacent to the column in addition to reinforcement required there for other loading cases.

The 1995 Code, Sec. 13.5.3.3, permits the value of γ_f to be increased to 1.0 (i.e., γ_v be reduced to zero) for edge columns when the moments act perpendicular to the edge of the structure and the applied shear $V_u \leq 0.75\phi V_c = v_c b_0 d$ and v_c is given by Eq. 10.5 or 10.6. The value of γ_f can be increased to 1.0 for corner columns when $V_u \leq 0.5\phi V_c$. For interior columns, γ_f may be increased by up to 25% when $V_u \leq 0.4\phi V_c$. If these changes in γ_v are invoked, the reinforcement ratio within the width $c_2 + 3h$, as described in the preceding paragraph, is limited to $0.375\rho_b$, where ρ_b is the balanced reinforcement ratio. These provisions apparently can be traced to an 1988 ACI Committee 352 report.[10.36,10.37]

The fraction of the unbalanced bending moment transferred by shear $\gamma_v M_u$ and the shear V_u are assumed to cause shear stresses that vary linearly around the critical section. The nominal shear strength of the connection is reached when the maximum shear stress at the critical section reaches v_c where for normal-weight concrete

$$v_c = \left(2 + \frac{4}{\beta}\right) \sqrt{f_c'} \quad \text{lb/in}^3 \qquad \text{but not greater than } 4 \sqrt{f_c'} \quad \text{lb/in}^2$$

(10.20)

where v_c is the ratio of the long side to the short side of column, and f_c' is the compressive cylinder strength of the concrete (psi). This value for v_c is to be multiplied by 0.75 for all-lightweight concrete or 0.85 for sand-lightweight concrete. The maximum shear stress due to the factored shear forces and moments must not exceed ϕv_c, where ϕ is the strength reduction factor for shear = 0.85 and v_c is given by Eq. 10.20. The stress limit from Eq. 10.6 must also be checked.

Calculation of the maximum shear stress at the critical section for various cases of connections is illustrated below. The calculations are based on the assumed distribution of shear stress and on section properties that follow the general form of equations developed by Di Stasio and van Buren.[10.10,10.38]

Interior Column Connection. Figure 10.18a shows the assumed critical section and the distribution of shear stress for an interior column connection. The shear stresses in the slab at the faces AB and CD of the critical section are given by

$$v_{AB} = \frac{V_u}{A_c} + \frac{\gamma_v M_u c_{AB}}{J_c}$$

(10.21)

$$v_{CD} = \frac{V_u}{A_c} - \frac{\gamma_v M_u c_{CD}}{J_c}$$

(10.22)

where c_{AB} is the distance from face AB of the critical section to centroidal axis cc, c_{CD} the distance from face CD of the critical section to centroidal axis cc, A_c the area of concrete at the critical section, and J_c the property of the section analogous to the polar moment of inertia.

$$A_c = 2d(c_1 + c_2 + 2d) \tag{10.23}$$

$$J_c = \frac{2d(c_1 + d)^3}{12} + \frac{2(c_1 + d)d^3}{12} + 2d(c_2 + d)\left(\frac{c_1 + d}{2}\right)^2 \tag{10.24}$$

In Fig. 10.19a, line ab represents the interaction relationship where the maximum shear stress (v_{AB} in this case) is limited to v_c from Eq. 10.20. Typical assumed shear stress distributions along the line are also indicated in the figure. The ordinate V_u/V_0 is the ratio of shear force transferred to column to the shear strength from Eq. 10.21 (or 10.22) when $M_u = 0$. It is evident that V_0 is equal to the V_u value given by Eq. 10.5. The abscissa, $\gamma_v M_u/\gamma_v M_0$, is the ratio of unbalanced bending moment transferred to column by shear

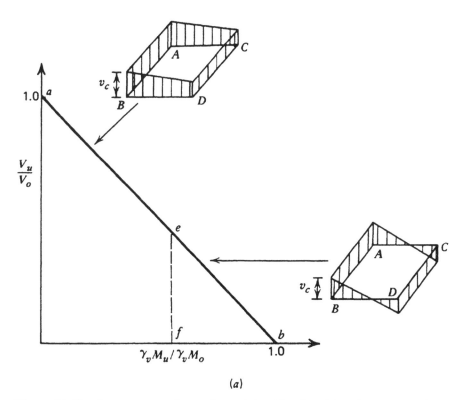

(a)

Figure 10.19 Shear–moment interaction relationships for slab–column connections: (a) interior column connection; (b) edge column connection; (c) corner column connection.

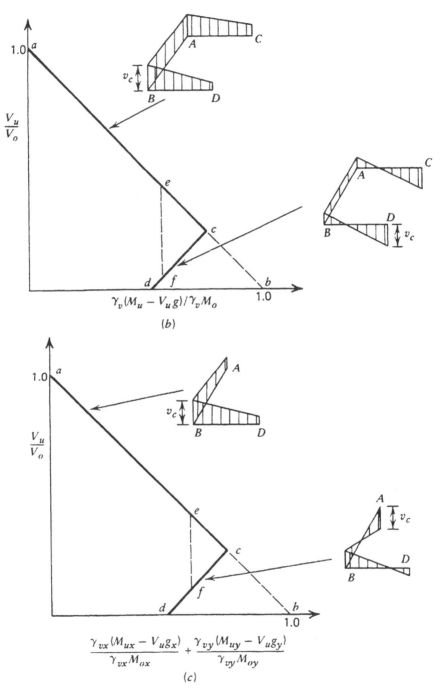

$$\gamma_v(M_u - V_u g)/\gamma_v M_o$$

(b)

$$\frac{\gamma_{vx}(M_{ux} - V_u g_x)}{\gamma_{vx} M_{ox}} + \frac{\gamma_{vy}(M_{uy} - V_u g_y)}{\gamma_{vy} M_{oy}}$$

(c)

Figure 10.19   (*Continued*).

to the moment strength from Eq. 10.21 (or 10.22) when $V_u = 0$. It is evident that $\gamma_v M_0$ equals $v_{AB} J_c / c_{AB}$, where v_{AB} is given by Eq. 10.20. The interaction line ab of Fig. 10.19a assumes that an unbalanced moment of at least $(1 - \gamma_v)M_u$ can be carried by flexure in the slab between the lines that are 1.5h each side of the column or column capital, where h is the slab thickness. If there is insufficient slab reinforcement within these lines, the interaction line will follow aef, where the position of ef depends on the flexural strength of the slab. Note that the dimensions of the connection and the concrete strength are the only variables influencing the position of the line ab. The slab steel content affects only the position of line ef.

Hoffman et al.[10.39] give derivations of J_c for several cases of circular interior, edge, and corner columns.

Edge Column Connection. Figure 10.18b shows the assumed critical section and the distribution of shear stress in the slab when moment is transferred normal to the edge of the slab. The dimension g is the distance between the centroidal axis of the critical section, cc, and the centroidal axis of the column. Since M_u is the unbalanced bending moment acting about the centroidal axis of the column, the unbalanced bending moment acting about the centroidal axis cc of the critical section is $M_u - V_u g$. The shear stresses in the slab at face AB and at points C and D of the critical section are given by

$$v_{AB} = \frac{V_u}{A_c} + \frac{\gamma_v(M_u - V_u g)c_{AB}}{J_c} \tag{10.25}$$

$$v_C = v_D = \frac{V_u}{A_c} - \frac{\gamma_v(M_u - V_u g)c_{CD}}{J_c} \tag{10.26}$$

where

$$A_c = d(2c_1 + c_2 + 2d) \tag{10.27}$$

$$J_c = \frac{2d[c_1 + (d/2)]^3}{12} + \frac{2[c_1 + (d/2)]d^3}{12} + (c_2 + d)dc_{AB}^2$$
$$+ 2[c_1 + (d/2)]d \left(\frac{c_1 + (d/2)}{2} - c_{AB} \right)^2 \tag{10.28}$$

$$c_{AB} = \frac{[c_1 + (d/2)]^2 d}{A_c} \tag{10.29}$$

$$c_{CD} = c_1 + \frac{d}{2} - c_{AB} \tag{10.30}$$

In Fig. 10.19b, line acd represents the interaction relationship when the maximum shear stress is limited to v_c from Eq. 10.20. For the portion ac the shear stress v_{AB} is critical, and for the portion cd the shear stresses v_C and v_D are critical. Typical assumed shear stress distributions along the interaction line are also shown in the figure. The ordinate V_u/V_0 is the ratio of shear force transferred to the column to the shear strength from Eq. 10.25 (or 10.26) when $M_u - V_u g = 0$. The abscissa $\gamma_v(M_u - V_u g)/\gamma_v M_0$ is the ratio of unbalanced bending moment transferred about the centroidal axis of the critical section by shear to moment strength from Eq. 10.26 when $V_u = 0$. The interaction line acd of Fig. 10.19b assumes that an unbalanced bending moment of at least $(1 - \gamma_v)(M_u - V_u g)$ can be transferred by flexure in the slab between the lines that are 1.5h each side of the column or column capital, where h is the slab thickness. If there is insufficient slab reinforcement within these lines, the interaction line will follow $aefd$, where the position of ef depends on the flexural strength of the slab.

Corner Column Connection. Figure 10.18c shows the assumed critical section and the distribution of shear stress in the slab when biaxial bending moment is transferred (bending in the x- and y-directions). The dimensions g_x and g_y are the distances in the x- and y-directions between the centroidal axes of the critical section, cc and dd, and the centroidal axes of the column. Since M_{ux} and M_{uy} are the unbalanced bending moments in the x- and y-directions acting about the centroidal axes of the column, the unbalanced bending moments acting about the centroidal axes cc and dd of the critical section are $M_{ux} - V_u g_x$ and $M_{uy} - V_u g_y$. The shear stresses in the slab at points A, B, and D of the critical section are given by

$$v_A = \frac{V_u}{A_c} + \frac{\gamma_{vx}(M_{ux} - V_u g_x)c_{AB}}{J_{cx}} - \frac{\gamma_{vy}(M_{uy} - V_u g_y)c_{AC}}{J_{cy}} \tag{10.31}$$

$$v_B = \frac{V_u}{A_c} + \frac{\gamma_{vx}(M_{ux} - V_u g_x)c_{AB}}{J_{cx}} + \frac{\gamma_{vy}(M_{uy} - V_u g_y)c_{BD}}{J_{cy}} \tag{10.32}$$

$$v_D = \frac{V_u}{A_c} - \frac{\gamma_{vx}(M_{ux} - V_u g_x)c_{CD}}{J_{cx}} + \frac{\gamma_{vy}(M_{uy} - V_u g_y)c_{BD}}{J_{cy}} \tag{10.33}$$

where

$$A_c = (c_x + c_y + d)d \tag{10.34}$$

$$J_{cx} = \frac{d[c_x + (d/2)]^3}{12} + \frac{[c_x + (d/2)]d^3}{12} + (c_y + d/2)dc_{AB}^2$$

$$+ [c_x + (d/2)]d \left(\frac{c_x + (d/2)}{2} - c_{AB} \right)^2 \tag{10.35}$$

$$J_{cy} = \frac{d[c_y + (d/2)]^3}{12} + \frac{[c_y + (d/2)]d^3}{12} + (c_x + d/2)dc_{BD}^2$$

$$+ [c_y + (d/2)]d \left(\frac{c_y + (d/2)}{2} - c_{BD}\right)^2 \qquad (10.36)$$

$$c_{AB} = \frac{[c_x + (d/2)]^2 d}{2A_c} \qquad (10.37)$$

$$c_{BD} = \frac{[c_y + (d/2)]^2 d}{2A_c} \qquad (10.38)$$

$$c_{CD} = c_x + \frac{d}{2} - c_{AB} \qquad (10.39)$$

$$c_{AC} = c_y + \frac{d}{2} - c_{BD} \qquad (10.40)$$

In Fig. 10.19c, line acd represents the interaction relationship when the maximum shear stress is limited to v_c from Eq. 10.20. For the portion ac, the shear stress v_C is critical, and for the portion cd, the shear stress v_A (or v_D) is critical. Typical assumed shear stress distributions along the interaction diagram are also shown in the figure. The ordinate V_u/V_0 is the ratio of shear force transferred to the column to the shear strength from Eq. 10.31 (or 10.32 or 10.33) when $M_{ux} - V_u g_x = M_{uy} - V_u g_y = 0$. The abscissa is expressed as the sum of the x- and y-direction ratios of unbalanced bending moment transferred about the centroidal axis of the critical section by shear to moment strength when $V_u = 0$. The interaction line acd of Fig. 10.19c assumes that the portions of the bending moments in the x- and y-directions not transferred by shear can be transferred by flexure in the slab between lines that are at the free edges of the slab and 1.5h on the other sides of the column or column capital, where h is the slab thickness. If there is insufficient slab reinforcement within these lines, the interaction line will follow aefd, where the position of ef depends on the flexural strength of the slab.

Example 10.1. A flat plate floor consists of a 7-in. (178-mm)-thick slab supported by 18-in. (457-mm) square columns on the lines of a square grid with 20 ft (6.10 m) between column centers. In the vicinity of each column the top steel in the slab consists of No. 6 [19 mm] bars placed both ways on 6-in. (152-mm) centers with an average effective depth of 5.5 in. (140 mm). The steel has a yield strength of 60,000 lb/in² (414 N/mm²). The concrete is normal weight with a compressive cylinder strength of 4000 lb/in² (27.6 N/mm²).

(a) Assuming that only gravity loads are acting and that the live load is placed uniformly over the entire floor, calculate the total uniform load per unit area of slab that would cause a punching shear failure at a symmetrically loaded interior slab–column connection.

(b) If the structure is subjected to lateral loading by earthquake motions when the factored gravity loading (dead plus live) of slab is 120 lb/ft² (5.75 kN/m²), calculate the unbalanced bending moment that would cause a shear failure at an interior slab–column connection. Check that the slab flexural reinforcement in the vicinity of the column is adequate if the negative moment due to the factored gravity loading is 5000 lb-ft/ft width (22.2 kNm/m).

SOLUTION. The critical section has sides of length $c + d = 18 + 5.5 = 23.5$ in. Therefore, the perimeter of the critical section $b_0 = 4(c + d) = 4 \times 23.5 = 94.0$ in.

(a) From Eq. 10.5, the nominal shear strength is

$$V_c = 4\sqrt{4000} \times 94 \times 5.5 = 130,800 \text{ lb}$$

From Eq. 10.4, the design shear strength is

$$\phi V_c = 0.85 \times 130,800 = 111,200 \text{ lb}$$

The loaded area of slab outside the critical section per column is

$$20^2 - \left(\frac{23.5}{12}\right)^2 = 396.2 \text{ ft}^2$$

Hence the total ultimate uniform load per unit area of slab when the design shear strength is reached is

$$w_u = \frac{112,200}{396.2} = 281 \text{ lb/ft}^2 \text{ (13.4 kN/m}^2\text{)}$$

(b) From Eq. 10.23,

$$A_c = 2 \times 5.5(18 + 18 + 2 \times 5.5) = 517 \text{ in}^2$$

From Eq. 10.24,

$$J_c = \frac{5.5(18 + 5.5)^3}{6} + \frac{(18 + 5.5)5.5^3}{6} + \frac{5.5(18 + 5.5)^3}{2} = 48,240 \text{ in}^4$$

From Eq. 10.19,

$$\gamma_f = \frac{1}{1 + \frac{2}{3}\sqrt{(18 + 5.5)/(18 + 5.5)}} = 0.60 \qquad \gamma_v = 1 - 0.6 = 0.4$$

From Eq. 10.20, the maximum shear stress at the design shear strength is

$$\phi v_c = 0.85 \times 4\sqrt{4000} = 215 \text{ lb/in}^2$$

Now

$$V_u = 120 \times 396.2 = 47,540 \text{ lb}$$

Hence, from Eq. 10.21,

$$215 = \frac{46,540}{517} + \frac{0.4M_u(23.5/2)}{48,240}$$

Therefore, the factored (ultimate) unbalanced bending moment at the design shear strength is

$$M_u = 1.263 \times 10^6 \text{ lb-in. (143 kN-m)}$$

CHECK. The slab flexure between lines 1.5 slab thicknesses each side of the column needs to transfer an unbalanced bending moment of $\gamma_f M_u$. That is, a slab strip of width $= 18 + (3 \times 7) = 39$ in. needs to transfer by flexure an unbalanced factored moment of $0.6 \times 1.263 \times 10^6 = 757,800$ lb-in. Also, the factored negative moment of 5000 lb-ft/ft width due to gravity loading amounts to a moment of $5000 \times 39 = 195,000$ lb-in. to be carried over the 39-in.-width slab strip. Hence, the design moment strength required of the 39-in.-wide strip is $757,800 + 195,000 = 952,800$ lb-in. Now for negative moment, $\rho = 0.44/(5.5 \times 6) = 0.0133$. Hence, the design negative-moment strength of the strip of width $b = 39$ in. is

$$\phi \rho b d^2 f_y \left(1 - 0.59\rho \frac{f_y}{f_c'}\right)$$

$$= 0.9 \times 0.0133 \times 39 \times 5.5^2 \times 60,000 \left(1 - 0.59 \times 0.0133 \times \frac{60,000}{4000}\right)$$

$$= 747,600 \text{ lb-in.}$$

Hence, spacing between centers of top steel needs to be made

$$\frac{747,600}{952,800} \times 6 = 4.7 \text{ in.}$$

Therefore, within the lines 1.5 slab thicknesses each side of the column, let the top steel in slab be No. 6 bars on 4.5-in. centers. (Note that according to the beam analogy described in Section 10.3.5, nominal slab steel would be required in the bottom of the slab in the vicinity of the column. Its quantity also would have to be checked against the steel required by the 1995 Code.)

10.3.4 ASCE-ACI Committee 426 Suggested Approach

The ASCE-ACI Committee 426 report,[10.5] published in 1977, contains suggested design provisions for the transfer of unbalanced moment and shear at slab–column connections. The suggested provisions provide a valuable addition to the rather brief recommendations given in the 1995 ACI Code,[10.1] and the more important ones will be summarized in the following.

General Considerations. The resisting capacity of a slab–column connection may be calculated by any procedure that takes into account the requirements of equilibrium and geometric compatibility and makes a rational assessment of the strength in combined bending, shear, and torsion of the critical section. Thus, the ACI Code approach described in Section 10.3.3 is applicable, as well as analyses based on thin plate theory and beam analogies.

An interior column is defined as one for which the distance from its periphery to any discontinuous edge of the slab exceeds the greater of four times the slab thickness or twice the development length of the slab reinforcement. This definition permits the effect of the edge on the shear strength of the slab transferring moment to be neglected when the distance from the column face to that edge exceeds this specified distance.

Approximate Procedure. For rectangular interior columns with a ratio of long side to short side of less than 2, the shear strength of the slab, without shear reinforcement, may be taken as

$$V_c = \frac{v_c b_0 d}{1 + [5.2(M_{u1} + M_{u2})/(b_0 V_u)]} \tag{10.41}$$

where v_c is given by Eq. 10.20, b_0 is the perimeter of the critical section defined in Sections 10.3.3 and 10.2.2, d the effective depth of the slab tension reinforcement, and M_{u1} and M_{u2} the unbalanced orthogonal moments to be transferred in directions 1 and 2. The proportion of M_{u1} and M_{u2} considered to be transferred by slab flexure is as defined using Eq. 10.19, and slab reinforcement should be provided as in Section 10.3.3 to transfer that proportion of the unbalanced bending moment. If for normal-weight concrete the term $5.2(M_{u1} + M_{u2})/b_0 V_u$ exceeds 0.6, slab reinforcement should also be present in the bottom of the slab placed continuously through the column region and anchored outside the critical section to control flexural cracking, as the 1995 Code requires.

Equation 10.41 was derived using the ACI Code approach and provides a simple conservative procedure for estimating the shear strength of interior connections. Presumably Eq. 10.6, added to the code after Ref. 10.5 was prepared, should also be satisfied.

Example 10.2. Calculate the unbalanced bending moment that would cause shear failure at the interior slab–column connection of Example 10.1(b) using the approximate procedure.

SOLUTION. The numerical values of V_u, v_c, b_0, and d have already been calculated for Example 10.1(b). $M_{u2} = 0$, since uniaxial column bending occurs and the ultimate unbalanced bending moment is $M_u = M_{u1}$. From Eq. 10.41, substituting ϕv_c for v_c to obtain the design shear strength,

$$47{,}540 = \frac{215 \times 94 \times 5.5}{1 + [5.2M_u/(94 \times 47{,}540)]}$$

Therefore, the factored unbalanced bending moment at the design shear strength is

$$M_u = 1.150 \times 10^6 \text{ lb-in. (130 kN-m)}$$

Note that Eq. 10.41 has given a slightly conservative moment capacity compared with Eq. 10.21. Note also that additional slab top steel is required for slab flexure, and nominal bottom steel, as in Example 10.1(b).

ACI Code Procedure. In cases where the ACI Code procedure is used for slabs without shear reinforcement, in which shear stresses are assumed to vary linearly around the critical section, a similar suggestion is also made to control cracking in the bottom of the slab at the connection. Such cracking is likely if, due to high unbalanced moment, the upward shear stress on one side of the connection is significant. It is recommended that when the algebraic difference in the shear stresses at the critical section in the slab on the opposite sides of the column exceeds $3\sqrt{f_c'}$ lb/in² ($0.25\sqrt{f_c'}$ N/mm²) for normal-weight concrete, slab reinforcement should be provided in the bottom face of the slab continuous through the column region and anchored outside the critical section, as the 1995 Code requires.

Beam Analogy Procedure. A beam analogy procedure may be used to determine the strength of slabs with or without shear reinforcement. The capacity of the slab must be checked at the critical section $d/2$ from the column face, and at critical sections where the shear reinforcement is reduced or terminated.

For a section in which the vertical shear carried by the concrete at the critical face of the critical section is considered to be added to the shear carried by the bent bar or stirrup shear reinforcement, the nominal ultimate shear stress of the concrete is taken to be one-half of that for slabs without shear reinforcement. The reason for this is the same as given in Section 10.2.3 for slabs with uniform shear. Figure 10.20 shows faces of the critical section of interior and exterior slab–column connections with square columns. Let M, V, and T without a subscript denote the strength for that action acting in combination with other actions, and with subscript 0 denote the strength for that action acting alone. Then for the analysis of the connections shown, the strength equations for normal-weight concrete for the faces of the critical section in pound and inch units (1 lb = 4.45 N and 1 in. = 25.4 mm) are:

Design moment capacity:

Fig. 10.20a:
$$M = \phi\rho(c + d)d^2 f_y \left(1 - 0.59\rho \frac{f_y}{f'_c}\right) \qquad (10.42)$$

where $\phi = 0.9$ and $\rho = A_b/s_b d$.

Design shear capacity without shear reinforcement:

Fig. 10.20a:
$$V \le V_0 = \phi(c + d)d4\sqrt{f'_c} \qquad \text{lb} \qquad (10.43)$$

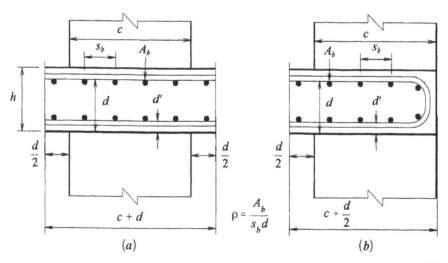

Figure 10.20 (a) (b)

$$\rho = \frac{A_b}{s_b d}$$

Figure 10.20 Faces of critical sections for strength equations for beam analogy for slab–column connections. (a) front, back, or side faces of an interior connection, or front face of an exterior connection; (b) side face of an exterior connection.

Fig. 10.20*b*: $$V \le V_0 = \phi \left(c + \frac{d}{2} \right) d4\sqrt{f'_c} \quad \text{lb} \qquad (10.44)$$

where $\phi = 0.85$.
   Design shear capacity with vertical stirrups as shear reinforcement:

Fig. 10.20*a*: $$V \le V_0 = \phi \left[(c + d)d \times 2\sqrt{f'_c} + A_v f_y \frac{d}{s} \right]$$
$$\le \phi(c + d)d \times 10\sqrt{f'_c} \qquad (10.45)$$

Fig. 10.20*b*: $$V \le V_0 = \phi \left[\left(c + \frac{d}{2} \right) d \times 2\sqrt{f'_c} + A_v f_y \frac{d}{s} \right]$$
$$\le \phi \left(c + \frac{d}{2} \right) d \times 10\sqrt{f'_c} \qquad (10.46)$$

where $\phi = 0.85$ and A_v is the area of shear reinforcement at spacing s at the face of the critical section.
   Design torsional capacity:

Fig. 10.20*a*: $$T \le T_0 = \phi \left[(c + d)h^2 \times 0.8\sqrt{f'_c} \right.$$
$$\left. + \alpha_t(c + d)(d - d') A_b \frac{f_y}{s_b} \right] \quad \text{lb-in.} \qquad (10.47)$$

where

$$\alpha_t = 0.66 + 0.33 \frac{c + d}{d - d'} \le 1.5$$

Fig. 10.20*b*: $$T \le T_0 = \phi \left[\left(c + \frac{d}{2} \right) h^2 \times 0.8\sqrt{f'_c} \right.$$
$$\left. + \alpha_t \left(c + \frac{d}{2} \right)(d - d')A_b \frac{f_y}{s_b} \right] \quad \text{lb-in.} \qquad (10.48)$$

where

$$\alpha_t = 0.66 + 0.33 \, \frac{c + d/2}{d - d'} \leq 1.5$$

Design combined shear and torsional capacity:

Fig. 10.20a:
$$\frac{T}{T_0} + \left(\frac{V}{V_0}\right)^2 \leq 1.0 \tag{10.49}$$

$$\left[\frac{3T}{\phi(c + d)^2 h \times 8\sqrt{f'_c}}\right]^2 + \left[\frac{V}{\phi(c + d)d \times 10\sqrt{f'_c}}\right]^2 \leq 1.0 \tag{10.50}$$

Fig. 10.20b:
$$\frac{T}{T_0} + \left(\frac{V}{V_0}\right)^2 \leq 1.0 \tag{10.51}$$

$$\left[\frac{3T}{\phi[c + (d/2)]^2 h \times 8\sqrt{f'_c}}\right]^2 + \left[\frac{V}{\phi[c + (d/2)]d \times 10\sqrt{f'_c}}\right]^2 \leq 1.0 \tag{10.52}$$

In the equations above, interaction between shear and torsion is assumed but not between torsion and moment or shear and moment. The torsion–shear interaction relationships, Eqs. 10.49 and 10.51, are based on, but are different from, the 1989 (and earlier) ACI Code[10.40] equations because the slab flexural reinforcement, but not the stirrups, has been assumed effective as torsion reinforcement in Eqs. 10.47 and 10.48. The torsional resistance provisions in the 1995 ACI Code[10.1] are completely different, and these provisions are based on the earlier codes. Equations 10.50 and 10.52 give another limiting relationship between torsion and shear, giving the upper limit on strength regardless of the quantity of shear and torsional reinforcement present, and are based on a limiting value of 8 $\sqrt{f'_c}$ lb/in² for the maximum torsional stress that can be applied without shear. Note also that the limiting value of 10 $\sqrt{f'_c}$ lb/in² for the maximum shear stress that can be provided without torsion (this also appears as the upper limit in Eqs. 10.45 and 10.46) is greater than the 1989 ACI Code[10.40] maximum value of 6 $\sqrt{f'_c}$ lb/in². Equations 10.49 and 10.50 appear in Ref. 10.5, and Eqs. 10.51 and 10.52 are obvious extensions.

In the analysis of a given slab–column connection using the strength equations given above, it is assumed that some redistribution of actions between critical faces can occur. Nevertheless, not all faces may reach all the limiting strengths, and the possible combinations of limiting strengths must be considered. For the case of an interior slab–column connection transferring unbalanced bending moment in one direction, the suggested beam analogy results in the eight possible limiting strength combinations shown in Fig.

10.21a. For an edge connection transferring unbalanced bending moment in one direction, the number of possible limiting strength combinations is two, as shown in Fig. 10.21b. For a corner connection transferring unbalanced bending moment in two directions, the number of possible limiting strength combinations is three, as shown in Fig. 10.21c. For an interior column the factored unbalanced moment strength and the shear strength may be written as

$$M_u \leq M_{AB} + M_{CD} + T_{BD} + T_{AC} + (V_{AB} + V_{CD})\frac{c_1 + d}{2} \quad (10.53)$$

$$V_u \leq V_{AB} + V_{BD} + V_{AC} - V_{CD} \quad (10.54)$$

where the subscripts AB, BD, and so on, indicate the capacities of those faces, and the direction of actions assumed are those shown in case 8 of Fig. 10.21a. For the other cases, the signs of shear and moment on face CD in the equations may need to be changed if the direction is different. Note that the shears on face AB, and the shears and bending moments on face CD, may not be the ultimate values.

For an edge connection the unbalanced moment strength perpendicular to the edge of the structure and the shear strength may be written as

$$M_u \leq M_{AB} + T_{BD} + T_{AC} + V_{AB}\left(\frac{c_1}{d} + \frac{d}{2}\right) + (V_{AC} + V_{BD})\frac{d}{4} \quad (10.55)$$

$$V_u \leq V_{AB} + V_{AC} + V_{BD} \quad (10.56)$$

where the direction of actions assumed are those in case 2 of Fig. 10.21b.

For an edge connection, the unbalanced moment strength parallel to the edge of the structure and the shear strength can also be written. For a corner connection, the unbalanced moment strength and shear strength may be written as

$$M_{ux} \leq M_{AB} + T_{BD} + V_{AB}\left(\frac{c_x}{d} + \frac{d}{2}\right) + \frac{V_{BD}d}{4} \quad (10.57)$$

$$M_{uy} \leq M_{BD} + T_{AB} + V_{BD}\left(\frac{c_y}{d} + \frac{d}{2}\right) + \frac{V_{AB}d}{4} \quad (10.58)$$

$$V_u \leq V_{AB} + V_{BD} \quad (10.59)$$

where the direction of actions assumed are those in case 3 of Fig. 10.21c.

To determine the strength of a given connection, all the limiting strength combinations (i.e., all the cases shown in Fig. 10.21 for the particular connection) need to be examined to determine the minimum external actions to

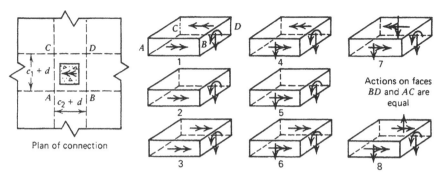

Failure conditions at faces *AB, BD, CD,* and *AC*

Cases 1, 4, 7 low *M/V* ratio; cases 3, 6, 8 high *M/V* ratio; cases 1, 2, 3 likely for low ρ and high c_1/d; cases 7, 8 likely for high ρ and low c/d.

(*a*)

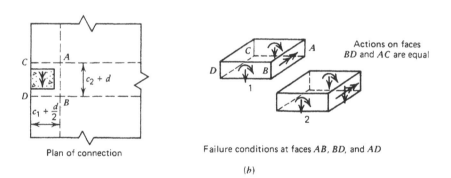

Failure conditions at faces *AB, BD,* and *AD*

(*b*)

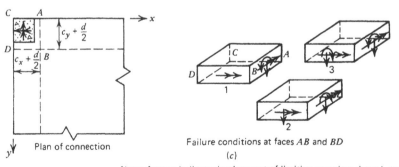

Failure conditions at faces *AB* and *BD*

(*c*)

Note: Arrows indicate development of limiting capacity: shear ↓, torsion ↷ bending ⇢ (clockwise looking in direction of arrow).

Figure 10.21 Limiting strength combinations for beam analogy for slab–column connections[10.5]: (*a*) interior connection; (*b*) edge connection; (*c*) corner connection.

cause failure. The different cases arise because it was considered by ACI Committee 426[10.5] that the full ultimate flexural, torsional, and shear capacities may not develop at the faces, and in some cases failure may occur with only a limited number of actions at full strength.

Reference 10.5 indicates that the beam analogy equations above give a more accurate prediction of the strength than the ACI Code approach based on a linear shear stress distribution. However, it is evident that there is difficulty in applying the beam analogy equations above to the case of interior columns, since there are so many (eight) cases of limiting strength combinations to be checked to determine the most critical case. Thus, for interior connections it is easier to use the ACI Code approach. However, for edge and corner connections, the beam analogy equations are easier to use than the ACI Code equations, since there are fewer cases of limiting strength combinations to examine.

For slabs with structural steel shearheads, the ACI procedure for slabs without unbalanced moment may be extended to slabs with unbalanced moment. Use of the beam analogy for determining the strength of exterior slab–column connections containing shearheads has been described by Hawkins and Corley.[10.41]

Torsion and Shear Transfer at Discontinuous Edges of Slabs. If a spandrel beam frames into an exterior column, the shear in the slab may be assumed to be transferred to the column via the beam. The spandrel beam must have torsional reinforcement to control torsional cracking, as well as shear reinforcement. If there is no spandrel beam, the slab edge will need slab reinforcement, adjacent to the column face and the slab edge, parallel to the slab edge in the top and bottom of the slab to control torsional cracks. This slab reinforcement should be anchored so as to be effective over at least four slab thicknesses from the column faces. In addition, the top reinforcement perpendicular to the edge of the slab should be hooked so that it bends back into the slab at the bottom, as shown in Fig. 10.20*b*, to form three sides of the reinforcement cage of torsion. The hooks should be on the outside, as shown, with the slab bars parallel to the edge of the slab inside the hooks.

10.3.5 Alternative Beam Analogy Approach for Interior Connections

The eight possible limiting strength combinations shown in Fig. 10.21*a* for interior connections make the application of that procedure relatively difficult because of the large number of failure cases to be considered. A simpler beam analogy has been developed which in effect assumes that case 6 of Fig. 10.21*a* is the critical limiting case[10.32] and that sufficient ductility in bending, torsion, and shear is available at the critical faces to allow development of the ultimate flexural, torsional, and shear capacities as required. Figure 10.22 shows the critical section lying at 0.5*d* from the column periphery and the direction of the actions assumed for the interior connection. The ultimate

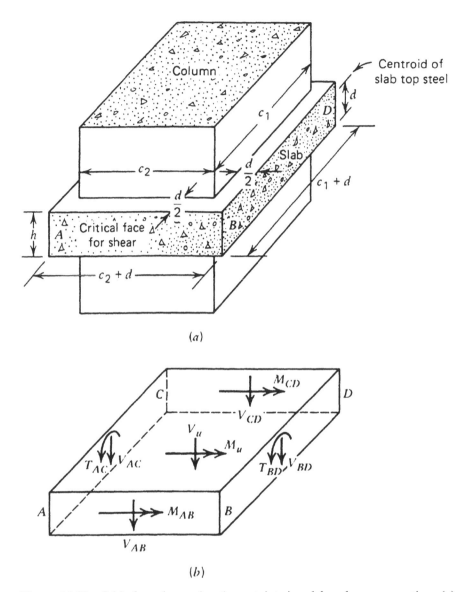

(a)

(b)

Figure 10.22 Critical section and actions at interior slab–column connection: (a) critical section; (b) actions at critical section.

(factored) unbalanced bending moment and ultimate shear force transferred by the connection are given by

$$M_u = M_{AB} + M_{CD} + T_{AC} + T_{BD} + (V_{AB} - V_{CD}) \frac{c_1 + d}{2} \qquad (10.60)$$

$$V_u = V_{AB} + V_{CB} + V_{AC} + V_{BD} \qquad (10.61)$$

The terms in these equations are explained in the following paragraphs.

Connections Without Shear Reinforcement. To derive the strength terms in Eqs. 10.60 and 10.61, the following simplifying assumptions will be made.

1. When the strength of the connection is developed, the slab tension reinforcement bars crossing faces *AB* and *CD* of the critical section reach the yield strength, and the flexural strength is reached in negative bending on face *AB* and positive bending on face *CD*. Compressive membrane (in-plane) forces will be present in the plane of the slab and may considerably enhance the flexural strength of the sections, but the order of flexural strength increase due to compressive membrane forces is difficult to assess and will be neglected. The assumption of the attainment of the positive moment flexural strength of face *CD* may be optimistic when the gravity load to horizontal load ratio is high (i.e., when V_u/M_u ratio or V_u/V_0 ratio is high), because then it is possible for negative moment to exist on face *CD* in the elastic range and a large redistribution of actions will be required at ultimate load to develop the positive-moment capacity of that face. However, the assumption is a convenient simplification. The estimate of strength given by the assumption when the gravity load to horizontal load ratio is high may not be unsafe, because the neglect of membrane action will mean that M_{AB} is underestimated. In any case, in most slabs the area of bottom steel adjacent to the column is smaller than the area of top steel, and thus the M_{CD} calculated may not be large compared with M_{AB}. Note that the bottom slab bars must be adequately anchored for the positive-moment capacity to be developed. The ultimate moment capacity will be assumed not to be reduced by the presence of shear on faces *AB* and *CD*.

2. The ultimate shear capacity is reached at face *AB* of the critical section. Face *AB* is the critical face for vertical shear because at that face the vertical shears due to the part of M_u transferred by shear stress and the part of V_u transferred by that face are additive, whereas at face *CD* these two shears act in opposite directions. That face *AB* is the critical face for shear has been assumed by the ACI 318 approach[10.1] and has been demonstrated in many tests (see, e.g., Ref. 10.30). The maximum vertical design shear stress on the critical face *AB* will be limited to $\phi 4\sqrt{f_c'}$ lb/in² [$\phi\sqrt{f_c'}/3$ N/mm²], provided that $0.5 \le c_1/c_2 \le 2$. Thus the limiting design shear force on face *AB* is

$$V_{AB} = \phi 4\sqrt{f_c'}(c_2 + d)d \qquad (10.62)$$

where the strength reduction factor $\phi = 0.85$. Note that the foregoing shear stress of $\phi 4\sqrt{f_c'}$ lb/in² for a slab is twice the $\phi 2\sqrt{f_c'}$ psi commonly used for beams. This considerably enhanced shear strength of slabs can be attributed to the location of the critical crack and the stress conditions at the apex of the crack, the presence of compressive membrane forces caused by the restraints provided by surrounding nonyielding portions of the slab, and other reasons (see Section 10.2.1). The ultimate shear capacity will be assumed not to be reduced by the development of the ultimate bending moment at the critical section.

3. The contributions of V_{AB}, V_{CD}, V_{AC}, and V_{BD} from the vertical shear V_u are assumed to be according to the tributary areas of the surrounding slab. The values of V_{AB} and V_{CD} will also receive a contribution from M_u due to the shear induced by moment transfer. If the portion of V_u transferred by face AB is $k_{AB}V_u$, the shear force induced on face AB by the moment transfer will be $\phi 4\sqrt{f_c'}(c_2 + d)d - k_{AB}V_u$, which by symmetry is also the shear force induced on face CD by moment transfer. These two moment-induced shears act in opposite directions. If the portion of V_u transferred by face CD is $k_{CD}V_u$, it is evident that

$$V_{CD} = k_{CD}V_u - [\phi 4\sqrt{f_c'}(c_2 + d)d - k_{AB}V_u] \qquad (10.63)$$

Thus, from Eqs. 10.60 and 10.61,

$$(V_{AB} - V_{CD})\frac{c_1 + d}{2} = [\phi 4\sqrt{f_c'}(c_2 + d)d - 0.5V_u(k_{AB} + k_{CD})](c_1 + d)$$

$$(10.64)$$

In symmetrical cases when equal portions of V_u are transferred to the four faces, $k_{AB} = k_{CD} = 0.25$.

4. The ultimate torsional capacity is assumed to be developed on the side faces, BD and AC, of the critical section. The torsional capacity will be calculated taking into account the reduction in torsional strength due to the vertical shear forces V_{BD} and V_{AC} caused by V_u using a circular interaction relationship between the torsional and vertical shear stresses. The ultimate torsional shear stress for the case of torsion without vertical shear will be assumed to be $\phi 4.8\sqrt{f_c'}$ lb/in² ($\phi 0.4\sqrt{f_c'}$ N/mm²), which is twice that recommended for beams by the 1989 ACI Code.[10.40] This enhanced torsional shear stress is assumed to be available for much the same reasons as the enhanced vertical shear stress described in assumption 2, particularly the presence of compressive membrane forces in the slab.[10.2] Test evidence[10.42] indicates that the ultimate torsional shear stress in the slab; without vertical shear, can be much greater than $4.8\sqrt{f_c'}$ lb/in²; measured values for this stress from the Japanese tests[10.42] were about five times that recommended here. The torsional strength of faces BD and AC can be obtained using the equations of the 1989 ACI Code[10.40] and its commentary.

In the case of symmetrically loaded connections ($k_{BD} = k_{AC}$), the vertical shear stresses on faces BD and AC induced by V_u may be written as

$$v_u = \frac{k_{BD}V_u}{d(c_1 + d)} = \frac{k_{AC}V_u}{d(c_1 + d)} \qquad \text{lb/in}^2 \qquad (10.65)$$

in which k_{BD} and k_{AC} give the portion of V_u being transferred to faces BD and AC. The maximum torsional shear stresses on faces BD and AC induced by T_{BD} and T_{AC} are

$$v_{tu} = \frac{3T_{BD}}{h^2(c_1 + d)} = \frac{3T_{AC}}{h^2(c_1 + d)} \qquad \text{lb/in}^2 \qquad (10.66)$$

The torsional strength of each face may be written as

$$T_{BD} = T_{AC} = \phi \frac{h^2(c_1 + d)}{3} \frac{4.8 \sqrt{f_c'}}{\sqrt{1 + (1.2v_u/v_{tu})^2}} \qquad \text{lb-in.} \qquad (10.67)$$

which when substituting v_{tu} from Eq. 10.66 and rearranging the terms gives

$$T_{BD} = T_{AC} = \phi \frac{1}{3} h^2(c_1 + d)\, 4.8\sqrt{f_c'}\, \sqrt{1 - \left(\frac{v_u}{4\sqrt{f_c'}}\right)^2} \qquad \text{lb-in.} \qquad (10.68)$$

in which v_u is given by Eq. 10.65.

In the case of unsymmetric loading, the vertical shear forces, $k_{BD}V_u$ and $k_{AC}V_u$ will not be equal and thus v_u will be different for faces BC and DA. In this case the values of T_{BD} and T_{AC} will need to be determined separately using the appropriate v_u values and summed.

Unbalanced Bending Moment Strength. Substituting Eqs. 10.64 and 10.68 into Eq. 10.60 gives the unbalanced bending moment design strength of the slab–column connection with symmetric loading $k_{BD} = k_{AC}$ as

$$M_u = (m_u + m_u')(c_2 + d) + [\phi 4\sqrt{f_c'}(c_2 + d)d - 0.5V_u(k_{AB} + k_{CD})](c_1 + d)$$

$$+ \phi \frac{2}{3} h^2(c_1 + d)4.8\sqrt{f_c'}\, \sqrt{1 - \left(\frac{v_u}{4\sqrt{f_c'}}\right)^2} \qquad \text{lb-in.} \qquad (10.69)$$

in which m_u and m_u' are the ultimate positive and negative resisting moments per unit width at faces CD and AB, respectively, including the strength reduction factor for flexure; V_u is the ultimate shear force to be transferred; and v_u is given by Eq. 10.65.

Note that the gravity loading on the slab will cause negative bending moments in the slab near the column. Hence, the slab flexural strength available to carry the unbalanced bending moment in Eq. 10.69 is the flexural strength remaining after the strength required to carry gravity loading has been subtracted.

Note also that the shear force to be transferred, V_u, cannot exceed

$$V_0 = \phi 4\sqrt{f_c'}\, d[2(c_1 + d) + 2(c_2 + d)] \qquad \text{lb} \qquad (10.70)$$

which is the ultimate shear capacity of the connection when the unbalanced bending moment, M_u, is zero. If the shear force to be transferred, V_u, equals V_0, it is evident from Eqs. 10.64 and 10.68 that no unbalanced bending moment can be transmitted by torsion or vertical shear but that unbalanced moment can be carried by the bending moment terms of Eq. 10.69, since they are unaffected by the magnitude of V_u. Thus, the theory implies that a slab–column connection loaded to its pure shear value is capable of sustaining in addition an unbalanced bending moment and gravity load moment equal to the sum of the negative moment strength of face AB and the positive moment strength of face CD of the critical section. This moment capacity when $V_u = V_0$ has been confirmed by test results, as for example those plotted in Fig. 10.26, which is discussed later.

An interaction diagram plotting values of M_u and V_u from Eqs. 10.69 and 10.70 which cause failure of a typical slab–column connection is shown in Fig. 10.23. Note that the ACI Code approach[10.1] results in an interaction diagram (see Fig. 10.19) consisting of a single straight line running between the shear strength V_0 without moment to the unbalanced moment strength without shear. The ACI and this beam analogy approaches both have the same V_0 value, but the unbalanced moment strengths with shear are different because of the empirical manner in which the ACI approach distributes the unbalanced moment between slab flexure and slab shear.

Example 10.3. Calculate the unbalanced bending moment that would cause shear failure at the interior slab–column connection of Example 10.1(b).

SOLUTION. Some of the required numerical quantities have already been calculated in Example 10.1. Since bottom steel is nominal, assume that $m_u = 0$ in Eq. 10.69. Other terms in Eq. 10.69 are:

Slab flexure, with top steel consisting of No. 6 bars on 4.5-in. centers, giving $\rho = 0.44/(5.5 \times 4.5) = 0.0178$:

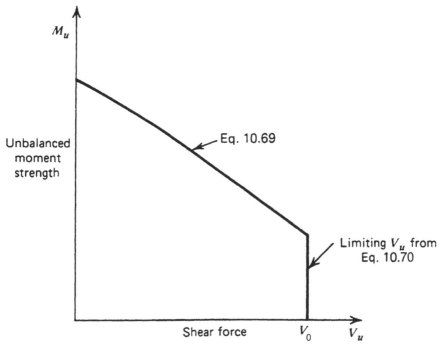

Figure 10.23   Interaction diagram for a slab–column connection.[10.32]

$$m'_u(c_2 + d) = \phi \rho d^2 f_y \left(1 - 0.59 \rho \frac{f_y}{f'_c} \right)(c_2 + d)$$

- factored gravity load negative moment over $(c_2 + d)$

$= 0.9 \times 0.0178 \times 5.5^2$

$\times 60{,}000 \left(1 - 0.59 \times 0.0178 \times \dfrac{60{,}000}{4000} \right) - 5000 \times 23.5$

$= 575{,}600 - 117{,}500 = 458{,}100$ lb-in.

Slab shear, with $k_{AB} = k_{CD} = 0.25$:

$[\phi 4\sqrt{f'_c}(c_2 + d)d - 0.25 V_u](c_1 + d)$

$= [0.85 \times 4\sqrt{4000} \times 23.5 \times 5.5 - 0.25 \times 47{,}540]23.5$

$= 373{,}800$ lb-in.

Slab torsion, with (from Eq. 10.65) $v_u = 0.25 \times 47{,}540/(5.5 \times 23.5) =$ 92.0 lb/in²:

$$\phi \frac{2}{3} h^2 (c_1 + d) 4.8 \sqrt{f_c'} \sqrt{1 - \left(\frac{v_u}{4\sqrt{f_c'}}\right)^2}$$

$$= 0.85 \times \frac{2}{3} \times 7^2 \times 23.5 \times 4.8\sqrt{4000} \sqrt{1 - \left(\frac{92.0}{4\sqrt{4000}}\right)^2}$$

$$= 184{,}500 \text{ lb}$$

Therefore, when $V_u = 47{,}540$ lb,

$$M_u = 458{,}100 + 373{,}800 + 184{,}500$$

$$= 1.016 \times 10^6 \text{ lb-in.}$$

This value of M_u is 20% less than the value given by Eq. 10.21 and 12% less than the value given by Eq. 10.41. Part of this conservatism no doubt arises from the greater torsional strength apparently available from the slab than has been allowed in the beam analogy equations.[10.40] Note that unlike the ACI Code approach, the beam analogy approach is capable of taking into account the actual contribution of slab reinforcement to the unbalanced moment strength and indicates that the unbalanced moment strength of the connection can be increased by increasing the content of the slab reinforcement. The ACI Code requires that two bottom bars be continuous and pass through the column core, but does not specify their area. The smallest practical area for this case is probably two-No. 4 bars, which would add 120,000 lb-in. to M_u. Two No. 6 bars would add about 240,000 lb-in. Thus, accounting for bottom reinforcement would bring the results of this beam analogy solution closer to the results obtained earlier.

Connections with Shear Reinforcement. The beam analogy method provides a convenient approach to take into account the effect of shear reinforcement.[10.32] Equations for the strength of slab–column connections transferring shear and unbalanced bending moment, with shear reinforcement in the form of either bent bars, vertical closed stirrups, or shearheads fabricated from structural steel shapes, can be derived by adding the contribution of the shear reinforcement to the strength of the connection without shear reinforcement, with an allowance made for the reduction in the strength of the concrete shear-resisting mechanism in accordance with the previous findings. The allowance for the reduced capacity of the concrete shear-resisting mechanism will be made by substituting $2\sqrt{f_c'}$ lb/in^2 [$\sqrt{f_c'}/6$ N/mm^2] for $4\sqrt{f_c'}$ lb/in^2 and $2.4\sqrt{f_c'}$ lb/in^2 [$0.2\sqrt{f_c'}$ N/mm^2] for $4.8\sqrt{f_c'}$ lb/in^2 into Eq. 10.69, which amounts to halving the vertical and torsional shear strengths of the concrete (these reduced stresses are, in fact, the vertical shear and torsional shear strength values recommended by the 1989 ACI Code[10.32] for beams).

The unbalanced moment design strength when $V_u \leq V_0$ therefore is

$$M_u = M_c + M_s \tag{10.71}$$

but not less than M_u from Eq. 10.69, in which M_s is the unbalanced moment strength due to the shear reinforcement and

$$M_c = (m_u + m_u')(c_2 + d) + [\phi 2\sqrt{f_c'}(c^2 + d)d - 0.5V_u(k_{AB} + k_{CD})](c_1 + d)$$

$$+ \phi \frac{2}{3} h^2(c_1 + d)2.4\sqrt{f_c'} \sqrt{1 - \left(\frac{v_u}{2\sqrt{f_c'}}\right)^2} \tag{10.72}$$

in which the notation is as in Eq. 10.69 and V_u is the ultimate shear force to be transferred. Values of M_s for various types of shear reinforcement are derived subsequently.

Bent Bars. Figure 10.24a shows a connection with bent bars as shear reinforcement. It is evident that the additional unbalanced bending moment capacity resulting from the bent bars will be due to the increase in shear strength on face *AB* (see Fig. 10.22), since the bars will have little influence on the torsional strength of faces *BD* and *AC*, and the shear stress on face *CD* is not critical. Thus, the additional unbalanced bending moment capacity of the connection due to the bent bars when $V_u \leq V_0$ may be written as[10.30,10.32]

$$M_s = \phi A_v f_y \sin \alpha(c_1 + d) \tag{10.73}$$

in which $\phi = 0.85$, A_v is the area of set of bent bars at face *AB*, f_y the yield strength of the bars, and α the angle of inclination of the bent portion to horizontal. Note that the bent bars must be effectively anchored at the bottom of the slab; otherwise, their efficiency will be greatly reduced.

Vertical Stirrups. Closed vertical stirrups placed around the slab bars passing both ways through the column (see Fig. 10.24b) will increase the unbalanced bending moment capacity due to the increase in vertical shear strength of face *AB* (see Fig. 10.22) and the increase in torsional strength of faces *BD* and *AC*. The additional moment due to the increase in vertical shear strength may be calculated from the stirrup force acting across the critical diagonal tension crack; this crack is assumed to be at 45° to the horizontal and therefore is crossed by d/s sets of stirrups, in which s is the spacing of the stirrup sets. The increase in torsional strength may be calculated using the approach of the 1989 ACI Code.[10.40] Thus, the additional unbalanced moment capacity of the connection due to closed stirrups on all four faces when $V_u \leq V_0$ may be written by summing the shear and torsional contributions as

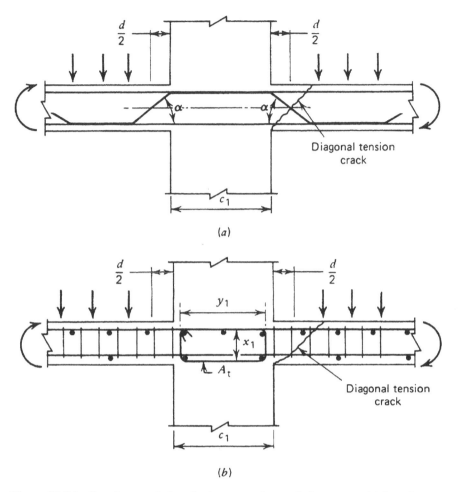

Figure 10.24 Bent bars and closed stirrups as shear reinforcement: (*a*) bent bars as shear reinforcement (only bars at faces *AB* and *CD* shown); (*b*) closed vertical stirrups as shear reinforcement.

$$M_s = \phi \left[\frac{A_v f_y d}{s} (c_1 + d) + \frac{2 \alpha_t x_1 y_1 A_t f_y}{s} \right] \qquad (10.74)$$

in which $\phi = 0.85$, A_v is the area of closed stirrups adjacent to face *AB* within a spacing s, f_y the yield strength of the stirrups, $\alpha_t = 0.66 + 0.33(y_1/x_1)$ but not greater than 1.5, x_1 the shorter center-to-center dimension of rectangle enclosed by stirrups on faces *BD* and *AC*, y_1 the longer center-to-center dimension of rectangle enclosed by stirrups on faces *BD* and *AC*, and A_t the area of one leg of the closed stirrups resisting torsion adjacent to faces *BC* and *DA* within a spacing s. Figure 10.24*b* shows x_1 and y_1 for a typical case.

Note that for the preceding torsional term to apply, the longitudinal steel in the slab passing within the area, x_1y_1, should satisfy the limits for A_l, given in the 1989 ACI Code.[10.40] The stirrup spacings should not exceed $0.5d$, to ensure that the critical diagonal tension cracks are effectively intersected by the stirrups. Typical arrangements of closed stirrups are shown in Fig. 10.25a.

Structural Steel Shearheads. A method for including the additional strength resulting from the presence of structural steel shearheads has also been developed.[10.30]

Shear Strength Outside Region of Shear Reinforcement. The strength of the region of the slab outside the shear reinforcement needs to be checked to ensure that it does not govern the strength of the connection. This means extending the shear reinforcement far enough to make the area outside shear reinforcement not critical.

Comparison of Theory with Test Results. The unbalanced moment strengths calculated by the simplified beam analogy given above have been shown to be conservative when compared with test results.[10.30]

10.3.6 Truss Analogy Approach Results

The truss analogy developed by Alexander and Simmonds[10.6,10.7] leads to a fundamentally different interaction relationship between moment and shear than is found using either the ACI Code approach or the beam analogy approach described earlier. Figure 10.26 gives an interaction diagram for a particular slab–column combination which was tested under several $M-V$ combinations by Stamenkovic and Chapman.[10.43] The test specimens included in this figure had identical top and bottom reinforcement. The truss analogy is somewhat conservative, but it reproduces the trends of the test results quite faithfully. The most striking difference between this interaction diagram and those presented earlier is that the shear capacity is quite insensitive to the presence of moment, at least until the moment is a large fraction of the moment capacity existing in absence of shear.

Moehle[10.44] suggested an essentially "square" interaction between moment and shear, that is, no interaction, as long as the applied $V_u \leq 0.75V_0$, where V_0 is the shear capacity in the absence of moment. This is consistent with Sec. 13.5.3.3 of the 1995 ACI Code. He also suggested a simplified truss concept for calculation of the moment capacity M_0, which is available in the absence of shear. This paper also suggests that the value of γ_v from Eq. 12.19 is significantly too large if the ACI Code's method of computing stresses on the critical section located at $d/2$ from the face of the column is to be used.

These studies strongly suggest that the linear stress distribution approach in the 1995 ACI Code[10.1] should be abandoned or greatly modified in order to make the predicted and observed results more closely related. The provi-

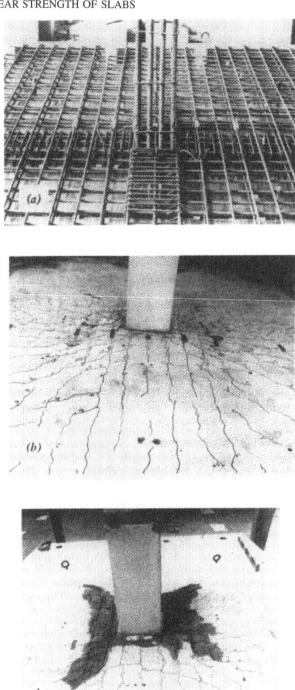

Figure 10.25   Reinforced concrete interior slab–column test specimen with closed stirrups as shear reinforcement[10.30]: (*a*) arrangement of reinforcement; (*b*) cracking in specimen after test; (*c*) broken concrete removed from specimen after test.

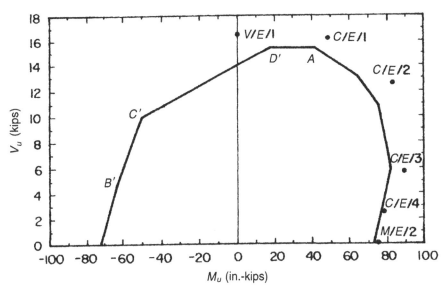

Figure 10.26 Moment–shear interaction diagram from truss analogy. (From Ref. 10.7. By permission from ACI International.)

sions of Code Sec. 13.5.3.3 accomplish part of this, but this change does not make the fundamental changes to the concepts that seem to be required.

10.3.7 Ductility of Slab–Column Connections

A punching shear failure of a slab–column connection without well-designed shear reinforcement is usually a brittle failure and can have disastrous consequences. Shear failure at a slab–column connection can result in progressive failures of adjacent connections of the same floor as the floor loading is transferred elsewhere, causing adjacent connections to become more heavily loaded. Also, lower floors may fail progressively as they become unable to support the impact of material dropping from above. Hence, attention should be given to the ductility associated with the shear strength of connections to avoid a brittle condition if possible. Ideally, the connection should contain reinforcement that holds the connection together after punching shear failure of the slab and prevents the slab from slipping down the column. Code provisions for connection design do not necessarily lead to ductile connections.

For slabs without shear reinforcement, some postfailure connection resistance can be obtained from the presence of sufficient quantities of bottom steel in the slab passing through the column. This steel can act as suspension steel, arresting movement of the slab down the column after punching shear failure, thus allowing redistribution of the gravity load to elsewhere in the floor. Bottom steel is more effective than top steel for this purpose, because the top steel has only a small thickness of concrete holding the bars in the slab, which is relatively easy to split off. The vertical forces carried by the slab steel are due to kinking of the slab bars and dowel action. Slab reinforcement to provide adequate postpunching resistance has been considered by Hawkins and Mitchell[10.13] and Mitchell and Cook.[10.14] The postpunching resistance of slabs can be improved by the presence of inclined flexural steel in the slab in the vicinity of the column or closed stirrups around the slab bars.

In earthquake-resistance design the ductility of the structure in the post-elastic range is also an important consideration. The present standards for seismic design assume that in the case of a major earthquake, the structure has sufficient ductility to absorb and dissipate energy by postelastic flexural deformation without collapse. During an earthquake, the slab–column connections of a flat plate structure will be subjected to repeated reversals of unbalanced bending moment, which may lead to a shear failure in the slab around the column due to a degradation of shear strength. It is not suggested that multistory flat plate buildings should normally be used as seismic-resistant structures without the presence of frames or walls in the structure to stiffen the building against excessive horizontal deflections due to seismic forces. Without such stiffening elements, considerable interstory deflections (drift) may occur, resulting in a serious nonstructural damage during a severe earthquake. However, even with the presence of such stiffening elements to reduce the interstory deflections, substantial unbalanced moments may need to be transferred at slab–column connections, and these need to be made adequately ductile. The use of closed stirrups in a slab around those slab bars that pass through the column (see Fig. 10.25a) has been shown by Carpenter et al.,[10.45] Islam and Park,[10.30] and others to result in a substantial increase in the ductility of the connection when subjected to cyclic unbalanced moments well into the inelastic range. The closed stirrups resulted in more ductile behavior at large deflections than a structural steel shearhead.[10.30] The success of the closed stirrups in producing a relatively ductile connection can be attributed not only to the stirrups providing torsional and flexural shear resistance at large deformations, but also to the stirrups holding the top and bottom slab reinforcement together in the vicinity of the column. This holding action prevented the top slab bars from splitting off the cover concrete and prevented the slab from moving down the column on the critical side. The tendency for a slab without shear reinforcement to drop down the column on the critical side of the column, and for the top reinforcing bars on that side of the slab to split off the cover concrete, is shown in Fig. 10.16b. The closed stirrup shear reinforcement of Fig. 10.25a meant that damage caused by cyclic unbalanced flexure in the inelastic range concentrated in the regions of the slab subjected to high torsion each side of the column (see Fig. 10.25b and c) and prevented significant movement of the slab down the column after shear failure had occurred. Note, however, that the tests demonstrated that although the ductility of the connection may be maintained by the shear reinforcement, the connection undergoes a large reduction in stiffness during cyclic loading due to diagonal tension cracking in the slab. Hence the connection would not contribute significantly to energy dissipation during a severe earthquake.

The ACI Committee 352 Report[10.36] and a background article by Moehle et al.[10.37] discuss ductility of slab–column connections. They concluded that an interstory drift of 1.5% typically requires a deflection ductility of about 2. They also concluded that in the absence of shear reinforcement, the typical beamless slab could reliably reach a deflection ductility of 2 only when the applied shear V_u was less than about $0.4V_0$, where V_0 is the shear capacity of the connection in the absence of unbalanced moment. Since an interstory drift of 1.5% may well be reached in a major earthquake even in a braced frame or in a frame with beams and columns providing the lateral force resisting system, this finding has serious implications for the designer of the beamless slab portions of the structure.

10.4 SLAB–WALL CONNECTIONS

Shear walls coupled with slabs are widely used in multistory apartment buildings. When lateral loading is applied to such a structure, due to wind or earthquake, the slabs tend to couple the shear walls, and as a result, shear forces and bending moments are transferred between the shear walls by the slabs. In particular, very high shear forces can exist in the slabs around the toes (ends) of the shear walls and punching shear failure can occur there. Only limited experimental work has been conducted on this problem.

From the observations and results obtained from one test on a pair of reinforced concrete shear walls coupled by a slab, Schwaighofer and Collins[10.46] have recommended that the shear force transferred from wall to wall by the coupling slab at punching shear failure of the slab may be assumed to act uniformly over a specified critical section. Thus, the ultimate shear force V_u (see Fig. 10.27a) is given by assuming a uniform ultimate shear stress of $\phi 4\sqrt{f_c'}$ lb/in² ($\phi\sqrt{f_c'}/3$ N/mm²) acting over the U-shaped critical section at $d/2$ from the faces of the wall so that the three faces of the U are of approximately the same length, where d is the effective depth of the slab.[10.46] The assumed critical section is shown in Fig. 10.27b. The design equation for ultimate shear force is given by

$$V_u = \phi 4\sqrt{f_c'}3d(t + d) \qquad \text{lb} \qquad (10.75)$$

where t is the thickness of the shear wall (1 lb = 4.45 N). Recommendations for flexural strength and stiffness of the coupling slab are also given in Ref. 10.46.

Taylor[10.47] has conducted tests on a number of model shear walls coupled by slabs and has investigated various types of slab reinforcement in the vicinity of the critical region of the walls. It was found that the provision of a

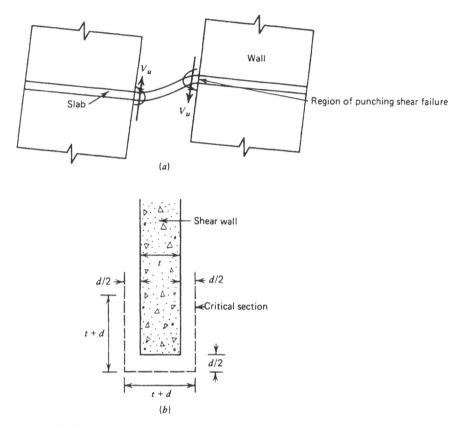

Figure 10.27   Shear strength of slab–wall connection[10.46]: (*a*) displacement of coupled walls and slab; (*b*) critical section of slab for shear strength.

structural steel member within the slab thickness running at right angles to the wall across the toe reduced damaging shear deformations in that region. However, as with slab–column connections, during cyclic loading the slab–wall connections were found to undergo a very large reduction in stiffness and it was concluded that the slab would not contribute significantly to energy dissipation during a severe earthquake.

10.5   SHEAR CAPACITY WITH HIGH-STRENGTH CONCRETE

High-strength concrete, sometimes referred to as high-performance concrete, can be defined as concrete with compressive strength greater than 8 to 10 kips/in² (55 to 70 N/mm²). Its use is somewhat restricted by the ACI Code in that there is general prohibition on taking the term $\sqrt{f_c'}$ greater than 100 lb/in² [8 N/mm²] for purposes of computing the shear strength of beams and slabs and the development length of reinforcing bars. This restriction is at least partially a result of a lack of definitive data rather than certain knowledge that the use of high-strength concrete is problematic.

Marzouk et al.[10.48] presented the results of a small series of tests of slabs with f'_c up to 74 N/mm² (10.7 kips/in²). Their tests indicated that the shear strengths computed following the ACI Code were safe for the strongest concretes, but that the excess beyond the code value was considerably smaller for the strongest concrete than for concretes of more normal strengths. In a pair of specimens subjected to shear only, doubling the concrete strength from about 32 N/mm² to 67 N/mm² (4.66 kips/in² to 9.74 kips/in²) increased the failure shear by only 7.5%, while the code equations would lead to an increase of about 45%.

Marzouk and Hussein[10.49] concluded from an earlier series of tests on interior slab–column connections transmitting only shear that the shear strength was better represented by a cube-root function of f'_c than by the square-root function used in the ACI Code. Concrete strengths ranged from 30 to 80 N/mm² (4.35 to 11.60 kips/in²) in this series. Three of the 17 slabs failed at loads lower than predicted by the ACI Code, with the lowest at 70% of the predicted value. This particular slab had a reinforcement ratio of 0.0049, which is quite low for the negative-moment region over a column in a beamless slab. All slabs with steel ratios equal to or larger than 0.0094 failed at loads greater than predicted by the ACI Code. The British Codes BS 8110[10.50] and the earlier CP 110[10.51] were both considerably more consistent than the ACI Code, but both also overpredicted the strength of the slab for which the ACI Code prediction was worst.

REFERENCES

10.1   *Building Code Requirements for Structural Concrete,* ACI 318-95, and *Commentary,* ACI 318R-95, American Concrete Institute, Farmington Hills, Mich., 1995, 371 pp.

10.2   R. Park and T. Paulay, *Reinforced Concrete Structures,* Wiley, New York, 1975, 769 pp.

10.3   M. Dragosvić and A. van den Beukel, "Punching Shear," *Heron,* Vol. 20, No. 2, 1974, 48 pp.

10.4   ASCE-ACI Committee 426, "The Shear Strength of Reinforced Concrete Members: Slabs," *J. Struct. Div., ASCE,* Vol. 100, No. ST8, August 1974, pp. 1543–1591.

10.5   ASCE-ACI Committee 426, "Suggested Revisions to Shear Provisions for Building Codes," *Proc. ACI,* Vol. 74, September 1977, pp. 458–469. Published in full as an ACI Committee Report (ACI 426.1R-77), American Concrete Institute, Detroit, 1979, 82 pp.

10.6   S. B. Alexander and S. H. Simmonds, "Ultimate Strength of Slab–Column Connections," *ACI Struct. J.,* Vol. 84, No. 3, May–June 1987, pp. 255–261.

10.7   S. H. Simmonds and S. B. Alexander, "Truss Model for Edge Column–Slab Connections," *ACI Struct. J.,* Vol. 84, No. 4, July–August 1987, pp. 296–303.

10.8   S. B. Alexander and S. H. Simmonds, "Bond Model for Concentric Punching Shear," *ACI Struct. J.,* Vol. 89, No. 3, May–June 1992, pp. 325–334.

10.9   S. B. Alexander and S. H. Simmonds, "Tests of Column–Flat Plate Connections," *ACI Struct. J.,* Vol. 89, No. 5, September–October 1992, pp. 495–502.

10.10 ACI-ASCE Committee 326, "Shear and Diagonal Tension," *Proc. ACI,* Vol. 59, January–March 1962, pp. 1–30, 277–334, and 352–396.

10.11 J. Moe, *Shearing Strength of Reinforced Concrete Slabs and Footings Under Concentrated Loads,* Development Department Bulletin D47, Portland Cement Association, Skokie, Ill., April 1961, 130 pp.

10.12 P. E. Regan, "Catenary Action in Damaged Concrete Structures," *Industrialization in Concrete Building Construction,* ACI SP-48, American Concrete Institute, Detroit, 1975, pp. 191–224.

10.13 N. M. Hawkins and D. Mitchell, "Progressive Collapse of Flat Plate Structures," *ACI J.,* Vol. 76, No. 7, July 1979, pp. 775–808.

10.14 D. Mitchell and W. D. Cook, "Preventing Progressive Collapse of Slab Structures," *J. Struct. Eng, ASCE,* Vol. 110, No. 7, July 1984, pp. 1513–1532.

10.15 *Design of Concrete Structures,* CSA Standard A23.3-94, Canadian Standards Association, Rexdale, Ontario, Canada, 1994.

10.16 R. C. Shilling, "Behavior of Shear Test Structure," M.S. thesis, Colorado State University, 1970, 66 pp. Also issued as Structural Research Report 5, Civil Engineering Department, Colorado State University, Fort Collins, Colo.

10.17 J. E. Carpenter, P. H. Kaar, and N. W. Hanson, "Discussion of Proposed Revision of ACI 318-63: Building Code Requirements for Reinforced Concrete," *Proc. ACI,* Vol. 67, September 1970, pp. 696–697.

10.18 *Shear Reinforcement for Slabs,* ACI 421.1R-92, American Concrete Institute, Detroit, 1992, 12 pp.

10.19 W. G. Corley and N. M. Hawkins, "Shearhead Reinforcement for Slabs," *Proc. ACI,* Vol. 65, October 1968, pp. 811–824.

10.20 A. Ghali and N. Hammill, "Effectiveness of Shear Reinforcement in Slabs," *Concrete International,* Vol. 14, No. 1, January 1992, pp. 60–65.

10.21 W. H. Dilger and A. Ghali, "Shear Reinforcement for Slabs," *J. Struct. Div.,* ASCE, Vol. 107, No. ST12, December 1981, pp. 2403–2420.

10.22 A. A. Elgabry and A. Ghali, "Tests on Concrete Slab–Column Connections with Stud-Shear Reinforcement Subjected to Shear-Moment Transfer," *ACI Struct. J.,* Vol. 84, No. 5, September–October 1987, pp. 433–442.

10.23 N. Hammill and A. Ghali, "Punching Shear Resistance of Corner Slab-Column Connections," *ACI Struct. J.,* Vol. 91, No. 6, November–December 1994, pp. 697–707.

10.24 J. D. Mortin and A. Ghali, "Connection of Flat Plates to Edge Columns," *ACI Struct. J.,* Vol. 88, No. 2, March–April 1991, pp. 191–198.

10.25 J. M. Hanson, *Influence of Embedded Service Ducts on Strength of Flat Plate Structures,* Research and Development Bulletin RDOOS, Portland Cement Association, Skokie, Ill., 1970, 16 pp.

10.26 *CRSI Handbook,* 8th ed., Concrete Reinforcing Steel Institute, Schaumberg, Ill., 1996.

10.27 J. L. Clarke, *Behaviour and Design of Pile Caps with Four Piles,* Technical Report 42.489, Cement and Concrete Association, London, November 1973, 19 pp.

10.28 P. Adebar, D. Kuchma, and M. P. Collins, "Strut-and-Tie Models for the Design of Pile Caps: An Experimental Study," *ACI Struct. J.,* Vol. 87, No. 1, January–February 1990, pp. 81–92.

10.29 W. B. Siao, "Strut-and-Tie Model for Shear Behavior in Deep Beams and Pile Caps Failing in Diagonal Splitting," *ACI Struct. J.,* Vol. 90, No. 4, July–August 1993, pp. 356–363.

10.30 S. Islam and R. Park, "Tests on Slab–Column Connections with Shear and Unbalanced Flexure," *J. Struct. Div., ASCE,* Vol. 102, No. ST3, March 1976, pp. 544–568.

10.31 N. M. Hawkins, "Shear Strength of Slabs with Moments Transferred to Columns," *Shear in Reinforced Concrete,* ACI SP-42, Vol. 2, American Concrete Institute, Detroit, 1974, pp. 817–846.

10.32 R. Park and S. Islam, "Strength of Slab–Column Connections with Shear and Unbalanced Flexure," *J. Struct. Div., ASCE,* Vol. 102, No. ST9, September 1976, pp. 1879–1901.

10.33 A. E. Long, "A Two-Phase Approach to the Prediction of the Punching Strength of Slabs," *Proc. ACI,* Vol. 72, February 1975, pp. 37–45.

10.34 A. Stamenkovic, "Flat Slab Construction: Column Head Strength Under Combined Vertical Load and Wind Moment," Ph.D. thesis, Imperial College, 1968.

10.35 N. W. Hanson and J. M. Hanson, "Shear and Moment Transfer Between Concrete Slabs and Columns," *J. Portland Cem. Assoc. Res. Dev. Lab.,* Vol. 10, No. 1, January 1968, pp. 2–16. Also issued as Development Department Bulletin D129, Portland Cement Association, Skokie, Ill., 1968, 15 pp.

10.36 "Recommendations for Design of Slab–Column Connections in Monolithic Reinforced Concrete Structures (ACI 352.1R-88),"*ACI Struct. J.,* Vol. 85, No. 6, November–December 1988, pp. 675–696.

10.37 J. P. Moehle, M. E. Kreger, and R. Leon, "Background to Recommendations for Design of Reinforced Concrete Slab–Column Connections," *ACI Struct. J.,* Vol. 85, No. 6, November–December 1988, pp. 636–644.

10.38 J. Di Stasio and M. P. van Buren, "Transfer of Bending Moment Between Flat Plate Floor and Column," *Proc. ACI,* Vol. 57, September 1960, pp. 299–314.

10.39 E. S. Hoffman, D. P. Gustafson, and A. J. Gouwens, *Structural Design Guide to the ACI Building Code,* Kluwer Academic Press, Boston, 1998, 466 pp.

10.40 *Building Code Requirements for Reinforced Concrete,* ACI 318-89 (Revised 1992), and *Commentary,* ACI 318R-89 (Revised 1992), American Concrete Institute, Detroit, 1992, 347 p.

10.41 N. M. Hawkins and W. G. Corley, "Moment Transfer to Columns in Slabs with Shearhead Reinforcement," *Shear in Reinforced Concrete,* ACI SP-42, Vol. 2, American Concrete Institute, Detroit, 1974, pp. 847–880.

10.42 Y. Kanoh and S. Yoshizaki, "Strength of Slab–Column Connections Transferring Shear and Moment," *Proc. ACI,* Vol. 76, March 1979, pp. 461–478.

10.43   A. Stamenkovic and J. C. Chapman, "Local Strength at Column heads in Flat Slabs Subjected to Combined Vertical and Horizontal Loadings," *Proc. Inst. Civil Eng. (London)*, Part 2, Vol. 57, June 1974, pp. 205–233.

10.44   J. P. Moehle, "Strength of Slab–Column Edge Connections," *ACI Struct. J.*, Vol. 85, No. 1, January–February 1988, pp. 89–98.

10.45   J. E. Carpenter, P. H. Kaar, and W. G. Corley, "Design of Ductile Flat Plate Structures to Resist Earthquakes," Paper 250, *Proceedings of the 5th World Conference on Earthquake Engineering,* Session SD, Rome, Italy, 1973.

10.46   J. Schwaighofer and M. P. Collins, "Experimental Study of the Behaviour of Reinforced Concrete Coupling Slabs," *Proc. ACI,* Vol. 74, March 1977, pp. 123–127.

10.47   R. G. Taylor, "The Nonlinear Seismic Response of Tall Shear Wall Structures," Ph.D. thesis, Department of Civil Engineering, University of Canterbury, 1977, 208 pp. and appendices.

10.48   H. Marzouk, M. Emam, and M. S. Hilal, "Effect of High-Strength Concrete Slab on the Behavior of Slab–Column Connections," *ACI Struct. J.*, Vol. 95, No. 3, May–June 1998, pp. 227–237.

10.49   H. Marzouk and A. Hussein, "Experimental Investigation on the Behavior of High-Strength Concrete Slabs," *ACI Struct. J.*, Vol. 88, No. 6, November–December 1991, pp. 701–713.

10.50   *Structural Use of Concrete,* BS 8110, Part 1:1985, *Code of Practice for Design and Construction,* British Standards Institution, London, 1985, 126 pp.

10.51   *Code of Practice for the Structural Use of Concrete,* CP 110, British Standards Institution, London, 1972, Part 1, 55 pp.

11 Prestressed Concrete Slabs

11.1 INTRODUCTION

Aside from possible economic advantages, the main reason that slabs are prestressed is to obtain a slab that is free, or nearly so, of cracks and deflection at the service load level. As a direct result of not having cracks at service loads, a prestressed slab is stiffer than a normal reinforced concrete slab having the same span and thickness, since the full cross-section rigidity is available rather than the rigidity of a cracked section. This higher rigidity resulting from having no cracks often allows the use of a thinner slab for the same span, with accompanying savings in the dead load of the structure. In addition, the use of draped post-tensioned tendons in continuous, cast-in-place slabs can result in a structure with approximately zero deflection under dead load plus some predetermined live load.

The prestressing operation also allows the designer to use very high strength steels, with breaking stresses from 240 to 270 kips/in² (1650 to 1860 N/mm²). Service load stresses will generally be in the range 120 to 160 kips/in² (830 to 1100 N/mm²) after losses, but since the steel has the initial tension applied with a jack and then locked in, the large steel strains are not translated into large deflections. High-strength concretes may also be used in prestressed slabs, but the strongest concretes are usually restricted to plant-cast pretensioned members, which are not discussed here.

The remainder of this chapter, which is purposefully kept short, deals with continuous post-tensioned slabs. The emphasis is on slabs spanning in two directions and one-way slabs are not considered. One-way slabs may generally be treated as wide beams (except for effects of concentrated loads, as discussed in Section 3.5.1) and design information is available in several textbooks.[11.1–11.3]

Most prestressed slabs spanning in two directions are supported directly on columns, with or without drop panels or column capitals, and are commonly called flat plates. In the United States it appears that most such slabs, which are economical because of the extremely simple formwork, are post-tensioned and then left unbonded. For a variety of reasons, with cost undoubtedly primary, the tendons are seldom grouted, and the lack of bond introduces some problems, which are discussed later.

Some prestressed slabs are built as lift slabs, in which all the floors are cast in a stack at ground level and then lifted to their final positions after being cured and post-tensioned. After lifting to the correct position, the slabs are anchored to the columns, using specialized hardware, and the slabs then differ from more conventional cast-in-place slabs primarily in the nature of the slab–column connection, which may not be able to transfer significant moment from slab to column.

11.2 BASIS FOR DESIGN

11.2.1 General Approach

The ACI Building Code[11.4] provides only general provisions for the design of prestressed concrete slabs. Additional guidance is given in the commentary to the ACI Code.[11.4] However, it is clear from the commentary that the principal design guide is the *Recommendations for Concrete Members Prestressed with Unbonded Tendons,*[11.5] which was prepared by ACI-ASCE Committee 423 on Prestressed Concrete, along with an earlier *Tentative Recommendations for Prestressed Concrete Flat Plates*[11.6] prepared by the same committee. The Post-Tensioning Institute's *Design of Post-Tensioned Slabs*[11.7] and the *Post-Tensioning Manual*[11.8] are also valuable references.

The design process is divided into the two distinct phases, as in most prestressed concrete design, of considering service load stresses and then considering the strength. The prestressing is normally applied to satisfy service load requirements, in combination with a set of allowable stresses. If the flexural failure capacity is then checked and found to be inadequate, additional steel in the form of deformed reinforcing bars is added to raise the strength to the necessary value. All unbonded prestressed flat plates will have auxiliary bar reinforcement at least in the negative moment regions over the columns.

11.2.2 Service Load Stresses

The analysis of slabs by general methods is recognized, but most of the emphasis is placed on the equivalent frame method described in Chapter 4, or a continuous beam analysis if there is no rigid connection to the columns, when considering the effects of dead and live load and the equivalent distributed loads from the prestressing forces. Consideration of the equivalent distributed loads from the prestressing forces introduces the concept of *load balancing,*[11.9] which enables the designer to choose a level of prestressing force that minimizes deflections.

Each slab strip is considered to be a wide beam with the moments uniformly distributed across the strip width, at this stage of the design. Stresses are then evaluated using the elastic equation

$$f_c = \frac{F_{se}}{A} \pm \frac{F_{se}(e')c}{I} \pm \frac{Mc}{I} \tag{11.1}$$

where F_{se} is the prestressing force, A the cross-sectional area, e' the effective eccentricity of prestressing force, including effects of secondary moments induced by prestressing, c the distance from centroidal axis to surface of member, I the moment of inertia of cross section, and M the moment due to service loads. Note that $F_{se} e'$ is the resulting prestressing moment, determined by applying the equivalent distributed loads from the prestressing forces to the slab and calculating the resulting moments as for the other imposed loads.

Stresses will need to be checked for the case of only dead load present and for dead load plus live load in the critical load patterns. These stresses are recognized as only rough approximations, since the moments are not uniformly distributed across the section. This is especially true in the negative-moment regions, where the local moments per unit width at columns may be several times the average moment per unit width.

Because of the real but uncomputed stress concentrations at the column, the average extreme fiber compression stress in the negative-moment region was limited to $0.30f'_c$ rather than the $0.45f'_c$ usually applicable to prestressed concrete in buildings by the earlier recommendations,[11.6] but the lower stress limit is not included in the recent recommendations and requirements. The tensile stress there is limited to $6\sqrt{f'_c}$ lb/in² [0.5 $\sqrt{f'_c}$ N/mm²]. According to the ACI Code,[11.4] bonded reinforcement, usually deformed reinforcing bars or welded wire fabric, having an area of at least $0.00075lh$ is provided directly over and adjacent to the columns, between lines that are $1.5h$ outside opposite column faces, extending to one-sixth of the clear span, where h is the slab thickness and l the span in the direction of reinforcement. This bonded reinforcement is to control cracking in the slab in the vicinity of the column due to the uncomputed stress concentrations.

The positive moments are more nearly uniformly distributed than the negative moments, and the normal allowable compressive stresses are used. The allowable tension ranges from $2\sqrt{f'_c}$ lb/in² [$\sqrt{f'_c}/6$ N/mm²] when there is no auxiliary reinforcement, to $6\sqrt{f'_c}$ lb/in² [0.5 $\sqrt{f'_c}$ N/mm²] when, according to the ACI Code, auxiliary reinforcement having the area $A_s = N_c/0.5f_y$ is present, where N_c is the tensile force in concrete under service dead plus live loads and f_y the yield stress of auxiliary reinforcement, but not over 60 kips/in² [420 N/mm²].

These stresses are applied to slabs with either bonded or unbonded tendons, and the allowable tensile stresses are specifically restricted to cases where the average prestress, F_{se}/A, exceeds 125 lb/in² (0.86 N/mm²) after all losses of prestress have occurred.[11.5]

It is the writers' opinion that application of the same allowable tensile stress criteria to bonded and unbonded slabs represents an unreasonable discrimination against slabs with bonded tendons. The load to initiate cracking would be the same, for the same areas of prestressed steel, but the behavior after cracking, in the absence of auxiliary reinforcement, would be appreciably better for a bonded slab than for an unbonded one.

After the required number of tendons is found for each orthogonal direction, their distribution across the width of the structure must be determined. The *Tentative Recommendations*[11.6] suggested that for continuous structures in which the ratio of the lengths of the sides of the panels is not over 1.33, 65 to 75% of the tendons should be in column strip, and the remainder uniformly distributed in the middle strip. However, design practice has changed greatly since that recommendation was made.

Most recent prestressed slab structures in North America have been built with all tendons in one direction grouped in narrow bands over and near the column lines, while the tendons in the orthogonal direction are uniformly distributed in what is termed a *banded* arrangement of tendons. This tendon arrangement leads to great simplification of the steel placement as compared with distributions based on column and middle strips.

The design concept is that the closely spaced tendons are thought of as forming beams in one direction, while the slab then spans as a one-way slab between these beams in the other direction. At some *balanced* load, normally the dead load plus a fraction of the service live load, stress conditions may be satisfactory. However, since the tendon distribution is drastically different than the elastic moment distribution, the cracking and deflection behavior at loads greater (or smaller) than the "balanced" load may not be satisfactory. This is a serviceability problem rather than a safety problem, however, as the flexural strength is primarily a function of the number of tendons and is not strongly influenced by their distribution. There may be detailing problems at edge columns because of the relatively wide spacing of the uniformly distributed tendons, and at least one test structure[11.10] had moment transfer deficiencies because of the lack of sufficient tendons through and near the columns.

It is thus required[11.4,11.5] that at least two tendons pass through or immediately adjacent to each edge or corner column. The ACI Code[11.4] specifies that the maximum spacing of tendons in column strips should not exceed the smaller of 5 ft [1.5 m] or eight times the slab thickness. The average prestress should be at least 125 lb/in^2 (0.9 N/mm^2), and normally should not exceed 500 lb/in^2 (3.4 N/mm^2), according to the *Recommendations*. The higher stresses lead to greater elastic and creep shortening of the slab, and this can be quite troublesome.

Structures with the tendons uniformly distributed across the full panel widths in both directions have been built, but the results were not entirely satisfactory in terms of cracking and deflections in one case.[11.11] Brotchie[11.12] infers from this and other tests[11.13] that the tendon distribution must follow the elastic distribution of moments reasonably closely, and suggests that the greater the span/depth ratio, the more closely the tendon distribution should match the elastic moment distribution if the service load behavior is to be satisfactory.

The load-balancing concept mentioned earlier deserves some comment. Its application to the design and analysis of beams is well documented.[11.2,11.9] Its application to slabs appears to need some additional explanation, although it will be helpful to start with a beam illustration. An interior span of a beam is shown in Fig. 11.1a, and an "ideal" post-tensioned tendon profile is shown. The depth is greatly exaggerated relative to the span.

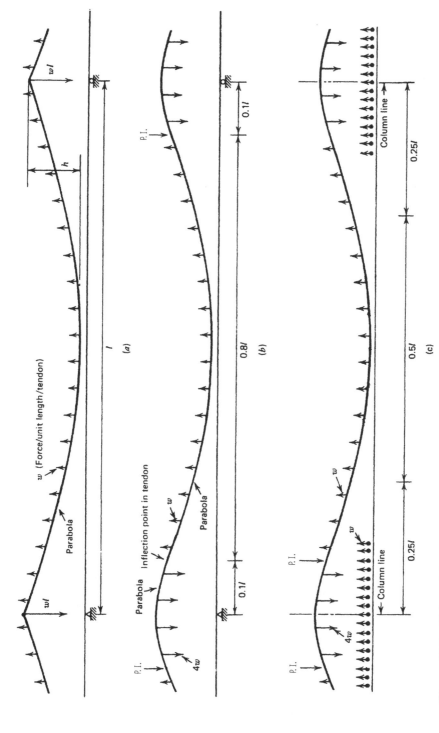

Figure 11.1 Slab-and-beam sections showing tendon profiles and forces applied by tendons to the concrete: (*a*) ideal tendon profile for beam; (*b*) practical tendon profile; (*c*) section at midspan of slab showing tendons in both directions.

The tendon has a force T that is assumed constant along its length. As a result of the parabolic curvature, the tendon applies a uniformly distributed upward force to the concrete. This uniformly distributed force is given by the expression

$$w = 8 \frac{Th}{l^2} \tag{11.2}$$

If the vertical force w, as shown in Fig. 11.1a, is exactly equal to the applied load (including dead load) acting downward, the load is said to be balanced, and it will be found that the bending moments due to applied loads are exactly canceled by moments caused by the upward forces, and the concrete in the member is subjected to neither moment nor shear but only uniform compressive stress, unless there are end moments present. There is a downward force on the concrete at the peak of the tendon, but the force goes straight to the reaction and does nothing to the beam.

The tendon force T changes with time because of creep and shrinkage of the concrete and relaxation of the steel stress. Consequently, the balanced load varies slightly with time, and the goal of zero deflection cannot realistically be met, although the deflections should remain very small.

The tendon profile shown is impractical, since the tendons cannot be placed and tensioned around the sharp corners occurring over each interior support. Consequently, a tendon profile similar to that in Fig. 11.1b will have to be used. The designer has control over the shape of the tendon and can establish the locations of the points of inflection of the tendon. With this profile, the tendon applies upward forces to the concrete in the central part of the span, but applies downward forces near the supports. The total upward and downward forces are the same, and if the points of contraflexure of the tendon profile are at 0.1l from the supports, the downward force per unit length is four times the upward unit force. In a beam this does not cause significant problems or seriously alter the load balancing since the large downward forces are applied near the supports. Khachaturian and Gurfinkel[11.1] developed extensive information on the effects of tendon profiles such as these, and the compound parabolas and other curves are easily taken into account.

A single square slab panel that is simply supported on nondeflecting supports on all four sides could be post-tensioned with two perpendicular sets of parabolic tendons having the shape shown in Fig. 11.1a. Both sets of tendons apply upward forces to the slab, and load will be balanced when the sum of the components from the crossing tendons equals the downward load. The downward component of the tendon reaction at the end of the tendon is transmitted directly to the support.

The tendon profile in a continuous slab will be basically the same as shown in Fig. 11.1b, with tendons in the two orthogonal directions. The action of a tendon passing directly over a column will be the same as in a beam. However, a tendon that is between column lines exerts the large downward force on the concrete in an area where there is no support reaction to resist the force.

This downward force from the tendon acts in the same direction as the applied load, and the two are capable of damaging the slab if they are not offset by other forces. These two forces can be resisted by the upward components of force from tendons placed in the perpendicular direction, provided that sufficient tendons are supplied.

The transfer of applied load to the supports can be visualized in terms of several steps. In the central region of a square symmetrical panel where the tendons in both directions have concave-upward curvature, each set of tendons balances half the load and transfers it back to the narrow strips near the column lines, where the concave-downward curvature exists in at least one set of tendons. In this region, the tendons parallel to the column lines, having concave-upward curvature, support the central part of the strip between two columns and transfer the load toward the columns. In the immediate vicinity of the columns, both sets of tendons have concave-downward curvature, and the entire load is transferred to the concrete over and near the column. It is evident that the *entire load* is transferred in *both directions* into the columns, a requirement that is obvious when the equilibrium of a slab segment is considered.

Design work is not done following these steps but achieves a similar end result. A set of tendons balancing *all* the load is determined for the *x*-direction, and then another set, again balancing *all* the load, is found in the *y*-direction. Only the central part of the tendon profile, where the curvature is concave upward, is normally considered when determining the number of tendons and the force in each. The end segments of the tendons are ignored, at least in the early stages of the design. The PTI's *Design of Post-Tensioned Slabs,*[11.7] and *Post-Tensioning Manual*[11.8] contain discussions of applying loading balancing to a slab with an irregular column layout, in addition to other design and behavior factors.

After the number of tendons is determined for each of the two directions, their spacings are determined. In panels that are nearly square, the earlier *Tentative Recommendations*[11.5] recommend placing 65 to 75% of the tendons in the column strips, as noted earlier. The tendons were usually uniformly distributed across each strip. Most current construction uses the banded tendon arrangement noted earlier.

Brotchie and Russell[11.14] and Brotchie and Robinson[11.15] developed an extension of the load-balancing concept to what they termed *moment balancing.* They suggest that a structure with zero deflection and in which the concrete is subjected to only axial compression forces (at some given applied load) results if the internal moment, $M = Te'$, is everywhere equal to the applied moment, where T is the tension force in the tendon and e' is the effective eccentricity of the tendon, including whatever secondary moment effects may exist.

In a beam, the result is exactly the same as that obtained using the load-balancing concept. In a slab, the force T may be interpreted as the tendon force attributable to the particular unit width of slab being considered, or if T is the force in a single tendon, it must be assumed to be acting on some tributary width of slab. Using a set of tendons having the shape shown in Fig. 11.1b, it is possible to reasonably satisfy the moment-balancing requirements in a flat plate slab, except in the immediate vicinity of the columns. The steel arrangement will be about like that given by the *Tentative Recommendations,* except that strict adherence to moment balancing would require different spacings for nearly every tendon. A practical solution might use three tendon spacings: (1) a very close spacing immediately over the columns, (2) a wider spacing in the rest of the column strip, and (3) a still wider spacing in the middle strip.

The entire discussion of load balancing and moment balancing is concerned with the behavior at service load or at dead load plus some fraction of the service live load, as was the initial discussion of allowable stresses. The comments apply equally to unbonded tendons and to those that are grouted after post-tensioning.

11.2.3   Flexural Strength

The structure must also be adequate for the factored ultimate loads. The strength analysis is relatively simple, especially in an unbonded slab where the moment capacity is not particularly sensitive to the tendon distribution. The applied moment acting on a critical section across the full width of a panel is found by multiplying the service dead and live load moments (M_d and M_l, respectively) by the appropriate load factors. Using the 1995 ACI Code,[11.4] this can be stated as

$$M_u = 1.4M_d + 1.7M_l \tag{11.3}$$

The resisting moment of a section containing both bonded reinforcing bars of area A_s and post-tensioned tendons of area A_{ps} can then be found with the following expression or some equivalent:

$$M_u = \phi \left[A_{ps}f_{ps} \left(d_p - \frac{a}{2} \right) + A_s f_y \left(d - \frac{a}{2} \right) \right] \tag{11.4}$$

where $\phi = 0.9$ is the strength reduction factor for flexure, A_{ps} the area of post-tensioned steel in a slab of width b, f_{ps} the stress in prestressing steel at a failure in flexure, d_p the effective depth of the prestressing steel, $a = (A_{ps}f_{ps} + A_s f_y)/(b \times 0.85f_c')$ the depth of an equivalent rectangular compression zone, b the width of the section being considered, A_s the area of bonded deformed bar reinforcement in width b, f_y the yield stress of bar reinforcement, and d the effective depth of bar reinforcement. Not all sections will have the auxiliary deformed reinforcement.

The only unknown is f_{ps}. For grouted tendons and ignoring any compression steel, this may be taken as Eq. 18-3 of the ACI Code:[11.4]

$$f_{ps} = f_{pu} \left(1 - \frac{\gamma_p}{\beta_1} \left[\rho_p \frac{f_{pu}}{f'_c} + \frac{d}{d_p} (\omega) \right] \right) \tag{11.5}$$

where f_{pu} is the breaking stress of tendons, $\gamma_p = 0.40$ for stress-relieved strands or 0.28 for low-relaxation strands, β_1 is the stress block parameter = 0.85 for concrete with $f'_c \leq 4,000$ lb/in^2 [30 N/mm^2], $\rho_p = A_{ps}/d_p b$ is the reinforcement ratio of prestressed steel, and $\omega = (A_s/bd)(f_y/f'_c)$ is the reinforcement index for bar reinforcement.

For unbonded tendons, the *Recommendations*[11.5] say that Eq. 18-5 of the ACI Code can be used, with all stresses in lb/in^2 units:

$$f_{ps} = f_{se} + 10,000 + \frac{f'_c}{300\rho_p} \quad \text{but not over } f_{py} \text{ or } f_{se} + 30,000 \tag{11.6}$$

where f_{se} = effective prestress at service load, after losses, and f_{py} is the yield stress of prestressed reinforcement. The SI equivalent expression for Eq. 11.6 is

$$f_{ps} = f_{se} + 70 + \frac{f'_c}{300\rho_p} \quad \text{but not over } f_{py} \text{ or } f_{se} + 200 \tag{11.6a}$$

where all stresses are in N/mm^2 units.

Equation 11.6, Code Eq. 18-5, was added to the code for thin members after it was found that the earlier Code Eq. 18-4 was significantly nonconservative for many slab cases. Code Eq. 18-4 is similar to Eq. 11.6, except that the divisor in the third term is 100 rather than 300, and the upper limits are higher. Code Eq. 18-4 was derived from the results of tests of beams and is a lower bound to that test data. Unfortunately, it had not been checked against the results of many slab tests before it was endorsed. A series of flat plate structures and accompanying one-way slabs were tested at the University of Texas at Austin,[11.10,11.16–11.19] and the results of those tests indicate that the stresses implied by Code Eq. 18-4 are not always attainable in thin slabs.

Most of the slabs tested had span/thickness ratios of 44, span/effective depth ratios of 53, and their service load behavior was generally adequate. Most of the slabs reached the required ultimate loads, but they did not always support the expected loads computed using the steel stresses given by Code Eq. 18-4.

Steel stresses were determined using both load cells at the ends of strands and from strains measured in the tendons at or near the points of maximum moment. The results of the tests of six slabs are summarized briefly in Table 11.1, which includes the results on one slab tested in Australia. The available data have been summarized by Mojtahedi and Gamble.[11.20]

The comparisons between computed and observed increases in stress Δf_s above the prestress level existing at the beginning of the tests are disturbing in some cases. Some of the slabs summarized in the table reached their required ultimate loads only because the design had been governed by service load considerations. If the amount of steel had been governed by strength considerations, most of these slabs would clearly have supported less load than the required minimum ultimate because of the lower-than-expected steel stress values.

A set of recommendations similar in many respects to the *Recommendations,* was developed in England.[11.21] This design guide also endorses the f_{ps} values for beams as being suitable for use with slabs. The Δf_s values, given in equation form in the British Code of Practice, BS 8110, for structural concrete,[11.22] are a function of the span/effective depth ratio and the prestressing steel index, $\rho_p f_{pu}/f_{cu}$, where f_{cu} is the cube strength. For a span/effective depth ratio of 30 the British Code gives Δf_s as ranging from $0.138 f_{se}$ to $0.200 f_{se,}$ with the lower values applying to high prestressing steel indices. For a span/effective depth ratio of 45 it gives Δf_s as ranging from $0.092 f_{se}$ to $0.133 f_{se}$. These specific values were computed assuming that $f_{se} = 0.6 f_{pu}$, where $f_{pu} = 1860 \text{ N/mm}^2 = 270 \text{ kips/in}^2$. Larger values of f_{se} lead to slightly smaller ratios of $\Delta f_s / f_{se}$.

The 1974 Australian Code[11.23] has a different expression for f_{ps}, which results in the predicted values of Δf_s as listed in the last column of Table 11.1. The equation, which was derived by Warwaruk et al.[11.24] from the results of beam tests, is

$$f_{ps} = f_{se} + 30,000 - \frac{\rho_p}{f'_c} \times 10^{10} \tag{11.7}$$

with all terms as previously defined and stresses having lb/in^2 units. The equivalent SI expression, from the Australian Code, is

$$f_{ps} = f_{se} + 210 - \frac{4.76\rho_p}{f'_c} \times 10^5 \tag{11.7a}$$

where the stresses are in N/mm^2 units.

This expression gives higher values of Δf_s than the current ACI Code Eq. 18-5, and they are larger than the measured changes in several cases. A code expression should lead to either the correct or to a conservative value.

The problem appears to be primarily one of geometry. The tendon stress increase is related directly to the total change in length of the concrete at the level of the tendon. The change in length, and hence Δf_s, can be shown to be nonlinearly and inversely related to the span/depth ratio of the slab or

TABLE 11.1   Changes in Stress in Unbonded Post-tensioned Strands in Several Slab and Shallow Beam Tests[a]

Slab	Observed Δf_s		ACI Code Δf_s (Eq. 18-5) (kips/in²)	Australian Code Δf_s (kips/in²)
	l/h	kips/in²		
Slab A[11.16]	44	19.5	19.9	26.6
Slab B[11.17]	44	21	24.5	27.7
Flat plate 1[11.18]	44	6–25[b]	14.5	22.5
Flat plate 2[11.10]	44	9–15[b]	22.4	27.3
Flat plate 3[11.19]	44	5–23[b]	14.7	22.9
Mark 4[11.13]	48	25.8	17.3[c]	25.4[c]

[a] 1 kip/in² = 6.89 N/mm²; Δf_s is the increase in steel stress above f_{se}.
[b] Range of values from strain and load cell measurements.
[c] Average of $+M$ and $-M$ section values.

beam, using simplistic models to represent the structure.[11.20] For a given deflection or for a deflection taken as some fraction of the span, it is apparent that the strain increase in a member with $l/h = 44$ is significantly less than in a member with $l/h = 28$, which is the shallowest beam considered in the development of Eq. 11.6.[11.25] This dependence of Δf_s on the l/h ratio has not been investigated systematically and has been only partially taken into account in the development of the ACI Code provisions.

Until this question is completely resolved, considerable prudence is advised. The simplest recommendation that can be made is to use the 1963 ACI Code[11.26] provisions, which give the following for unbonded members:

$$f_{ps} = f_{se} + 15,000 \text{ lb/in}^2 \qquad (11.8)$$

The equivalent SI expression is

$$f_{ps} = f_{se} + 103 \text{ N/mm}^2 \qquad (11.8a)$$

The actual f_{ps} may often exceed this value, but apparently will not often be much less than this value.

11.2.4   Shear Strength

The shear stresses near the column in a flat plate may be a problem, and prestressing the slab may increase the shear strength somewhat but does not eliminate the problem. The 1995 Code[11.5] uses nominal shear stresses on a perimeter located $d/2$ outside the column. These stresses, computed using the factored ultimate loads and divided by the strength reduction factor, $\phi = 0.85$, for shear, are then compared with the concrete capacity. The shear stress is computed as

$$v_u = \frac{V_u - V_p}{b_0 d\phi} \tag{11.9}$$

where V_u is the factored shear force acting on that perimeter, V_p the vertical component of the prestressing force computed at the perimeter of the critical section and b_0 the length of perimeter $d/2$ away from column. The force term V_p is likely to be very small in most practical post-tensioned slabs because the tendon slopes will be small.

The concrete shear capacity, in force units, for interior columns and other columns where the slab extends at least four times its thickness beyond the column, is given by

$$V_c = (\beta_p \sqrt{f_c'} + 0.3 f_{pc}) b_0 d + V_p \tag{11.10}$$

with stresses in lb/in^2 units, and where β_p = the smaller of 3.5 or ($\alpha_s d/b_0$ + 1.5), where α_s is 40 for interior columns, 30 for edge columns, and 20 for corner columns; $f_{pc} = f_{se} A_{ps}/A_g$ the average compression stress in slab; $A_g = bh$; b is the width of slab containing prestressing steel area A_{ps} (normally either the entire width of the structure or of a full panel); and h the slab thickness. The equivalent SI expression, with stresses in N/mm^2 terms, is

$$V_c = (\beta_p \sqrt{f_c'} + 0.3 f_{pc}) b_0 d + V_p \tag{11.10a}$$

where β_p is the smaller of 0.29 or ($\alpha_s d/b_0$ + 1.5)/12, where α_s is 40 for interior columns, 30 for edge columns, and 20 for corner columns. The use of this expression is to be limited to $f_c' < 5000$ psi (35 N/mm^2) and $f_{pc} <$ 500 psi (3.5 N/mm^2), because of lack of test data from slabs with higher values. No guidance is given for a case where f_{pc} is substantially different in the two directions. The expression was originally developed for web-shear (principal tension) cracking in prestressed beams, but is adequately conservative relative to the test data for slabs.

Most of the test data are for specimens where the slab-column connection is subjected to shear loading only. The 1995 ACI Code uses the same approach for both reinforced and prestressed concrete slabs for cases where a significant moment is to be transferred, using the same A_c and J_c properties for the critical sections of the slab as are considered extensively in Chapter 10. The requirement that the stress values implied in Eq. 11.10 be restricted to cases where the slab extends at least four times its thickness outside a column seems unduly conservative. Test data reported by Sunidja[11.27] and Foutch et al.[11.28] clearly show that the stress enhancements due to the prestressing force are also available at edge columns with no slab extension beyond the column. It is also clear that the high local f_{pc} values associated with banded tendons are applicable rather than the f_{pc} value averaged over the panel width.

11.2.5 Concluding Comments

The successful design and construction of prestressed slabs requires considerable knowledge of and attention to the smallest details of the anchorage systems, stressing equipment, strand protection, sheathing, and other items. The engineer undertaking the design of a post-tensioned flat plate should ideally already be familiar with (or be willing to spend the effort required to become familiar with) the design, analysis, and behavior of post-tensioned concrete beams before adding the complexities accompanying the additional dimension.

11.3 CORROSION CONCERNS

Although prestressed concrete structures are generally relatively free of cracking, this has not always been enough to protect the reinforcement from corrosion. This has been especially true for parking structures in cold climates, where cars carry significant amounts of deicing chemicals into the structures. The early unbonded, post-tensioned tendons often had little corrosion protection of the steel, and time has proven that coating a tendon with grease and wrapping it in kraft paper will not protect the strand or wires in the tendon.

The report *Corrosion and Repair of Unbonded Single Strand Tendons,*[10.29] prepared by ACI/ASCE Committee 423, contains an extensive review of the problem and, as the title implies, illustrations and examples of repair methods that have been used. Corrosion of a prestressing strand or wire differs from that of an ordinary reinforcing bar in at least two ways. Each wire is relatively small in diameter, and the loss of a unit thickness due to corrosion causes a greater relative reduction in area than in a larger bar. Because of the combination of the chemistry and heat treatment of the prestressing steel and the high sustained stress, a strand or prestressing wire is much more susceptible to stress corrosion than is a reinforcing bar. The presence of chlorides increase the likelihood of stress-corrosion failures.

Single-strand tendons are now encased in tight-fitting plastic sheaths, with careful attention paid to ensuring that all the space not filled with the wires in the strand is filled with a grease that has been selected for its corrosion-inhibiting qualities. These protected strands are generally produced following a specification published by the Post-Tensioning Institute.[11.30]

Protecting the strand in this manner produced a big improvement in durability but is not in itself adequate. The early uses of the plastic-sheathed strands often left small unprotected lengths of strand at the anchorages. The sheath could not pass through the anchorage, and so had to be removed, and this often left a short length of strand exposed and vulnerable to corrosion. In addition, the grouting of the exposed pockets left at the anchorages after the strands were stressed sometimes provided entry points for water and chlorides. More recent installations have generally made provisions to protect the junction between the strand sheath and the anchorage, and to protect the anchorage and the end of the strand more completely. In some instances the anchorages are coated with plastic and a sealed end cap is provided so that the entire strand anchorage is both sealed and electrically isolated. This is often referred to as an *encapsulated system*. Details are given in Ref. 10.29 and in the various reports, standards, and papers referred to in it.

REFERENCES

11.1   N. Khachaturian and G. Gurfinkel, *Prestressed Concrete,* McGraw-Hill, New York, 1969, 460 pp.

11.2   T. Y. Lin and Ned H. Burns, *Design of Prestressed Concrete Structures,* 3rd ed., Wiley, New York, 1981, 646 pp.

11.3   Michael Collins and Denis Mitchell, *Prestressed Concrete Structures,* Prentice Hall, Upper Saddle River, N.J., 1991, 766 pp.

11.4   *Building Code Requirements for Structural Concrete,* ACI 318-95 and *Commentary,* ACI 318R-95, American Concrete Institute, Farmington Hills, Mich., 1995, 371 p.

11.5   ACI-ASCE Committee 423, *Recommendations for Concrete Members Prestressed with Unbonded Tendons,* ACI 423.3R-96, American Concrete Institute, Farmington Hills, Mich., 1996, 19 pp.

11.6   ACI-ASCE Committee 423, *Tentative Recommendations for Prestressed Concrete Flat Plates, Proc. ACI,* Vol. 71, February 1974, pp. 61–71.

11.7   *Design of Post-Tensioned Slabs,* 2nd ed., Post-Tensioning Institute, Phoenix, Ariz., 1984, 54 pp.

11.8   *Post-Tensioning Manual,"* 5th ed., Post-Tensioning Institute, Phoenix, Ariz., 1990, Sec. 5.3.

11.9   T. Y. Lin, "Load-Balancing Method for Design and Analysis of Prestressed Concrete Structures," *Proc. ACI,* Vol. 60, June 1963, pp. 719–742.

11.10   N. H. Burns and R. Hemakom, *Strength and Behavior of Post-Tensioned Flat Plates with Unbonded Tendons,"* Department of Civil Engineering, University of Texas at Austin, Austin, Texas, September 1975, 267 p.

11.11   J. E. Ferris, "Prestressed Parking Deck: Sydney," *Proceedings of the 3rd Australian Building Research Congress,* Sec. C., Melbourne, Victoria, Australia, 1967, pp. 283–286.

11.12   J. F. Brotchie, *On the Design of Prestressed Flat Plates,* Division of Building Research Technical Paper (Second Series) 3, Commonwealth Scientific and Industrial Research Organization, Melbourne, Victoria, Australia, 1974, 39 pp.

11.13   J. F. Brotchie and F. D. Beresford, "Experimental Study of a Prestressed Concrete Flat Plate Structure," *Civ. Eng. Trans., Inst. Eng. Aust.,* Vol. CE9, October l967, pp. 267–282.

11.14   J. F. Brotchie and J. J. Russell, "Flat Plate Structures," *Proc. ACI,* Vol. 61, August 1964, pp. 959–996, plus Part 2 Suppl.

11.15   J. F. Brotchie and N. I. Robinson, "The Action of Unbonded Prestressing Tendons," *Archit. Sci. Rev.,* Vol. 9, No. 4, December 1966, pp. 115–117.

11.16 W. R. Vines, "Strength and Behavior of a Post-Tensioned Concrete Slab with Unbonded Tendons," M.S. thesis, Department of Civil Engineering, University of Texas at Austin, May 1976, 176 pp.

11.17 F. A. Charney, "Strength and Behavior of a Partially Prestressed Concrete Slab with Unbonded Tendons," M.S. thesis, Department of Civil Engineering, University of Texas at Austin, May 1976, 179 pp.

11.18 N. H. Burns and R. Hemakom, "Test of a Scale Model Post-Tensioned Flat Plate," *J. Struct. Div., ASCE,* Vol. 103, No. ST6, June 1977, pp. 1237–1255.

11.19 N. H. Burns and C. V. Winter, "Test of Four Panel Post-Tensioned Flat Plate with Unbonded Tendons," Preliminary Report, Department of Civil Engineering, University of Texas at Austin, September 1976, 22 pp.

11.20 S. Mojtahedi and W. L. Gamble, "Ultimate Steel Stresses in Unbonded Prestressed Concrete," *J. Struct. Div., ASCE,* Vol. 104, No. ST7, July 1978, pp. 1159–1165. (Discussion Vol. 105, No. ST9, September 1979, pp. 1862–1864.)

11.21 *The Design of Post-Tensioned Concrete Flat Slabs in Buildings,* Recommendations of the Concrete Society Working Party on Post-Tensioned Flat Slab Construction, The Concrete Society, London, 1974, 27 p.

11.22 *Structural Use of Concrete,* BS 8110: Part 1:1997, *Code of Practice for Design and Construction,* British Standards Institution, London, 1997, 121 pp.

11.23 *SAA Prestressed Concrete Code,* (Metric Units), Australian Standards 1481–1974, Standards Association of Australia, North Sydney, New South Wales, Australia, 1974, 68 pp.

11.24 J. Warwaruk, M. A. Sozen, and C. P. Siess, *Investigation of Prestressed Reinforced Concrete for Highway Bridges, Part III, Strength and Behavior in Flexure of Prestressed Concrete Beams,* Bulletin 464, University of Illinois Engineering Experiment Station, Urbana, Ill., 1962, 105 p.

11.25 A. H. Mattock, J. Yamakazi, and B. T. Kattula, "Comparative Study of Prestressed Concrete Beams, with and Without Bond," *Proc. ACI,* Vol. 68, February 1971, pp. 116–125.

11.26 *Building Code Requirements for Reinforced Concrete,* ACI 318-63, American Concrete Institute, Detroit, 1963, 144 pp.

11.27 H. Sunidja, "Response of Prestressed Concrete Plate–Edge Column Connections," Ph.D thesis, University of Illinois at Urbana–Champaign, 1982, 232 pp. Also issued as *Civil Engineering Studies,* Structural Research Series 498, University of Illinois, Urbana, Ill., March 1982.

11.28 D. A. Foutch, W. L. Gamble, and H. Sunidja, "Tests of Post-Tensioned Concrete Slab–Edge Column Connections," *ACI Struct. J.,* Vol. 87, No. 2, March–April 1990, pp. 167–179; "Closure to Discussion" Vol. 88, No. 1, pp. 116–117.

11.29 ACI/ASCE Committee 423, *Corrosion and Repair of Unbonded Single Strand Tendons,* ACI 423.4R-98, American Concrete Institute, Farmington Hills, Mich., 1998, 20 p.

11.30 *Specification for Unbonded Single Strand Tendons,* Revised 1993, Post-Tensioning Institute, Phoenix, Ariz., 1993, 20 pp.

12 Membrane Action in Slabs

12.1 INTRODUCTION

The yield line theory due to Johansen considers the presence of only moments and shear forces at the yield lines in the slab and gives a good indication of the ultimate load when the yield line pattern can form without the development of membrane (in-plane) forces in the slab. However, membrane forces are often present in reinforced concrete slabs at the ultimate load as a result of the boundary conditions and the geometry of deformations of the slab segments. If the edges of slabs are restrained against lateral movement by stiff boundary elements, compressive membrane forces are induced in the plane of the slab when, as the slab deflects, changes of geometry cause the slab edges to tend to move outward and to react against the bounding elements. The compressive membrane forces so induced enhance the flexural strength of the slab sections at the yield lines (provided that the slab is not over-reinforced), which will cause the ultimate load of the slab to be greater than the ultimate load calculated using Johansen's yield line theory. At larger deflections the slab edges tend to move inward and, if the edges are laterally restrained, tensile membrane forces are induced that may enable the slab to carry significant load by catenary action of the reinforcing steel. Even in slabs that are not intentionally restrained against lateral movement at the edges, the deflection of the slab at the ultimate load may result in a more favorable distribution of internal forces in the slab, which can enhance the load-carrying capacity as indicated in the experimental studies of reinforced concrete slabs discussed at the ends of Chapters 6 and 7.

A number of research studies on membrane action in reinforced concrete slabs have been conducted. However, only approximate ultimate strength theories have yet been developed, and the studies have pointed to difficulties in incorporating membrane action in design. Nevertheless, there is no doubt that membrane action will significantly increase the ultimate load of many reinforced concrete slab systems, even if membrane action has not been considered in the design.

12.2 UNIFORMLY LOADED LATERALLY RESTRAINED REINFORCED CONCRETE SLABS

12.2.1 General Behavior and Review of Past Research

Figure 12.1 shows the load-central deflection curve of a uniformly loaded two-way rectangular reinforced concrete slab with laterally restrained edges.

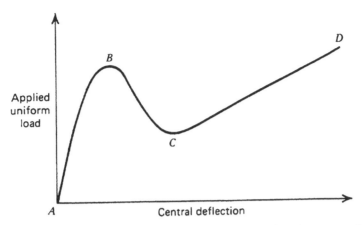

Figure 12.1 Load–deflection relationship for two-way reinforced concrete slab with edges restrained against lateral movement.

As the load is increased from A to B, the yield line pattern develops, and with the help of compressive membrane forces, the slab reaches its enhanced ultimate load at B. The introduction of compressive membrane forces in the slab can be thought of as being due to jamming of the slab segments between the boundary restraints, which causes the slab strips to arch from boundary to boundary. The induced compressive membrane force in the slab results in an enhancement of the flexural strength of the slab sections, as typical moment-axial force interaction diagrams show (see, e.g., Fig. 12.2). In the slab the compressive membrane forces are never great enough for the tension steel not to yield and therefore will always result in an increase in the ultimate moment of resistance at the yield lines. Figure 12.2 illustrates that the enhancement of moment for a given compressive membrane force will be particularly high in the case of lightly reinforced slabs.

It should be noted that the compressive membrane forces that develop at small deflections are a result of cracking in the slab. In a clamped elastic slab, there is no tendency for the ends of the span to move outward. If one integrates the bottom surface strain along the length of the span, the sum is zero. After cracking, this integration leads to a net gain in length.

As the deflection increases beyond B (see Fig. 12.1), the load carried by the slab decreases rapidly because of a reduction in the compressive membrane force. As C is approached, the membrane forces in the central region of the slab change from compression to tension. For slabs with rigid boundaries, the central deflection of the slab at C has been found to approximately equal the slab thickness. Beyond C the slab carries load by the reinforcement acting as a plastic tensile membrane with full-depth cracking of the concrete over the central region of the slab due to the large stretch of the slab surface. The slab continues to carry further load with an increase in deflection until at D the reinforcement begins to fracture or some other com-

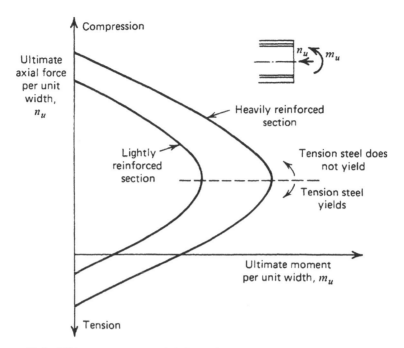

Figure 12.2   Ultimate moment-axial force interaction of symmetrically reinforced concrete sections.

ponent in the system fails. The load at B will be referred to as the ultimate load. Tensile membrane action is useful in preventing a catastrophic failure when the ultimate load is reached, and in heavily reinforced slabs the load carried at large deflections in the tensile membrane range can actually exceed the ultimate load. Unlike the compression membrane phase, tensile membrane forces can develop even in a thin elastic plate as the deflections become large.

Compressive membrane action has been observed in several full-scale tests. Ockleston[12.1] and Liebenberg[12.2] have tested uniformly loaded panels of full-scale reinforced concrete slab-and-beam floors. The loading was not applied to the whole floor but only to particular interior panels. It was found that the lateral restraint provided by the stiffness of the surrounding beams and panels was sufficient to enforce considerable membrane action in the loaded panels. The ultimate loads of the three panels tested by Ockleston were more than twice the ultimate loads predicted by Johansen's yield line theory. Ockleston correctly attributed this enhancement of strength to the presence of compressive membrane forces. Also, in a test conducted by Gamble et al.[12.3] on a $\frac{1}{4}$-scale model of a nine panel (three by three) reinforced concrete slab-and-beam floor, when the interior panel alone was subjected to a uniformly distributed live load, the supporting beams failed when the load on the panel was approximately twice the ultimate load predicted by Johansen's yield line theory. This slab was in the tensile membrane stage, as the midspan deflection was about twice the slab thickness, and tensile cracks penetrated the full thickness of the slab in the central portion of the panel.

Tests and analytical studies of uniformly loaded reinforced concrete single panels with the boundaries laterally restrained have been reported by Powell,[12.4] Wood,[12.5] Christiansen,[12.6] Schlaich,[12.7] Park,[12.8–12.11] Sawczuk,[12.12] Brotchie et al.,[12.13,12.14] Millington,[2.15] Keenan,[12.16] Roberts,[12.17] Girolami et al.,[12.18] Hung and Nawy,[12.19] Hopkins and Park,[12.20] Black,[12.21] Desayi and Kulkarni,[12.22] and others. These tests and analytical studies have demonstrated that the ultimate load of single panels may be significantly (many times) higher than that given by Johansen's yield line theory, particularly if the boundary restraint is stiff, the span/depth ratio of the panel is high, and the reinforcing steel ratio is small. The load carried by tensile membrane (catenary) action at large deflections has also been shown to be significant in many slabs, provided that the reinforcing bars are anchored adequately.

Two $\frac{1}{4}$-scale models of a reinforced concrete nine-panel (three by three) flat slab system supported by four interior columns and by a continuous wall around the perimeter have been tested under uniform static and dynamic loading by Criswell,[12.23] and showed a 30% increase in ultimate load due to compressive membrane forces from the restraint of the boundary walls. Comparison of this test result with the load enhancement obtained by Ockleston,[12.1] Liebenberg,[12.2] and Gamble et al.[12.3] illustrates that the increase in ultimate load due to membrane action is more significant for slab-and-beam systems than for flat slab systems. To investigate the possibility of incorporating membrane action in design, a $\frac{1}{4}$-scale model of a reinforced concrete nine-panel (three by three) slab-and-beam floor system was tested by Hopkins and Park.[12.24] The center panel had been designed for an ultimate load of twice Johansen's yield line theory ultimate load, and the beams were designed to carry the enhanced ultimate load, and the resulting axial forces. The floor reached the anticipated ultimate load, but it was found that in such a design it could be difficult to ensure that crack widths and deflections at service load were not excessive.

A large number of restrained slabs have been tested by the Army Corps of Engineers Waterway Experiment Station and others in investigations into the design and behavior of protective construction for various purposes. Woodson[12.25] has summarized much of the earlier work in addition to presenting data from his own investigations of the effects of various arrangements of shear reinforcement on the behavior of restrained one-way slabs. The cases reviewed included both static and dynamic loading of one- and two-way slabs in addition to boxlike structures. The box structures were often buried, in various fill materials. These slabs were generally thicker than found in normal civil construction, but a wide range of span/depth ratios was included. The recently increased need to design to resist terrorist attack has made understanding the full range of behavior of restrained slabs and walls even more important than previously.

12.2.2   Behavior in the Compressive Membrane Range

The compressive membrane stage of behavior covers two ranges of deflections, A to B and B to C of Fig. 12.1. As the load is increased from A to B, the slab behavior initially is elastic, combined with inelastic behavior at the critical sections at higher loads until at B the yield line pattern for the slab is fully developed. As the deflection increases from B to C, the slab deformation is due mainly to further plastic rotation at the yield lines. Thus, in the range BC the slab can be considered to be deforming as a mechanism due to plastic rotations at the yield lines. Plastic theory can be developed to follow the load–deflection path BC at and after ultimate load with compressive membrane forces acting. The theory can be developed first for a restrained strip and then extended to a two-way slab. To determine the theoretical ultimate load of the slab from this load–deflection relationship, the deflection at the ultimate load needs to be known. This deflection can be determined from theoretical considerations of elastic and plastic deformations at maximum load or empirically from test results. In the following sections theory is developed for the load–deflection relationship for region BC, and then consideration is given to the deflection at ultimate load with reference to test results.

Plastic Theory for Load–Deflection Behavior of a Restrained Strip at and After Ultimate Load. Many of the research workers listed previously have developed theories to follow the load–deflection relationship of strips. The approach adopted here is that due to Park,[12.8,12.9] which has been used as a basis by many of the other studies.

A fixed-end strip with plastic hinges developed is shown in Fig. 12.3. The strip is initially of length l and is fully restrained against rotation and vertical translation at the ends. The ends of the strip are considered to be partially restrained against lateral displacement, and the outward lateral movement at each end is t.

It is necessary to include the effect of any lateral displacement that may occur at the ends of the strip, because compressive membrane action is dependent on the restriction of small lateral displacements, and the behavior of the strip is sensitive to any lateral displacements that may occur. The lateral

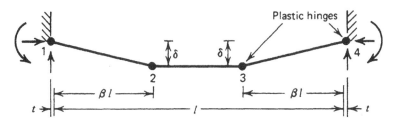

Figure 12.3   Plastic hinges of restrained strip.

displacement t may be calculated from the movement of the boundary system under the action of the membrane force. It is to be noted that the behavior of the strips is much less sensitive to imperfect restraint against rotation and vertical translation than to small lateral displacements.

The strip in Fig. 12.3 is considered to have symmetrically positioned plastic hinges as shown in the figure. It will be assumed that at each plastic hinge the tension steel has yielded, the compressed concrete has reached its strength with the stress distribution as defined by the ACI 318-95[12.26] equivalent rectangular concrete stress block, and the tensile strength of the concrete can be neglected. It will also be assumed that the top steel at opposite supports has the same area per unit width, that the bottom steel is constant along the length of the strip, but that the top and bottom steel may be different.

The portions of the strip in Fig. 12.3 between the critical (plastic hinge) sections are assumed to remain straight. However, because of the sensitivity of the theory to axial shortening, the axial strains in the strip due to elasticity, creep, and shrinkage will be taken into account. The sum of the elastic, creep, and shrinkage axial strain, ε, will be assumed to have a constant value along the length of the strip, since the membrane force is constant along the length. Because of ε, the shortening of middle portion 23 of the strip will be $\varepsilon(1 - 2\beta)l$, and because of symmetry, the ends of portion 23 will approach the center of the strip by $0.5\varepsilon(1 - 2\beta)l$. The outward lateral displacement of each boundary is t, and hence the horizontal distance from each end of the portion 23 of the strip to the adjacent boundary is $\beta l + 0.5 \varepsilon(1 - 2\beta)l + t$. Also, owing to ε, the lengths of end portions 12 and 34 will decrease to $(1 - \varepsilon)\beta l$. Figure 12.4 shows the change in dimensions of end portion 12 due to ε and t. The thickness of the strip is h and the neutral axis depths at yield sections 1 and 2 are c' and c, respectively. These two neutral-axis depths may be different because the top and bottom steel may not be similar. If ϕ is the inclination of portion 12, the distance between points A and B of Fig. 12.4 is shown by the geometry of the deformations to be

$$[\beta l + 0.5\varepsilon(1 - 2\beta)l + t] \sec \phi = (h - c') \tan \phi + (1 - \varepsilon)\beta l - c \tan \phi$$

$$h - c' - c = \frac{2\beta l \sin^2 (\phi/2) + \varepsilon \beta l \cos \phi + 0.5\varepsilon(1 - 2\beta)l + t}{\sin \phi} \quad (12.1)$$

For this equation, since ϕ and ε are small,

$$\sin \phi = 2 \sin \frac{\phi}{2} = \frac{\delta}{\beta l} \quad \text{and} \quad \cos \phi = 1$$

Therefore,

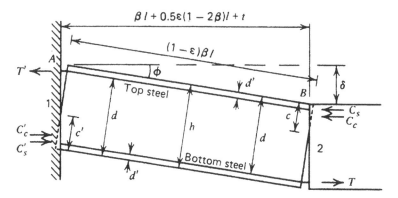

Figure 12.4   Portion of strip between yield sections 1 and 2 of Fig. 12.3.

$$c' + c = h - \frac{\delta}{2} - \frac{\beta l^2}{2\delta}\left(\varepsilon + \frac{2t}{l}\right) \tag{12.2}$$

Also, for equilibrium, the membrane forces acting on sections 1 and 2 of portion 12 of the strip are equal, and therefore

$$C'_c + C'_s - T' = C_c + C_s - T \tag{12.3}$$

where C'_c and C_c are the concrete compressive forces, C'_s and C_s the steel compressive forces, and T' and T the steel tensile forces, acting on sections 1 and 2, respectively. Using the ACI concrete compressive stress block (see Fig. 12.5), the concrete compressive forces can be written for a strip of unit width as

$$C'_c = 0.85f'_c\beta_1c' \tag{12.4}$$

$$C_c = 0.85f'_c\beta_1c \tag{12.5}$$

where f'_c is the concrete cylinder strength and β_1 the ratio of the depth of the equivalent rectangular stress block to the neutral-axis depth, as defined in ACI 318-95[12.26] (i.e., $\beta_1 = 0.85$ for $f'_c \leq 4000$ lb/in² and for $f'_c > 4000$ lb/in², β_1 reduces linearly by 0.05 for each 1000 lb/in² greater than 4000 lb/in², but β_1 must not be less than 0.65 [$\beta_1 = 0.85$ for $f'_c \leq 30$ N/mm² and for $f'_c > 30$ N/mm², β_1 reduces linearly by 0.05 for each 7 N/mm², but β_1 must not be less than 0.65]). Substituting Eqs. 12.4 and 12.5 into Eq. 12.3 and rearranging gives

$$c' - c = \frac{T' - T - C'_s + C_s}{0.85f'_c\beta_1} \tag{12.6}$$

Solving Eqs. 12.2 and 12.6 simultaneously gives

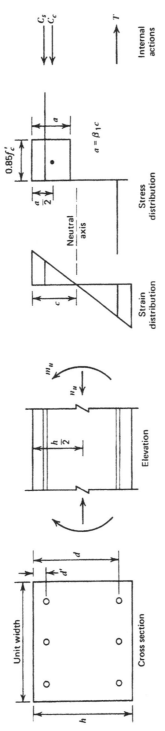

Figure 12.5 Conditions at positive moment yield section.

$$c' = \frac{h}{2} - \frac{\delta}{4} - \frac{\beta l^2}{4\delta}\left(\varepsilon + \frac{2t}{l}\right) + \frac{T' - T - C_s' + C_s}{1.7f_c'\beta_1} \qquad (12.7)$$

$$c = \frac{h}{2} - \frac{\delta}{4} - \frac{\beta l^2}{4\delta}\left(\varepsilon + \frac{2t}{l}\right) - \frac{T' - T - C_s' + C_s}{1.7f_c'\beta_1} \qquad (12.8)$$

Hence, use of the geometry of deformations and the equilibrium conditions has enabled the neutral-axis depths at the critical section to be determined. Note from Eqs. 12.7 and 12.8 that if the steel is similar top and bottom ($T' = T = C_s' = C_s$) and $\varepsilon = 0$ and $t = 0$, the neutral-axis depth at the yield sections varies between $0.5h$ and $0.25h$ for vertical deflections δ at the yield sections, which vary between zero and h. Note also that the effect of strip shortening ($\varepsilon > 0$) and outward movement at the boundaries ($t > 0$) is to reduce the neutral-axis depths at the yield sections. Note, however, that if the deflection is very small ($\delta \rightarrow 0$), and if $\varepsilon > 0$ and $t > 0$, the neutral-axis depths given by Eqs. 12.7 and 12.8 tend to $-\infty$ because of the third term on the right-hand side of those equations. These equations apply properly only when the deflection of the slab is significant. The initial deflections are, in fact, governed by elastic behavior. The plastic theory discussed above applies only at and after the ultimate load is reached.

In the typical lightly reinforced concrete slab used in practice, the neutral axis at ultimate load found by equating the tension in the steel to the compression in the concrete is small, the neutral axis lying quite close to the compressed surface of the concrete in most cases. Hence, if the lateral restraint at the edges causes the neutral axis at the yield sections at the ultimate load to be at a greater depth, an overall compressive membrane force will exist in the strip. The presence of this compression will result in an enhancement in the moment capacity of the sections, as the interaction diagram shown in Fig. 12.2 indicates.

Figure 12.5 shows conditions at a positive-moment yield section of unit width. The stress resultants at the section C_c, C_s, and T are statically equivalent to the membrane force n_u, acting at middepth, and the resisting moment m_u, summed about the middepth axis. Therefore, for a strip of unit width

$$n_u = C_c + C_s - T = 0.85f_c'\beta_1c + C_s - T \qquad (12.9)$$

$$m_u = 0.85f_c'\beta_1c(0.5h - 0.5\beta_1c) + C_s(0.5h - d') + T(d - 0.5h) \qquad (12.10)$$

where c is given by Eq. 12.8. For a negative-moment yield section of unit width, m_u' is given by an equation similar to Eq. 12.10, and $n_u' = n_u$ for equilibrium.

Considering end portion 12 or 34 of the strip, the sum of the moments of the stress resultants at the yield sections about an axis at middepth at one end is $m_u' + m_u - n_u\delta$. Shear forces have been neglected, since their net contribution to the analysis by virtual work will be zero. On substituting c' and c from Eqs. 12.7 and 12.8 into the equations for m'_u, m_u, and n_u, it is found that

$$
\begin{aligned}
m'_u + m_u - n_u\delta &= 0.85f'_c\beta_1 h \left[\frac{h}{2}\left(1 - \frac{\beta_1}{2}\right) + \frac{\delta}{4}(\beta_1 - 3) \right. \\
&\quad + \frac{\beta l^2}{4\delta}(\beta_1 - 1)\left(\varepsilon + \frac{2t}{l}\right) + \frac{\delta^2}{8h}\left(2 - \frac{\beta_1}{2}\right) \\
&\quad \left. + \frac{\beta l^2}{4h}\left(1 - \frac{\beta_1}{2}\right)\left(\varepsilon + \frac{2t}{l}\right) - \frac{\beta_1\beta^2 l^4}{16h\delta^2}\left(\varepsilon + \frac{2t}{l}\right)^2 \right] \\
&\quad - \frac{1}{3.4f'_c}(T' - T - C'_s + C_s)^2 + (C'_s + C_s)\left(\frac{h}{2} - d' - \frac{\delta}{2}\right) \\
&\quad + (T' + T)\left(d - \frac{h}{2} + \frac{\delta}{2}\right)
\end{aligned}
$$

$$(12.11)$$

If portion 12 or 34 of the strip of unit width is given a virtual rotation θ, the virtual work done by the actions at the yield sections of the portion is

$$
(m'_u + m_u - n_u\delta)\theta \tag{12.12}
$$

By equating the work done by the actions at the yield sections of the portions of the strip given by Eq. 12.12 to the work done by the loading on the strip in undergoing the virtual displacement, an equation can be obtained that relates the deflection of the strip to the load carried.

Note, however, that the load–deflection relationship so obtained will assume that the critical sections have reached their strength from the onset of deflections. Hence, the initial part of the load–deflection curve plotted from such a relationship will not be accurate because of the assumed plastic behavior. Thus, the load–deflection relationship so obtained will not apply at small deflections when the slab is acting elastically, nor at greater deflections before ultimate load when the critical sections are acting only partly plastically The derived load–deflection relationship can be expected to apply accurately only when sufficient deformation has occurred to allow full plasticity to develop at the critical sections.

Example 12.1. A fixed-end reinforced concrete strip of unit width and span l carries a uniformly distributed load per unit length w. The strip is restrained against lateral movement at its ends by an elastic surround. The surround at each end has a lateral stiffness S, where S is the load per unit outward displacement of the support at each end (lb/in.). The strip has a modulus of elasticity E_c and depth h. Shortening of the strip due to creep and shrinkage of the concrete may be neglected. Calculate the uniform load–central deflection relationship for the strip using plastic theory.

SOLUTION. For uniform load per unit length w present over the entire length of the strip, the positive-moment plastic hinge occurs at midspan (i.e., $\beta = 0.5$ in Fig. 12.3), and therefore Eq. 12.11 with $\beta = 0.5$ applies to both end portions of the strip. If each end of the strip is given a virtual rotation θ about the supports, the virtual work done by the loading on each end portion is $(wl/2)(\theta l/4)$. Hence, incorporating Eq. 12.2, the virtual work equation may be written as

$$\frac{wl}{2}\frac{\theta l}{4} = (m'_u + m_u - n_u\delta)\theta$$

Therefore,

$$\frac{wl^2}{8} = m'_u + m_u - n_u\delta \tag{12.13}$$

where the right-hand side of Eq. 12.13 is given by Eq. 12.11. The value of $\varepsilon + (2t/l)$ required for Eq. 12.11 is found as below, where n_u has been substituted from Eq. 12.9.

$$\varepsilon + \frac{2t}{l} = \frac{n_u}{hE_c} + \frac{2}{l}\frac{n_u}{S}$$

$$= \left(\frac{1}{hE_c} + \frac{2}{lS}\right)\left\{0.85f'_c\,\beta_1\left[\frac{h}{2} - \frac{\delta}{4} - \frac{\beta l^2}{4\delta}\left(\varepsilon + \frac{2t}{l}\right)\right.\right.$$

$$\left.\left. - \frac{T' - T - C'_s + C_s}{1.7f'_c\beta_1}\right] + C_s - T\right\}$$

Therefore,

$$\varepsilon + \frac{2t}{l}$$

$$= \frac{\left(\dfrac{1}{hE_c} + \dfrac{2}{lS}\right)\left[0.85f'_c\beta_1\left(\dfrac{h}{2} - \dfrac{\delta}{4} - \dfrac{T' - T - C'_s + C_s}{1.7f'_c\beta_1}\right) + C_s - T\right]}{1 + 0.2125\dfrac{f'_c\beta_1\beta l^2}{\delta}\left(\dfrac{1}{hE_c} + \dfrac{2}{lS}\right)}$$

$$\tag{12.14}$$

The load–central deflection relationship is given by Eqs. 12.13, 12.11, and 12.14 for the part of the deflection range where plastic theory is applicable.

Example 12.2. A series of fixed-end concrete strips was tested by Roberts[12.17] within a stiff surrounding frame (see Fig. 12.6a). The load was applied to each strip by a system of point loads that simulated uniformly distributed loading. Each strip had a clear span of $l = 56.5$ in. (1.435 m). The strip

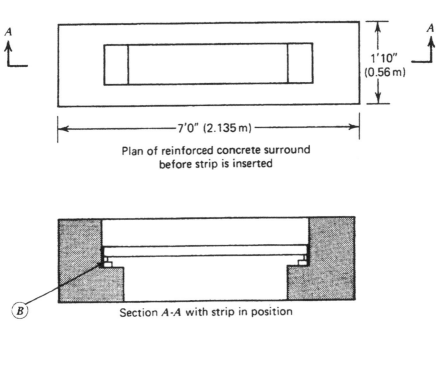

Plan of reinforced concrete surround
before strip is inserted

Section *A-A* with strip in position

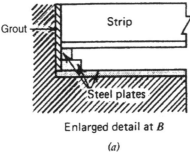

Enlarged detail at *B*

(a)

Figure 12.6 Laterally restrained reinforced concrete strip—tests and theory, Example 12.2: (*a*) details of surround and strip support[12.17]; (*b*) experimental and theoretical load–deflection relationships.

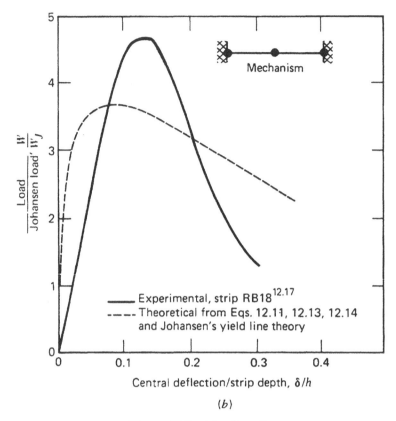

(b)

Figure 12.6   (*Continued*).

designated RB18 had an overall depth of 3 in. (76.2 mm), and the concrete cylinder strength was 3130 lb/in² (21.6 N/mm²), assuming the cylinder strength to be 0.8 of the cube strength measured by Roberts. The strip contained 0.578% of tension reinforcement at the positive-moment plastic hinge with an effective depth of 2.66 in. (67.6 mm) and a yield strength of 35,000 psi (241 N/mm²). There was no other reinforcement in the plastic hinge regions. The modulus of elasticity of the concrete may be taken to be 3.36 × 10⁶ lb/in² (23,200 N/mm²), and creep and shrinkage may be neglected since the loading was short-term and the strip was placed in the surround for insufficient time for significant differential shrinkage between the strip and the surround to take place. Load tests on the elastic surround showed the surround stiffness at each end to be $S = 3.29 \times 10^6$ lb/in. (0.576 × 10⁶ N/mm) for a beam of unit width. Calculate the load–central deflection relationship given by plastic theory for the strip and compare the load carried by the laterally restrained strip with that if the strip were not laterally restrained.

SOLUTION. Consider a unit width of strip. From reinforcement details, $C'_s = 0$, $T' = 0$, $C_s = 0$, and $T = 0.00578 \times 2.66 \times 35,000 = 538$ lb. Since $f'_c < 4000$ lb/in², $\beta_1 = 0.85$. From Eq. 12.13, $Wl/8 = m'_u + m_u - n_u\delta$, where W is the total load on the strip. If the strip is not laterally restrained, $m'_u = 0$, $n_u = 0$, and total ultimate load is W_j.

$$m_u = T\left(d - 0.58\frac{T}{f'_c}\right) = 538\left(2.66 - 0.59 \times \frac{538}{3130}\right) = 1377 \text{ lb-in.}$$

Therefore,

$$\frac{W_j l}{8} = 1377 \text{ lb-in.} \tag{i}$$

If the strip is laterally restrained; from Eqs. 12.13 and 12.11;

$$\frac{Wl}{8} = 0.85 \times 3130 \times 0.85 \times 3\left\{0.8625 - 0.5375\delta - 59.85\left[\frac{\varepsilon + (2t/l)}{\delta}\right]\right.$$

$$+ 0.06568\delta^2 + 76.48\left(\varepsilon + \frac{2t}{l}\right) - 45,110\left.\left(\frac{\varepsilon + (2t/l)}{\delta}\right)^2\right\}$$

$$- \frac{538^2}{3.4 \times 3130} + 538\left(2.66 - 1.5 + \frac{\delta}{2}\right)$$

$$= 6448 - 3378\delta + 4458\delta^2 + 518,900\left(\varepsilon + \frac{2t}{\delta}\right)$$

$$- 406,040\left[\frac{\varepsilon + (2t/l)}{\delta}\right] - 306,000,000\left[\frac{\varepsilon + (2t/l)}{\delta}\right]^2 \tag{ii}$$

Now

$$\frac{1}{hE_c} + \frac{2}{lS} = \frac{10^{-6}}{3 \times 3.36} + \frac{2 \times 10^{-6}}{56.5 \times 3.29} = 0.1100 \times 10^{-6} \text{ in./lb}$$

and from Eq. 12.14,

$$\varepsilon + \frac{2t}{l}$$

$$= \frac{0.11 \times 10^{-6}\left[0.85 \times 3130 \times 0.85\left(\frac{3}{2} - \frac{\delta}{4} + \frac{538}{1.7 \times 3130 \times 0.85}\right) - 538\right]}{1 + 0.2125[(3130 \times 0.85 \times 0.5 \times 56.5^2)/\delta]0.11 \times 10^{-6}}$$

$$= \frac{343.6 - 62.19\delta}{10^6 + 99,260/\delta} \tag{iii}$$

Thus, Eq. ii with Eq. iii substituted gives $Wl/8$ as a function of the central deflection δ. The theoretical enhancement in load-carrying capacity due to membrane action is given by the ratio W/W_J obtained from the ratio of Eq. ii to Eq. i. The theoretical value of W/W_J so obtained is plotted against δ/h in Fig. 12.6b. Figure 12.6b also shows the ratio of experimental W to theoretical W_J plotted against the experimental δ/h ratio, as measured by Roberts,[12.17] and the maximum (ultimate) load predicted by the theory to be conservative. This is due primarily to the theory assuming that the concrete strength is the uniaxial value, whereas in the test the concrete at the ends of the strip was confined transversely by the friction between the ends of the strip and the surround. For example, Roberts conducted some supplementary tests[12.17] which showed that the stress in the concrete at the ends of the strip was probably 2000 lb/in² (13.8 N/m²) greater than the cylinder strength. Figure 12.6b also shows that after maximum load, the load carried decreased more quickly than predicted by the theory. This is not surprising, since the theory assumes that the concrete compressive stress block parameters (mean stress and β_1) remain at the ACI values, whereas in fact as the deflection increases the stress block parameters will change at high extreme fiber concrete compression strains and crushing of concrete will occur; also, the theory assumes that the concrete stress–strain properties are reversible when the neutral axis decreases as the deflection increases, whereas in fact permanent set occurs on reversal of strain. The discrepancy between theoretical and experimental curves before maximum load is reached is also to be expected since, as discussed previously, the initial behavior is in fact elastic, whereas plastic theory has been used, which cannot be justified until near maximum load.

Both the test and theoretical results in Fig. 12.6b indicate the large enhancement in load-carrying capacity, over the Johansen value, which occurs due to compressive membrane action.

Example 12.3. A fixed-end reinforced concrete slab strip of unit width has a tension steel ratio at the critical negative- and positive-moment sections of 0.003, and the ratio of effective depth of tension steel to overall depth is 0.85. The effect of compression steel may be ignored. The steel has a yield strength of 60,000 lb/in² (414 N/mm²). The concrete cylinder strength is 4000 lb/in² (27.6 N/mm²) and hence $\beta_1 = 0.85$. The modulus of elasticity of the concrete is 3.8×10^6 lb/in² (26,000 N/mm²). The ends of the strip are restrained against lateral movement by an elastic surround, which has a lateral stiffness S at each end, where S is the membrane force to produce outward displacement (lb/in.). The strip is uniformly loaded over its length. Calculate the uniform load–central deflection relationship for the strip using plastic theory for span/depth ratios l/h of 20 and 40 for a range of surround stiffnesses varying from infinitely rigid to very flexible.

SOLUTION. Consider a unit width of strip. From reinforcement details, $C'_s = 0$ and $C_s = 0$. $T' = T = 0.003 \times 0.85h \times 60{,}000 = 153h$ lb. From Eq. 12.13, $wl^2/8 = m'_u + m_u - n_u \delta$. If the strip is not laterally restrained ($n_u = 0$),

$$m'_u = m_u = 153h \left(0.85h - 0.59 \times \frac{153h}{4000} \right) = 126.6h^2 \text{ lb-in.}$$

Therefore, the Johansen ultimate load per unit length is

$$w_J = 2 \times 125.6h^2 \times \frac{8}{l^2} = 2026 \left(\frac{h}{l} \right)^2 \text{ lb/in.}$$

If the strip is laterally restrained, from Eqs. 12.13 and 12.11,

$$w = 7504 \left(\frac{h}{l} \right)^2 - 11{,}203 \frac{\delta}{h} \left(\frac{h}{l} \right)^2 + 4552 \left(\frac{\delta}{h} \right)^2 \left(\frac{h}{l} \right)^2$$

$$+ 1662 \left(\varepsilon + \frac{2t}{l} \right) - 433.5 \frac{h}{\delta} \left(\varepsilon + \frac{2t}{l} \right)$$

$$- 307.0 \left(\frac{h}{\delta} \right)^2 \left(\frac{l}{h} \right)^2 \left(\varepsilon + \frac{2t}{l} \right)^2 \text{ lb/in.}$$

where, from Eq. 12.14,

$$\varepsilon + \frac{2t}{l} = \frac{(340.0 - 190.1\delta/h)(1 + 2hE_c/lS)}{10^6 + 47.53(l/h)^2(h/\delta)(1 + 2hE_c/lS)}$$

Figure 12.7 shows the ratio w/w_J given by the equations above plotted against δ/h for $l/h = 20$ and 40 and a range of S values. It is evident that the effect of the stiffness of the lateral restraint on both the maximum load carried and the shape of the load–deflection relationship given by plastic theory is very significant. If the surround is very stiff, the ultimate load is much higher than the Johansen load and is reached at a relatively small central deflection (0.1 to 0.25 times the strip depth in this example, with the smaller proportion of the depth applying to the case of the thicker slab). If the surround is very flexible, the ultimate load may not be much higher than the Johansen load and is reached at a relatively large deflection (0.8 to 0.9 times the strip depth in this example).

To obtain some feeling for the lateral stiffness required to achieve significant membrane action, the lateral stiffness of the surround, S (i.e., load per unit outward displacement of the support at each end) can be compared with the axial stiffness of the beam over each half-span, S_b (i.e., load per unit shortening of the half-span). When $S = S_b$,

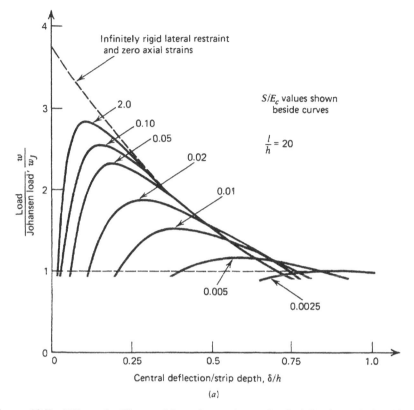

Figure 12.7 Effect of stiffness of lateral restraint on load–deflection relationship for laterally restrained strip of Example 12.3: (a) span/depth ratio $l/h = 20$; (b) span/depth ratio $l/h = 40$.

$$\frac{n_u}{S} = \frac{n_u \cdot 0.5l}{hE_c}$$

or

$$S = \frac{2hE_c}{l}$$

Hence, for this example, when the total outward displacement of the surround ends equals the shortening of the beam (i.e., $S = S_b$), $S = 0.1E_c$ when $l/h = 20$ and $S = 0.05E_c$ when $l/h = 40$. Figure 12.7a shows that when $l/h = 20$, very significant membrane action can be enforced when the surround and the beam have the same stiffness (i.e., $S = 0.1E_c$) and the increase in membrane action to be attained making the surround much more stiff is not great. On the other hand, Fig. 12.7b shows that when $l/h = 40$, the membrane action

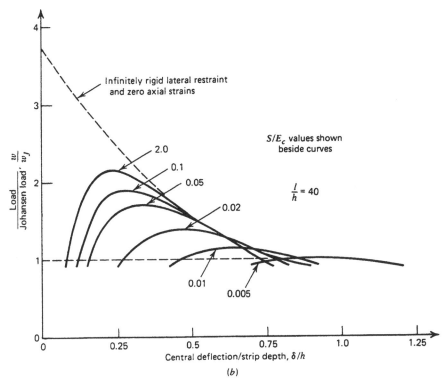

Figure 12.7 (*Continued*).

enforced when the surround and the beam have the same stiffness (i.e., $S = 0.05E_c$) is significant, but more is to be gained by making the surround stiffer. Thus, strips with small l/h ratios are less sensitive to outward displacements at their ends than are strips with large l/h ratios. It is also evident that the surround stiffness need not be enormous to achieve membrane action close to that for an infinitely rigid surround.

The dashed curves in Fig. 12.7a and b are for the hypothetical case of a rigid–plastic strip ($E_c = \infty$) with infinitely stiff surround. It is evident that elastic shortening of the strip causes a greater reduction in the membrane action for thinner strips than for thicker strips.

Central Deflection at the Ultimate Load of a Uniformly Loaded Rectangular Slab with All Edges Restrained. The plastic theory for the load–deflection relationship developed for restrained strips can be extended to two-way slabs. However, to determine the theoretical ultimate load of the slab from the load–deflection equation, the value of the central deflection at ultimate load, δ_u, must be known. The determination of δ_u has caused difficulty in the past, and both empirical and semiempirical approaches have been adopted by investigators. It is to be noted that δ_u is not necessarily the deflection at the peak of the load–deflection curve given by plastic theory. For example, δ_u cannot always be assumed to be the deflection at maximum load in the curves of Fig. 12.7. This is because the plastic theory curves do not hold at small deflections when the slab is acting elastically, or at greater deflections when the slab is acting partly elastically and partly plastically before the yield line pattern has fully formed. Ideally, a method for calculating the central deflection at ultimate load from the elastic and plastic deformations is required. Some of the investigators listed in Table 12.1, and Desayi and Kulkarni,[12.22] have devised semiempirical analytical methods for calculating the δ_u/h ratio. However, these methods require further verification because they do not appear to be generally applicable. The most acceptable approach at present is probably to examine the range of experimentally determined values for δ_u/h and to select a safe empirical value.[12.8, 12.9]

Reinforced concrete slabs have been tested in a number of investigations with lateral restraint provided at all edges by very stiff surrounding frames. The range of ratios of measured central deflection at the ultimate load to slab thickness, δ_u/h, are shown in Table 12.1. The slabs tested by Hung and Nawy were within a more flexible test frame than the other slabs, and hence their slabs reached ultimate load at central deflections that were a greater proportion of the slab thickness (between $0.62h$ and $0.89h$) than for the other slabs. For slabs within stiff surrounding frames, it would seem that a reasonably conservative estimate (i.e., an underestimate) of the ultimate load would be obtained at a central deflection of 0.5 of the slab thickness. It should be borne in mind that the ultimate load is not reached at a sharp peak in the load–

TABLE 12.1   Measured Central Deflection/Slab Thickness at Ultimate Load of Uniformly Loaded Laterally Restrained Slabs

Investigator	Number of Slabs	$\dfrac{l_x}{l_y}$	$\dfrac{l_y}{h}$	$\dfrac{\delta_u}{h}$	$\dfrac{\delta_u}{l_y}$
Powell[12.4]	15	1.75	16	0.33–0.44	0.021–0.028
Wood[12.5]	3	1.0	30	0.5	0.017
Park[12.8]	5	1.5	20	0.39–0.50	0.020–0.025
	1	1.5	27	0.48	0.018
	3	1.5	40	0.37–0.50	0.009–0.013
Brotchie and Holley[12.14]	4	1.0	20	0.36–0.57	0.018–0.029
	3	1.0	10	0.10–0.11	0.010–0.011
Keenan[12.16]	4	1.0	24	0.33–0.51	0.014–0.021
	1	1.0	15	0.20	0.013
	1	1.0	12	0.18	0.018
Hung and Nawy[12.19]	7	1.0	24	0.81–0.89	0.034–0.038
	5	1.43	17	0.62–0.74	0.037–0.044
Black[12.21]	4	1.0	33	0.34–0.71	0.010–0.022

deflection curve, and hence that the load does not differ significantly from the ultimate load over a small range of deflections (see Figs. 12.7 and 12.13). Hence, exact precision in the determination of the deflection at ultimate load is unnecessary. Also, a conservative value for the ultimate load will be obtained from the theoretical load–deflection curve if the deflection at ultimate load is overestimated, since the theoretical plastic theory curve reaches its peak at smaller deflections. Thus, a value of δ_u/h of 0.5 should lead to a reasonable estimate of the ultimate load of slabs with span/depth ratios in the range 20 to 40, but as Table 12.1 shows, it may be overly conservative for slabs with smaller span/depth ratios, which tend to reach ultimate load at significantly smaller δ_u/h ratios than 0.5.

Table 12.1 also lists values of deflection at ultimate load divided by the short span, δ_u/l_y. The values are about as variable as the δ_u/h values, but the thicker slabs do not necessarily have the lower δ_u/l_y values. The slabs reported by Hung and Nawy[12.19] have the largest values, again apparently in response to the lower restraining frame stiffness. The δ_u/l_y variable is probably less useful than the δ_u/h value in trying to describe and understand the behavior of restrained slabs.

The load–central deflection curves of Fig. 12.7 for strips tend to indicate the attainment of the maximum plastic theory load at central deflections considerably smaller than $0.5h$. However, it should be noted that the central deflection at the ultimate load of a two-way slab with compressive membrane action will not be the same as that of a strip (or one-way slab) with the same span/depth ratio and stiffness of lateral restraint, because in a two-way slab a large proportion of the positive-moment yield lines lie in the regions of the slab that have a smaller deflection than at the slab center, whereas in the strip the entire positive-moment yield section has the same deflection as the strip center. Thus, strips (and one-way slabs) will reach ultimate load at a smaller central deflection than two-way slabs, and care should be taken when interpreting deflection results for them.

When the surrounding frame is less stiff, the ultimate load is reached at greater central deflections than when the surrounding frame is very stiff (see the central deflections of Hung and Nawy in Table 12.1 and the strip results in Fig. 12.7). Thus, calculation of the ultimate load using a δ_u/h value of 0.5 may at first sight appear to be a crude approximation for these cases. However, examination of Fig. 12.7, and experimentally measured load–deflection curves, shows that when the surround is less stiff, and/or the axial strains in the slab are high, the load–deflection curve rises slowly to the ultimate load and then falls away gradually. That is, although the peak of the load–deflection curve may occur at much larger central deflections than $0.5h$ in this case, the load–deflection curve is fairly flat near ultimate load and the load is near ultimate load over a good range of deflections. Hence, when the surround is less stiff and/or the axial strains are high, provided that the effect of surround flexibility and slab axial strains are taken into account in calculating the load–deflection relationship, the theoretical ultimate load calculated at a central deflection of 0.5 of the slab thickness should still give a reasonably accurate value for the ultimate load.

Plastic Theory for Load–Deflection Behavior of a Uniformly Loaded Rectangular Slab with All Edges Restrained at and After Ultimate Load. Slabs that have edges parallel to the x- and y-axes and are reinforced in the x- and y-directions will be considered. To analyze such slabs, the following assumptions will be made:[12.8,12.9]

1. The slab is composed of strips running in the x- and y-directions which have the same depth as the slab. The x-direction strips contain only the x-direction steel and the y-direction strips contain only the y-direction steel.
2. The yield line pattern of the slab is as shown in Fig. 12.8. The yield sections of the strips lie on the yield lines and have the same deflection as the actual slab. (The simplification of assuming corner lines at 45° to the slab edges was shown in Chapter 8 to result in not more than 3% error in theoretical ultimate load for slabs with all edges fixed against restraint.)
3. The yield sections of the strips are at right angles to the direction of the strips. At the yield sections the torsional moments are zero, the

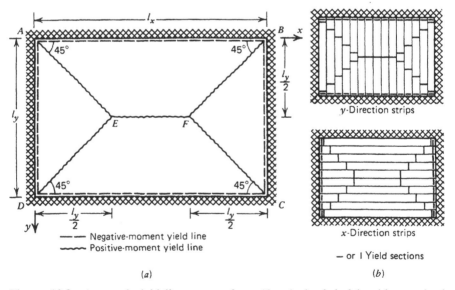

Figure 12.8 Assumed yield line pattern for uniformly loaded slab with restrained edges: (*a*) actual slab; (*b*) systems of strips.

tension steel has yielded, and the compressed concrete has reached its strength as defined by the ACI rectangular concrete stress block.[12.26] The strength of concrete in tension is neglected.

4. The portions of the strips between yield sections remain straight.

5. The top steel at opposite supports is the same but may be different from the top steel in the other direction. The top steel extends far enough into the slab from the edges to ensure the formation of the type of yield line pattern shown in Fig. 12.8. The bottom steel is placed over the entire area of the slab. The steel area, per unit width, is constant for each layer of steel but may be different for steel in each direction and for top and bottom steel.

6. The sum of the elastic, creep, and shrinkage axial strains in each strip, ε, is the same for all strips running in the same direction, but may be different for x- and y-direction strips.

7. The outward lateral displacement that occurs at each boundary, t, is the same for all strips running in the same direction, but may be different for x- and y-direction strips.

8. The slab reaches its ultimate load at a central deflection of one-half the slab thickness.

To determine the load–central deflection relationship, the slab is given a virtual displacement in the direction of the loading. If the virtual displacement at the yield line EF is unity, the end portions of the strips of the segments of the slab will undergo virtual rotations of $2/l_y$ about the lines of yielding at the edges of the slab. The middle portion of those strips that are divided into three portions will remain horizontal. Hence, the virtual work equation may be written as

$$
\iint w_u \Delta \, dx \, dy = 4 \int_0^{0.5l_y} (m'_{ux} + m_{ux} - n_{ux}\delta) \frac{2}{l_y} \, dy
$$
$$
+ 4 \int_0^{0.5l_y} (m'_{uy} + m_{uy} + m_{uy} - n_{uy}\delta) \frac{2}{l_y} \, dx
$$
$$
+ 2(m'_{uy} + m_{uy} - n_{uy}\delta) \frac{2}{l_y} (l_x - l_y) \tag{12.15}
$$

where w_u is the ultimate uniform load per unit area, Δ is the virtual displacement at element of slab with sides dx by dy, and the right-hand side is given by Eq. 12.12. The x and y subscripts refer to x- and y-direction quantities, respectively.

The virtual work done by the external loading, given by the left-hand side of Eq. 12.15, is

$$\frac{1}{3} w_u l_y^2 + \frac{1}{2} w_u (l_x - l_y) l_y = \frac{w_u l_y}{6} (3l_x - l_y)$$

The value of $m'_{ux} + m_{ux} - n_{ux} \delta$ for the first term on the right-hand side of Eq. 12.15 can be found by substituting into Eq. 12.11 the appropriate x-direction quantities and also $\delta = 0.5hy/0.5l_y = yh/l_y$ and $\beta = y/l_y$. The values of $m'_{uy} + m_{uy} - n_{uy} \delta$ for the second and third terms on the right-hand side of Eq. 12.15 can be found substituting into Eq. 12.11 the appropriate y-direction quantities and also $\delta = xh/l_y$ and $\beta = x/l_y$ for the second term and $\delta = 0.5h$ and $\beta = 0.5$ for the third term. On performing the foregoing substitutions and integrations on Eq. 12.15, the following equation for slabs with $l_x \geq l_y$ is obtained:

$$\frac{w_u l_y^2}{24} \left(3\frac{l_x}{l_y} - 1 \right)$$

$$= 0.85 f'_c \beta_1 h^2 \left\{ \frac{l_x}{l_y} (0.188 - 0.141\beta_1) + (0.479 - 0.245\beta_1) \right.$$

$$+ \frac{\varepsilon'_x}{16} \left(\frac{l_y}{h} \right)^2 \frac{l_x}{l_y} (3.5\beta_1 - 3) + \frac{\varepsilon'_y}{16} \left(\frac{l_y}{h} \right)^2 \left[2 \frac{l_x}{l_y} (1.5\beta_1 - 1) + (0.5\beta_1 - 1) \right]$$

$$\left. - \frac{\beta_1}{16} \frac{l_x}{l_y} \left(\frac{l_y}{h} \right)^4 \left[(\varepsilon'_x)^2 \frac{l_x}{l_y} + (\varepsilon'_y)^2 \right] \right\} - \frac{1}{3.4 f'_c} [(T'_x - T_x - C'_{sx} + C_{sx})^2$$

$$+ \frac{l_x}{l_y} (T'_y - T_y - C'_{sy} + C_{sy})^2] + (C'_{sx} + C_{sx}) \left(\frac{3h}{8} - d'_x \right)$$

$$+ (T'_x + T_x) \left(d_x - \frac{3h}{8} \right) + (C'_{sy} + C_{sy}) \left[\frac{l_x}{l_y} \left(\frac{h}{4} - d'_y \right) + \frac{h}{8} \right]$$

$$+ (T'_y + T_y) \left[\frac{l_x}{l_y} \left(d_y - \frac{h}{4} \right) - \frac{h}{8} \right] \tag{12.16}$$

where

$$\varepsilon'_x = \varepsilon_x + \frac{2t_x}{l_x} \tag{12.17}$$

$$\varepsilon'_y = \varepsilon_y + \frac{2t_y}{l_y} \tag{12.18}$$

and where the x and y subscripts refer to the x- and y-direction quantities, respectively.

In Eqs. 12.16 to 12.18, the strains ε_x and ε_y are the sums of the axial elastic, creep, and shrinkage strains in the strips and may be calculated from the membrane forces, the modulus of elasticity for the concrete, and the creep and shrinkage coefficients. The lateral boundary displacements t_x and t_y may be calculated from the membrane forces and the stiffness of the surround, and for interior panels surrounded by wide exterior panels the main contribution will be from the stretch of the sides of the surround rather than lateral bowing of the sides. In Eq. 12.16 the simplifying assumption is made that the x- and y-direction strips have constant values for ε_x and ε_y and for t_x and t_y, although ε_x and ε_y may be unequal and t_x and t_y may be unequal.

The membrane forces per unit width in each direction at ultimate load, n_{ux} and n_{ux}, are required to establish the axial strains and the lateral edge displacements. The membrane force is constant along each strip but varies from strip to strip since β and δ are variables. However, in most cases the mean membrane force per unit width will not differ much from the actual value, and it can be assumed that use of the mean value will give sufficient accuracy. The total membrane force in the x-direction of the slab of Fig. 12.8 is given by substituting the x-direction steel quantities, $\delta = yh/l_y$ and $\beta = y/l_x$ (for $y < l_y/2$), and c from Eq. 12.8, into Eq. 12.9, and integrating the resulting expression over the width of the slab. The mean n_{ux} is found by dividing the total membrane force in that direction by l_y. Thus,

$$\text{mean } n_{ux} = \frac{2}{l_y} \int_0^{0.5l_y} n_{ux} \, dy$$

$$= 0.85 f'_c \beta_1 h \left[\frac{7}{16} - \frac{\varepsilon'_x}{4} \frac{l_x}{l_y} \left(\frac{l_y}{h} \right)^2 \right]$$

$$+ \frac{1}{2} (C'_{sx} + C_{sx} - T'_x - T_x) \tag{12.19}$$

Similarly, it may be shown that

$$\text{mean } n_{uy} = \frac{1}{l_x} \left[2 \int_0^{0.5l_x} n_{uy} \, dx + n_{uy}(l_x - l_y) \right]$$

$$= 0.85 f'_c \beta_1 h \left[\frac{3}{8} + \frac{1}{16} \frac{l_y}{l_x} - \frac{\varepsilon'_y}{4} \left(\frac{l_y}{h} \right)^2 \right]$$

$$+ \frac{1}{2} (C'_{sy} + C_{sy} - T'_y - T_y) \tag{12.20}$$

The variation of the actual membrane force from the mean depends on the particular slab. For a square slab with equal top and bottom steel, if $\varepsilon'_x = \varepsilon'_y = 0$, the maximum and minimum membrane forces per unit width differ by $\pm 14\%$ from the mean. For large values of ε'_x and ε'_y the difference becomes greater, but the membrane forces become smaller and hence have less effect.

Large axial strains and lateral displacements can cause a considerable reduction in the compressive membrane action in some cases. To show the extent of this decrease in strength, reduction coefficients have been plotted against ε'_x and ε'_y for various cases in Fig. 12.9, where the reduction coefficient, R, is defined as

$$R = \frac{w_u \text{ from Eq. 12.16 with values for } \varepsilon'_x \text{ and } \varepsilon'_y}{w_u \text{ from Eq. 12.16 with } \varepsilon'_x = \varepsilon'_y = 0} \qquad (12.21)$$

In Fig. 12.9, R has been plotted for various l_x/l_y, l_y/h and $\varepsilon'_x = \varepsilon'_y$ values. In the figure $\varepsilon'_x = \varepsilon'_y$ has been assumed, since in most cases this will be approximately correct. It is evident that high values of ε'_x and ε'_y cause a negligible reduction in the strength of thick slabs, but for thin slabs the reduction is large and may be great enough to reduce the load to the Johansen value. The values for ε'_x and ε'_y likely to be met in practice depend very much on the loading history of the slab. Elastic strains can be calculated easily, but the creep and shrinkage strains are more difficult to estimate since they depend very much on the magnitude and the duration of the long-term loading and the differential shrinkage between slab and surround.

Plastic Theory for Load–Deflection Behavior of a Uniformly Loaded Rectangular Slab with Some Edges Laterally Restrained. The edge panel of a slab-and-beam floor will have one or more edges that are laterally unrestrained. According to the strip approximation, membrane forces will not develop in the direction of the span at right angles to the unrestrained edge. Such slabs can be analyzed by this theory by taking the simple ultimate moments in the direction perpendicular to laterally unrestrained edge or edges, but taking the enhanced ultimate moments with membrane action in the direction between laterally restrained edges.[12.8]

Plastic Theory for Uniformly Loaded Slab-and-Beam Floor Systems. If compressive membrane action is to be utilized in the design of slab-and-beam floors, the lateral restraint at the edges of each panel must be provided by the stiffness of the surrounding beams and panels. The lateral stiffness available has to be examined very closely because membrane action is dependent on the restriction of very small horizontal translations and large horizontal forces are involved.

Consider a continuous reinforced concrete slab-and-beam floor with stiff beams (see Fig. 12.10a), which is loaded uniformly over the entire area until the ultimate load is reached. Let the design of the floor be such that failure occurs over the entire at the same load. If the supporting beams have a low ultimate flexural strength, the mechanism formed will involve failure of the beams and of the panels (Fig. 12.10b), the floor folding along yield lines that

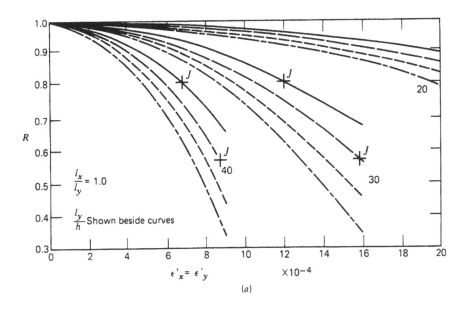

(a)

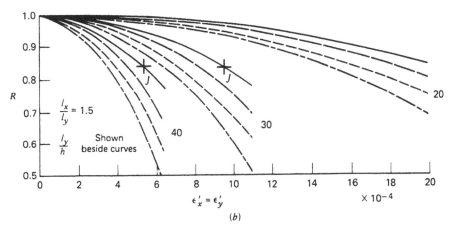

(b)

Figure 12.9 Reduction coefficients for rectangular slabs with restrained edges and equal steel top and bottom in each direction.

run across the whole width. Compressive membrane action cannot develop in the panels in this mechanism since restraint against lateral translation is not available from the beams. Alternatively, if the beams have a high ultimate strength, the failure will be confined to the panels (Fig. 12.10c). Compressive membrane action can develop in most of the panels of this mechanism. For an interior panel of the type marked A, membrane forces could develop in the direction of each span, owing to the horizontal restraint of the surrounding beams and panels. For an exterior panel of the type marked B, the membrane forces that develop in the direction perpendicular to the outside edge would

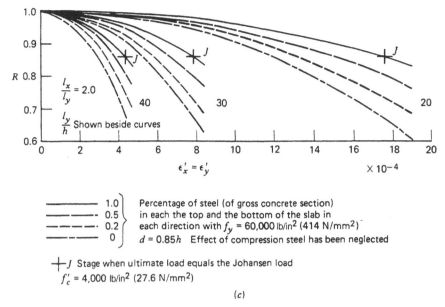

—————— 1.0 ⎫   Percentage of steel (of gross concrete section)
————·— 0.5 ⎪   in each the top and the bottom of the slab in
——— — 0.2 ⎬   each direction with f_y = 60,000 lb/in² (414 N/mm²)
———— — 0   ⎭ $d = 0.85h$ Effect of compression steel has been neglected

╋–J Stage when ultimate load equals the Johansen load
f_c' = 4,000 lb/in² (27.6 N/mm²)

(c)

Figure 12.9   (*Continued*).

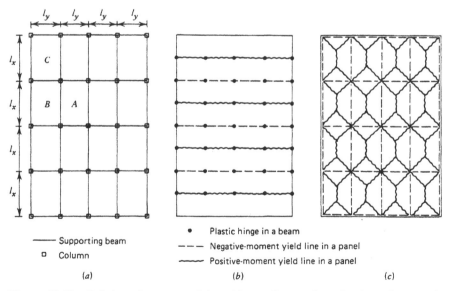

—————— Supporting beam
□   Column

•   Plastic hinge in a beam
– – –   Negative-moment yield line in a panel
———   Positive-moment yield line in a panel

(a)   (b)   (c)

Figure 12.10   Reinforced concrete slab-and-beam floor and mechanisms due to uniform loading: (*a*) floor; (*b*) beam and panel failure; (*c*) panel failure.

be small, since an edge beam alone would not be stiff enough to restrain lateral movement, but membrane forces could develop in the direction parallel to the outside edge. For a corner panel of the type marked C, the membrane forces developed in the directions of both spans would be small.

Figure 12.11 shows the horizontal forces acting on the supporting beams of the floor when the mechanism of Fig. 12.10c develops with compressive membrane action. The membrane forces in the edge panels acting normal to the outside edges of the floor are neglected in the figure. For the membrane action shown in Fig. 12.11 to take place, the requirements are as follows. If the panels of the floors are the same size, only the beams marked AB, CD, EF, and GH in Fig. 12.11 will tend to bow laterally, since the remaining interior beams will be loaded by equal and opposite horizontal forces. (Bowing of the edge beams is neglected since they are assumed to be unloaded.) The bowing of these four interior beams is restricted by the deep beam action of the adjacent panels, the effectiveness of which will depend on the l_x/l_y

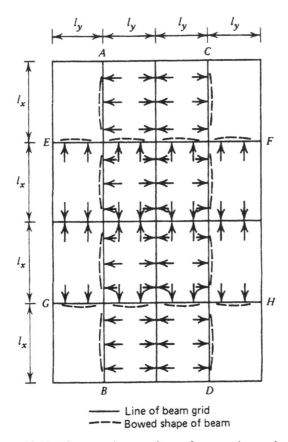

Figure 12.11 Compressive membrane forces acting on beams.

ratio of the exterior panels. If $l_x/l_y = 1.0$, there is no doubt that the lateral bowing will be reduced to negligible quantities, but narrow exterior panels may obviously allow a considerable reduction in the compressive membrane action of adjacent interior panels.

All the beams of the floor will have to act as ties in order to balance the membrane forces acting on beams AB, CD, EF, and GH of Fig. 12.11. If it is assumed that at the ultimate load of the panels, the beams carry the membrane forces of the adjacent panels equally, the tie force P_{ux} in each interior beam in the x-direction is approximately

$$P_{ux} = l_y \times \text{mean } n_{ux} \tag{12.22}$$

In the edge beams in the x-direction, the tie force will be approximately one-half of this value. Similarly, in the y-direction interior beams, the tie forces P_{uy} (in the edge beams $P_{uy}/2$) are approximately

$$P_{uy} = l_x \times \text{mean } n_{uy} \tag{12.23}$$

In these equations, mean n_{ux} and mean n_{uy} are the mean membrane forces, per unit width, in the panels given by Eqs. 12.19 and 12.20. More accurate values for the tie forces, especially for the forces in the edge beams, could be found by an analysis that takes into account the stiffness of the panels as deep beams and the extensibility of the ties.

In addition to the reinforcement placed for bending in the vertical plane, the beams will require additional reinforcement to provide for the tie forces. The extension of the beams will allow the edges of the panels to move laterally. For example, if at the ultimate load of an interior panel the tie steel is stressed uniformly along its length to 30,000 lb/in² (207 N/mm²), the resulting tensile strain (0.001) would cause a very large reduction in the ultimate strength of thin slabs, as Fig. 12.9 shows. This may make the design of thin slabs using compressive membrane action impracticable. In practice, however, it is unlikely that the tie steel will have a high uniform tensile stress along its whole length.

Economics of Utilizing Compressive Membrane Action in Design. The use of compressive membrane action allows the designer of continuous slab-and-beam floors to reduce the amount of reinforcement in the panels to less than that required by Johansen's yield line theory. For the same ultimate loads and concrete dimensions, there will be more steel in the beams supporting the panels designed to include membrane action than in the beams supporting the panels designed by Johansen's theory, however, since extra steel will be required for tie reinforcement. For economical use of compressive membrane action, the resulting reduction in the steel content of the panels should be greater than the extra reinforcement placed in the beams. It can be shown,[12.27] however, that for a slab-and-beam floor designed to include compressive membrane action, more steel is required in the beams for ties than can be saved in the panels if all the membrane force is resisted by tie steel placed continuously around the panels. This follows because the lever arm of the additional concrete force in the panel due to compressive membrane force is smaller than that of the steel it replaces, and hence the compressive membrane force has to be larger than the force in the replaced steel if the ultimate load is to be the same. This would appear to make compressive membrane action of academic interest only.

Where compressive membrane action theory is of use, however, is in showing that a floor that has been designed by Johansen's yield line theory to carry a particular ultimate load on all the panels can, in fact, carry on some panels an ultimate load much greater than the design ultimate load, provided that the surrounding panels are lightly loaded, since the steel in the lightly loaded adjacent beams and panels can be used to develop the compressive membrane action in the heavily loaded panels. This means, for example, that slab-and-beam floors with only the alternate panels loaded could carry extremely high ultimate loads on the loaded panels. Also, steel placed in beams for earthquake or wind loading could be utilized to carry some additional gravity loading on the panel when the structure is subjected to gravity loading alone.[12.28] That is, the incorporation of membrane action in design is most efficient when use can be made of steel placed in the beams for alternative loading cases to provide the tie reactions for the membrane forces in the panels.

Experimental Results from Uniformly Loaded Single Panels. A number of investigators have reported test results from uniformly loaded single panels tested with all edges restrained or partially restrained against rotation and vertical and horizontal translation. Most of these tests have involved panels with very rigid surrounding frames and subjected to short-term loading. A few long-term loading tests have also been conducted. Some of these results are reviewed and compared with theoretical predictions in the following paragraphs.

Short-Term Loading Tests on Slabs with Stiff Surrounds. The main investigators who have conducted short-term loading tests on slabs in stiff surrounding frames are listed in Table 12.1. In the investigation conducted by Park,[12.8] uniformly loaded rectangular slabs were tested with either three or four edges restrained against rotation and translation in an extremely stiff surrounding steel frame. The test frame and edge support details are shown in Fig. 12.12. Slabs have also been tested in extremely stiff surrounding frames by Powell,[12.4] Wood,[12.5] Brotchie and Holley,[12.14] and Keenan.[12.16] The details of the slabs tested with all edges restrained, and the measured ultimate loads, are shown in Table 12.2. In the case of the slabs of Powell, Wood, and Park, the concrete strengths were measured from 6-in. (150-mm) cubes, and these strengths were multiplied by 0.8 to estimate the concrete cylinder strength.

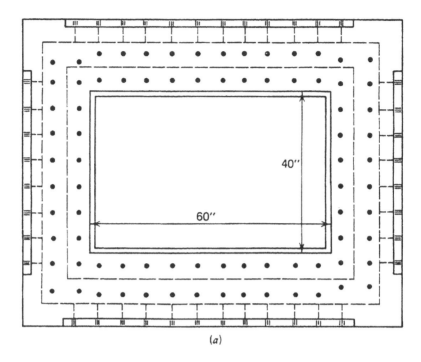

(a)

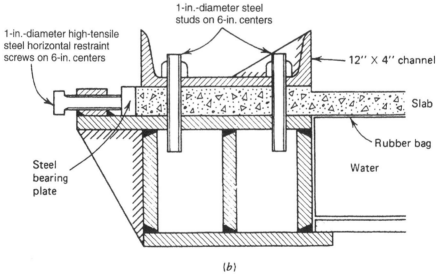

(b)

Figure 12.12   Steel test frame and edge support details (1 in. = 25.4 mm)[12.8,12.9,12.11]: (a) base of test frame; (b) restrained edge; (c) simply supported edge.

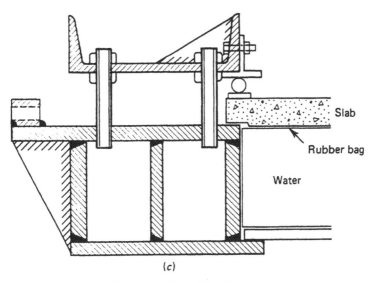

(c)

Figure 12.12   *(Continued)*.

Table 12.2 also shows the theoretical ultimate loads calculated using Johan-sen's yield line theory and using Eq. 12.16, assuming that $\varepsilon_x' = \varepsilon_y' = 0$. The assumption of zero axial strains and zero lateral edge displacements for Eq. 12.16 was made in view of the short-term loading and the great stiffness of the surrounding frames. In the ultimate load calculations reported in Ref. 12.8, the concrete compressive stress block parameters used were those recom-mended by Hognestad et al.[12.29] (the concrete stress block parameters in the ACI Code[12.26] are based on Ref. 12.29). However, in Table 12.2 the theoretical ultimate loads have been calculated using the ACI Code concrete stress block parameters directly, as in Eq. 12.16, and hence there are slight differences between these and the ultimate loads calculated in Ref. 12.8.

Table 12.2 shows that the Johansen load was greatly exceeded in the tests, especially in the case of the lightly reinforced slabs. Unreinforced slabs, which have zero ultimate load according to Johansen, in fact have consider-able strength due to compressive membrane action. The theory, including the effect of compressive membrane action (Eq. 12.16), gives a reasonable pre-diction for the experimental ultimate load in the case of most of the slabs in the table. For the 40 slabs, the mean value for $w_{\text{test}}/w_{\text{theory}}$ is 1.02, with a standard deviation of 0.125. It is evident that the values of $w_{\text{test}}/w_{\text{theory}}$ would have been higher if the actual values for ε_x' and ε_y' had been included in the calculation of w_{theory}, showing that in general the theory is conservative. The conservatism arises from the use of the uniaxial concrete strength rather than the actual confined values in the slab, and the use of an approximate deflection at ultimate load.

Figure 12.13 shows the measured load–central deflection curves of Park's series *A* slabs compared with the theoretical load–deflection curves given by

TABLE 12.2 Tests and Theory: Short-Term Uniform Loading on Slabs with All Edges Restrained by a Stiff Surround[a]

Investigator	Slab Mark	Dimension $l_x \times l_y \times h$ (in.)	$\dfrac{l_x}{l_y}$	$\dfrac{l_y}{h}$	Reinf. Short Span Top	Short Span Bottom	Long Span Top	Long Span Bottom	Concrete Cylinder Strength (lb/in²)	(a) Experimental Ultimate Load, w_{test} (lb/in²)	(b) Theoretical Ultimate Load from Eq. 12.16, with $\varepsilon'_x = \varepsilon'_y = 0$, w_{theory} (lb/in²)	(c) Theoretical Johansen Load, w_J (lb/in²)	$\dfrac{(a)}{(c)}$ $\dfrac{w_{test}}{w_J}$	$\dfrac{(a)}{(b)}$ $\dfrac{w_{test}}{w_{theory}}$
Powell[12.4]	S46	36 × 20.57 × 1.286	1.75	16	0.25	0.25	0.25	0.25	5810	45.0	39.8	5.47	8.22	1.13
	S47	36 × 20.57 × 1.286	1.75	16	0.25	0.25	0.25	0.25	6500	38.6	43.8	5.48	7.05	0.88
	S50	36 × 20.57 × 1.286	1.75	16	0.45	0.45	0.45	0.45	5400	48.1	40.5	9.81	4.90	1.19
	S54	36 × 20.57 × 1.286	1.75	16	0.71	0.71	0.71	0.71	5940	52.9	47.7	15.4	3.43	1.11
	S55	36 × 20.57 × 1.286	1.75	16	0.71	0.71	0.71	0.71	5340	55.0	44.2	15.4	3.57	1.25
	S58	36 × 20.57 × 1.286	1.75	16	0.97	0.97	0.97	0.97	5800	49.6	54.0	25.0	1.98	0.92
	S59	36 × 20.57 × 1.286	1.75	16	0.97	0.97	0.97	0.97	5700	50.8	53.5	25.0	2.03	0.95
	S62	36 × 20.57 × 1.286	1.75	16	1.53	1.53	1.53	1.53	5950	61.8	65.4	38.6	1.60	0.95
	S63	36 × 20.57 × 1.286	1.75	16	1.53	1.53	1.53	1.53	5270	67.3	61.4	38.3	1.76	1.10
	S48	36 × 20.57 × 1.286	1.75	16	0	0	0	0	5950	36.9	36.8	0	—	1.00
	S53	36 × 20.57 × 1.286	1.75	16	0	0	0	0	5450	42.0	33.8	0	—	1.24
	S56	36 × 20.57 × 1.286	1.75	16	0	0	0	0	5540	37.5	34.4	0	—	1.09
	S57	36 × 20.57 × 1.286	1.75	16	0	0	0	0	5740	30.2	35.6	0	—	0.85
	S60	36 × 20.57 × 1.286	1.75	16	0	0	0	0	5750	32.5	35.6	0	—	0.91
	S64	36 × 20.57 × 1.286	1.75	16	0	0	0	0	5760	35.0	35.7	0	—	0.98
Wood[12.5]	FS12	68 × 68 × 2.25	1.0	30.2	0	0.26		0.26	4720	16.88	16.7	1.50	11.22	1.01
	FS13	68 × 68 × 2.25	1.0	30.2	0.26	0.26	0.26	0.26	3840	12.27	14.7	3.00	4.09	0.84
	FS14	68 × 68 × 2.25	1.0	30.2	0	0		0	4140	9.31	13.3	0	—	0.70
Park[12.8]	A1	60 × 40 × 2	1.5	20	0.38	0.19	0.41	0.20	4780	30.8	27.6	8.41	3.66	1.12
	A2	60 × 40 × 2	1.5	20	0.84	0.42	0.43	0.21	4280	31.3	27.4	11.5	2.72	1.14

Slab	$l_x \times l_y \times h$												
A3	60 × 40 × 2	1.5	20	1.44	0.72	0.45	0.22	5000	37.8	37.0	20.5	1.84	1.02
A4	60 × 40 × 2	1.5	20	2.42	1.21	0.47	0.23	4020	37.3	36.9	25.9	1.44	1.01
D1	60 × 40 × 2	1.5	26.7	0	0	0	0	5020	24.6	23.2	0	—	1.06
D2	60 × 40 × 1.5	1.5	40.8	0	0	0	0	4960	12.9	12.9	0	—	1.00
D3	60 × 40 × 0.98	1.5	39.6	0	0	0	0	5140	4.61	5.69	0	—	0.81
D4	60 × 40 × 1.01	1.5	40.4	0	0	0	0	4440	4.14	5.23	0	—	0.79
D5	60 × 40 × 0.99	1.5		0	0	0	0	3550	4.14	4.02	0	—	0.96
Brotchie and Holley[12.14] 42	15 × 15 × 0.75	1.0	20	0	0	0	0	5060	35.6	37.0	0	—	0.96
44	15 × 15 × 0.75	1.0	20	0	0	0	0	4320	25.9	31.7	0	—	0.82
46	15 × 15 × 0.75	1.0	20	1.0	1.0	1.0	1.0	5490	59.5	50.6	18.8	3.17	1.18
48	15 × 15 × 0.75	1.0	20	2.0	2.0	2.0	2.0	4870	55.0	50.5	34.3	1.60	1.09
45	15 × 15 × 1.5	1.0	10	0	0	0	0	4270	144.0	125.4	0	—	1.15
47	15 × 15 × 1.5	1.0	10	1.0	1.0	1.0	1.0	4570	209.0	188.9	81.1	2.58	1.11
49	15 × 15 × 1.5	1.0	10	2.0	2.0	2.0	2.0	4620	220.0	221.1	150.1	1.47	1.00
Keenan[12.16] 3S1	72 × 72 × 3	1.0	24	0.82	0.82	0.82	0.82	3550	32.2	30.7	17.7	1.82	1.05
3S2	72 × 72 × 3	1.0	24	0	0	0	0	4140	23.6	21.1	0	—	1.12
3S3	72 × 72 × 3	1.0	24	0.82	0.82	0.82	0.82	4120	34.6	33.6	17.9	1.94	1.03
3S4	72 × 72 × 3	1.0	24	0.82	0.82	0.82	0.82	3300	32.1	26.3	17.6	1.83	1.22
4.75S1	72 × 72 × 4.75	1.0	15.2	0.89	0.89	0.89	0.89	3170	85.0	80.8	50.6	1.68	1.05
6S1	72 × 72 × 6	1.0	12	1.33	1.33	1.33	1.33	3620	182.0	190.6	130.9	1.39	0.96
										Mean			1.02
										Standard deviation			0.125

ᵃThe reinforcement ratio was determined using the actual effective depths in the case of Park's slabs and using the mean effective depths for the other slabs. The steel yield strengths were: 30.6 or 37.0 kips/in² for Powell's slabs, 33.8 kips/in² for Wood's slabs, 40 to 51 ksi for Park's slabs, 55 or 60 kips/in² for the slabs of Brotchie and Holley, and 47.4 or 49.6 kips/in² for Keenan's slabs. The concrete compressive strengths for the slabs of Powell, Wood, and Park were measured from 6-in. cubes, and it is assumed above that the concrete cylinder strength = 0.8 × cube strength; the concrete strengths for the slabs of the other investigators were measured from cylinders. Dimensions l_x and l_y are the clear spans and h is the slab thickness. 1 lb/in² = 0.00689 N/mm²; 1 kip/in² = 6.89 N/mm², 1 in. = 25.4 mm.

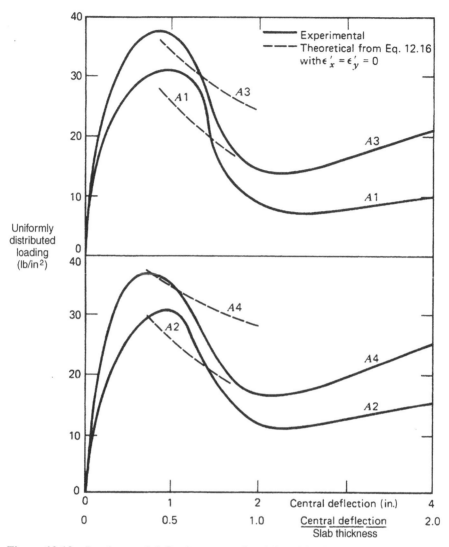

Figure 12.13 Load-central deflection curves for slabs with all edges restrained (1 lb/in² = 6.89 kN/m², 1 in. = 25.4 mm).[12.8]

a form of Eq. 12.16.[12.8] In this form of Eq. 12.16 the central deflection was left in the equation as a variable (rather than being put equal to $0.5h$ as in Eq. 12.16), $\varepsilon_x' = \varepsilon_y' = 0$ was assumed, and the concrete stress block parameters were as given in Ref. 12.29. Figure 12.14 illustrates a slab after the ultimate load was reached and shows the clearly developed yield line pattern.

Tests were also conducted by Park[12.8] on eight slabs with similar dimensions to Al to A4 but with only three edges restrained. These slabs showed experimental ultimate loads of 1.5 to 2.9 times the Johansen load. The ratio

Figure 12.14   Yield line pattern of uniformly loaded slab with restraint at all edges.[12.9]

of experimental ultimate load to theoretical ultimate load for these slabs[12.8] varied between 0.83 and 1.27 when it was assumed that only one-way membrane action occurred, that the central deflection at ultimate was $0.4h$, that the corner yield lines were at 45° to the slab edges, and that the stress block parameters were as given in Ref. 12.29.

The results from slabs with all edges restrained tested by Hung and Nawy[12.19] have not been shown in Table 12.2. Equation 12.16 with $\varepsilon'_x = \varepsilon'_y = 0$ gave an average value for the ratio of experimental ultimate load to theoretical ultimate load of only 0.58 for the slabs of Hung and Nawy, which had all edges clamped. The relatively low experimental ultimate load appears to have been due to the surrounding frame used in those tests not being very stiff, resulting in ε'_x and ε'_y values that are too large to be neglected in the calculation of the theoretical ultimate loads from Eq. 12.16. The lack of a very stiff surrounding frame in the tests by Hung and Nawy is also demonstrated by the large deflections at which ultimate load was reached for those slabs, as indicated in Table 12.1. No attempt has been made here to calculate the theoretical ultimate loads of the slabs of Hung and Nawy, including the effect of ε'_x and ε'_y because the displacement of the panel edges were not reported with their test results.

Long-Term Loading Tests on Slabs with Stiff Surrounds. It is evident that the enhanced strength due to compressive membrane action is reduced by the long-term effects of creep and shrinkage. To check long-term effects experimentally, eight unreinforced concrete slabs with all edges restrained against rotation and translation were tested by Park[12.9] by applying a sustained uniform load for a period of six weeks (except in one case). During this period, measurements of creep and shrinkage strains were made on specimens. At the end of the period the slabs (with one exception) were loaded to failure and the experimental ultimate loads compared with the theoretical ultimate loads from Eq. 12.16 with the measured values for axial strains and lateral boundary displacements substituted. It was found that at very high sustained loads the creep strains were very significant. For instance, one slab with a span/depth ratio of 40 failed at a load of 0.78 of the short-term ultimate load after that load had been applied for less than one day. However, slabs with sustained loads of approximately one-third of the short-term ultimate load showed no visible cracking during the sustained loading and it was evident that the axial creep strains were negligible. The tests confirmed the theory in that the ultimate load of slabs with a span/depth ratio of 40 was greatly reduced by axial strains, but for thicker slabs the reduction was negligible. The theoretical ultimate loads calculated for the test slabs were mainly conservative when compared with the experimental values and were influenced by the difficulty of estimating the axial strains and lateral boundary displacements accurately. At the ultimate load a well defined yield line pattern formed, as observed for the slab in Fig. 12.15.

Surround Stiffness Tests. To investigate the stiffness and strength of the exterior panels of slab and beam floors when resisting the membrane action of the interior panels, a series of 20 idealized small-scale nine-panel concrete floors were tested by Park.[12.10] Each model floor was three panels wide in each direction and the interior panel [which was 12 in. (305 mm) square] was loaded uniformly to failure with the eight surrounding exterior panels subjected only to the resulting reactive forces. Beams were not cast with the floor, but the floor was supported on rollers around the interior panel and around the outside edges of the floor. The exterior panels were reinforced in various ways and were of various sizes. It was found that tie reinforcement should be placed around the interior panel if membrane action was to be enforced without the failure of the exterior panels. The ultimate loads of the interior panels showed good agreement with the theory, which included the axial strains in the panel and the lateral displacement at the edges due to the stretch of the tie steel once the concrete had cracked. It was found that the exterior panels of the floor should be almost square if lateral bowing is to be avoided. Figure 12.16 shows one of the slabs after testing; the white square indicates the edges of the interior panel.

Tests on Panels with Controlled Horizontal Reactive Forces at Edges. Six reinforced concrete panels have been tested by Girolami et al.[12.18] Each panel was 6 ft (1.83 m) square and 1.75 in. (44 mm) thick and was cast monolith-

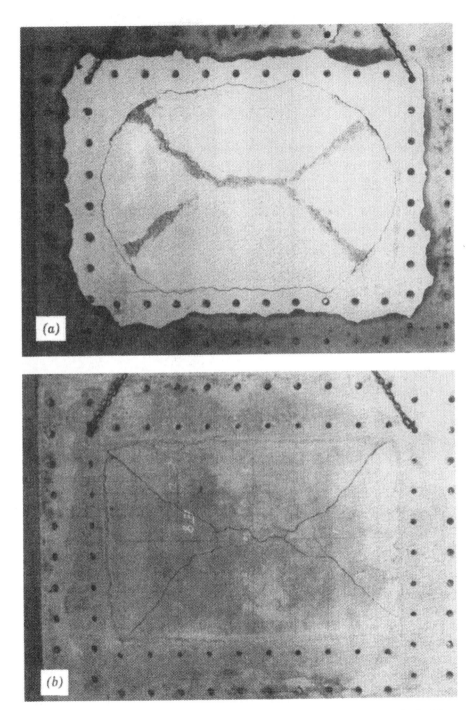

Figure 12.15 Yield lines of a slab at failure[12.9]: (*a*) loaded face; (*b*) unloaded face.

Figure 12.16   Yield lines of an interior panel.[12.10]

ically with 6-in. (152-mm)-deep by 3-in. (76-mm)-wide spandrel beams. The panel and spandrel beams contained both negative- and positive-moment reinforcement. Three of the test structures were supported only at the intersections of the spandrel beams, while three were supported at several points along the spandrel beams to investigate the effect of nondeflecting beams. The panels were subjected to a number of point loads over their surface simulating uniform vertical loading, and vertical loads were also applied to cantilever extensions of the beams to maintain a certain amount of restraint at the corners. Equal horizontal compressive loads were applied at five equally spaced points on each side of the slab. The three corner-supported structures initially developed a typical panel type of yield line pattern with positive-moment yield lines along the diagonals of the panel, but ultimately failed in a folding type of mechanism with yield lines running only parallel to the panel edges and with the beams participating in the failure mechanism. Failure occurred within the panel in the three structures with nondeflecting beams.

The measured ultimate load of the panels was 1.7 to 2.1 times the load calculated by Johansen's yield line theory. The load carried when the central deflection of the panel was 0.5 of the slab thickness was close to the ultimate load. A simple iterative procedure was used to calculate the theoretical ultimate load, including the effect of membrane forces. This simple procedure was possible because the applied membrane forces were known, being held constant by horizontal jacks during the tests. Each step in this iterative procedure involved estimating the vertical deflection of the panel and then calculating the ultimate load of the panel corresponding to this deflection from the equilibrium of the slab segments bounded by the yield lines, taking into account the enhanced moment strengths determined from the section properties and the known membrane forces. The procedure was repeated until the load giving the estimated deflection, calculating using elastic plate theory with fully cracked sections, agreed with the load given by the equilibrium requirements of the slab segments. For the tests involving nondeflecting beams, the equilibrium requirements for one of the triangular segments of the panel bounded by the yield lines may be written as

$$\frac{W}{4}\frac{l}{6} = m_u'l + m_ul - \frac{2}{3}P\delta_{cs} \qquad (10.24)$$

where W is the total load on the panel; l the panel span; m_u' and m_u the negative- and positive-moment strengths of slab per unit width, respectively; P the total applied horizontal force on one side of panel; and δ_{cs} the central deflection of slab. The deflections were assumed to vary parabolically across the slab, and therefore $\frac{2}{3}\delta_{cs}$ is the mean slab deflection along a positive-moment yield line. The ratio of experimental ultimate load to theoretical ultimate load estimated by this procedure varied between 1.02 and 1.17. For the slabs with deflecting beams, a different equilibrium equation from Eq. 12.24 was required because of the different failure mechanism, and very good agreement between experimental and theoretical ultimate loads were also obtained for these slabs. These test results indicate clearly that the ultimate load of a panel can be estimated with very good accuracy if the membrane forces acting on the panel are known.

Ghoneim and MacGregor[12.30–12.32] reported the results of tests of 19 simply supported square or rectangular slabs, many of which were subjected to concentric in-plane compression forces in one or both directions. In all cases without in-plane compression forces acting, the distributed loads at failure were significantly larger than the computed yield line (Johansen) load as a result of tension membrane forces that developed at large deflections. The short span was 1829 mm (72.0 in.) in all cases, and thicknesses ranged from 64.7 to 70.1 mm (2.55 to 2.76 in.) except for two square slabs which were

92.8 and 92.7 mm (3.65 in.) thick. The short span/thickness ratios were thus about 28 or about 19.7. The ratio of long span/short span was 2.33 (four slabs) or 1.50 (four slabs) for the rectangular cases. (There is a conflict between Refs. 12.30 and 12.31 about the ratio of long span/short span for the slabs noted here as having this ratio at 2.33, but it is believed that 2.33 is the correct value.)

In every case, application of in-plane compression forces to the thinner slabs resulted in a reduction in the ultimate load as compared with the slabs without these forces. The effects of deflection (P–Δ effects) outweighed the benefits of adding a compression force to the cross section. The thicker slab, with span/thickness = 19.7, gained 11% load capacity when the in-plane force was added.

The slabs with in-plane compression forces were appreciably stiffer than those without, as a result of significant delays in the onset of cracking. In most cases the in-plane compression stress was in the range 0.36 to $0.4f'_c$, with three cases ranging down to $0.21f'_c$. The higher stresses corresponded to thrusts slightly higher than the balanced thrust, from the moment–thrust interaction diagram. In all cases, applying the in-plane compression forces greatly reduced the deflection at maximum load, with the extent of the reduction depending on panel shape and on the magnitude and direction(s) of the in-plane forces.

Reference 12.32 has an analysis of the slabs that were tested, with good agreement reported between the test and analytical results. Although the tests are interesting and useful in aiding the understanding of slab behavior, they are not closely related to the main topic of this chapter because the in-plane forces were concentrically applied rather than developing or being applied with a significant eccentricity.

Experimental Results from a Slab-and-Beam Floor Designed to Incorporate Compressive Membrane Action. Tests on multipanel reinforced concrete slab-and-beam floors by Ockleston,[12.1] Liebenberg,[12.2] Gamble et al.,[12.3] and others have demonstrated the enhancement of the ultimate load of interior panels due to membrane action when subjected to uniform loading when the surrounding panels were lightly loaded. In no case, however, was membrane action allowed for in the design of the floor. A $\frac{1}{4}$-scale nine-panel (three by three) reinforced concrete slab-and-beam floor was designed to incorporate membrane action and tested to check the serviceability and ultimate load.[12.24] The equations derived by Park[12.9] were used to determine the ultimate uniform load of the panels. The design ultimate load the floor was 800 lb/ft^2 (38.3 kN/m^2) and the design required an enhancement factor (ultimate load/Johansen load) of 2.00 for the interior panel, 1.35 for the center edge panels, and 1.00 for the corner panels. The panels were lightly reinforced, the top and bottom steel of all panels being 0.16% and 0.15% of the gross concrete area, respectively; the steel had a yield strength of 52 kips/in^2 (359 N/mm^2). All panels were 1.94 in. (49.2 mm) thick and the spans between beam centers were 66 in. × 66 in. (1.68 m × 1.68 m) for the interior panel, 66 in. × 48 in. (1.68 m × 1.22 m) for the center edge panels, and 48 in. × 48 in. (1.22 m × 1.22 m) for the corner panels. The floor was supported at the beam junctions. From the known steel quantities of the panels and the required ultimate load the maximum allowable lateral movement at the panel edges was estimated and the membrane forces in the panels at ultimate load were calculated. These membrane forces were then considered to be uniformly distributed outward in-plane forces acting on the surrounding beams and panels, and the beams were designed for the required strength for bending due to gravity loads and tensions due to the membrane forces. The lateral deformations of the floor due to axial stretch of the beams under tension, the bending and shear deformations of the edge panels under membrane forces, and the axial shortening of interior and center edge panels under membrane forces were estimated and the beam sizes and reinforcement adjusted until the outward movement of the panel boundaries was approximately equal to the maximum allowed. This calculation necessitated some iterations and approximations. For example, the outward movement of panel edges was calculated on the basis of uncracked sections without allowance for creep, and then for the interior panel was increased by a factor of 4 to allow for loss of stiffness due to cracking and creep. A full account of the design may be seen elsewhere.[12.33] During the test loading, a range of patterns of uniformly distributed loads were applied in the service load range, and then the floor was loaded to failure with equal uniform loading on all panels. Subsequent to failure of the interior panel, the center edge and corner panels were loaded to failure.

The behavior of the floor up to a uniform load of 375 lb/ft² (18.0 kN/m²) applied over the whole floor was entirely satisfactory. However, loading up to 450 lb/ft² (21.5 kN/m²) led to a significant increase in deflections and cracking. The cracking present in the bottom of the floor at a load of 450 lb/ft² (21.5 kN/m²) is shown in Fig. 12.17. The crack widths shown in the figure were those measured when the load was reduced to 375 lb/ft² (18.0 kN/m²). When scaled up to obtain the prototype crack widths, they would be unacceptable for a service load of this magnitude. The load–central deflection relationship for the interior panel is shown in Fig. 12.18. Failure of the floor occurred in the interior panel at 850 lb/ft² (40.7 kN/m²), when the central deflection of the interior panel was almost equal to the panel thickness. This ultimate load was 6% greater than the design ultimate load. At this stage the yield line patterns in the center edge and corner panels were well advanced. The high steel strains measured at the midspan and support sections of the interior and exterior beams at ultimate load was evidence that the additional longitudinal steel placed in the center spans of the beams had been utilized in developing the strength of the floor. In later tests on the outer panels, with the load on the interior panel kept constant at 600 lb/ft² (28.7 kN/m²), the center edge panels failed at 966 lb/ft² (46.2 kN/m²) and the corner panels at 1170 lb/ft² (56.0 kN/m²), there being evidence of membrane enhancement

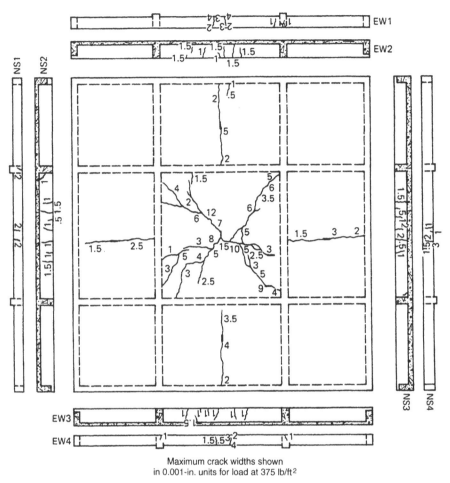

Figure 12.17 Crack pattern of slab and beam floor at 450 lb/ft² (21.5 kN/m²).[12.24]

in both cases. Figure 12.19 shows top and bottom views of the floor after the test.

The design method was necessarily a simplification, but it did provide a reasonable assessment of panel compressions and beam tensions due to membrane action. The test illustrated that design allowing for membrane action is possible. However, the magnitude of crack widths and deflections under high sustained service loads must be cause for concern. The doubt concerning the long-term behavior of the floor, and the requirements that the design loads must be high before membrane action can be fully exploited, would appear to limit the applicability of membrane action in design to relatively thick, heavily loaded slabs with reliable lateral restraint.

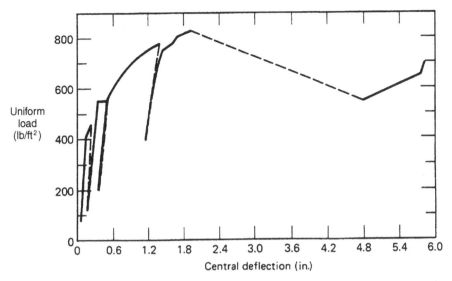

Figure 12.18 Load–central deflection relationship measured for interior panel (1 lb/ ft² = 47.9 N/m², 1 in. = 25.4 mm).[12.24]

12.2.3 Behavior in the Tensile Membrane Range

Near the end of the compressive membrane action range, in the central region of the slab, the large stretch of the slab surface causes the cracks there to penetrate the entire thickness of the slab depth, and the load then is carried mainly by the reinforcing bars acting as a tensile membrane. For a slab with a very stiff surround, as point C in Fig. 12.1 is approached the membrane forces change from compression to tension in the central region of the slab and the boundary restraints begin to resist inward movement of the slab edges. Initially, the outer regions of the slab will act with the surround as part of the compression ring supporting the tension membrane action in the inner region of the slab. With further deflection beyond point C, the region of tensile membrane action gradually spreads throughout the slab and the load carried by the yielding reinforcement acting as a tensile membrane (with full-depth concrete cracking), like a soap bubble, increases until the steel starts to fracture at point D. For a slab with a less stiff surround that reaches ultimate load (point B in Fig. 12.1) at a large deflection, some tensile membrane action will have developed by the stage point B is reached, in which case the portion BC is less steep, and further development of tensile membrane action develops rapidly after point B. Knowledge of the region CD of Fig. 12.1 is important because as soon as the ultimate load of the slab is reached in the practical case of gravity loading (which once applied remains unchanged as the slab deflects), the load will drop suddenly through the slab unless the tensile membrane strength of the reinforcement is great enough to "catch" the load.

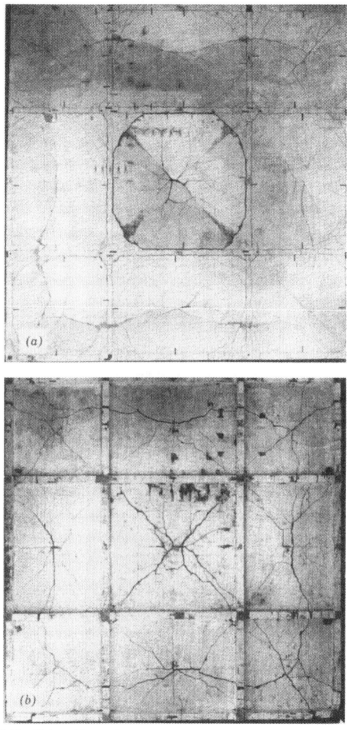

Figure 12.19 Slab-and-beam floor after testing[12.24]: (*a*) loaded surface after testing; (*b*) unloaded surface after testing.

Tests by Powell,[12.4] Wood,[12.5] Park,[12.11] Brotchie and Holley,[12.14] Keenan,[12.16] Black,[12.21] and others have indicated that for heavily reinforced slabs the load at point D in Fig. 12.1 can significantly exceed the ultimate load at point B. Therefore, in many cases tensile membrane action provides a useful means of preventing catastrophic failure.

It is to be noted that significant tensile membrane action is possible only for a slab on beams or other stiff supports that provide edge restraint. Tensile membrane action cannot be a significant feature of the large-deflection behavior of beamless slabs, except for pattern loading and in connection with a suspension system that may occur after shear failure around a column if there is adequate, well-anchored bottom steel over the column.

Analysis of Tensile Membrane Behavior. The general case of a slab fixed against all translation at the edges and containing different amounts of steel in the x- and y-directions can be analyzed assuming that at the stage when the load carried by tensile membrane action becomes important: (1) all the concrete has cracked throughout its depth and hence is incapable of carrying any load, (2) all the reinforcement has reached the yield strength and hence acts as a plastic membrane, (3) no strain hardening of steel occurs, and (4) only the reinforcement that extends over the entire area of the slab contributes to the membrane.

In the general case of orthotropic reinforcement, the yield force of the reinforcement per unit width will be different in the x- and y-directions (i.e., $T_x \neq T_y$). Consider the small rectangular element with sides of length dx and dy carrying uniformly distributed load per unit area of w in Fig. 12.20. There are no shear forces acting on the element since the reinforcement mesh has no shear resistance. Equilibrium of forces in the z-direction requires that

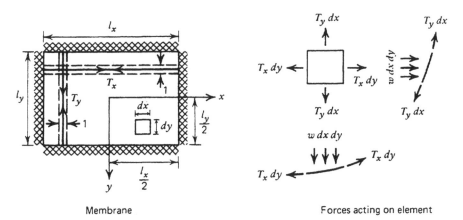

Membrane   Forces acting on element

Figure 12.20   Uniformly loaded plastic tensile membrane.

$$0 = w \, dx \, dy - T_x dy \frac{\partial z}{\partial x} + T_x dy \left(\frac{\partial z}{\partial x} + \frac{\partial^2 z}{\partial x^2} \, dx \right)$$

$$- T_y dx \frac{\partial z}{\partial y} + T_y dx \left(\frac{\partial z}{\partial y} + \frac{\partial^2 z}{\partial y^2} \, dy \right)$$

from which

$$\frac{T_x}{T_y} \frac{\partial^2 z}{\partial x^2} + \frac{\partial^2 z}{\partial y^2} = \frac{w}{T_y} \tag{12.25}$$

To reduce this equation to the standard plastic membrane theory equation with equal yield forces per unit width in each direction, transform the x-axis by substituting

$$X = x \sqrt{\frac{T_y}{T_x}} \tag{12.26}$$

Then it may be shown that

$$\frac{\partial^2 z}{\partial X^2} + \frac{\partial^2 z}{\partial y^2} = - \frac{w}{T_y} \tag{12.27}$$

Thus, the actual membrane of Fig. 12.20 is replaced by the equivalent simple membrane of Fig. 12.21, which has the same uniform force per unit width T_y in each direction and is governed by Eq. 12.27. Standard membrane theory[12.11,12.34] can then be used to solve Eq. 12.27, and the following relationship

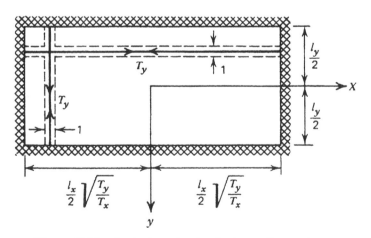

Figure 12.21 Equivalent simple plastic tensile membrane.

between uniform load per unit area w and the central deflection of the plastic tensile membrane δ results.

$$\frac{wl_y^2}{T_y\delta} = \frac{\pi^3}{4\displaystyle\sum_{n=1,3,5,\ldots}^{\infty}\frac{1}{n^3}(-1)^{(n-1)/2}\left\{1 - \dfrac{1}{\cosh[(n\pi l_x/2l_y)\sqrt{T_y/T_x}]}\right\}} \qquad (12.28)$$

Equation 12.28 gives a linear relationship between w and δ and is an approximation for the portion of the load–deflection curve between C and D of Fig. 12.1. The limiting point D will depend on the ductility of the steel. Figure 12.22 shows Eq. 12.28 plotted for various l_x/l_y and T_y/T_x values. Only the practical case of $T_y \geq T_x$ for slabs with $l_x \geq l_y$ has been plotted, since economy of steel will always require more steel per unit width in the direction of the short span than in the direction of the long span. The curves illustrate that as the l_x/l_y ratio increases, the load carried by the long span rapidly becomes negligible and the slab approaches the one-way case. The yield forces per unit width in Eq. 12.28 have been taken as T_x and T_y because it is usual for the bottom steel to be placed over the entire area of the panel, and sufficient lap with top steel normally occurs around the panel edges to provide adequate anchorage. Should the bottom steel be anchored adequately into the surround and the top steel extend over the entire area of the panel, T_x and T_y could be replaced by $T_x + T_x'$ and $T_y + T_y'$, respectively. Note also that the bars must be effectively anchored at the supports to carry the yield force.

Experimental Results. If laterally restrained slabs are loaded uniformly by bags filled with water under pressure, the falling and rising branches of the load–deflection curve after ultimate load (the region BCD of Fig. 12.1) can be followed in tests. Figure 12.23 shows a slab (slab A3 of Table 12.2) at the end of its test run at the completion of the tensile membrane stage when the bars begin to fracture.[12.11] A typical load–deflection relationship (for slab A4 of Table 12.2) is shown in Fig. 12.24; for this slab the top steel existed only around the edges and that steel has not been included in the assumed plastic membrane. On the whole, comparison of theoretical curves given by Eq. 12.28 and experimental curves for portion CD has shown the theory to be conservative.[12.11] This was evidently because a pure plastic tensile membrane did not develop over the entire slab. For the lightly reinforced slabs particularly, the cracking that was present at the end of the test runs was not a great deal more extensive than the cracking that was present when the yield line pattern developed, and thus at very high deflections the load was in fact carried by a stronger combined bending and tensile membrane action. The assumption of no strain hardening of steel also makes the theory conservative. Thus, Eq. 12.28 gives a lower limit to the load carried by tensile membrane action. It was concluded from the test results of Powell[12.4] and Park[12.11] that a safe

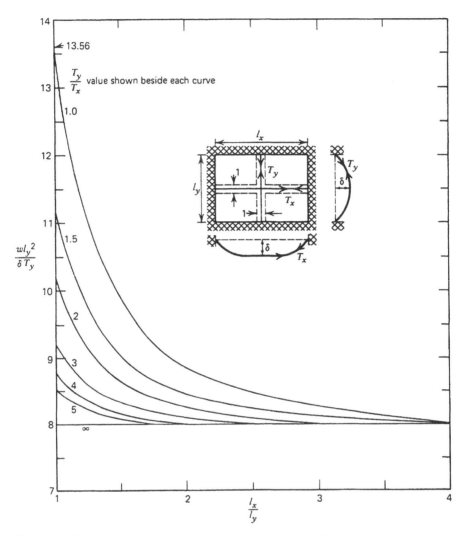

Figure 12.22   Load–central deflection relationships for uniformly loaded rectangular plastic tensile membranes.[12.11]

value for the central deflection after tensile membrane action could be taken to be 0.1 of the short span. Any greater deflection was likely to cause bar fracture. Thus, the ultimate load at the end of the tensile membrane range (at point D in Fig. 12.1) can be calculated from Eq. 12.28 with $\delta = 0.1 l_y$ substituted.

Tests by Keenan[12.16] and Black[12.21] on square slabs have confirmed that $0.1 \times$ span gives a safe maximum value for the central deflection after tensile membrane action. In fact, Black found that fracture of reinforcing steel oc-

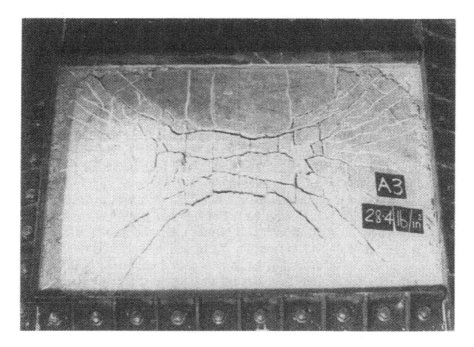

Figure 12.23 Crack pattern at unloaded face of slab A3 at end of tensile membrane stage.[12.11]

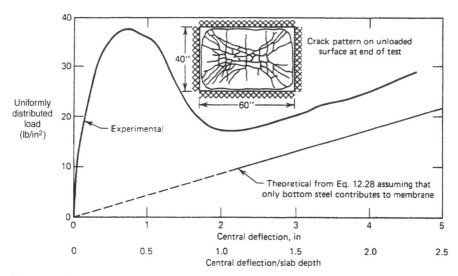

Figure 12.24 Load–deflection curves and cracking for slab A4 (1 lb/in² = 6.89 kN/m², 1 in. = 25.4 mm).[12.11]

curred at 0.14 or 0.15 times the span in his tests. To make Eq. 12.28 less conservative, it was suggested by Keenan and Black that for square slabs the value of $wl^2/\delta T_y = 13.56$ given theoretically by Eq. 12.28 (see Fig. 12.22) could be taken as $wl^2/\delta T_y = 20$.

Significance of Tensile Membrane Action. It is of interest to compare the theoretical tensile membrane load at the suggested maximum deflection with the theoretical ultimate load in the compression membrane range. Consider a square slab with fully restrained edges that is reinforced by steel with a yield strength of 60,000 lb/in² (414 N/mm²) placed in both directions in the top around the edges only and in the bottom over the whole area of the slab. Let the areas of steel top and bottom be the same and the distances between the centroids of top and bottom steel be 0.7 of the slab depth. Let the concrete cylinder strength be 4000 lb/in² (27.6 N/mm²). For the theoretical ultimate load (at point *B* of Fig. 12.1) from Eq. 22.16 with $\varepsilon'_x = \varepsilon'_y = 0$ to equal the tensile membrane load from Eq. 12.28 at a central deflection of 0.1 of the span with only the bottom steel contributing to the tensile membrane, the steel ratio for both the top and the bottom steel required in each direction for a span/depth ratio of 40 is 0.73%; for a span/depth ratio of 30 this figure increases to 1.29%. In fact, the steel percentages actually required will be less than these because, as has already been noted, Eq. 12.28 is conservative and also because the actual ultimate load will be smaller than that given by Eq. 12.16 with $\varepsilon'_x = \varepsilon'_y = 0$ because some axial strains in the slab and lateral boundary displacements will actually occur. It is evident that many reinforced concrete slabs will have a load-carrying capacity due to tensile membrane action that will exceed the ultimate load reached in the compression membrane range.

In a slab-and-beam floor when tensile membrane action develops in the panels at large deflections, the beams will need to carry compression to provide the reactive forces for the tensile forces in the panels. If the supporting system is not very stiff, tensile membrane forces can still develop in the central region of the panel, since the outer regions of the panel can act as a compression ring to provide the reactive forces. The load–deflection curve of Fig. 12.18 shows that tensile membrane action was capable of arresting the decrease in the load-carrying capacity of the interior panel of the slab-and-beam floor at large deflections during the test. Figure 12.19*a* shows considerable concrete crushing near the corners of the interior panel across the diagonals, indicating large compressive membrane forces there, and wide full-depth cracking in the central region of the panel due to tensile membrane forces there. It is also possible that if a panel is loaded several times to successively higher loads, and with different combinations of panels loaded, it may not develop much compressive membrane effect because of residual deformations and a reduction in the surround stiffness. So much residual deformation may accumulate that the panel may slip past the compressive membrane stage into the tensile membrane stage. Slab T-1 of Ref. 12.3

seemed to exhibit the effects of residual deformation from various combinations of pattern loading, in that no "snap through" of the interior panel was observed during the tests.

12.3 CONCENTRATED LOADS ON LATERALLY RESTRAINED REINFORCED CONCRETE SLABS

Tests and analytical studies of laterally restrained reinforced concrete slabs subjected to concentrated loads have been reported by Taylor,[12.35] Aoki and Seki,[12.36] Batchelor et al.,[12.37,12.38] and others. These studies have shown that compressive membrane stresses induced by lateral restraint at the slab boundaries as the slab deflects will increase the ultimate concentrated load when failure occurs in a flexural mode with the formation of a yield line pattern. Lateral restraint has also been found to result in an increase in the ultimate concentrated load when failure occurs in a shear mode due to punching. Semiempirical approaches for the determination of the ultimate concentrated load of laterally restrained panels have been developed in the studies noted above. Design codes for bridge decks could now include an allowance for the enhancement in strength due to compressive membrane action.

12.4 SLABS WITH EDGES FREE TO MOVE LATERALLY

Even if the support conditions of slabs are such that lateral movement at the edges can occur freely, such as the case of slabs with edges simply supported on rollers, the geometry of deformations as the slab deflects will result in the development of some membrane forces in the slab. This occurs in uniformly loaded slabs at relatively large deflections when the slab regions at the edges tend to move inward but are restrained from doing so by the adjacent outer regions. The result is an outer ring of compression supporting tensile membrane forces in the inner (central) region of the slab.

Taylor[12.39] has conducted a preliminary analytical study of the effect of membrane action on slabs with edges free to move laterally. Figure 12.25 illustrates a uniformly loaded simply supported square slab at large deflections after the yield line pattern has formed. In the central region of the slab, the cracks will have penetrated through the full depth of the slab. If the equilibrium of segment *A* is considered, it is evident that the total tension in the reinforcement shown in the elevation must be balanced by the total compression in the compression zones at each end of the segment. That is, these forces must balance taking the yield line overall, and will not be in balance along each part of the yield line as is assumed in Johansen's yield line theory. In other words, the lower region of the yield line is in tension and the upper outer region will be in compression. Hence, the effect of the change of geometry due to deflections is simply to increase the effective lever arm of the

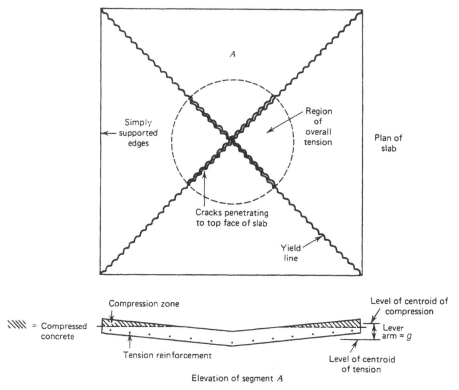

Figure 12.25 Tensile membrane action in uniformly loaded slab without lateral restraint at slab edges.[12.39]

internal forces. For a square slab of side l carrying a uniformly distributed load per unit area w, if the equilibrium of segment A is considered by taking moments about the support line, the following equation can be written for the slab at a particular deflection:

$$\frac{wl^2}{24} = g \, \Sigma \, T \tag{12.29}$$

where g is the distance from centroid of steel to centroid of concrete compression and $\Sigma \, T$ is the total tensile force in reinforcing steel. Taylor found that load–deflection characteristics obtained from such calculations agree fairly well with actual load–deflection curves measured for simply supported slabs after development of the yield line pattern. For the most part, the measured curves give a higher load-carrying capacity, due largely to strain hardening of the reinforcement at large deflections.

12.5   RECENT COMPUTATIONAL APPROACHES

Meamarian et al.[12.40] derived an equation similar to Eq. 12.11, but with some additional terms to account for the possible presence of prestressed reinforcement. They also developed a computer program to complete the iterative solutions often necessary with this equation. They compared the results of analysis with results of tests of 16 slabs conducted at the U.S. Army Waterways Experiment Station and reported by Guice.[12.41] These one-way slabs had spans of 24 in. (610 mm) and had span thickness ratios of 10.4 or 14.8, and all contained shear reinforcement. The ratios of computed/measured values of maximum load, and axial force, moment deflection at maximum load, were fairly close to 1 for the average of all 16 tests, but the scatter was large. The agreement on maximum load was best, and was worst for axial force and deflection. The largest measured deflections were underestimated by the analyses.

At least two investigations into the behavior of restrained slabs have used finite element analysis methods. Famiyesin and Hossain[12.42,12.43] and Welch[12.44] both used nonlinear material and geometric models to explore various aspects of behavior. Famiyesin and Hossain used layered shell elements to model the thickness of slabs, using 10 layers to represent the concrete and four to represent the reinforcement. Nonlinear stress–strain curves were adopted for both concrete and steel, with the effects of biaxial stresses considered for the concrete. The tensile stress–strain curve for concrete included a tension stiffening component so that the loss of tension capacity was gradual as cracking progressed. Their analysis was tested by comparison with various test results already referred to, including tests by Powell,[12.4] Park,[12.8] Keenan,[12.16] Hung and Nawy[12.19] and others. All the graphical comparisons of measured and computed load–deflection curves were with Powell's data.

The material properties used in the analyses included the measured values of Young's modulus, and the initial slopes of the computed load–deflection curves matched the measured slopes. However, as the loads increased, the computed deflections soon were too small and remained so until the peak load in the compression membrane phase had been reached. There may be several reasons for this difference, including at least: (1) The compression stress–strain curve for the concrete did not have a descending branch and/or its stiffness decayed too slowly as the strain increased; (2) the concrete did not creep; (3) tension stiffening may been too large; and (4) the restraint against lateral displacement was absolute, while the experimental slabs must have had at least some displacement.

Analyses were done using both *load control,* in which the load was increased in small steps, and *deflection control,* in which small deflection increments were imposed. The deflection control mode led to load–deflection curves of the same general nature as shown in Fig. 12.1, but as noted previously, the peak loads were reached a deflections considerably smaller than the measured deflections, and the tension membrane phase of behavior was much stiffer than the measured values. The deflection control calculations were very computational intensive compared to load control calculations. The load control calculations could not lead to the decrease in load shown by the range $B-C$ in Fig. 12.1 but rather, reached a point comparable to point B, and then the curve went nearly horizontally to intersect the line $C-D$. The distinction between the two papers is that the first deals with slabs that are clamped against rotation and deflection on all four sides, while the second deals with slabs clamped on three sides and simply supported on the fourth.

Welch[12.44] conducted an extensive investigation of the effects of restraint of one-way slabs using the finite element program ABAQUS. He considered slabs over a wide range of span/thickness ratios and found $l/t = 18$ to be a useful dividing line between "thick" and "thin" slabs when comparing analytical and test results. Both material and geometric nonlinearities were considered in the analyses, and both beam and continuum elements were used in various trials. Only the beam elements produced meaningful results, where nine layers of elements were used to represent the member thickness. It was found that contrary to the usual experience with finite element analyses, increasing the number of elements in a span did not necessarily improve the results but rather, in many cases led to an earlier termination of the analysis due to numerical convergence problems.

Reasonable results were obtained for some thin slabs. The predicted loads were too high and the predicted deflections were too low, often much too low, for the thick slabs. The slabs were modeled as fully restrained in the axial direction, which clearly was too much restraint for the thick slabs. The addition of longitudinal and rotational springs at the supports to try to make the restraints more nearly like those in the physical tests introduced numerical instabilities which often led to early termination of the analyses as compared to the fully restrained cases.

Welch also derived an equation similar to Eq. 12.11 but made slightly different assumptions about the axial strains in the slab segments. An equation equivalent to Eq. 12.14 was also derived, and the different assumptions led to a formulation that is considerably more complex than Eq. 12.14. For thick slabs, the two analyses give nearly identical results. For thin slabs, Welch's analysis gives larger peak loads that occur at smaller deflections, with the differences increasing as the slabs become thinner. This was an unexpected result, since the change in the analysis should have increased the axial flexibility, or decreased the axial stiffness, of the slab, and this was expected to lead to slightly smaller peak loads and larger deflections.

Welch analyzed test data from more than 100 one-way restrained slab tests, using both his modified analysis and the analysis presented in this chapter. He found it helpful to divide the slabs into three classes (rather than the two divisions used earlier), based on their span/depth ratios. He called $l/t < 18$ thick slabs, $18 < l/t < 22$ thin slabs, and $l/t > 22$ very thin slabs. While various investigators have used the equations developed in this chapter, in conjunction with an estimate of the deflection (often about slab thickness/2) at peak load, Welch had greater success correlating the computed and measured peak loads when he computed the load corresponding to the maximum thrust in the slab. He referred to this as a *thrust-indexed* load calculation, as opposed to the earlier efforts using *deflection-indexed* load calculations.

Various methods of computing deflections corresponding to the peak load were tried. The agreement between test and calculation varied widely but was generally better for thick slabs than for thin slabs. Part of the difference between the thickness cases was attributed to the fact that the thicker slabs were generally governed by material strength properties, while the thinner slabs were often governed by geometric instability, that is, by snap-through.

Welch also considered the tensile membrane phase of behavior and again found it useful to consider different ranges of span/thickness ratios to define different behaviors, including the problem of whether the ultimate stress or yield stress was most appropriate in computing the maximum load.

REFERENCES

12.1   A. J. Ockleston, "Arching Action in Reinforced Concrete Slabs," *Struct. Eng.,* Vol. 36, No. 6, June 1958, pp. 197–201. See also *Loading Tests on Reinforced Concrete Slabs Spanning in Two Directions,* Paper 6, Portland Cement Institute, Johannesburg, South Africa, October 1958.

12.2   A. C. Liebenberg, "Arch Action in Reinforced Concrete Slabs," *Proceedings of the 1963 Diamond Jubilee Convention of the South African Institution of Civil Engineers,* 1963, pp. 99–102. See also *Arch Action in Concrete Slabs,* Research Report 234, C.S.I.R., Johannesburg, South Africa, 1963.

12.3   W. L. Gamble, M. A. Sozen, and C. P. Seiss, "Tests of a Two-Way Reinforced Concrete Slab," *J. Struct. Div., ASCE,* Vol. 95, No. ST6, June 1969, pp. 1073–1096.

12.4   D. S. Powell, "Ultimate Strength of Concrete Panels Subjected to Uniformly Distributed Loads," M.Sc. thesis, Cambridge University, 1956.

12.5   R. H. Wood, *Plastic and Elastic Design of Slabs and Plates,* Thames and Hudson, London, 1961, pp. 225–261.

12.6   K. P. Christiansen, "The Effect of Membrane Stresses on the Ultimate Strength of the Interior Panel in a Reinforced Concrete Slab," *Struct. Eng.,* Vol. 41, No. 8, August 1963, pp. 261–265.

12.7   J. Schlaich, "Die Gewolbewirkung in durchlaufenden Stahlbetonplatten (Vault Action in Continuous Reinforced Concrete Slabs)," Thesis, Technische Hochschule, Stuttgart, 1963.

12.8   R. Park, "Ultimate Strength of Rectangular Concrete Slabs Under Short-Term Uniform Loading with Edges Restrained against Lateral Movement," *Proc. Inst. Civ. Eng.,* Vol. 28, June 1964, pp. 125–150.

12.9   R. Park, "The Ultimate Strength and Long-Term Behaviour of Uniformly Loaded Two-Way Concrete Slabs with Partial Lateral Restraint at All Edges," *Mag. Concr. Res.,* Vol. 16, No. 48, September 1964, pp. 139–152.

12.10   R. Park, "The Lateral Stiffness and Strength Required to Ensure Membrane Action at the Ultimate Load of a Reinforced Concrete Slab and Beam Floor," *Mag. Concr. Res.,* Vol. 17, No. 50, March 1965, pp. 29–38.

12.11   R. Park, "Tensile Membrane Behaviour of Uniformly Loaded Rectangular Reinforced Concrete Slabs with Fully Restrained Edges," *Mag. Concr. Res.,* Vol. 16, No. 46, March 1964, pp. 39–44.

12.12   A. Sawczuk, "Membrane Action in Flexure of Rectangular Plates with Restrained Edges," *Flexural Mechanics of Reinforced Concrete,* ASCE-1965-50, ACI SP-12, ASCE-ACI, Detroit, November 1964, pp. 347–358.

12.13   J. F. Brotchie, A. Jacobson, and S. Okubo, *Effect of Membrane Action on Slab Behaviour,* Research Report R65-25, Department of Civil Engineering, Massachusetts Institute of Technology, Cambridge, Mass., 1965, 101 pp.

12.14   J. F. Brotchie and M. J. Holley, "Membrane Action in Slabs," *Cracking, Deflection, and Ultimate Load of Concrete Slab Systems,* ACI SP-30, American Concrete Institute, Detroit, 1971, pp. 345–377.

12.15   C. F. Millington, *Comparison of the Ultimate Load Carrying Capacity of Laterally Restrained/Unrestrained, Reinforced/Unreinforced Cement Mortar Model Beams and Two Way Panels,* Atomic Weapons Research Establishment Report 0-87/65, United Kingdom Atomic Energy Authority, Aldermaston, Berkshire, England, January 1966, 158 pp.

12.16   W. A. Keenan, *Strength and Behavior of Restrained Reinforced Concrete Slabs Under Static and Dynamic Loadings,* Technical Report R621, U.S. Naval Civil Engineering Laboratory, Port Hueneme, Calif., April 1969.

12.17   E. H. Roberts, "Load-Carrying Capacity of Strips Restrained Against Longitudinal Expansion," *Concrete,* Vol. 3, No. 9, September 1969, pp. 369–378.

12.18   A. G. Girolami, M. A. Sozen, and W. L. Gamble, "Flexural Strength of Reinforced Concrete Slabs with Externally Applied In-Plane Forces," and W. L. Gamble, H. Flug, and M. A. Sozen, "Strength of Slabs Subjected to Multiaxial Bending and Compression," *Civil Engineering Studies,* Structural Research Series 369, Department of Civil Engineering, University of Illinois, Urbana, Ill., 1970.

12.19   T. Y. Hung and E. G. Nawy, "Limit Strength and Serviceability Factor in Uniformly Loaded, Isotropically Reinforced Two-Way Slabs," in *Cracking, Deflection and Ultimate Load of Concrete Slab Systems,* ACI SP-30, American Concrete Institute, Detroit, 1971, pp. 301–324.

12.20   D. C. Hopkins and R. Park, "Compressive Membrane Action in a Laterally Restrained Circular Reinforced Concrete Slab," *Proceedings of the 3rd Australasian Conference on the Mechanics of Structures and Materials,* Vol. 1, Session A2, Auckland, New Zealand, August 1971.

12.21   M. S. Black, "Ultimate Strength of Two-Way Concrete Slabs," *J. Struct. Div., ASCE,* Vol. 101, No. ST1, January 1975, pp. 311–324.

12.22   P. Desayi and A. B. Kulkarni, "Load-Deflection Behaviour of Restrained R/C Slabs," *J. Struct. Div., ASCE,* Vol. 103, No. ST2, February 1977, pp. 405–419.

12.23 M. E. Criswell, *Design and Testing of a Blast-Resistant Reinforced Concrete Slab System,* Technical Report N-72-10, Weapons Effects Laboratory, U.S. Army Engineer Waterways Experiment Station, Vicksburg, Miss., 1972, 312 pp.

12.24 D. C. Hopkins and R. Park, "Test on a Reinforced Concrete Slab and Beam Floor Designed with Allowance for Membrane Action," in *Cracking, Deflection and Ultimate Load of Concrete Slab Systems,* ACI SP-30, American Concrete Institute, Detroit, 1971, pp. 223–250.

12.25 S. C. Woodson, *Effects of Shear Reinforcement on the Large-Deflection Behavior of Reinforced Concrete Slabs,* Technical Report SL-94-18, U.S. Army Corps of Engineers, Waterways Experiment Station, Vicksburg, Miss., September 1994, 319 pp. Also submitted as Ph.D. thesis, University of Illinois at Urbana–Champaign, Urbana, Ill., 1992, 319 pp.

12.26 *Building Code Requirements for Structural Concrete,* ACI 318-95 and *Commentary,* ACI 318R-95, American Concrete Institute, Farmington Hills, Mich., 1995, 371 pp.

12.27 R. Park, "The Ultimate Strength of Uniformly Loaded Laterally Restrained Rectangular Two-Way Concrete Slabs," Ph.D. thesis, University of Bristol, 1964, 294 pp.

12.28 D. C. Hopkins and R. Park, Closure on discussion on "Test of a Reinforced Concrete Slab and Beam Floor Designed with Allowance for Membrane Action," *Proc. ACI,* Vol. 70, January 1973, pp. 68–69.

12.29 E. Hognestad, N. W. Hanson, and D. McHenry, "Concrete Stress Distribution in Ultimate Strength Design," *Proc. ACI,* Vol. 52, December 1955, pp. 455–479.

12.30 M. G. Ghoneim and J. G. MacGregor, "Tests of Reinforced Concrete Plates Under Combined Inplane and Lateral Loads," *ACI Struct. J.,* Vol. 91, No. 1, January–February 1994, pp. 19–30.

12.31 M. G. Ghoneim and J. G. MacGregor, "Behavior of Reinforced Concrete Plates Under Combined Inplane and Lateral Loads," *ACI Struct. J.,* Vol. 91, No. 2, March–April 1994, pp. 188–197.

12.32 M. G. Ghoneim and J. G. MacGregor, "Prediction of the Ultimate Strength of Reinforced Concrete Plates Under Combined Inplane and Lateral Loads," *ACI Struct. J.,* Vol. 91, No. 6, November–December 1994, pp. 688–696.

12.33 D. C. Hopkins, "Effects of Membrane Action on the Ultimate Strength of Reinforced Concrete Slabs," Ph.D. thesis, University of Canterbury, 1969, 358 pp.

12.34 S. Timoshenko and J. N. Goodier, *Theory of Elasticity,* 2nd ed., McGraw-Hill, New York, 1951, 506 pp.

12.35 R. Taylor, "Some Tests on the Effect of Edge Restraint on Punching Shear in Reinforced Concrete Slabs," *Mag. Concr. Res.,* Vol. 17, No. 50, March 1965, pp. 39–44.

12.36 Y. Aoki and H. Seki, "Shearing Strength and Cracking in Two-Way Slabs Subjected to Concentrated Load," in *Cracking, Deflection, and Ultimate Load of Concrete Slab Systems,* ACI SP-30, American Concrete Institute, Detroit, 1971, pp. 103–126.

12.37 P. Y. Tong and B. deV. Batchelor, "Compressive Membrane Enhancement in Two-Way Bridge Slabs," in *Cracking, Deflection and Ultimate Load of Concrete Slab Systems,* ACI SP-30, American Concrete Institute, Detroit, 1971, pp. 271–286.

12.38 B. deV. Batchelor and I. R. Tissington, "Shear Strength of Two-Way Bridge Slabs," *J. Struct. Div., ASCE,* Vol. 102, No. ST12, December 1976, pp. 2315–2331.

12.39 R. Taylor, "A Note on a Possible Basis for a New Method of Ultimate Load Design of Reinforced Concrete Slabs," *Mag. Concr. Res.,* Vol. 17, No. 53, December 1965, pp. 183–186.

12.40 N. Meamarian, T. Krauthammer, and J. O'Fallon, "Analysis and Design of Laterally Restrained Structural Concrete One-Way Members," *ACI Struct. J.,* Vol. 91, No. 6, November–December 1994, pp. 719–725.

12.41 L. K. Guice, *Behavior of Partially Restrained Reinforced Concrete Slabs,* Technical Report SL-86-32, Structures Laboratory, U.S. Army Waterways Experiment Station, Vicksburg, Miss., 184 pp.

12.42 O. O. R. Famiyesin and K. M. A. Hossain, "Optimized Design Charts for Fully Restrained Slabs by FE Predictions," *J. Struct. Eng., ASCE,* Vol. 124, No. 5, May 1998, pp. 560–569.

12.43 O. O. R. Famiyesin and K. M. A. Hossain, "Development of Charts for Partially Clamped Slabs by Finite-Element Predictions," *J. Struct. Eng., ASCE,* Vol. 124, No. 11, November 1998, pp. 1339–1349.

12.44 R. W. Welch, "Compressive Membrane Capacity Estimates in Laterally Edge Restrained Reinforced Concrete One-Way Slabs," Ph.D. thesis, University of Illinois at Urbana–Champaign, 1999, 422 p.

13 FIRE RESISTANCE OF REINFORCED CONCRETE SLABS

13.1 INTRODUCTION

Most buildings are subject to fire resistance requirements in addition to the strength and serviceability requirements that are given in the ACI and other material-specific codes. The required fire resistance or fire endurance is given in the general building code governing a specific geographic or political area. The required endurance is a function of at least the intended use of the building and its size, where the size includes height and floor area or areas of compartments that are separated by adequate barriers. Additional factors include the presence or absence of installed fire detection equipment, alarms, smoke control systems, and fire-extinguishing equipment such as sprinkler systems. The required endurance will be stated in hours, ranging from $\frac{1}{2}$ to 4 hours in North American codes, and the time may be different for different elements in the same building, such as slabs, beams, girders, columns, and walls. The required fire endurance of exterior walls will be partially governed by the separation of the building from its neighbors.

The criteria for the determination of the fire resistance of a specific member are given (in the United States) in the ASTM E119[13.1] standard, which is generally written based on the expectation of experimental verification of designs. The intent is that a member or system be subjected to its full unfactored service load, that is, 1.0 DL + 1.0 LL, and then subjected to a thermal loading following a specified time–temperature curve, which is shown in Fig. 13.1. At the end of the desired endurance period, two hours for instance, a member or system must:

1. Not suffer structural collapse, and if it functions as a barrier between two fire compartments, it must neither
2. Experience a temperature rise on the side away from fire of more than 250°F (139°C) as the average of several measurements, nor
3. Permit passage flame or hot gases through the floor or wall sufficient to ignite cotton waste that is held near the floor or wall.

While the ASTM E119 requirements have their origins in experimental work, it is possible to compute the structural capacity and the temperature

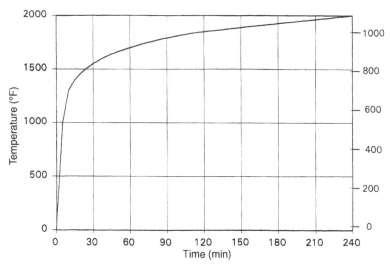

Figure 13.1   Standard time–temperature curve from ASTM E 119.

rise in many practical cases. Many, if not most slab structures will be critical in heat transmission, item 2, before they are approaching structural collapse, item 1. They will seldom be damaged to the point of permitting the passage of flames or hot gases, item 3, until long after the other two limits have been passed.

The ACI Code[13.2] contains no provisions relating to fire resistance except a note that the cover of the reinforcement may have to be increased because of fire resistance requirements. However, a recent standard prepared by ACI Committee 216[13.3] contains code provisions for determining the fire resistance of many common reinforced and prestressed concrete members and various masonry components such as walls and fire protection for structural steel columns.

Reinforced concrete structures have significant inherent fire resistance. Concrete has a relatively low coefficient of thermal conductivity, a relatively large specific heat, and is relatively tolerant of elevated temperatures. The steel reinforcement is much more sensitive to elevated temperatures than is the concrete, but the concrete is often able to protect the steel from excessive temperatures for significant time periods. A fully loaded simply supported reinforced concrete member that is also free to grow in length as it is heated will be at or near the collapse stage when the reinforcement temperature is around 1000°F (540°C), at which temperature the yield stress of the reinforcement will have been reduced to approximately the service load stress. Continuity and restraint of the expansion accompanying heating may greatly increase the fire endurance of a member compared to a simply supported member of the same cross section and span.

In the following sections in this chapter we consider briefly the thermal resistance of slabs, the structural resistance of slabs and other reinforced concrete members, and finally, touch on some special problems associated with the behavior of very high strength and saturated concretes at fire temperatures. References are cited that will lead the interested reader to other specialized literature, as there is no intention in this chapter to provide a complete treatment of the topic.

13.2 THERMAL RESISTANCE

Figure 13.2 shows the experimentally determined thicknesses required for various thermal endurances for slabs (or walls) made of different concrete materials. For a given slab thickness, the endurance is a function of thermal conductivity and specific heat of the materials, which are in turn functions of the specific gravity and mineral composition of the concrete materials. The lighter-weight materials are better insulators and thus require less thickness for a given level of protection. The carbonate aggregate concretes are somewhat more efficient insulators than the siliceous aggregate concretes because of differences in the thermal properties at room temperature. In addition, heating limestone causes an endothermic reaction accompanying decomposition of some of its components, and this further delays the transfer of heat through the slab. The carbonate aggregate concretes are also somewhat less likely to spall during heating than are the siliceous aggregate concretes because of expansive phase transformations that occur in crystalline quartz components of the siliceous aggregates. The lightweight aggregates are generally

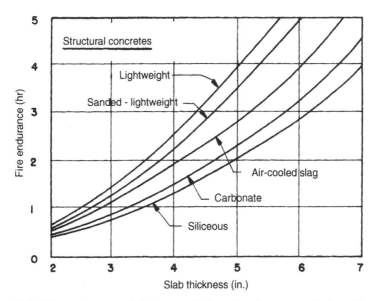

Figure 13.2 Fire endurance of slabs or walls based on heat transmission. (From Ref. 13.4. By permission from the Concrete Reinforcing Steel Institute.)

stable at fire temperatures because their processing temperatures during manufacture were higher than the fire temperature.

It is possible to compute the thermal regime within a slab or other reinforced concrete member. However, the governing partial differential equation is highly nonlinear because of both the transient nature of the imposed heat cycle and because the thermal conductivity and specific heat are both temperature dependent. Time-step analyses have been conducted using both finite difference and finite element methods. Details of these analysis methods are given by Harmathy.[13.5] The thermal and mechanical properties of various concretes at elevated temperatures are given in a paper by Abrams.[13.6]

In those cases in which the thermal resistance of a slab is not adequate, either the slab may be thickened or various insulating materials may be added to the lower (heated) side. The CRSI book,[13.4] an ACI Committee 216 report[13.7] and a PCI manual[13.8]all contain information on such cases. These reports also contain information on the thermal resistance of two-course floors where part of the thickness is of light-weight material and the remainder is of normal-weight concrete.

13.3   STRUCTURAL FIRE RESISTANCE

13.3.1   Members Unrestrained Against Length Change

As noted above, a fire attacks the structure by reducing the strengths of the materials by heating them. The moment capacities are gradually reduced until the structure can no longer resist the applied load, which is normally taken to be the unfactored service load. For a slab structure, the load capacity can be evaluated by means of the yield line theory discussed earlier, with the moment capacities as reduced by the effects of the heat substituted for the appropriate positive and negative moments. The discussion in this section assumes that the change in length of the member, which extends upon heating, is unrestrained by its supports, other parts of the structure, or other external means. Some of the effects of this growth in length are considered in the next subsection.

The moment capacities can be calculated by means of the usual equations, modified for the effects of temperature. The nominal moment capacity, per unit width, can be stated as

$$m_n = A_s f_y d \left(1 - 0.59\rho \frac{f_y}{f'_c}\right) = A_s f_y (d - a/2) \qquad (13.1)$$

where the second statement of the equation is in terms of the equivalent rectangular stress block. The moment capacity of the section as affected by elevated temperatures can be stated as

$$m_{n\theta} = A_s f_{y\theta}(d_\theta - a_\theta/2) \tag{13.2}$$

where the θ subscripts denote temperature-dependent values. The depth of the equivalent rectangular stress block is computed as

$$a_\theta = \frac{A_s f_{y\theta}}{0.85 f'_{c\theta}\, b} \tag{13.3}$$

where A_s and b must be consistent. Either or both the concrete and steel strengths may be affected by the elevated temperature profile existing in the member.

The application of Eqs. 13.2 and 13.3 requires information about the distribution of temperature through the thickness of the slab and about the remaining strengths of the concrete and steel at the relevant temperatures. Some information will be given, after a few comments about the generally expected results.

The positive moment capacity will be governed almost entirely by a reduced yield stress of the steel resulting from the elevated temperature. The top surface of the slab, which is the compression zone, will remain cool enough that the compressive strength of the concrete will not be affected. Thus, only f_y changes in Eqs. 13.2 and 13.3. Since the clear cover to the bottom reinforcement is often only 0.75 in. (20 mm), the cover to the center of the lowest reinforcement is about 1.0 in. (25 mm) in many practical cases. Critical steel temperatures may be reached in about 1.5 hours, depending on the type of concrete, and this will determine fire endurance of a simply supported member that is not restrained against heat-induced growth in length.

The negative moment capacity determination is somewhat more complex. The steel near the top surface of the slab will be only slightly affected within any slab that has adequate thermal resistance. However, the bottom concrete layers, the compression zone, will be affected by the elevated temperatures. Thus, in Eq. 13.3, a reduced compressive strength of $f'_{c\theta}$ will have to be used. In many cases, the value of $f'_{c\theta}$ will have to be an averaged value because the temperature and remaining strength gradients will be very large near the heated surface, even when the depth a_θ is fairly small. In addition, the $a_\theta/2$ fraction in Eq. 13.2 may need to be replaced when a_θ becomes large and the stress distribution cannot be approximated satisfactorily as uniform. An analysis approach that considers the compression zone to be formed of several thin layers may be necessary to make a good estimate of the stress block parameters rather than using Eq. 13.2 directly. In addition, it is possible for the effective depth d_θ to decrease because of sloughing or spalling of the surface layers of the concrete.

Figure 13.3 is one depiction of the temperature distributions in the lower parts of solid slabs made with three different aggregates. These graphs are

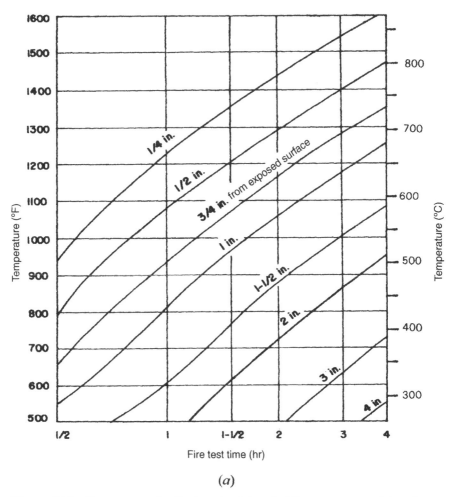

(a)

Figure 13.3   Temperature distributions in slabs of different concrete: (a) carbonate aggregate concrete; (b) siliceous aggregate concrete; (c) sand-lightweight aggregate concrete. (From Ref. 13.4. By permission from the Concrete Reinforcing Steel Institute.)

specifically for slabs that had been naturally, as opposed to kiln, dried until the internal middepth relative humidity was 75%, the standard test condition. Damper concrete will remain slightly cooler; dryer concrete will be slightly hotter at the same time and distance from the heated surface. The nature of the heat transfer process is such that the lower 2 or 3 in. (50 or 75 mm) of concrete will be about the same temperature regardless of whether the slab is 4 in. or 7 in. (100 or 175 mm) thick.

From these graphs one can see that if the objective is to prevent the steel temperature from exceeding 1000°F (540°C) at 2 hours into a standard fire,

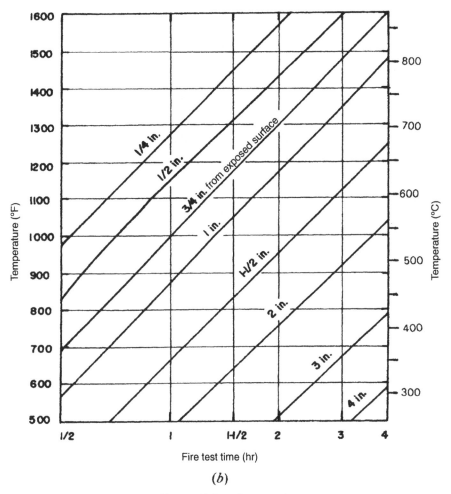

Fire test time (hr)

(*b*)

Figure 13.3   (*Continued*).

about 1.2 in. (30 mm) of cover to the bar center is required for carbonate aggregate concrete, about 1.4 in. (35 mm) for siliceous aggregate, and about 1.2 in. (30 mm) for sanded lightweight concrete having a density of about 113 lb/ft³ (1800 kg/m³). The presence of a bar in the concrete causes only negligible disturbances to the temperature distribution, and the concrete temperature at the same distance from the heated surface as the center of the steel layer is normally used as the steel temperature.

As noted earlier, the yield strength of steel is gradually reduced as its temperature increases. Harmathy and Stanzak,[13.10] conducted an extensive series of tests, and their paper includes information on the yield and ultimate stresses, the changing nature of the stress–strain curves, and on creep of steel at elevated temperatures. Figure 13.4 has been used by nearly all North Amer-

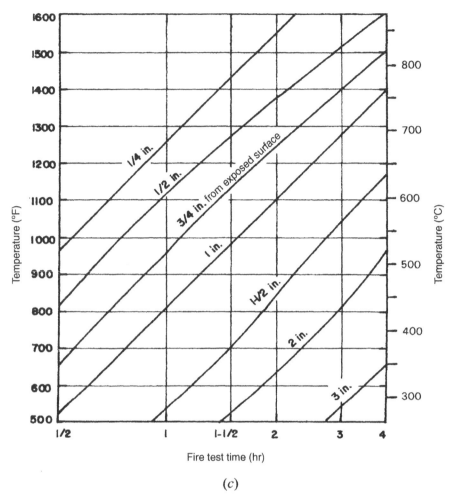

(c)

Figure 13.3   (*Continued*).

ican documents on fire resistance of concrete structures, for example in Ref. 13.4, and is assumed to apply to both grades 40 [grade 300] and 60 [420] bars. It can be seen that 1000°F (540°C) leads to about 65% of the room-temperature yield strength, 1100°F (590°C) to about 54%, and 1200°F (650°F) to about 37%. Since the normal combined load and ϕ-factors will be somewhat less than 2 for most reinforced concrete buildings, it can be expected that collapse of an unrestrained simply supported beam or one-way slab will occur when the steel temperature reaches between 1000 and 1100°F (540 and 590°C).

It should be noted that prestressing steel is considerably more sensitive to elevated temperatures than hot-rolled reinforcing bars. The PCI report[13.8] contains temperature-strength data for prestressing steels. A cold-worked (strain-

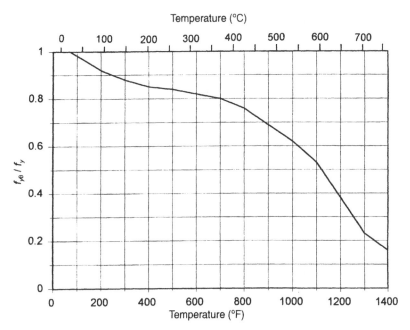

Figure 13.4 Yield stress versus temperature for reinforcing bars. (From Ref. 13.11. By permission from R. L. Brockenbrough & Assoc., Inc., Pittsburgh, Pa.)

hardened) bar would also be expected to be more sensitive, as would any heat-treated steel, such as high-strength structural bolts. It is not known whether Thermex processed steel, recently introduced in North America, is unduly sensitive to elevated temperatures. It is heat-treated to some extent and at least would appear to have the potential for being more sensitive than the more ordinary hot-rolled bar.

Although concrete is more tolerant of elevated temperatures than steel, the surface layers of concrete on the fire-exposed side of a slab or wall will be much hotter than the steel, which is protected by an inch or more of the same concrete. Just as the specific heat and thermal conductivity depend on the type of aggregate, the remaining strength at elevated temperatures also depends on the aggregate type and also on the test conditions. Figure 13.5 gives temperature-strength curves for three different concretes. These curves were obtained with specimens that were loaded to about $0.4f'_c$ while they were being heated, and after the desired temperature was reached they were loaded to failure. Tests in which the concrete is heated in the unstressed state and then loaded to failure in general lead to lower strengths, presumably because the applied stress helps delay the onset of microcracking. Concrete that has been heated in the unloaded state, cooled, and then tested is even weaker.

Most other material properties, including Young's modulus, the coefficient of thermal expansion, and the creep properties, are nonlinearly temperature dependent. Information on these properties may be found in the ACI 216

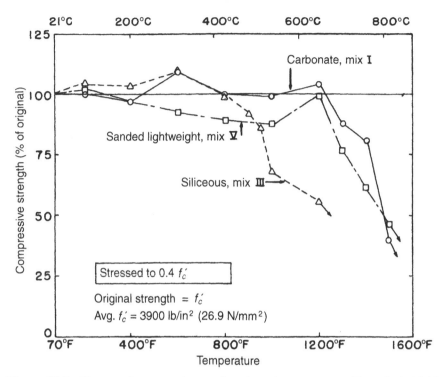

Figure 13.5   Compressive strength as a function of temperature. (From Ref. 13.12. By permission from ACI International.)

report,[13.7] the PCI manual,[13.8] the CRSI book,[13.4] Harmathy's book,[13.5] and in a paper by Abrams.[13.6]

Example 13.1. Assess the remaining moment capacity at 2 hours into a standard fire exposure of a limestone aggregate slab that is 7 in. (178 mm) thick, reinforced with grade 60 [420] No. 4 [13 mm] bars spaced at 12 in. (305 mm) with 1 in. (25 mm) of cover from the center of the bar to the surface of the concrete. The compressive strength of the concrete is 4 kips/in² (27.6 N/mm²). Make the determination for the cases of positive and negative moment, that is, with the steel near the fire-exposed lower surface and with the steel located 6 in. from the fire-exposed face.

SOLUTION. First find the moment capacity of the slab at room temperature.

$$a = \frac{0.2 \text{ in}^2 \times 60 \text{ kips/in}^2}{0.85 \times 4 \text{ kips/in}^2 \times 12 \text{ in.}} = 0.294 \text{ in. (7.47 mm)}$$

$$m_n = m_n' = 0.2 \text{ in}^2/\text{ft} \times 60 \text{ kips/in}^2 (6 - 0.294/2) \text{ in.} = 70.2 \text{ kip-in./ft}$$

$$= 5.85 \text{ kip-ft/ft (26.02 kN-m/m)}$$

Next find the positive moment capacity at 2 hours. The top of slab temperature will be much less than 250°F (120°C), and therefore the concrete compressive strength will be unaffected. From Fig. 13.3 the temperature at 1.0 in. (25 mm) from the fire-exposed surface is about 1060°F (570°C). From Fig. 13.4, the yield stress of the reinforcement has been reduced to about 58% of its original value. From this:

$$a_\theta = \frac{0.2 \text{ in}^2 \times 60 \text{ kips/in}^2 \times 0.58}{0.85 \times 4 \text{ kips/in}^2 \times 12 \text{ in.}} = 0.171 \text{ in. (4.34 mm)}$$

$$m_{n\theta} = 0.2 \text{ in}^2/\text{ft} \times 60 \text{ kips/in}^2 \times 0.58 \,(6 - 0.171/2) \text{ in.}$$

$$= 41.2 \text{ kip-in./ft} = 3.43 \text{ kip-ft/ft (15.26 kN-m/m)}$$

Note that $m_{n\theta}/m_n = 3.43/5.85 = 0.586$, approximately the same as $f_{y\phi}/f_y$.

Now find the negative moment capacity at 2 hours. More approximations are needed in this calculation than for the positive moment capacity. The slab surface temperature is nearly the same as the fire temperature, 1850°F (1010°C), and some of the slab concrete must be assumed to be worthless at this temperature. The top reinforcement temperature cannot be determined from Fig. 13.3 but is probably not over 200°F (95°C) and the strength is not less that about $0.95f_y$, from Fig. 13.4. The concrete has a strong thermal gradient near the fire-exposed (compression) side. Temperatures can be found from Fig. 13.3, and remaining strengths from Fig. 13.5.

Depth (in.)	Temp. (°F)	$\%f_c'$
0	1850	0
0.25	1440	80
0.50	1300	90
0.75	1170	100
1.00	1060	100

The first 0.25-in. (6.4-mm) layer will be ignored, leading to $d_\theta = 5.75$ in. (146 mm). Since the room temperature value of a was about 0.3 in. (7.5 mm), it will be assumed that the concrete in the compression zone more than 0.25 in. (6.4 mm) from the surface can be represented by $f_{c\theta}' = 0.85f_c'$. This then leads to

$$a_\theta = \frac{0.2 \text{ in}^2 \times 60 \text{ kips/in}^2 \times 0.95}{0.85 \times 4 \text{ kips/in}^2 \times 0.85 \times 12 \text{ in.}} = 0.328 \text{ in. (8.33 mm)}$$

$$m_{n\theta}' = 0.2 \text{ in}^2/\text{ft} \times 60 \text{ kips/in}^2 \times 0.95(5.75 - 0.328/2) \text{ in.}$$

$$= 63.67 \text{ kip-in./ft} = 5.31 \text{ kip-ft/ft (23.62 kN-m/m)}$$

Since $m_{n\theta}'/m_n' = 5.31/5.85 = 0.91$, slightly less than $f_{y\theta}/f_y$, it can be seen

that the negative-moment capacity has been decreased by the fire exposure at
a much smaller rate than the positive-moment capacity. There are more approximations involved in this calculation than for the positive moment, but it
is believed that they are on the conservative side and that the internal lever
arm for the steel couple is perhaps slightly larger than assumed here.

For a continuous one-way slab supporting a uniformly distributed load, the
nominal collapse load can be computed from

$$m_n + m_n' = \frac{wl^2}{8} \tag{13.4}$$

Thus, the loss of strength due to the fire exposure is dependent on the loss
of strength of both positive and negative moment sections, and can be evaluated simply in terms of the sum of the moment capacities. For the example
case, the sum of $m_n + m_n'$ at room temperature is $5.85 + 5.85 = 11.70$ kip-
ft/ft, and the sum $m_{n\theta} + m_{n\theta}'$ at 2 hours is $3.43 + 5.31 = 8.74$ kip-ft/ft. The
ratio of capacity at 2 hours to that at room temperatures is $8.74/11.70 =$
0.747, indicating that about three-fourths of the capacity is retained 2 hours
into the fire. Continued fire exposure would gradually reduce the capacity,
with the positive-moment resistance diminishing significantly faster than the
negative moment, since the bottom steel temperature is high enough at 2 hours
to be in the temperature range where small increases in temperature lead to
disproportionate decreases in $f_{y\theta}$.

In a more typical slab that has m_n' considerably larger than m_n, the structural
capacity would diminish more slowly than in this example. In the case of a
two-way slab, the same concept as used in Eq. 13.4 is applicable, except that
the "8" in the equation would be replaced by some other, larger, number. For
a square slab fixed at all edges, the "8" would be replaced with "24" unless
it is deemed necessary to determine the remaining capacity of the two layers
of positive-moment reinforcement separately, since their temperatures could
be somewhat different.

It is obvious that heating the member will cause it to grow in length. Other
heat-induced deformations may be less obvious but also very important. Because there is a strong temperature gradient through the thickness of a slab
(or beam), the differential expansion induces curvature. The member deflects,
or at least tries to deflect, toward the source of the heat. That is, it tries to
deflect toward the fire, or downward for a horizontal member heated from
below.

If the member is simply supported, the only effect of the differential heating is deflection. However, if the member is continuous, the end slopes that
accompany the deflection of a simply supported member will be restrained,
leading to an increase in the negative moments at the ends of the span. This
growth in negative moment can occur early in the fire[13.13] and in beam tests

has led to negative moments approximately equal to the yield moment long before the concrete and steel in the member were hot enough to be affected by heat. Since the loading is constant, an increase in the negative moments is accompanied by an equal decrease in the positive moment and a potentially very large shift in the points of contraflexure toward midspan.

If the length of the top reinforcement has been selected considering only the moment diagram(s) accompanying vertical load, the top steel may be much too short for the moment diagram changes induced by the fire. It seems possible that the best single action one can take to improve the fire resistance of a reinforced concrete structure would be to provide at least some full-length top reinforcement in all beams.

13.3.2   Members Restrained Against Length Change

As discussed in Chapter 12, a slab in which the free lateral movement of the boundaries is restrained may in some cases resist much more load than can be accounted for by flexural forces alone. At small deflections, compressive membrane forces may become important; at larger deflections tensile membrane forces may also become important. The fire environment adds additional considerations, since heating a slab or beam causes a growth in length, and if this growth is restrained, in-plane compressive forces are induced. The forces can become very large when the restraint is quite stiff. The induced compression will generally be helpful to a slab, and indeed the methodology given in the CRSI book[13.4] (and several earlier PCA reports) shows that it is often possible to demonstrate that adequate structural capacity exists several hours into a fire without even taking the reinforcement into account.

That book outlines an empirical method for accounting for the stiffness of the restraining parts of a slab structure in those cases where it can be assumed, because of isolation of a fire by fire separation walls or because of the nature of the fuel distribution, that only part of a slab floor is heated and that the surrounding parts can provide restraint against elongation. The method of evaluating the restraint stiffness is much different than any described in Chapter 12, although the intent is the same.

Although the compressive forces will generally be beneficial to the restrained slab, the equilibrating forces and deformations may be very harmful to the surrounding structure. The unrestrained growth in length at 4 hours in a fire can approach 1% of the original length.[13.14] The supports can seldom accommodate such motion, and if the supports attempt to restrain the elongation, the forces can easily damage the supports. In one fire in a very long [728 ft (222 m)] slab structure built without expansion joints,[13.15–13.17] the slab grew in length by more 3 ft (1 m). This resulted in the shear failures of many supporting columns, since no single-story column can possibly accommodate a relative displacement of 18 in. (500 mm).

It is possible for extension in length of a heated member to cause damage to a supporting member, say a stiff column, even when the particular stiff column is remote from the heated area. This can occur because the axial stiffness in compression of a typical reinforced concrete beam or slab is extremely high, and if a pair of interior spans extend in length by 2 in. (50 mm) each, the 2 in. of motion will be transmitted almost undiminished to the ends of the members. If there is a member so stiff that 2 in. of deflection will damage it, it will be damaged.

13.4   SPECIAL CONSIDERATIONS FOR HIGH-STRENGTH CONCRETE

Most properties of concrete improve as the strength increases, but this has not been true of the fire resistance. Very high strength concretes, and more ordinary strength concretes that are damp or saturated, have often been subjected to explosive spalling as they are heated. The spalling is the result of expanding water in the pore structure, and if the water and steam cannot escape as rapidly as pressure is generated, internal pressures can build up until the tensile strength of the concrete is exceeded and spalling occurs. Lower-strength concretes are less vulnerable to this damage because of their greater porosity, but will spall when the moisture content is high enough and heating rapid enough. The phenomenon has the name *moisture clog spalling,* and an early description was given by Harmathy.[13.18] Although the failure might be characterized as a steam explosion, it is more complex than that and can occur at relatively low temperatures, quite early in a fire.

A comprehensive review of the problem is contained in two reports from the U.S. National Institute of Standards and Technology (NIST).[13.19,13.20] One of the presentations summarized in Ref. 13.20 described tests of high-strength [60 N/mm^2 (8.7 kips/in^2)] lightweight aggregate concrete intended for use in an offshore oil drilling platform, an environment in which drying is impossible. One paragraph, by M. P. Gillen, describing the tests and the conclusions from them, will be quoted completely:

The test results indicated the important role of moisture in the spalling phenomenon. Explosive spalling was observed in specimens predried at 60°C (moisture was not completely eliminated), and in moist specimens with steel fibers (steel fibers increased the tensile strength of the concrete but resulted in spalling larger pieces of concrete). However, no spalling was observed in the specimen that was completely predried at 105°C and in moist specimens that contained polypropylene fibers. Polypropylene fibers with lengths between 150 and 200 mm were found to be most effective for the Heidrun concrete. The study concluded that:

1. The primary culprit for spalling is moisture.
2. Stress and restraint have an influence on the spalling behavior.
3. Polymeric fiber addition reduces spalling and offers a tremendous financial advantage over fire protection by passive coatings or other means.

It would appear that the polypropylene fibers melted or burned and in doing so created a void system adequate to dissipate the internal pore pressure accompanying heating of the concrete. One of the questions still to be answered is whether very high strength concretes can ever dry enough, under normal exposure conditions, to reduce or eliminate the spalling risk. These materials ordinarily have very low porosities, and they dry very slowly, if at all.

REFERENCES

13.1   ASTM E119-88, "Standard Test Methods for Fire Tests of Building Construction and Materials," *ASTM Annual Book of Standards,* Vol. 04.07, West Conshohocken, Pa., 1998.

13.2   *Building Code Requirements for Structural Concrete,* ACI 318-95 and *Commentary,* 318R-95, American Concrete Institute, Farmington Hills, Mich., 1995, 371 pp.

13.3   Joint Committee 216 of the American Concrete Institute and the Masonry Society, *Standard Method for Determining Fire Resistance of Concrete and Masonry Construction Assemblies,* ACI 216.1-97/TMS 0216.1-97, American Concrete Institute, Farmington Hills, Mich., 1997, 26 p.

13.4   *Reinforced Concrete Fire Resistance,* Concrete Reinforcing Steel Institute, Chicago, 1980.

13.5   T. Z. Harmathy, *Fire Safety Design and Concrete,* Longman Scientific & Technical (co-published with John Wiley & Sons), 1993, 412 pp.

13.6   M. S. Abrams, "Behavior of Inorganic Materials in Fire," *Design of Buildings for Fire Safety,* STP 685, American Society for Testing and Materials, Philadelphia, 1979, pp. 14–75.

13.7   ACI Committee 216, *Guide for Determining the Fire Endurance of Concrete Elements,* ACI 216R-89, American Concrete Institute, Detroit, 1989, 44 pp. (lists 142 references).

13.8   A. H. Gustaferro and L. D. Martin, *Design for Fire Resistance of Precast Prestressed Concrete,* 2nd ed., PCI MNL-124-89, Prestressed Concrete Institute, Chicago, 1989, 85 pp.

13.9   M. S. Abrams and A. H. Gustaferro, "Fire Endurance of Slabs as Influenced by Thickness, Aggregate Type, and Moisture," *J. PCA Res. Dev. Lab.,* Vol. 10, No. 2, May 1968, pp. 9–24. Reprinted as Research Department Bulletin 223, Portland Cement Association, Skokie, Ill.

13.10 T. Z. Harmathy and W. W. Stanzak, "Elevated-Temperature Tensile and Creep Properties of Some Structural and Prestressing Steels," *Fire Test Performance,* ASTM STP 464, American Society for Testing and Materials, Philadelphia, 1970, pp. 186–208.

13.11 R. L. Brockenbrough and B. G. Johnston, *Steel Design Manual,* United States Steel Corporation, Pittsburgh, Pa., 1974, 257 pp.

13.12 M. S. Abrams, "Compressive Strength of Concrete at Temperatures to 1600°F," *Temperature and Concrete,* ACI SP-25, American Concrete Institute, Detroit, 1971, pp. 33–58.

13.13   T. D. Lin, A. H. Gustaferro, and M. S. Abrams, *Fire Endurance of Continuous Reinforced Concrete Beams,* PCA Research and Development Bulletin RD072.01B, Portland Cement Association, Skokie, Ill., 1981, 23 pp.

13.14   S. L. Selvaggio and C. C. Carlson, "Restraint in Fire Tests of Concrete Floors and Roofs," *Fire Test Methods: Restraint and Smoke,* STP 422, American Society for Testing and Materials, Philadelphia, 1966, pp. 21–39. Reprinted as Research Department Bulletin 220, Portland Cement Association, Skokie, Ill.

13.15   E. B. Cohn and W. A. Wall, "Military Personnel Records Center Built Without Expansion Joints," *J. ACI,* Vol. 54, No. 12, June 1958, pp. 1103–1110.

13.16   J. A. Sharry, C. Culver, R. Crist, and J. P. Hillelson, "Military Personnel Records Center Fire, Overland, Missouri (Part 1)," *Fire J.,* Vol. 68, No. 3, May 1974, pp. 5–9.

13.17   E. Walker, W. W. Stender, and H. E. Nelson, "Military Personnel Records Center Fire, Overland, Missouri (Part 2)," *Fire J.,* Vol. 68, No. 4, July 1974, pp. 65–70.

13.18   T. A. Harmathy, "Effect of Moisture on the Fire Endurance of Building Elements," *Moisture in Materials in Relation to Fire Tests,* STP 385, American Society for Testing and Materials, Philadelphia, 1965, pp. 74–95.

13.19   L. T. Phan, *Fire Performance of High-Strength Concrete: A Report on the State of the Art,* NISTIR 5934, NIST, Gaithersburg, Md., December 1996, 105 pp.

13.20   L. T. Phan, N. J. Carino, D. Duthinh, and E. Garboczi, *Proceedings of the International Workshop on Fire Performance of High-Strength Concrete,* NIST, Gaithersburg, Md., February 13–14, 1997, NIST Special Publication 919, U.S. Department of Commerce, Washington, D.C., 1997, 164 pp.

INDEX

ACI code provisions:
  crack control, 545
  deflection control, 515, 526, 626
  direct design method, *see* Direct design
    method
  equivalent frame method, *see* Equivalent
    frame method
  holes, 133, 254
  shear strength, 560, 564, 566, 584, 594,
    631
  slab serviceability, 146, 515
  slab strength, 146
Advanced strip method, *see* Strip method
Affinity theorem, 354
Analysis approaches, 11
Approximate yield line patterns, 407, 507
Approximation methods:
  elastic theory, 49
  yield line theory, 407, 415, 507

Banding of reinforcement, 242, 254, 471, 624
Beam analogy, 49, 582, 585, 601
Beam-and-girder floor, 5, 499
Beam-and-slab floor, 3, 499, 660
Beamless slabs, 3, 10, 68, 101,117, 125, 138,
    277, 377, 494, 621
Beam moments, 68, 236, 499
Beam relative flexural stiffness, 60, 68, 103
Beam relative torsional stiffness, 63
Beams, *see* Supporting beam design
Bent bars, 567, 609
Boundary conditions, 31, 210, 307

Causes of cracking, 532
Checkerboard loadings, *see* Pattern loadings
Choice of slab type, 10
Circular fans, 341
Circular slabs, 40, 399
Classical elastic plate theory, *see* Elastic
    theory analysis
Clear span concept, 148, 153
Column capitals, 80, 94
Column moment, 157, 200
Column size, 61, 68

Column stiffness, 64, 200
Column strips, 61
Comparison with test results, 155, 296, 424,
    665, 676, 683
Complete load-deflection behavior of slabs,
    13, 434, 637
Composite beam-slab collapse mechanisms,
    371, 440, 499
Compressive membrane action, 216, 556, 603,
    636, 640
Concentrated loads, 122, 339, 360, 687
Corner effects, 54, 260, 349, 453
Corner reinforcement, 260, 349, 453
Corrosion protection, 531
Crack control, *see* Cracking
Cracking:
  ACI code provisions, 515, 545
  aesthetic considerations, 531
  causes, 532
  computation of width of cracks in service
    load range:
    one-way slabs, 533
    two-way slabs, 540
  corrosion protection, 531
  design examples, 539, 543, 547
  need for crack control, 261, 454, 530
  permissible service load crack widths, 531,
    545
  types of cracking, 532

Deflection control, *see* Deflections
Deflections:
  ACI code provisions, 516, 526
  computation of deflections in service load
    range:
    immediate, 516, 531
    long-term, 516
    limiting thicknesses, 527
  need for deflection control, 261, 453, 515
  permissible service load deflections, 518
  ultimate load range with membrane action,
    637
Design approach for slabs:
  ACI code methods, 16, 18, 144

Design approach for slabs (*Continued*)
  elastic theory method, 13
  general lower bound method, 14, 223, 232
  prestressed concrete, 621
  strip method, 15, 232, 258, 277
  yield line theory, 15, 449
Design of reinforcement in accordance with
        predetermined field of moments:
  mixed moment fields, 227
  negative moment fields, 227
  positive moment fields, 225
  rules for placing reinforcement, 228
Direct design method:
  clear span concept, 148, 153
  column moments, 195, 200
  column stiffness, 200
  column strips, 61, 180
  conditions, 145
  design examples, 148, 196, 202
  distribution of moments:
    to beam and slab in column strip, 180,
        187
    to column and middle strip, 180
    to positive and negative moment regions,
        195, 198
  general approach, 16, 146
  middle strips, 61, 180
  shear forces in beams, 188
  static moment, 148
Discontinuity lines, 238, 247
Drop panels, 3, 528
Ductility:
  slab-column connections, 613
  slab sections, 207, 304

Effective width of beams, 62, 376
Elastic models, 42
Elastic theory analysis:
  approximate methods, 49
  basis, 13, 21
  beam relative flexural stiffness, 60, 102
  beam relative torsional stiffness, 63
  boundary conditions, 31
  classical plate theory, 13, 21, 37
  column relative stiffness, 50, 64, 67
  column size, 80
  concentrated loads, 122
  elastic models, 42
  equilibrium, 22
  equivalent column relative flexural stiffness,
        50, 64, 67
  examples, 51
  finite difference methods, 42
  finite element methods, 46

holes, 133
Lagrange's equation, 21
linearly varying loads, 130
line loads, 122
method of solution, 37
moment-deformation relationships, 24
moments acting at angle to coordinate axes,
        see Transformation of moments to
        different axes
moments in corner panels, 117
moments in edge panels:
  span parallel to edge of structure, 103
  span perpendicular to edge of structure,
        107
moments in interior panels:
  effect of pattern loadings, 81
  effect of relative stiffness of supporting
        beam, 68
  effect of size of supporting column or
        capital, 80
openings, *see* Holes
plate analog, 45
Poisson's ratio, 34
principal moments, 36
reactions, 32, 52, 76
shear-deformation relationships, 30
static moment, 148
torsional moments, 22, 24, 28
Elevated temperature response, *see* Fire
        resistance
Equilibrium equation, 24, 207, 321
Equilibrium method, 205, 232, 321, 332
Equivalent column, *see* Equivalent frame
        method
Equivalent column flexural stiffness, 64
Equivalent frame, *see* Equivalent frame
        method
Equivalent frame method:
  column moments, 157, 168
  column strips, 61, 182
  design aids, 186
  design examples, 164, 188
  distribution of moments to column and
        middle strips, 180
  equivalent column, 18, 64, 163
  equivalent frame, 18, 157
  general approach, 18, 146
  middle strips, 61, 182
  plane frame idealizations, 174
    for lateral loadings, 178
    for vertical loadings, 175
  positive and negative moments, 157
  shear forces in beams, 188
Extent of top reinforcement, *see* Length of
        top reinforcement

Finite difference methods, 42
Finite element methods, 46
Fire endurance, *see* Fire resistance
Fire resistance, 695
   concrete properties, 703
   examples, 704
   high strength concrete, 708
   moment capacities, 698
   restraint of deformations, 707
   steel properties, 701
   temperature distributions, 700
   thermal resistance, 697
Flat plates, 3, 10, 377, 494, 551
Flat slabs, 3, 7, 10, 377, 494, 551
Flexural strength, 155, 211, 628

General lower bound limit analysis and
         design, *see* Lower bound method
Grouted tendons, 621, 623, 628

High strength concrete:
   fire resistance, 708
   shear resistance, 615
Hillerborg's strip method, *see* Strip method
History of development of slabs, 7
Holes, 133, 390, 566
Hollow slab, 6

Immediate deflections, 515, 521

Johansen's yield criterion, *see* Yield criterion
Johansen's yield line theory, *see* Yield line
         theory
Joists, 5, 6

Kinking of reinforcement, 212

Lagrange's equation, 21
Lateral restraint, 636, 645, 660, 687
Length of top reinforcement, 260, 454, 458
Lift slabs, 621
Limit analysis, 14
Linearly varying loads, 130
Line loads, 122
Load balancing, 624
Loading transferred to beams, 32, 76, 502
Long-term deflections, 516
Lower bound method:
   advanced strip method, *see* Strip method
   boundary conditions, 210
   design approach, 223, 230
   design examples, 218, 261

design of reinforcement in accordance with
      a predetermined field of moments,
      223
   equilibrium equation, 207
   general approach, 14, 205, 217, 232
   strip method, 232, 234, 277
   transformation of moments to different
      axes, 209
   yield criterion, 211

Membrane action:
   comparison with test results, 647, 665, 676,
      683
   compressive membrane action, 216, 636,
      640
   concentrated loads, 555, 687
   deflection behavior, 636, 653
   effect on flexural strength of sections, 216
   examples, 645, 647
   lateral restraint, 636, 645, 660, 687
   load-deflection behavior, 637
   one-way laterally restrained slabs, 640
   slab with edge(s) free to move laterally,
      660, 687
   surround stiffness and strength, 645, 655,
      660
   tensile membrane action, 636, 679, 687
   two-way laterally restrained slabs, 656
Middle strips, 61
Minimizing steel volume, 258, 468, 664
Minimum load principle, 314
Minimum weight design, *see* Minimizing
      steel volume
Moment-deformation relationships, 24, 206,
      304
Moments acting an angle to coordinate axes,
      see Transformation of moments to
      different axes
Moments in corner panels, 117, 363, 369,
      407, 413
Moments in edge panels, 101, 363, 369, 407,
      409
Moments in interior panels, 68, 363, 407

Negative moment reinforcement:
   length of top reinforcement, 259, 454, 458
   variation across slab, 281, 453, 496
Nodal forces:
   breakdown cases, 330
   at free edges, 329
   at intersection of yield lines, 328
   magnitude, 323

One-way slabs, 5, 9, 78, 122, 342, 640
Openings, *see* Holes

Partial loadings, *see* Pattern loadings
Pattern loadings, 81, 200, 461
Pile caps, 574
  truss analogy, 578
Plane-frame analysis, *see* Equivalent frame
    method
Poisson's ratio, 34
Prestressed concrete slabs:
  basis for design, 9, 622
  flexural strength, 628
  grouted tendons, 621, 629
  load balancing, 624
  service load stresses, 620, 623
  shear strength, 631
  tendon distribution, 623
  unbonded tendons, 621, 629
Principal moments, 36
Punching shear, 552

Reactions, 32, 52, 76, 188, 211, 235, 499,
    502
Re-entrant corners, 255, 277
References, 19, 57, 139, 202, 230, 300, 445,
    513, 547, 617, 634, 691, 709
Reinforcement:
  banded, 242, 254, 471
  concentration at columns, 193, 283, 496
  corner of slabs, 229, 260, 453, 590, 596
  design moment strength, 241, 224, 258,
    308, 451
  distribution to positive and negative
    moment sections, 157, 180, 195,
    261, 454
  holes in slabs, 138
  isotropic, 213, 310
  kinking, 215
  length of top steel, 260, 454, 458
  maximum spacing, 258, 551
  maximum steel ratio, 224, 258, 451
  minimum steel ratio, 224, 258, 451
  orthotropic, 310
  rules for placing reinforcing for general
    field of moments, 223
  skew reinforcement, 230, 255, 403
  temperature and shrinkage reinforcement,
    224, 258, 451
  torsional reinforcement, 229, 260, 453, 596
Reinforcing steel, *see* Reinforcement
Ring slabs, 401

Secondary beams, 5, 509
Segment equilibrium method, 287
Serviceability:
  crack widths, 530
  deflections, 515, 624
Serviceability checks, *see* Serviceability
Service ducts, 574
Shear-deformation relationships, 30
Shear failure, *see* Shear strength
Shear forces in beams, 188
Shear forces at yield lines, 322
Shear reinforcement:
  bent bars, 566, 609
  stirrups, 566, 610
  structural steel shearheads, 569, 610
  stud shear reinforcement, 572
Shear strength:
  ACI code provisions, 560, 566
  beam-action, 551
  ductility, 613
  free edges, 572
  lateral restraint, 601, 687,
  mechanism of shear failure, 554
  openings, 572
  pile caps, 574
  prestressed concrete slabs, 631
  punching shear, 552
  resistance provided by concrete, 554
  resistance provided by shear reinforcement,
    see Shear reinforcement
  service ducts, 574
  slab-column connections with unbalanced
    bending moment:
    ACI code provisions, 584
    beam analogy, 582, 595, 601
    behavior, 579
    corner column connection, 590
    design example, 591, 595, 607
    ductility, 601, 613
    edge column connection, 589
    interior column connection, 521, 535
    truss models, 583
  slab-wall connections, 615
  truss models, 554, 564, 574, 583
  two-way action, 552
  uniform shear, 554
Simple strip method, *see* Strip method
Skewed slabs, 230, 255, 403
Slab-and-beam floor, 5
Slab-column connections:
  shear strength, *see* Shear strength
  transfer of unbalanced moment, 384, 579
  yield line patterns, 379
Slab types, 3, 7

Slab-wall connections, 615
Static moment, 17, 50, 68, 132, 148
Stirrups, 566, 610
Strip action, 234
Strip loadings, *see* Pattern loadings
Strip method:
  advanced strip method:
    design applications, 285
    design examples, 285
    general approach, 277
    types of slab element, 278
  comparison with test results, 296
  simple strip method:
    banding of reinforcement, 242, 254
    column supports, 254, 255
    comparison with yield line theory
      ultimate load, 256
    corner reinforcement, 260
    design applications, 258
    design examples, 261, 265, 269
    design method, 258
      discontinuity lines originating from
        slab corners, 238
      discontinuity lines originating from
        slab sides, 247
    distribution of reinforcement to positive
      and negative moment zones, 253,
      261
  free edges, 255
  holes, 254
  length of top reinforcement, 260
  minimizing the steel volume, 236, 259
  re-entrant corners, 255
  skewed slabs, 255
  strip action, 234
  strong bands, 254, 296
  superposition of loading cases and moment
    strengths, 261, 346, 455
  triangular slabs, 255
Strong bands, 254, 296
Structural steel shearheads, 569, 610
Superposition of loading cases and moment
  strengths, 261, 346, 455
Supporting beam design, 32, 76, 188, 371,
  499, 660

Tensile membrane action, 636, 679
Thickness of slab, 146, 526
Torsion, reinforcement for, 229, 260, 453,
  595, 606
Torsional moments, 22, 27, 35, 54, 260, 596,
  606
Torsional stiffness, 63, 66

Transformation of moments to different axes,
  36, 209
Trial and error methods, 415
Triangular slabs, 255
Truss methods:
  pile caps, 578
  slab-column connections with unbalanced
    bending moment, 583
  uniform shear, 564
Two-way slabs, 5, 8, 10, 68, 101, 117
Twisting moments, *see* Torsional moments

Ultimate moments of resistance:
  isotropic reinforcement, 213, 310
  orthotropic reinforcement, 310
  prestressed slabs, 628
Unbonded tendons, 621, 629
Upper bound method, 14

Virtual work method, 311

Waffle slabs, 5
Width of cracks, *see* Cracking

Yield criterion, 211, 308
Yield line patterns, 306, 308
Yield lines, 306
Yield line theory:
  affinity theorem, 354
  approximate yield line patterns, 407, 507
  axes of rotation, 306
  banded reinforcement, 471
  beamless slabs:
    distribution of reinforcement, 495
    folding yield line patterns, 377, 494
    local yield line patterns at columns, 379
    unbalanced moment transfer at slab-
      column connections, 384
  checkerboard loading, 461
  circular fans, 341
  circular slabs, 337, 399
  comparison with test results, 424
  components of internal work done, 314
  composite beam-slab collapse mechanisms,
    371, 499
  compressive membrane action, 656
  concentrated loads, 339, 360
  conditions at ultimate load, 305
  corner effects, 349, 453
  corner reinforcement, 349, 453
  crack control, 453
  deflection control, 453
  design examples, 474, 477, 480, 484, 486,
    497

Yield line theory (*Continued*)
  design method, 450
  distribution of reinforcement, 452, 471, 496
  ductility of sections, 304
  effective width of beams, 376
  equilibrium method, 321, 332
  extent of top reinforcement, *see* Length of
      top reinforcement
  general cases of uniformly loaded
      rectangular slabs:
    all edges supported, 363, 407
    three edges supported, 366, 409
    two edges supported, 369, 413
  length of top reinforcement, 454, 458
  loading transferred to beams, 502
  membrane action, 656
  minimizing the reinforcement, 468
  minimum load principle, 314
  nodal forces:
    breakdown cases, 330
    at free edges, 329
    at intersection of yield lines, 328
    magnitude, 323
  openings, 390
  regular polygon shaped slabs, 335
  reinforcement, 303, 310, 451
  ring slabs, 401
  secondary beams, 509
  serviceability checks, 449, 453, 476, 479,
      482, 485, 492
  shear forces at yield lines, 322
  skew slabs, 403
  strength requirements, 450
  superposition of loading cases and moment
      strengths, 346, 455
  supporting beam design, 374, 499
  trial-and-error method, 415
  ultimate moments of resistance, 209, 211,
      308, 451
  virtual work method, 311
  yield line patterns, 306, 308
  yield lines, 306

Lightning Source UK Ltd.
Milton Keynes UK
UKOW06n1842160315

247965UK00001B/101/P